# Performance-Based Seismic Engineering: Vision for an Earthquake Resilient Society

GEOTECHNICAL, GEOLOGICAL AND EARTHQUAKE ENGINEERING

Volume 32



For further volumes:
http://www.springer.com/series/6011

Matej Fischinger
Editor

# Performance-Based Seismic Engineering: Vision for an Earthquake Resilient Society

Springer

*Editor*
Matej Fischinger
Faculty of Civil and Geodetic Engineering
University of Ljubljana
Ljubljana, Slovenia

ISSN 1573-6059 ISSN 1872-4671 (electronic)
ISBN 978-94-017-7930-2 ISBN 978-94-017-8875-5 (eBook)
DOI 10.1007/978-94-017-8875-5
Springer Dordrecht Heidelberg New York London



Printed on acid-free paper

Springer is part of Springer Science+Business Media (www.springer.com)

*In honour of Peter Fajfar and Helmut Krawinkler, the founders of the Bled Workshops.*

# Preface

The desire for good performance is inherently built into the human mind, so that performance based design has always existed in one form or another. But the perception of performance has frequently been vague and insufficiently quantified. Even today the occurrence of major earthquakes continues to confirm that there are fundamental differences between the expectations of stakeholders and the performance which is actually provided by traditional design. Only about two decades ago increased public awareness and the simultaneous development of advanced engineering tools and methodologies matured enough to trigger activities leading towards the formulation of an up-to-date concept of performance based design.

Since the very beginning, Peter Fajfar and Helmut Krawinkler were in the forefront of these new ideas. They initiated and organized three famous workshops (those which were held in 1992, 1997, and 2004), which became known simply as the Bled Workshops – Bled is a small town in Slovenia, next to the nice Lake Bled, where the events were organized. These workshops produced widely cited reference books, which provided visions about the future development of earthquake engineering, as foreseen by leading researchers in the field. There are very few scientific events which can repeatedly bring together the best and leading researchers from all over the world, and thus provide a forum with a strong impact and authority for important developments in a particular scientific field. During Bled 1 (1992) the new emerging tools of nonlinear seismic analysis and design were discussed. These tools were, at the time, and still, are a prerequisite for modern performance-based earthquake engineering, a burgeoning idea that was incubated in the minds of the participants. During Bled 2 (1997) it became clear that performance-based design had become one of the leading new ideas in earthquake engineering. By the time that Bled 3 was convened, in 2004, the procedures and methods of performance-based design and evaluation, which had been developed during extensive research, were being gradually adopted into everyday practice.

Now, 20 years after the foundation of the tradition of the Bled workshops, we are witnesses to a world-wide breakthrough of this idea, with many different implementations and applications. The major research activities in the field of performance-based earthquake engineering have been supported and coordinated by large networks of research institutions and laboratories. However, even if this significant progress is taken into account, the earthquake engineering community still faces many big challenges. Over just the last 5 years, several devastating earthquakes have reminded us that these destructive events still threaten the lives of millions of people, and very large amounts of property, as well as the social structure and economic well-being of individuals, communities, and countries all over the world. These events have clearly demonstrated that some of the traditional concepts of performance based design are becoming out-of-date. First of all it has become clear that our research interest should go beyond the narrow technical aspects, and that the seismic resilience of society as a whole should become an essential part of the planning and design process. The Bled 4 workshop was organized in order to discuss, develop and promote this idea in the light of the state-of-the-art achievements in the field, and this book presents the outcomes of this event. The workshop started exactly 20 years after the day when Slovenia had declared independence, 40 years after the Institute of Structural Engineering, Earthquake Engineering and Construction IT (IKPIR) had been established at the University of Ljubljana, and 500 years after the strongest earthquake to ever hit Slovenian lands, which occurred in 1511.

First of all, the participants of the 2011 event built on the tradition of the earlier Bled workshops, which were initiated by Professors Fajfar and Krawinkler, in order to honour their important research contributions. To our great sorrow, soon after the workshop the earthquake engineering community had to face the loss of Helmut Krawinkler, even while he was still actively contributing to the finalization of this book, which meant a lot to him. I will never forget Helmut's communication in January 2012, telling me "To put it bluntly, Bled 2011 was my last very good and lasting memory". Today this sentence fills me with both sadness and happiness. But first of all it committed me to fulfil Helmut's wish, and to get this book published, in spite of the problems which I had to face. In order to honour Helmut's memory, Gregory Deierlein prepared the introductory chapter of the book, based on Helmut's Power Point presentation, which was presented at the beginning of the 2011 workshop. So the book includes Helmut's last and priceless address to the engineering community, together with his vision and advice for the future development of performance based design and earthquake engineering. I am very grateful to Greg for undertaking this extremely difficult but most important task.

Our joint aim has been to develop a common global vision for earthquake engineering and seismic risk management, while at the same time recognizing the unique regional traditions which do exist. This book therefore consists of three major parts (IV–VI), presenting the vision of the three world regions which lead in earthquake engineering – Japan and Asia, Europe, and the Americas. Whereas the majority of the chapters in the Americas group were contributed by authors from the western US and Canada, Part VI also presents the views and visions which are held in the eastern US, Mexico and Chile. In order to make sure that New Zealand, as one of the leading schools in earthquake engineering, was not missed out, Nigel Priestley contributed two chapters to the book. By doing so, in spite of the serious condition of his health, Nigel proved his great energy and devotion to research, and – I can dare say – also his friendship to me. I am therefore very grateful to him for ensuring that his views could be given in this book, thus providing a more complete picture of the vision of future code developments. And primarily, I express my gratitude to the regional coordinators Masayoshi Nakashima (Japan/Asia) and Peter Fajfar (Europe), as well as Jack Moehle and Andrei Reinhorn, who together coordinated the Americas group. The regional coordinators proposed invited participants and contributors, defined the regional concept of the presentation, and served as one of the two reviewers of each chapter required by the publisher. Without their unswerving support I would not have been able to finish this task. I am particularly obliged to the Japanese researchers, who participated in spite of the enormous commitments and day-and-night work which they had to perform in the months immediately following the 2011 Tohoku Earthquake. Here I would like to express my special thanks to Masayoshi Nakashima, who gave the final initiative for the Japan/Asia group to participate.

After Helmut's introduction (Part I) the book starts with Part II – Global Vision – which first includes three chapters contributed by three distinguished researchers from the three participating regions, giving a broad introduction to the problems to be discussed and considered. The first chapter was contributed by Stephen Mahin, the director of the Pacific Earthquake Engineering Research Centre (PEER). The PEER Centre has always been among the leading institutions which have been involved in the development and promotion of performance-based-design (PBD) methodologies. The "PEER methodology", which is used by many authors in this book, is frequently considered to be synonymous with PBD procedures in general. In this chapter, entitled "Engineering Challenges on the Way to Resilient Structures and Communities", the engineering aspects of resilient communities are discussed, focusing on the question of how to increase the post-earthquake operationality of those structures and on the lifelines which are critical to a community's needs in the aftermath of a major earthquake, and the ability of occupants to "shelter-in-place" during repairs. Hiroshi Akiyama, Professor Emeritus of the University of Tokyo, a close friend of both Peter Fajfar and Helmut Krawinkler, contributed the chapter on the use of energy principles in earthquake engineering. The importance of this contribution is best described in the review written by Masayoshi Nakashima: "A legendary design concept developed by Professor Akiyama is summarized in this chapter. The importance of cumulative structural damage is emphasized, and the concept of energetic equilibrium is the plausible answer to allow for the damage. The chapter should be published as a historical note to 'energy-based seismic design'." While this concept has not, recently, been sufficiently addressed, I am convinced that many performance objectives and goals on the path towards resilient structures will be more efficiently achieved using energy principles. The third chapter was written by Žiga Turk, who served both as Minister for Economic Development, and as Minister for Education, Research, Culture and Sport in the past governments of the Republic of Slovenia, as well as acting as Secretary General of the Reflection Group on the Future of Europe. Žiga Turk analyses the profound changes that the world is going through, and how civil engineering should respond to these challenges. Concluding with the statement that "the essence of technology is nothing technical" he supports one of the main observations in this book, that PBD should go beyond narrow technical interests, and should focus on the resilience of communities and society. Three more papers in the Global Vision part of the book address important developments in the codification of direct displacement-based seismic design, and the earthquake resistant design/retrofit of bridges with advanced materials.

As mentioned above, several devastating earthquake disasters (Haiti, Chile, L'Aquila, Tohoku, and Christchurch) occurred shortly before the 2011 workshop. Most workshop participants were involved in the post-earthquake reconnaissance and disaster-relief efforts. This valuable experience has been incorporated into all chapters of the book, and in particular into Part III – New Vision after Recent Earthquakes. These disasters occurred in very different, and very differently developed, parts of the world. However, they all had consequences that were far beyond those expected, and they all revealed significant weaknesses in the expected performance evaluation and earthquake preparedness plans. The main message of this part is best described by Masayoshi Nakashima: "If 'resiliency' is defined as the ability to recover to normal conditions as quickly as possible, then true resiliency cannot be obtained by focusing on individual components separately. . . . As long as building performance is investigated on only an individual basis, a full picture of the community performance cannot be obtained." There is also one very important message to be given. We too often concentrate on earthquake engineering procedures which are only suitable for developed countries. However, out of all the above-mentioned events, the Haiti earthquake was the worst, if not the worst earthquake catastrophe in modern history. As pointed out by Eduardo Miranda (Chap. 9): "Resilience encompasses on the one hand a measure of the impact of earthquake on society and on the other the capacity to recover from the disaster." Consistently with this, Sergio Alcocer (Chap. 32 in Part VI) has analysed the specifics of developing countries which determine the earthquake preparedness activities that are suitable for this environment.

At a time very soon after the Tohoku earthquake, we were honoured by the presence of His Excellency Toshimitsu Ishigure, the Ambassador of Japan in the Republic of Slovenia, at the opening session of the Bled 4 workshop. The Ambassador talked about his own broad personal experience of earthquakes, particularly when he was involved in several rescue activities as the Head of the Overseas Disaster Assistance Division at the Ministry of Foreign Affairs. He led the disaster relief team after the Earthquake of North Afghanistan in 2002, and after the Tsunami disaster in Phuket in 2004, and he was personally involved in the rescue operations after the 2003 Algeria earthquake. In Kobe 2005 he was involved in the establishment of the International Recovery Platform, which is the worldwide conference on disaster prevention under the auspices of the UN. As a guest of honour, he addressed the participants of the Bled 4 workshop with the following words: "First of all I would like to express my sincere gratitude to all Slovenians and citizens from other countries for their heartfelt sympathy and solidarity with Japan, which is now facing difficulties due to the huge earthquake and tsunami disaster on March 11 this year. However, Japan will not simply rebuild, but rather reshape itself into a more dynamic country. Today, I am really grateful for being able to take part in the Bled 4 Workshop: Performance Based Seismic Engineering-Vision for an Earthquake Resilient Society. Especially, at this moment after the disastrous earthquakes which happened this year, I think we have a great opportunity to learn from these experiences in order to minimize casualties and to prevent secondary disasters, and the need for this kind of study is highly regarded among the people as well. ... Having seen with my own eyes the aftermath of earthquake disasters, I am really well aware of the importance of preventive measures for potential natural disasters, and the importance of developments in the technology of seismic engineering. I am therefore firmly convinced that, from your research and discussions which will be exchanged at this conference, new knowledge and technology to prevent disasters and minimize earthquake casualties will emerge, and so contribute to saving as many lives as possible in potential earthquake disasters all around the world. I wish great success to the Bled 4 workshop."

The second guest of honour at the Bled 4 workshop was Professor Matjaž Mikoš, the Dean of the Faculty of Civil and Geodetic Engineering of the University of Ljubljana, who, as a hydraulic engineer and as a hydrologist active in the fields of landslide mitigation and flood prevention, fully understands how important performance-based seismic engineering is in order to build an earthquake resilient society. In his welcome speech he said: "It is a special privilege to be in a position to work together with Professor Fajfar in the same faculty, and therefore I will take the opportunity of this opening address and express my personal and our faculty's sincere thanks for the contributions of Professors Krawinkler and Fajfar, who have contributed so much to the field of seismic engineering, and who are the founders of these scientific workshops at Bled. The International Decade for Natural Disaster Reduction, in the last century, intensified international cooperation and initiated new ways of thinking in this field ... . Different natural hazards such as tsunamis, earthquakes, volcanic eruptions or floods are inevitable on this Earth, but we can build up our capacities, prepare early warning plans, raise levels of preparedness, and work hard on prevention. And this is precisely what you will be working on during these days at Bled."

Significant speeches were also given by Peter Fajfar's former post-graduate students, who had achieved high-ranking positions in Slovenian society, and in the institutions which are responsible for natural disaster prevention. Roko Žarnić addressed the audience as the Minister of the Environment and Spatial Planning of the Republic of Slovenia. He presented the efforts for upgrading the disaster resilience of the Slovenian community by introducing the newly established Slovenian Council for Measures of Seismic Resilience, and described the recovery efforts after recent earthquakes in Slovenia. Črtomir Remec, the President of the Slovenian Chamber of Engineers and the President of the European Council of Engineering Chambers, emphasized the importance of PBD methodologies for the development of design practice.

Browsing through this book, which has emerged as the main result of the Bled 4 workshop, I hope that it will continue the tradition of the excellent "Bled

**Fig. 1** Bled 4 workshop participants

publications", which have served as reference books in earthquake engineering. There are many people who have contributed to this success. Firstly, I would like to express my gratitude to Atilla Ansal, the Secretary General of the European Association of Earthquake Engineering and Springer's Geotechnical, Geological and Earthquake Engineering Series Editor, for his kind and encouraging consideration of this book, and Petra Steenbergen, Springer's Senior Publishing Editor, for her help and patience with the delay in the preparation of the manuscript. And I am, of course, deeply grateful to the invited authors (the first authors of all the chapters, as well as Patricio Bonelli, Gian Paolo Cimellaro, Gregory Deierlein and Gaetano Manfredi) and their co-authors (please see the List of contributors), who put a lot of effort and care into preparing the 32 chapters of this book in spite of their very busy schedules. I equally thank the other invited participants – Boštjan Brank, Mehmed Čaušević, Vojko Kilar, Vladimir Sigmund and Roko Žarnić – who participated in the interesting and fruitful discussions. I am particularly obliged to Božidar Stojadinović, with whom we planned this wonderful event for several years. I conclude this introduction with a group photo of the Bled 4 workshop participants, as a lasting memory of this event (Fig. 1).

Ljubljana, Slovenia
June 2013

Matej Fischinger

# Contents

## Part III New Vision After Recent Earthquakes



## Part IV Vision in Japan and Asia

## Part VI Vision in Americas

# About the Editor

Matej Fischinger (born in 1954) is a Professor of Earthquake Engineering and Reinforced Concrete Structures at the University of Ljubljana in Slovenia. He is a Member of the Slovenian Academic Society of Technical and Natural Sciences and the Vice-president of the Slovenian Association for Earthquake Engineering.

His research has been concerned with earthquake resistant design of RC structures and the inelastic design procedures. He is the co-author of the N2 method, proposed in 1989 in his Ph.D. dissertation supervised by Professor Fajfar. For the related research work they got the highest research award in the Republic of Slovenia. Since then the N2 method, which was recently incorporated into Eurocode 8, has become one of the leading push-over methods in earthquake engineering. His current interest is in the seismic resistance of bridges, RC industrial buildings and structural walls, performance-based design methodologies, the Eurocode, and the use of information technology in education. Using inelastic macro-models the research group led by Matej Fischinger made several successful benchmark predictions of the response of RC structural walls.

As a designer, consultant or reviewer, he has participated in many design projects (in particular of high-rise apartment buildings, bridges, industrial buildings, precast buildings, NPP Krško and related buildings). He has been very active in the development of the EC8 and its introduction as a National code in Slovenia. Slovenia was the first country to adopt Eurocodes as the national code on January 1, 2008. He wrote the commentary for the RC section. He has actively participated in the recent modifications of the design rules for prefabricated structures in EC8. These results are based on extensive research within several EU research projects, where Matej Fischinger served as the Slovenian co-ordinator.

# Contributors

**Hiroshi Akiyama** Professor Emeritus, Faculty of Engineering, University of Tokyo, Tokyo, Japan

**Sergio M. Alcocer** Instituto de Ingeniería, Universidad Nacional Autónoma de México, UNAM, México DF, Mexico

**M. Nuray Aydınoğlu** Department of Earthquake Engineering, Kandilli Observatory and Earthquake Research Institute, Boğaziçi University, Istanbul, Turkey

**Chadi Ayoub** C. Engineering, Inc., Houston, TX, USA

**Jack W. Baker** Department of Civil and Environmental Engineering, John A. Blume Earthquake Engineering Center, Stanford University, Stanford, CA, USA

**Tracy C. Becker** Disaster Prevention Research Institute (DPRI), Kyoto University, Uji, Kyoto, Japan

**Patricio Bonelli** Department of Structural Engineering, Universidad Técnica Federico Santa María, Valparaíso, Chile

**Rubén Boroschek** Department of Structural Engineering, University of Chile, Santiago, Chile

**Gian Paolo Cimellaro** Department of Structural, Geotechnical and Building Engineering (DISEG), Politecnico di Torino, Torino, Italy

**Víctor Contreras** Rubén Boroschek and Associates, Santiago, Chile

**Carlos Cruz** Department of Civil and Environmental Engineering, University of Alberta, Edmonton, AB, Canada

**Luc Davenne** University Paris 10, Paris, France

**Gregory G. Deierlein** Department of Civil and Environmental Engineering, John A. Blume Earthquake Engineering Center, Stanford University, Stanford, CA, USA

**Matjaž Dolšek** Faculty of Civil and Geodetic Engineering, University of Ljubljana, Ljubljana, Slovenia

**Norberto Dominguez** Ecole Polytechnique, Civil Engineering Master Program, Mexico City, Mexico

**Kenneth J. Elwood** Civil and Environmental Engineering Department, University of Auckland, Auckland, New Zealand

**Peter Fajfar** Faculty of Civil and Geodetic Engineering, University of Ljubljana, Ljubljana, Slovenia

**Michael N. Fardis** Department of Civil Engineering, University of Patras, Patras, Greece

**Matej Fischinger** Faculty of Civil and Geodetic Engineering, University of Ljubljana, Ljubljana, Slovenia

**Paolo Franchin** Department of Structural and Geotechnical Engineering, University of Rome "La Sapienza", Rome, Italy

**Hiroshi Fukuyama** Building Research Institute, Tsukuba-shi, Ibaraki-ken, Japan

**Curt B. Haselton** Department of Civil Engineering, California State University, Chico, CA, USA

**Adnan Ibrahimbegovic** Ecole Normale Supérieure, LMT-Cachan, Cachan, France

**Iunio Iervolino** Dipartimento di Ingegneria Strutturale, Università degli Studi di Napoli Federico II, Naples, Italy

**Takahito Inoue** Hyogo Earthquake Engineering Research Center, National Research Institute for Earth Science and Disaster Prevention, Shijimi, Miki, Hyogo, Japan

**Tatjana Isaković** Faculty of Civil and Geodetic Engineering, University of Ljubljana, Ljubljana, Slovenia

**Toshikazu Kabeyasawa** Building Research Institute, Tsukuba-shi, Ibaraki-ken, Japan

**Toshimi Kabeyasawa** Earthquake Research Institute, University of Tokyo, Tokyo, Japan

**Koichi Kajiwara** Hyogo Earthquake Engineering Research Center, National Research Institute for Earth Science and Disaster Prevention, Miki City, Hyogo-ken, Japan

**Weng Yuen Kam** Beca, Auckland, New Zealand

**Fatemeh Kavianipour** TTG Corporation, Pasadena, CA, USA

**Kazuhiko Kawashima** Department of Civil Engineering, Tokyo Institute of Technology, Meguro, Tokyo, Japan

**Yousok Kim** Earthquake Research Institute, University of Tokyo, Tokyo, Japan

**Helmut Krawinkler** Department of Civil and Environmental Engineering, John A. Blume Earthquake Engineering Center, Stanford University, Stanford, CA, USA

**Koichi Kusunoki** Division of Disaster Mitigation Science, Earthquake Research Institute, The University of Tokyo, Tokyo, Japan

**Jui-Liang Lin** NARLabs, National Center for Research on Earthquake Engineering, Taipei, Taiwan

**Pao-Chun Lin** NARLabs, National Center for Research on Earthquake Engineering, Taipei, Taiwan

**Ting Lin** Department of Civil, Construction and Environmental Engineering, Marquette University, Milwaukee, WI, USA

**Stephen Mahin** Department of Civil and Environmental Engineering, University of California, Berkeley, Berkeley, CA, USA

**Gaetano Manfredi** Dipartimento di Ingegneria Strutturale, Università degli Studi di Napoli Federico II, Naples, Italy

**Damijan Markovic** EDF, SEPTEN, Lyon, France

**Tomohiro Matsumiya** Department of Architecture and Building Engineering, Kinki University, Higashi-Osaka, Osaka, Japan

**Roberto Meli** Instituto de Ingeniería, Universidad Nacional Autónoma de México, UNAM, México DF, Mexico

**Eduardo Miranda** Department of Civil and Environment Engineering, Stanford University, Stanford, CA, USA

**Jack Moehle** Department of Civil and Environmental Engineering, University of California, Berkeley, Berkeley, CA, USA

**Sarira Motaref** Department of Civil and Environmental Engineering, University of Connecticut, Storrs, CT, USA

**Takuya Nagae** Hyogo Earthquake Engineering Research Center, National Research Institute for Earth Science and Disaster Prevention, Miki, Hyogo, Japan

**Masayoshi Nakashima** Disaster Prevention Research Institute (DPRI), Kyoto University, Uji, Kyoto, Japan

**Manabu Nakayama** Hyogo Earthquake Engineering Research Center, National Research Institute for Earth Science and Disaster Prevention, Miki, Hyogo, Japan

**Melissa O'Brien** Department of Civil and Environmental Engineering, University of Nevada, Reno, NV, USA

**Stefano Pampanin** College of Engineering, Civil and Natural Resources Engineering, University of Canterbury, Christchurch, New Zealand

**Alexandra Papailia** Department of Civil Engineering, University of Patras, Patras, Greece

**Paolo Emilio Pinto** Department of Structural and Geotechnical Engineering, University of Rome "La Sapienza", Rome, Italy

**Maria Polese** Dipartimento di Ingegneria Strutturale, Università degli Studi di Napoli Federico II, Naples, Italy

**Nigel Priestley** Jacobs School of Engineering, Department of Structural Engineering, University of California, San Diego, CA, USA

ROSE School European School for Advanced Studies in Reduction of Seismic Risk, Pavia, PV, Italy

Priestley Structural Engineering, Diamond Harbour, New Zealand

**Andrea Prota** Dipartimento di Ingegneria Strutturale, Università degli Studi di Napoli Federico II, Naples, Italy

**Andrei M. Reinhorn** Department of Civil Structural and Environmental Engineering, University at Buffalo – State University of New York, Buffalo, NY, USA

**Klemen Rejec** Faculty of Civil and Geodetic Engineering, University of Ljubljana, Ljubljana, Slovenia

**José I. Restrepo** Department of Structural Engineering, University of California at San Diego, San Diego, CA, USA

**Rodrigo Retamales** Rubén Boroschek and Associates, Santiago, Chile

**Saiid M. Saiidi** Department of Civil and Environmental Engineering, University of Nevada, Reno, NV, USA

**Tomohiro Sasaki** Hyogo Earthquake Engineering Research Center, National Research Institute for Earth Science and Disaster Prevention, Miki City, Hyogo-ken, Japan

**Majid Baradaran Shoraka** Civil Engineering Department, University of British Columbia, Vancouver, BC, Canada

**Akira Tasai** Department of Architecture and Urban Culture, Yokohama National University, Yokohama, Kanagawa, Japan

**Masaomi Teshigawara** Department of Architecture, Yokohama National University, Nagoya, Aichi, Japan

**Miha Tomaževič** Slovenian National Building and Civil Engineering Institute, Department of Structures, Ljubljana, Slovenia

**Ching-Yi Tsai** Department of Civil Engineering, National Taiwan University, Taipei, Taiwan

**Keh-Chyuan Tsai** Department of Civil Engineering, National Taiwan University, Taipei, Taiwan

**Georgios Tsionis** Department of Civil Engineering, University of Patras, Patras, Greece

**Žiga Turk** Faculty of Civil and Geodetic Engineering, University of Ljubljana, Ljubljana, Slovenia

**Gerardo M. Verderame** Dipartimento di Ingegneria Strutturale, Università degli Studi di Napoli Federico II, Naples, Italy

**Victor K. Victorsson** Global Engineering, Swiss Reinsurance Company Ltd., Mythenquai, Zürich, Switzerland

**Ashkan Vosooghi** AECOM, Sacramento, CA, USA

**John W. Wallace** Department of Civil Engineering, University of California, Los Angeles, Los Angeles, CA, USA

**An-Chien Wu** NARLabs, National Center for Research on Earthquake Engineering, Taipei, Taiwan

**Tony Y. Yang** Civil Engineering Department, University of British Columbia, Vancouver, BC, Canada

**Richelle G. Zafra** Department of Civil Engineering, University of the Philippines Los Baños, Laguna, Philippines

# Part I
# Helmut's Vision

# Chapter 1
# Challenges Towards Achieving Earthquake Resilience Through Performance-Based Earthquake Engineering

**Helmut Krawinkler and Gregory G. Deierlein**

**Abstract** Much has been accomplished in performance-based earthquake engineering over the past two decades. Processes have been established that facilitate probabilistic seismic hazard analysis, evaluation of relevant engineering demand parameters through advanced modeling and nonlinear response history analysis, quantification of damage measures and associated repair/replacement costs at the component level, and aggregation of losses for structural and nonstructural systems. The outcome is a probabilistic assessment of direct economic loss and collapse safety due to earthquakes. In contrast to assessment of structural collapse and direct losses, comparatively less has been accomplished in quantifying factors that affect downtime, business interruption, and community functions. These issues are critically important to bridge between performance of a single structure and the earthquake resilience of a community or region or country. A key aspect of resilience is looking beyond direct damage and losses to their implications on disaster response and recovery. From a societal perspective, resilience is the key challenge to mitigate the lasting effects of earthquakes. Drawing upon relevant research and recent initiatives in California to create more earthquake resilient communities, this paper explores challenges to improve performance-based engineering to address specific aspects of resilience.

---

(Helmut Krawinkler) Author was deceased at the time of publication.

H. Krawinkler • G.G. Deierlein (✉)
Department of Civil and Environmental Engineering, John A. Blume Earthquake Engineering Center, Stanford University, Stanford, CA, USA
e-mail: ggd@stanford.edu

M. Fischinger (ed.), *Performance-Based Seismic Engineering: Vision for an Earthquake Resilient Society*, Geotechnical, Geological and Earthquake Engineering 32,
DOI 10.1007/978-94-017-8875-5_1, 

**Keywords** Performance-based earthquake engineering • Resilience • Risk • Safety • Loss assessment

## 1.1 Introduction

Over the past 20 years, performance-based earthquake engineering (PBEE) has developed from the conceptual framework to a workable set of procedures and enabling technologies. As described in SEAOC's *Vision 2000* report (SEAOC 1995), "*the intent of performance-based earthquake engineering is to provide methods for siting, designing, constructing and maintaining buildings, such that they are capable of providing predictable performance when affected by earthquakes.*" Here the key distinction from traditional earthquake engineering is the emphasis on *predictable performance* – implying the need for methods to determine the expected response of structures and to relate this to meaningful performance metrics. In first generation implementations of PBEE, such as *FEMA 273* (1997), performance is quantified by approximate relationships between structural component deformations and qualitative performance measures of Immediate Occupancy, Life Safety and Collapse Performance. In contrast, the current second-generation procedures, most notably those embodied in *FEMA P-58 Seismic Performance Assessment of Buildings* (2012a), quantify performance in terms of direct economic losses and collapse risk. Other performance measures, including risks of building closure, repair times and casualties are also included in the *FEMA P-58* procedures, though admittedly with more reliance on judgment.

Whereas the primary developments in PBEE have focused on the performance of individual buildings and facilities, from a societal view, it is ultimately the aggregate performance of the built environment and resilience of communities that are most important. The United Nations International Strategy for Disaster Reduction defines resilience as follows: "*The capacity of a system, community or society potentially exposed to hazards to adapt, by resisting or changing in order to reach and maintain an acceptable level of functioning and structure. This is determined by the degree to which the social system is capable of organizing itself to increase this capacity for learning from past disasters for better future protection and to improve risk reduction measures* (UNISDR 2004)." Implied in this statement is awareness, planning, improved protection, leadership, and resource allocation. PBEE can contribute to each of these aspects, but major contributions can be made to improved awareness, protection and planning. The paper discusses the role of PBEE in quantifying earthquake risks and facilitating better informed planning and design of the built environment. In taking a broader view of performance, a key challenge is to move beyond evaluation of direct losses from earthquakes to emphasize factors that are most important to recovery and rebuilding.

## 1.2 PBEE: Background and Status

### *1.2.1 PBEE Framework*

The high level objectives of PBEE are to develop scientifically-based transparent engineering methods and tools that can:

1. Facilitate decision making of cost-effective risk management of the built environment in areas of high seismicity
2. Facilitate the implementation of performance-based design and evaluation by the engineering profession
3. Provide a foundation on which code writing bodies can base the development of transparent performance-based provisions
4. Facilitate the development and implementation of innovative systems (response modification devices, rocking/self-centering systems, etc.)

The underlying framework for the current generation of performance-based approaches is shown in Fig. 1.1. This framework was developed by the Pacific Earthquake Engineering Research (PEER) Center (Cornell and Krawinkler 2000; Moehle and Deierlein 2004; Krawinkler and Miranda 2004) and has since been implemented in the *FEMA P58* (2012a). The framework provides a clearly articulated procedure to relate quantitative measures of the earthquake hazard to system performance metrics. While this overall framework is well-established, details of

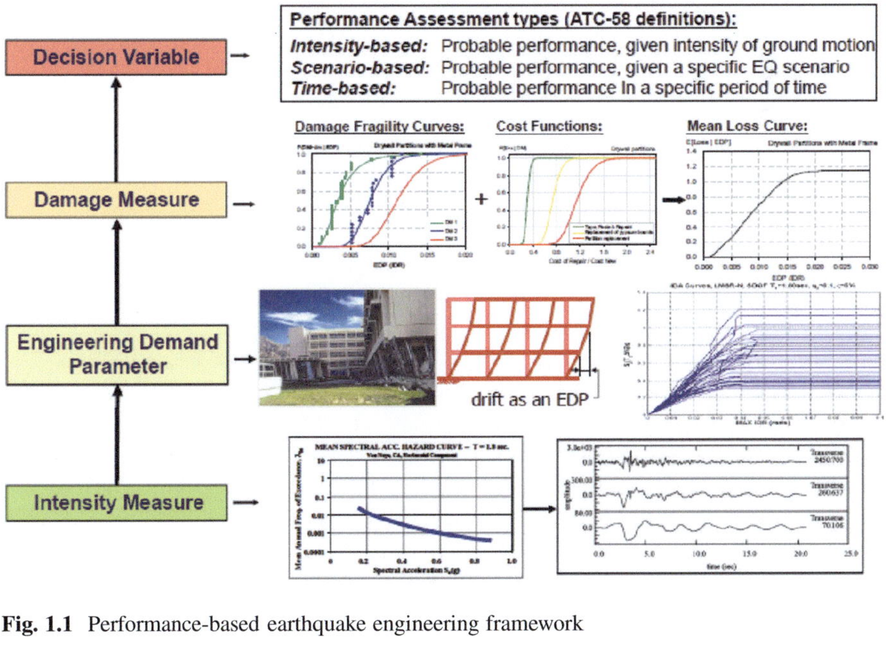


**Fig. 1.1** Performance-based earthquake engineering framework

the procedures are still being further developed and refined. Brief highlights of methodology components and their current status are as follows:

***Earthquake Hazard***: For use in nonlinear dynamic analyses, the earthquake hazard is characterized by input ground motions, which may be obtained by scaling or spectrally matching recorded motions or through earthquake simulations. While it is generally accepted to characterize the ground motions based on their spectral acceleration intensity, there is continued exploration on ways to incorporate frequency content, duration, and other aspects of the earthquake hazard in the input ground motions. The concept of Conditional Spectra, which accounts for correlation of ground motion intensities at multiple periods, has been proposed as a more appropriate target than Uniform Hazard Spectra to characterize the spectral intensity (e.g., Baker 2011; Bradley 2010), and research is ongoing to address near-fault directivity pulses, duration, and other effects (e.g., Champion and Liel 2012; Chandramohan et al. 2013; Shahi and Baker 2011). For a comprehensive summary and recommendations on this topic the reader is referred to a recent report, *Selection and Scaling Earthquake Ground Motions for Performing Response-History Analyses* (NIST 2011).

***Structural Analysis***: Nonlinear dynamic (response history) analysis is arguably the most mature component of PBEE, but many challenges remain to validate and improve the reliably of technologies to simulate the response of realistic structures from the initiation of damage up to the onset of collapse. Commercially available analysis software with capabilities to simulate elastic and moderately nonlinear response of three-dimensional models are becoming used in practice (Deierlein et al. 2010); however, the ability of these to model large inelastic deformations is questionable. Even in research, where models have been developed to capture strength and stiffness degradation up to the onset of collapse (e.g., Ibarra et al. 2005; Haselton et al. 2010), the modeling capabilities are limited to certain behavioral effects and by calibration of phenomenological parameters. Moreover, the accuracy of models to determine demand parameters, such as local deformations, residual drifts, and floor accelerations has not been fully validated. As other components of the PBEE process mature, the limitations in nonlinear structural analysis will become more important to address.

***Damage Assessment***: Perhaps the most unique new feature of PBEE is the formalization of damage assessment models, where the damage states and demand parameter limits are defined in terms of repair thresholds that have specific costs and consequences. For example, the limiting drift criteria for partition walls correspond to repair states that increase from (1) patching and repainting, to (2) replacement of gypsum wallboards, to (3) complete replacement of the wall and its embedded electrical and mechanical components (Taghavi and Miranda 2003). These repair limits can then be related to the cost, duration and other implications of repair. The *FEMA P-58* (2012a) development effort created many new damage fragility curves for a wide range of structural and nonstructural components and facilitated the practical implementation of damage assessment. Nevertheless, to fully realize the full potential of PBEE, further work remains to validate and expand the library of damage data and fragility functions.

***Performance Calculations***: Translating damage into appropriate performance metrics is the most important stage of PBEE, though probably the least well-developed. Performance measures have been coined "death, dollars and downtime", referring to risk of casualties, economic losses, and loss of function, but quantifying these seemingly straightforward metrics remains the most elusive. To date, most emphasis has been on calculating direct costs associated with repair of damage. *FEMA P58* provides repair costs, developed by professional cost estimators, for each component damage function. *FEMA P58* also includes consequence functions to calculate casualties, repair time, and building placard tagging (denoting safety for occupancy), though with relatively little data or hard science to determine these, their development relies heavily on judgment. As will be expanded on later, in addition to the need to validate and improve these existing performance models for individual facilities, more thought must be given to measures of communities (e.g., cities and urban regions comprised of large building inventories) and to relate building-specific measures to community-wide concerns.

### *1.2.2 Benchmarking Building Performance*

Some of the first applications of the PBEE tools have been to evaluate the performance of buildings designed according to current building codes. The studies are intended to provide a basis against which to judge the performance of other new or existing buildings and to evaluate the effectiveness of building code provisions. In companion studies, Haselton et al. (2010) and Ramirez et al. (2012) evaluated the performance of a set of modern concrete-framed buildings, designed for a high-seismic region near Los Angeles. They reported rates of collapse risk that range from 0.4 to 3.6 % in 50 years and expected annual losses (direct costs) on the order of about 1 % of the building replacement cost. With such data, the more important question becomes whether this level of performance is appropriate or optimal (in a cost-benefit sense) for individual building owners or society at large.

In an extension to this study, Ramirez and Miranda (2012) examine the breakdown of losses associated with repair versus building replacement. As shown in Fig. 1.2, their results reveal that over half of the expected loss is from damage that is deemed non-repairable (residual drifts in excess of 1.5 %), leading to building demolition. Their results also confirm that building collapse is a small contributor to direct losses for modern building designs. However, whether building replacement arises from collapse or demolition, apart from the cost of replacement, the complete replacement of the building has important long-term consequences on displacement of occupants and loss of function. This is in contrast to direct losses associated with damage of non-structural components, which accrue rapidly under modest ground motion intensities, but could be repaired faster and, possibly, while the building

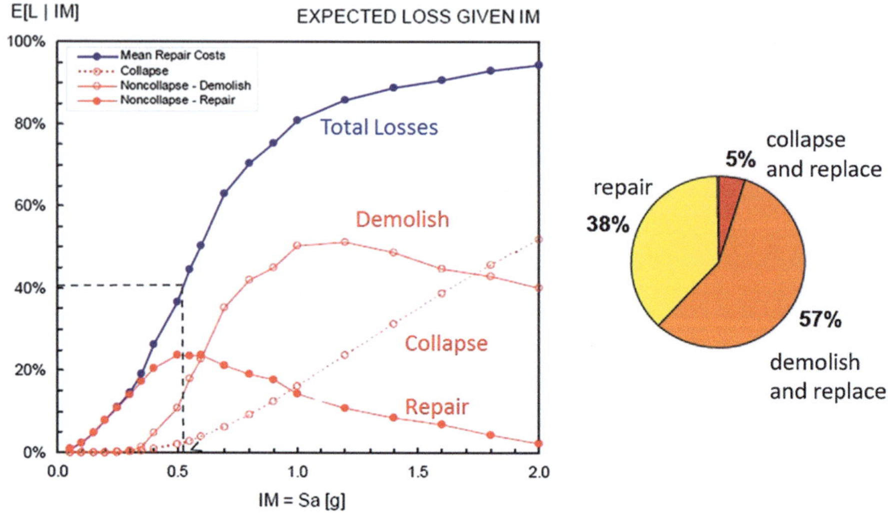


Fig. 1.2 Components of expected loss for a low-rise office building (Ramirez and Miranda 2012)

remains occupied. Thus, direct economic losses due to these repairs may have significantly less impact on indirect losses than direct losses associated with major structural repairs or building replacement.

In a related study, Liel et al. (2010) and Liel and Deierlein (2013) examine the collapse safety and losses of non-ductile concrete buildings, representative of buildings constructed before ductile detailing provisions were introduced to practice in the mid-1970s. The reported collapse risks for the non-ductile concrete buildings are on the order of 30 to 40 times higher than for modern code-conforming buildings, whereas direct economic losses (due to repair and replacement) are only twice those for modern buildings. This data helps confirm that it is the collapse and casualty risks, rather than direct economic losses, which are the primary consideration for existing non-ductile concrete buildings. Questions related to the safety of non-ductile concrete buildings and what, if any, government policies or other measures should be implemented to address the risk, are the focus of the Concrete Coalition (http://www.concretecoalition.org/) and related efforts in California.

### *1.2.3 Implementation of PBEE Framework*

The PBEE framework described above is influencing the development of guidelines and standards in the United States. Three significant developments are briefly summarized below.

***FEMA P58***: The development of *FEMA P58 Seismic Performance Assessment of Buildings* (2012a) represents a comprehensive implementation of PBEE. The *FEMA P58* procedures allow for evaluating the risks of (1) collapse and casualties, (2) direct economic losses to repair damage or replacement of collapsed or demolished buildings, (3) repair time, which is indexed off of repair costs, and (4) building closure, which is defined in terms of criteria defined for an "unsafe" (red) post-earthquake building inspection placard. *FEMA P58* incorporates these performance measures in three approaches that are referred to as intensity-based, earthquake scenario-based, or time-based assessments. The intensity-based assessment, where performance is calculated for a specified spectral acceleration response spectrum, is the most basic of the approaches and a subset component of the other two. Results of the scenario-based assessment, defined by an earthquake fault rupture magnitude and distance to the building site, reflects both the expected value of ground motion spectral intensity and the dispersion of this intensity for the specified scenario. The time-based assessment is the most comprehensive of the approaches, considering all earthquakes affecting a site and their risk of occurrence over a specified period of time.

In addition to assessment procedures, *FEMA P58* provides a library of damage and consequence functions, to evaluate losses in common building systems. Software called PACT (Performance Assessment Toolkit) is also available to apply the procedures and facilitate their practical use by design professionals.

***FEMA P695* and new MCE Maps**: The *FEMA P695 Quantification of Building Seismic Performance Factors* (2009) outlines a procedure to determine seismic force reduction factors (e.g., R, $\Omega_o$ and $C_d$ factors) that are used to define the minimum seismic base shear requirements in US building codes, such as the *ASCE 7* (ASCE 2010). The underlying approach of *FEMA P695* entails quantifying the collapse risk using nonlinear dynamic analysis, combined with judgment-based factors to account for uncertainties. Nonlinear dynamic analyses are used to assess the median value of notional collapse fragility curves, and the dispersion (uncertainty) in the collapse fragility is determined by variability in nonlinear response due to alternative ground motion records along with judgments of uncertainties arising due to the quality of (1) design and construction, (2) nonlinear analysis models, and (3) knowledge of structural behavior. While *FEMA P695* was conceived for the specific purpose of establishing response parameters for design, the collapse assessment procedures follow a performance-based approach that can be modified for more general use. Perhaps the most remarkable aspect of *FEMA P695* is to establish a minimum collapse risk, defined as a conditional collapse probability of 10 % under the Maximum Considered Earthquake (MCE) intensity. This collapse risk is based on judgments informed by benchmark studies of representative buildings designed according to current building code provisions.

In the United States, the MCE ground motion intensity has traditionally been defined in terms of ground motion exceedance rates, typically a 2 % chance of exceedance in 50 years. Building on the collapse fragilities defined in *FEMA P695*, the MCE seismic design maps for the United States have recently been revised to provide more consistent collapse safety over the entire United States (Luco et al. 2007). These new MCE design maps are targeted based on a maximum risk of collapse with a 1 % chance of exceedance in 50 years. This "risk targeted" approach is in contrast to previous MCE maps that were based on ground motion exceedance rates. Similar to the permissible collapse risk criteria of *FEMA P695*, the target risk of 1 % in 50 years is based on a combination of judgment and benchmark building studies. The new MCE design map intensities were obtained by integrating site ground motion hazard information with a generic collapse fragility curve that has an assumed lognormal dispersion of 0.6 and a 10 % probability of collapse at the MCE intensity (as specified in the *FEMA P695* procedures). Thus, given the default collapse fragility and the ground motion hazard for a specific location, the MCE intensity was determined for each map location so as to yield a target collapse risk of 1 % in 50 years. These uniform risk MCE maps have been adopted into the latest *ASCE 7* (2010) seismic design standard.

**Tall Building Guidelines**: As an alternative to traditional prescriptive design requirements for tall buildings, new guidelines have recently been developed to assess the adequacy of tall buildings based on nonlinear dynamic analysis (PEER 2010; LATBSDC 2011). The guidelines are intended to provide equivalent performance to that provided by prescriptive building code requirements, while providing a more transparent design basis that can be modified to provide enhanced performance. By focusing attention on the intended performance, they highlight important questions as to whether tall buildings, with high occupancies and potential consequences from earthquake damage, should be designed to higher performance targets than conventional low-rise buildings.

### *1.2.4 PBEE of Distributed Systems*

Whereas the current implementations of PBEE are primarily geared towards evaluating the performance of individual facilities, there are obvious cases where PBEE approaches only make sense to apply at the system level. For example, in transportation systems the performance of the overall highway system must consider network interactions between individual bridges. Thus, except for bridge collapse safety, which has direct implications on the safety of drivers, the functional performance of individual bridges is only important as it relates to functionality of the overall highway system, whose performance is typically measured in terms of traffic delay time (e.g., Kiremidjian et al. 2007; Chang et al. 2000). The same sort of argument could be made for other utility systems, such as water distribution systems, where the water service level depends on the performance and interactions between various network components associated with water supply, storage, treatment, and pipeline transmission (e.g., Davis et al. 2012; Romero et al. 2010).

Conceptually, extension of the PBEE framework from component to system performance is straightforward, but, implementation of the framework presents several challenges. As most systems are geographically distributed, performance assessment requires earthquake scenario-based approaches, which consider earthquake damage and functionality of components across the distributed network. Thus, the ground motion hazard assessment requires consideration of spatial correlations between ground motion intensities for scenario earthquakes (e.g., Han and Davidson 2012). While the seismic demands and physical damage can generally be evaluated discretely for each component, the consequence of damage on system performance requires a comprehensive system analysis, considering network interactions between the components. Evaluation of the system performance itself may be further complicated by exogenous effects of the earthquake on the functional demands on the systems. For example, travel times and delays on a transportation system depend on both the physical condition of the highway network and on the demand for transportation. As the travel demand is a function of economic or other activity, it is likely to be impacted by earthquake damage to non-transportation facilities and systems. Similarly, service level demands for water and other utilities may be impacted by earthquake damage to other systems. Therefore, to the extent that the changes in demand and interdependencies between systems depend on socio-economic factors impacted by the earthquake, these factors should be considered in assessing their earthquake performance.

## 1.3 From PBEE to Earthquake Resilience

While the performance-based methods described previously are a major step forward towards quantifying and managing earthquake risks of individual buildings, a much broader interpretation of performance is needed to understand how communities will be impacted and recover from devastating earthquakes. Consideration of recovery, including its dependence on available resources and the human workforce, raises important new questions that go beyond the traditional PBEE metrics. As illustrated in Fig. 1.3, resilience relates to the loss in functionality in a community that depends on the amount of damage caused by the earthquake disaster and the rate at which the functionality is recovered. The total loss is represented by the "loss triangle" which is the integration of the reduced system function over time to recovery (NRC 2011). This loss can be reduced by (1) pre-disaster mitigation to reduce earthquake damage and its consequences, and (2) planning and taking appropriate measures to hasten recovery and rebuilding. Thus, a key component of resilience is to incorporate post-disaster recovery and rebuilding considerations into the pre-disaster evaluation and planning. There is a large body of published work on resilience to earthquakes and other natural hazards, ranging from theoretical to applied and from socio-economic and political aspects to engineering oriented (e.g., UNISDR 2004; NRC 2011; Bruneau et al. 2003; Cutter et al. 2010; Poland 2012). Common to most of these are four dimensions to resilience from earthquakes:

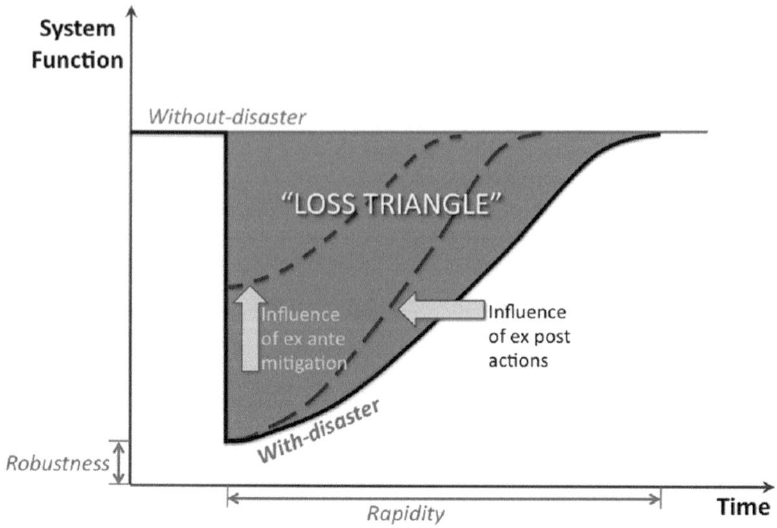


**Fig. 1.3** Idealized concept of resilience (NRC 2011)

***Technical*** – concerning the physical characteristics of the built environment including (1) evaluation of the expected seismic performance of buildings, lifeline systems, etc. and implications on post-earthquake functionality, and (2) planning and designing ways to improve performance through retrofit of existing facilities and enhancements to new facilities. As recovery and rebuilding is central to resilience, the technical engineering considerations must go beyond evaluation of expected damage to address post-earthquake functionality (e.g., safety to aftershocks) and repair of the buildings and infrastructure.

***Organizational*** – concerning governance and organizations that have responsibility to plan and lead post-earthquake response, recovery and rebuilding. While the natural emphasis in organizations is on preparations for emergency response, resilience planning requires emphasis on longer-term considerations, such as natural hazards considerations in land use planning and development of streamlined post-earthquake decision-making procedures that can facilitate repair and rebuilding.

***Social*** – concerning individual residents and non-governmental community organizations and (1) how these groups are likely to be impacted by the earthquake, (2) measures that can be taken to lessen these impacts on these groups, and (3) ways to enhance the capability of these groups to participate in recovery and rebuilding. One of the most important social factors concerns the availability of housing or shelters to help ensure that communities will not be displaced and can function after the earthquake. The social component also involves the effectiveness of civic and religious organizations to help coordinate local recovery and rebuilding.

***Economic*** – relating to (1) the economic consequences of the earthquake, including direct economic losses and indirect losses associated with business interruption, lost jobs, etc. and (2) the availability of resources to rebuild after a disaster, including insurance, availability of financing, government grant programs, and savings of individuals or business. An important related factor affecting the earthquake impact and recovery is the economic profile of the community.

While there is general consensus as to the overall goals and definition of resilience, one of the major challenges is to measure resilience, since this is an essential step towards identifying and overcoming weaknesses. As one research group notes regarding resilience measures, "qualitative models tend to be more comprehensive than quantitative models, which are instead more discipline-oriented. This observation demonstrates the marked disconnect between what is thought to be an ideal understanding of resilience versus what is actually measurable" (Verrucci et al. 2012). Studies that attempt to comprehensively quantify resilience metrics in all four of its dimensions generally resort to indexed ratings across a broad range of topics, such as (1) population and building density in areas of high expected ground shaking, (2) typical age and quality of building stock, (3) availability of emergency response and shelter facilities, (4) prevalence of earthquake insurance and financial resources of communities, and (5) strength of community organizations, etc. (Verrucci et al. 2012; Cutter et al. 2010). Studies that are more quantitative, such as examination of restoration of water service following the Northridge earthquake (Davis et al. 2012) or critical lifeline and support systems (Bruneau et al. 2003), tend to be more case- and discipline-specific.

Notwithstanding the challenges in measuring resilience, there is no question that efforts to measure and improve resilience must consider its multiple dimensions. This is not to say that specific steps to improve resilience cannot be discipline-specific, since most improvements are usually developed and implemented within a discipline. But, in order to be effective, all individual efforts to improve resilience must be devised and integrated through a larger overarching plan that helps establish performance requirements for the individual components.

Experiences from large earthquakes and other natural disasters demonstrate that community resilience cannot be evaluated solely in terms of the performance of individual buildings or lifeline system components. The February 2011 earthquake in New Zealand is an obvious example where the damage to individual buildings has had a disproportionate effect in the social and economic devastation of the central business district of Christchurch. This situation is at odds with the fact that current building code requirements in New Zealand, and most other countries, do not distinguish between design requirements for buildings in a densely populated urban region, which can be impacted by a single earthquake, and buildings in outlying suburban areas (Liu 2012). The new "risk targeted" MCE maps in the *ASCE 7* (2010) are another example, where efforts to make building codes risk consistent across the United States may be at odds with risks to specific urban regions. Similar comparisons could be made to design requirements for levees and other flood protection, and whether components of a network that are essential to a city or region (such as levees around New Orleans) should be designed to higher standards than ones where the consequences of isolated failure are less.

## 1.4 San Francisco Resilient City Initiative

To mark the 2006 centennial of the 1906 San Francisco earthquake and fire, an earthquake scenario study was conducted to consider what would happen to modern day San Francisco if the 1906 M7.9 earthquake were to reoccur. The study predicted a disaster with up to 3,400 deaths, 10,000 buildings destroyed, 250,000 households displaced, and $120 billion in losses (Kircher et al. 2006). This study, together with increased awareness of risks from the 1989 Loma Prieta earthquake and other disasters, prompted the San Francisco Planning and Urban Research Association (SPUR) to undertake an initiative to evaluate ways to make San Francisco more resilient to earthquakes. Spearheaded by earthquake engineers, this "resilient city" initiative involves a broad range of design and emergency professionals, city government officials, and urban planners (Poland 2009; SPUR 2009). It provides a focused example to promote resilience through pre-earthquake mitigation and planning for post-earthquake recovery, and it illustrates ways that PBEE can help inform the process and for earthquake engineers to engage with a broader constituency. This resilient city initiative (Fig. 1.4) has been an integrating mechanism for other related efforts, including the CAPPS project (Community Action Plan for Seismic Safety, http://sfcapss.org/) to identify vulnerabilities in the San Francisco and ways to mitigate these so as to preserve the city's diverse communities. The CAPPS project identified comparable overall damage and losses as for the 1906 earthquake scenario study but with more specifics on the vulnerable building stock in San Francisco. It also makes recommendations on steps to mitigate

**Fig. 1.4** San Francisco resilient city initiative (SPUR 2009)

damage risks through seismic retrofit and to facilitate post-earthquake recovery by establishing governance plans and repair standards for rebuilding.

The SPUR initiative embraces the goal that "Resilient communities have an ability to govern after a disaster has struck. The communities adhere to building standards that allow power, water and communication networks to begin operating again shortly after a disaster and allow people to stay in their homes, travel to where they need to be, and resume a fairly normal living routine within a few weeks. They are able to return to a *new* normal within a few years." (Poland 2009). The resilient city initiative is built around a realistic assessment of damage from an "expected earthquake" and its impact on response and rebuilding. Seismic mitigation and recovery strategies are then identified and evaluated to enable an appropriate timetable for recovery. The concept of an "expected earthquake" (scenario earthquake) is important to establish a common basis for evaluation and planning over geographically distributed facilities, systems and organizations. The "expected earthquake" is defined as a M7.2 event on a nearby portion of the San Andreas fault. This is not the most extreme earthquake that can affect San Francisco, but it is judged to be the most appropriate for overall assessment and planning purposes. Presumably, scenarios that are more or less severe could be evaluated in follow up studies to fine tune the planning. Resilience assessment is based on transparent performance measures of facilities and systems, considering direct earthquake damage and its implications on the city-wide recovery effort.

Seismic performance targets for facilities and systems are defined based on the implications of damage on post-earthquake functionality and repairs. Building performance is characterized by the following performance categories:

*A* – *Safe and operational*: Essential facilities such as hospitals and emergency operations centers
*B* – *Safe and usable during repair*: "shelter-in-place" residential buildings and buildings needed for emergency operations
*C* – *Safe and usable after repair*: current minimum design standard for new, non-essential buildings
*D* – *Safe but not repairable*: below standard for new, buildings; often used as a performance goal for existing buildings undergoing voluntary rehabilitation
*E* – *Unsafe* – partial or complete collapse: damage that will lead to casualties in the event of the "expected" earthquake

Targets for performance of utility and transportation systems are organized into the following three categories, depending on how quickly their level of service can be restored following the expected earthquake:

*Category I* – resume 100 % service within 4 h
*Category II* – resume 90 % service within 72 h, 95 % service within 30 days and service 100 % within 4 months
*Category III* – resume 90 % service within 72 h, 95 % service within 30 days, and 100 % service within 3 years

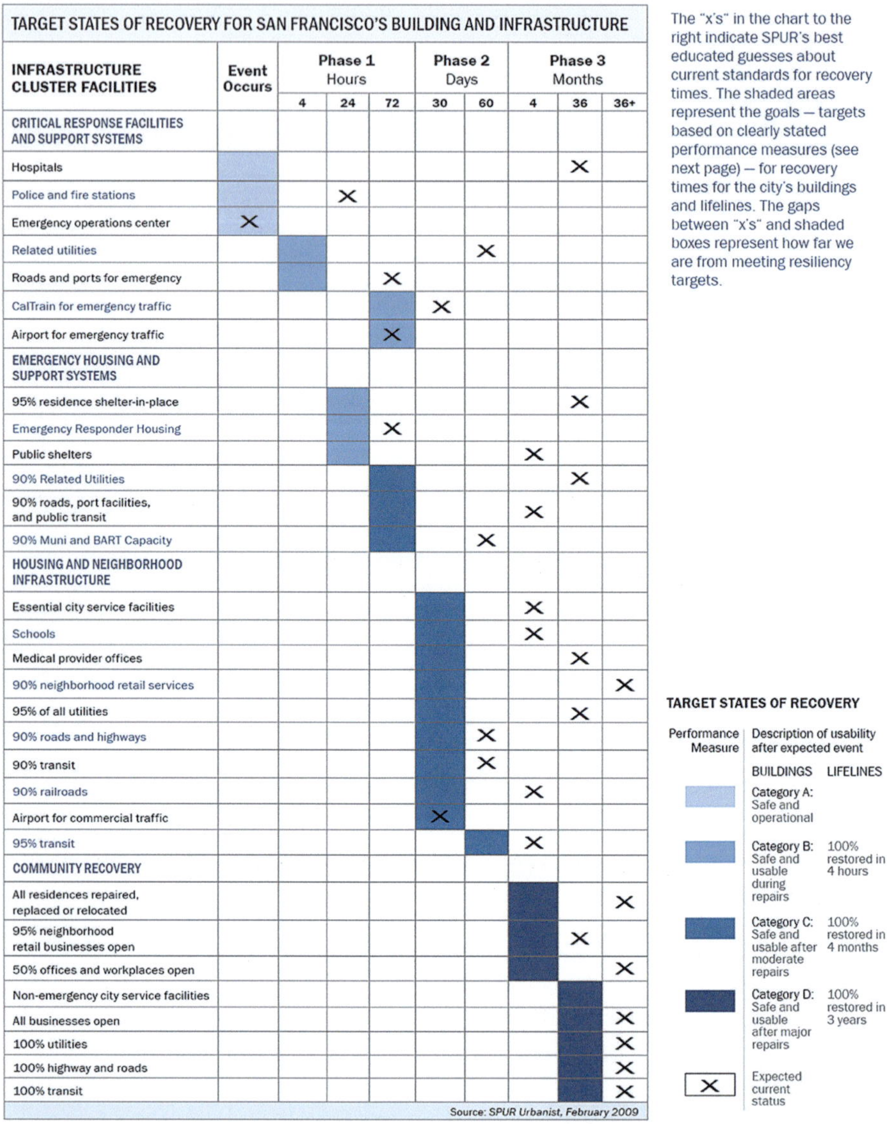


**Fig. 1.5** Target recovery states for San Francisco's buildings and infrastructure (SPUR 2009)

Using these categories, specific target goals for building and infrastructure are established, considering city-wide needs. These are illustrated in Fig. 1.5, where specific performance goals are identified for buildings based on their occupancy type and usage and for lifeline systems (designated by shading corresponding to building categories A through D and systems categories I through III). The "X" markers in Fig. 1.5 are estimates of performance for the current inventory of facilities,

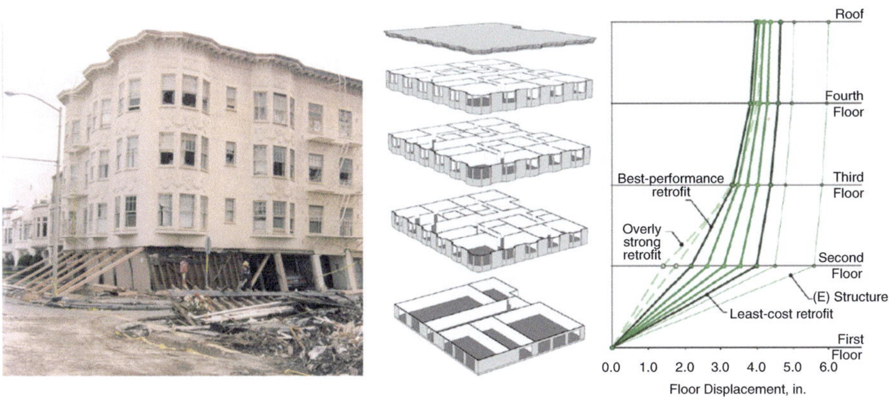


**Fig. 1.6** Assessment and retrofit for soft-story wood-framed buildings (FEMA 2012b)

indicating where measures are needed to upgrade buildings and other facilities. It should be noted that while there is some data to support the performance targets and inventory estimates in Fig. 1.5, these are based largely on judgments from the professional participants of the SPUR resilient city initiative and related CAPSS project.

While buildings in category E, deemed to pose a significant life safety risk, are a primary concern, another important focus is to determine whether buildings can provide for post-earthquake occupancy, including "shelter-in-place" for residential buildings (SPUR 2011). This emphasis on post-earthquake performance is an important new consideration since performance-based research and developments have traditionally focused on collapse (life-safety risk) and repair cost (economic losses). Comparatively less attention has been paid to quantifying post-earthquake occupancy and function, in part due to the lack of specified performance targets. In this regard, the specific targets defined by the building performance categories (A through E) and specified in Fig. 1.5 are a major step forward to quantifying the performance targets for individual buildings to ensure community resilience.

In addition to outlining a framework for community resilience, the resilient city initiative has captured the attention of civic leaders and prompted earthquake mitigation legislation to address an important weakness that was brought to light. The CAPPS project identified soft-story wood-framed apartment buildings (see Fig. 1.6) as a significant weakness, where scenario earthquake damage posed a significant collapse risk (category E) and would displace a large number of residents. This prompted the development of performance-based guidelines to assess and retrofit soft-story wood-frame buildings (FEMA 2012b) and to recent legislation by City of San Francisco to require mandatory of these buildings (SFGate 2013). This is an excellent example where seismic mitigation policies resulted from (1) identifying the risks to both the building occupants and broader community, and (2) providing cost-effective engineering solutions to assess and mitigate the risks through retrofits designed by performance-based methods.

| CoRE Rating | Safety | Reparability | Functionality |
|---|---|---|---|
| ★★★★★ | Life Safe | Loss <5% | Occupiable Immediately<br>Functional < 72 hours |
| ★★★★ | Life Safe | Loss <10% | Occupiable Immediately<br>Functional < 1 month |
| ★★★ | Life Safe | Loss <20% | Occupiable < 1 month<br>Functional < 6 months |
| Certified | Life Safe | Not estimated | Not estimated |
| Not Certified | Life Safe Hazard | Not estimated | Not estimated |

**Fig. 1.7** Building seismic rating system of the US Resiliency Council (Reis et al. 2012)

Another noteworthy development catalyzed by the resilient city initiative involves the development and implementation of a seismic rating system for buildings. Seismic building ratings have long been suggested as a mechanism to raise awareness of the expected building performance by building owners, occupants, and other stakeholders, but previous efforts to develop rating systems have languished. Building on the momentum of the resilient city initiative, the existing buildings committee of the Structural Engineers Association of Northern California has proposed a seismic rating system that reflects performance metrics similar to the A to E categories identified previously (SEAONC 2012). More recently, this rating system has been embraced by the U.S. Resiliency Council (http://usrc.org/), which is a new nonprofit organization that has been created to institutionalize implementation of the rating system. The U.S. Resiliency Council follows an approach of voluntary ratings, similar to how the LEED program is applied to evaluate green building performance (http://new.usgbc.org/leed). Shown in Fig. 1.7 is the proposed building rating system metrics, which are defined based on performance during the "expected earthquake". The performance categories of safety, reparability, and functionality are defined along the lines of building performance targets identified in SPUR's resilient city plan.

## 1.5 PBEE as a Facilitator Towards Seismic Resilience

Performance-based methods and technologies clearly have an important role in assessing and designing for community resilience. However, to effectively serve this role, PBEE research and development needs to expand beyond the current emphasis on calculating direct losses (collapse risk and repair costs) and place greater attention on post-earthquake functionality and repair. Referring to Fig. 1.8, SPUR's five building performance categories (A through E) can be described in terms of the resilience loss triangle, introduced previously in Fig. 1.3. For comparison, characteristic values of direct losses due to repair are also shown in Fig. 1.8. The

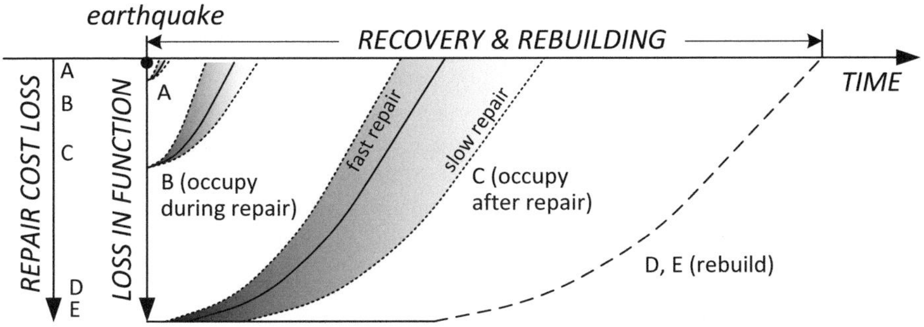


**Fig. 1.8** Direct repair cost and loss of function for alternative building performance categories

figure highlights several important distinctions between the performance for each building category:

- Whereas the direct repair costs for building categories A, B and C are relatively close (5, 10 and 20 %, per Fig. 1.7), the difference in post-earthquake functionalities are dramatically different. In particular, buildings in category C, which requires repairs prior to reoccupancy, have essentially the same initial loss in function as buildings in categories D (damaged beyond repair) and E (collapsed). On the other hand, buildings in category B (safe to reoccupy during repairs) have a much smaller loss in functionality.
- Beyond the initial loss in functionality, the speed with which repairs can commence and be completed can have a major effect on the total functional loss. Repairs for category B buildings, which are safe to reoccupy immediately after an earthquake, are likely to begin and be completed much sooner than those in category C buildings. The duration of repairs for both categories B and C will, of course, depend upon the details of the repairs and whether the original building designs included provisions to facilitate repairs, e.g., by isolating inelastic action in structural elements that are easy to replace.
- Once buildings are damaged to the point to be technically or economically prohibitive to repair (category D), the buildings have essentially the same loss in functionality as collapsed buildings (category E). As illustrated previously by the example of Fig. 1.2 and as has been observed in damaged buildings in Christchurch (from the 2011 earthquake), existing buildings may be far more likely to experience losses in category D that is generally recognized.

These considerations from Fig. 1.8 highlight the critical importance of two damage thresholds to community resilience: (1) the threshold damage for building closure, which differentiates between building category B and C, and (2) the threshold of damage that makes repairs prohibitive and demolition inevitable, which differentiates between building category C and D. While these thresholds are generally related to the amount of damage and repair costs, more so than the cost of damage, they may depend heavily on the nature of the damage and implications on repair. Foremost of these considerations is whether there is significant structural damage that jeopardizes the building safety and triggers building closure. This point runs counter to observations that damage to non-structural elements is a major contributor to "expected losses". While damage to nonstructural components is disruptive and can be expensive to repair, it is usually not the major driver to trigger building closure. A related consideration is whether there are significant residual story drifts, which are one of the primary triggers for building closure and, potentially, demolition. Data from the 1995 Kobe earthquake suggest a residual story drift threshold of ~1.4 % for demolition of steel-framed buildings (Iwata et al. 2006), and values for other systems are likely to be lower.

Looking beyond the performance of individual buildings, the San Francisco resilient city study highlights the importance of evaluating potential damage to the overall community – taking into account the region's inventory of buildings and the utility and transportation systems. As illustrated in Fig. 1.5, in addition to differentiating between building performance targets for critical versus non-critical facilities, the targets should also consider specific community needs for housing and commerce. While individual residential or office buildings are typically considered to be non-critical, it is critical to maintain functionality (occupancy) for a sufficient number of buildings in order to preserve community functions that are necessary for human welfare, recovery and rebuilding.

## 1.6 Concluding Remarks

While tremendous advancements have been made in PBEE methods and enabling technologies, many important challenges remain. Certainly, there is continuing need for improvements and refinements in all aspects of the methodology, from the characterization of ground motions through to evaluation of performance. However, from the standpoint of community resilience, the authors would venture that the most important research needs include the following:

- Improved analysis technologies to enable more reliable evaluations of residual drift and the collapse safety of structures. This is motivated by the need to (1) identify existing buildings that pose a significant life-safety risk (category E buildings), (2) differentiate between buildings that are safe or unsafe to occupy after an earthquake (category B versus C buildings), and (3) differentiate between buildings that are or are not likely to be demolished after an earthquake (category C versus D buildings).
- Improved evaluation of the economic loss and functional performance of large inventories (portfolio's) of buildings and implications on socio-economic factors for communities. These data are important to establish appropriate performance targets for buildings to ensure that communities can survive, rebuild and flourish again after a large earthquake. The critical need is to provide quantitative

measures to substantiate and refine the judgment-based targets proposed by the SPUR project in Fig. 1.5.

- Improved technologies to enable comprehensive regional earthquake scenario studies that reliably simulate ground motions, damage and reduced function of facilities and systems, and the process of rebuilding and restoration of functions. Ideally, these technologies would be based on more scientific (fundamental) models of phenomena that would reduce reliance on empirical models and judgment to assess both the immediate damage and the rebuilding process. In addition to modeling the physical building and infrastructure systems, the simulations to assess the reduction and restoration of functions should, to the extent possible, consider socio-economic factors that connect the physical and human elements.
- Development of innovative structural systems, devices and materials for buildings and infrastructure that can improve resilience by (1) decreasing the damage potential on functional performance and (2) facilitating repair and rebuilding. Design innovations are needed for new facilities as well as retrofit and repair of existing facilities. The value and effectiveness of these innovations should be judged in the context of how they reduce direct losses and improve resilience.

**Acknowledgments** The concept for this paper was conceived of by Helmut Krawinkler as a collaborative effort with the second author. This collaboration was cut short by Helmut's unexpected death on April 16, 2012, leaving the second author to fulfill Helmut's vision for the paper. Apart from the specific references cited in the paper, the primary sources for this paper are the inspiration, knowledge and insights that Helmut shared over his career and at the Bled 4 workshop in June 2011.

The authors gratefully acknowledge the contributions of many colleagues and former students who have helped advance the state-of-art in PBEE and for financial support for their work by the National Science Foundation, the Pacific Earthquake Engineering Research (PEER) Center, the Applied Technology Council (funded by FEMA and NIST), and the John A. Blume Earthquake Engineering Center at Stanford University. They are also grateful to the organizers and sponsors of the Bled 4 workshop, and in particular the tremendous contribution of Matej Fischinger in planning and hosting the workshop and producing the conference proceedings.

## References

ASCE (2010) Minimum design loads for buildings and other structures, ASCE/SEI standard ASCE 7-10. ASCE, Reston

Baker JW (2011) Conditional mean spectrum: tool for ground motion selection. J Struct Eng 137(3):322–331

Bradley BA (2010) A generalized conditional intensity measure approach and holistic ground-motion selection. Earthq Eng Struct Dyn 39:1321–1342

Bruneau M, Chang SE, Eguchi RT, Lee GC, O'Rourke TD, Reinhorn AM, Shinozuka M, Tierney K, Wallace WA, von Winterfeldt D (2003) A framework to quantitatively assess and enhance the seismic resilience of communities. Earthq Spectra 19(4):733–752

Champion C, Liel A (2012) The effect of near-fault directivity on building seismic collapse risk. Earthq Eng Struct Dyn 41(10):1391–1409

Chandramohan R, Lin T, Baker JW, Deierlein GG (2013) Influence of ground motion spectral shape and duration on seismic collapse risk. In: Proceedings of the 10th international conference on urban earthquake engineering, Tokyo, 9 p

Chang S, Shinozuka M, Moore J (2000) Probabilistic earthquake scenarios: extending risk analysis methodologies to spatially distributed systems. Earthq Spectra 16(3):557–572

Cornell A, Krawinkler H (2000) Progress and challenges in seismic performance assessment. PEER Newsletter, University of California, Spring 2000

Cutter SL, Burton CG, Emrich CT (2010) Disaster resilience indicators for benchmarking baseline conditions. J Homel Secur Emerg Manag 7(1). Article 51

Davis CA, O'Rourke TD, Adams ML, Rho MA (2012) Case study: Los Angeles water services restoration following the 1994 Northridge Earthquake. In: Proceedings of 15th world conference earthquake engineering, Lisbon

Deierlein GG, Reinhorn AM, Willford MR (2010) Nonlinear structural analysis for seismic design. NEHRP Seismic design technical brief no 4, NIST GCR 10-917-5, NIST, Gaithersburg

FEMA (1997) NEHRP guidelines for the seismic rehabilitation of buildings. FEMA report 273, Washington, DC

FEMA (2009) Quantification of building seismic performance factors. FEMA P695, Washington, DC

FEMA (2012a) Guidelines for seismic performance assessment of buildings. FEMA P58/Pre-Release, Aug 2012, Washington, DC

FEMA (2012b) Seismic evaluation and retrofit of multi-unit wood-frame buildings with weak first stories. FEMA P-807, Washington, DC

Han Y, Davidson RA (2012) Probabilistic seismic hazard analysis for spatially distributed infrastructure. Earthq Eng Struct Dyn 41:2141–2158

Haselton CB, Liel AB, Deierlein GG, Dean BS, Chou JH (2010) Seismic collapse safety of reinforced concrete buildings: I. Assessment of ductile moment frames. J Struct Eng 137(4):481–491

Ibarra L, Medina R, Krawinkler H (2005) Hysteretic models that incorporate strength and stiffness deterioration. Earthq Eng Struct Dyn 34:1489–1511

Iwata Y, Sugimioto K, Kuwamura H (2006) Reparability limit of steel structural buildings: based on the actual data of the Hyogoken-Nanbu earthquake. In: Cauffman SA (ed) Wind and seismic effects: proceedings of the 38th joint panel meeting, National Institute of Standards and Technology (NIST) special publication 1057, Gaithersburg, MD, pp 23–32

Kircher CA, Seligson HA, Bouabid J, Morrow GC (2006) When the big one strikes again – estimated losses due to a repeat of the 1906 San Francisco earthquake. Earthq Spectra 22(S2):S297–S339

Kiremidjian A, Moore J, Fan YY, Yazlali O, Basoz N, Williams M (2007) Seismic risk assessment of transportation network systems. J Earthq Eng 11:371–382

Krawinkler H, Miranda E (2004) Chapter 9: Performance-based earthquake engineering. In: Bertero VV, Bozorgnia Y (eds) Earthquake engineering: from engineering seismology to performance-based engineering. CRC Press, Boca Raton

LATBSDC (2011) An alternative procedure for seismic analysis and design of tall buildings located in the Los Angeles Region. Los Angeles Tall Buildings Structural Design Council, http://www.tallbuildings.org

Liel AB, Deierlein GG (2013) Cost-benefit evaluation of seismic mitigation alternatives for older reinforced concrete frame buildings. Earthq Spectra 29(4):1391–1411

Liel AB, Haselton CB, Deierlein GG (2010) Seismic collapse safety of reinforced concrete buildings: II. Comparative assessment of non-ductile and ductile moment frames. J Struct Eng ASCE 137(4):492–502

Liu A (2012) Integrated design for achieving building seismic resilience. In: Proceedings of 15th world conference earthquake engineering, Lisbon

Luco N, Ellingwood BR, Hamburger RO, Hooper JD, Kimball JK, Kircher CA (2007) Risk-targeted versus current seismic design maps for the conterminous United States. In: SEAOC

2007 convention proceedings, Squaw Creek, CA, 26–29 September 2007. Structural Engineers Association of California, pp 163–175
Moehle J, Deierlein GG (2004) A framework methodology for performance-based earthquake engineering. In: Proceedings 13th world conference on earthquake engineering, Vancouver, paper 679
NIST (2011) Selection and scaling earthquake ground motions for performing response-history analyses. NIST GCR 11-917-15, National Institute of Standards and Technology, Gaithersburg
NRC (2011) National earthquake resilience: research, implementation and outreach. National Research Council, National Academies Press, Washington, DC
PEER (2010) Seismic design guidelines for tall buildings. Pacific Earthquake Engineering Research Center, University of California at Berkeley, Berkeley
Poland C (2009) The Resilient City: defining what San Francisco needs from its seismic mitigation polices. http://www.spur.org/initiative/resilient-city
Poland C (2012) Guidelines for creating disaster-resilient communities. In: Proceedings of 15th world conference earthquake engineering, Lisbon
Ramirez CM, Miranda E (2012) Significance of residual drifts in building earthquake loss estimation. Earthquake Eng Struct Dyn 41:1477–1493
Ramirez CM, Liel AB, Mitrani-Reiser J, Haselton CB, Spear AD, Steiner J, Deierlein GG, Miranda E (2012) Expected earthquake damage and repair costs in reinforced concrete frame buildings. Earthq Eng Struct Dyn 41:1455–1475
Reis E, VonBerg E, Stillwell K, Mayes R (2012) The U.S. Resiliency Council – principles of formation. In: SEAOC-SEANM 2012 convention proceedings, Santa Fe, New Mexico, 12–15 September 2012. Structural Engineers Association of California, Sacramento, pp 82–88
Romero N, O'Rourke TD, Nozick LL, Davis CA (2010) Seismic hazards and water supply performance. J Earthq Eng 14:1022–1043
SEAOC (1995) Vision 2000, performance-based seismic engineering of buildings. Structural Engineers Association of California, Sacramento
SEAONC Existing Buildings Comm (2012) SEAONC's earthquake performance rating system: translating ASCE 31–03. SEAOC-SEANM 2012 convention proceedings, Santa Fe, New Mexico, 12–15 September 2012. Structural Engineers Association of California, Sacramento, pp 68–81
SFGate (2013) SF mayor wants to mandate earthquake upgrades. San Francisco Chronicle, 16 Feb 2013. http://www.sfgate.com
Shahi SK, Baker JW (2011) An empirically calibrated framework for including the effects of near-fault directivity in probabilistic seismic hazard analysis. Bull Seismol Soc Am 101(2):742–755
SPUR (2009) The Resilient City, part 1: before the disaster. SPUR Urbanist, Issue 479, San Francisco, Feb 2009. www.spur.org
SPUR (2011) Safe enough to stay. SPUR Shelter-In-Place Task Force Report. http://www.spur.org/initiative/resilient-city
Taghavi S, Miranda E (2003) Response assessment of nonstructural building elements. PEER report 2003/05, Pacific Earthquake Engineering Research Center (PEER), University of California, Berkeley
UNISDR (2004) Living with risk: a global review of disaster reduction initiatives, United Nations Pub. http://www.unisdr.org/we/inform/publications/657
Verrucci E, Rossetto T, Twigg J (2012) Multi-disciplinary indicators for evaluating the seismic resilience of urban areas. In: Proceedings of 15th world conference earthquake engineering, Lisbon

# Part II
# Global Vision

# Chapter 2
# Engineering Challenges on the Way to Resilient Structures and Communities

**Stephen Mahin**

**Abstract** Safety of occupants is of paramount importance in the design of any structure or system. However, in the aftermath of recent major earthquakes, worldwide attention has increasingly focused on the need for resilient structures and lifelines that are able to return to service quickly following a major earthquake. In this paper, engineering aspects of resilient communities are discussed, focusing on increasing the post-earthquake operability of those structures and lifelines critical to a community's needs in the aftermath of a major earthquake and the ability of occupants to "shelter-in-place" during repairs

**Keywords** Design criteria • Continued occupancy • Buildings • Bridges • Seismic isolation • Protective structures • Resiliency • Resilient communities

## 2.1 Introduction

Engineers specializing in the field of earthquake engineering are constantly reminded of the importance of their work by the worldwide occurrence of earthquakes, both great and small. Earthquakes pose a threat to nearly every country in the world. While experimental and analytical research and design studies provide important sources of information for continual improvement of building codes and construction practices, the largest impetus for change comes from the disparity between the observed and expected behavior. Sometimes a region is inadequately prepared for an earthquake, and efforts are made quickly to improve

S. Mahin (✉)
Department of Civil and Environmental Engineering, University of California, Berkeley, Berkeley, CA 94720-1710, USA
e-mail: mahin@berkeley.edu

M. Fischinger (ed.), *Performance-Based Seismic Engineering: Vision for an Earthquake Resilient Society*, Geotechnical, Geological and Earthquake Engineering 32,
DOI 10.1007/978-94-017-8875-5_2, 

the quality and enforcement of building codes. In other cases, earthquake damages identify shortcomings in fundamental theory or detailing and analysis requirements, and research is carried out to determine appropriate remedies for these problems. In other cases, public expectations regarding the performance of the constructed environment have evolved, and design criteria should change in keeping with these expectations.

Architectural styles, construction types, building codes, quality of engineering and construction, seismic hazard, and public expectations differ significantly from region to region around the world. As such, the performance of structures and lifelines observed in one region may be acceptable to those who live there, even though those living in another region might not view similar behavior were it to occur near them positively. Nonetheless, because of the substantial time between the occurrence of design level or greater seismic events in any one region of the world, engineers and public officials can analyze and draw valuable lessons from the performance of structures and systems that occur in other regions of the world. However, there is increasing recognition that the impact of an earthquake on the occupants and owner of a structure depends on more than the degree of structural damage to just that structure, and that an damaging earthquake can transform from a survivable disaster to a catastrophic event if the extent of even moderate damage to structures and lifelines results in the closing of a critical number of key businesses, services and housing units.

Moreover, in many industrialized countries, public and government officials have come to expect high levels of safety in a citizen's everyday life. For examples, spalling and cracking associated with formation of plastic hinges are pointed out as structural failures by public media, while they are expected according to *ductile* design principles. It appears that there is a growing discrepancy between the public's expectations and the focus of many engineers on providing the minimum seismic design needed to protect life safety.

Recent earthquakes in China, Chile, New Zealand and Japan have been as large or larger than the maximum levels of excitation considered in the design of typical structures. These events have tested many structures to their limits. In most cases, modern structures have done well. However, it is clear from all of these events that large earthquakes trigger other disasters that compound the disaster and may elevate its status to that of a catastrophe. For example, earthquake ground shaking in China not only leads to the collapse of many individual buildings, but also to an extraordinary number of landslides that caused additional damages and hampered rescue and recovery operations. Both Chile and Japan suffered from the effects of tsunami following the main earthquake ground shaking. The severity of ground shaking in New Zealand, Japan and Chile were large enough to trigger large-scale liquefaction. While such issues are often considered in the design of an individual structure, the cascading effects of them all occurring in a region as a result of the same event and transforming a natural disaster into a catastrophic regional event have only recently been recognized.

## 2.2 Lessons Learned and Relearned from Recent Earthquakes

Recent earthquakes in Chile, New Zealand and Japan reiterate many lessons learned in past earthquakes. For example, recent earthquakes remind us of the substantial risk posed by existing structures and lifelines constructed when earthquake engineering concepts were not as advanced as they are today. Tremendous damage was observed in New Zealand in unreinforced masonry structures, and damage was observed in all three countries in reinforced concrete constructed prior to the adoption of modern ductile detailing requirements. Other lessons re-learned include the need for providing a continuous load path, avoiding discontinuities in the lateral-load resisting systems, designing connections to develop the capacities of the elements framing into them, protecting systems from potential failures of the supporting soil, and so on. There are extraordinary opportunities to learn from these recent earthquakes. Substantial amount of instrumentation was installed throughout Japan, Chile and New Zealand to measure the movement of the ground, water and structures. Moreover, many efforts have been undertaken to document in high detail direct damages and the economic, social, political, medical and other impacts of this damage.

### *2.2.1 Effects of Ground Motions on Engineered Facilities*

Earthquakes in Chile, New Zealand and Japan are often noted as being larger than expected events. While these events were indeed large, generally predictions of future earthquakes are probabilistic in nature. These events should serve as a warning that median probabilistic estimates are not deterministic bounds on what may occur in future events. Specific ground motion observations are provided below related to subduction zone events, the importance of aftershocks and special issues raised regarding near and far field ground shaking.

### *2.2.2 Subduction Zone Events*

The recent subduction zone ruptures in Chile and Japan illustrate the potential for the damaging effects of such earthquakes to be experienced over a vast geographic area. Strong motion records are often very intense (up to 2.9 g) and very long (3 min). The motions in Japan are particularly complex, showing strong evidence of multiple segments rupturing sequentially along the fault. There are several thousands of records from these earthquakes and their aftershocks, especially in Japan, including ones at free field sites, down-hole arrays and in buildings.

Efforts are underway to better characterize these features and their effects on structures and soils. The potential of subduction zone events to damage a large numbers of buildings spread over a large geographic region may suggest need for design criteria that mitigate the severity of this damage.

The occurrence of a larger than expected subduction zone event in Japan has prompted considerable discussion of the potential of other large subduction zone events in Japan and other locations around the Pacific Ocean. The possibility of these "larger than expected events" has a substantial influence on predictions of future tsunami, both locally, but also in transoceanic tsunami waves from sources such as the Cascadia and Aleutian subduction zones in North America, as well as similar zones in Asia and South America.

### 2.2.3 Near and Far Field Motions

While subduction zone events in Chile and Japan provide new challenges to engineers, damages in Christchurch, New Zealand, and in many instances in the Wenchuan earthquake in China were due to the intensity of shaking that occurred in the near field. Because Japan (and Chile) are susceptible to such near-field fault ruptures, as in the 1995 Hyogo-Ken Nambu earthquake, studies to better understand the special features of off-shore subduction ground motions and on-shore fault ruptures that lead to damage are needed. In Japan and China, substantial evidence exists for tall buildings and other long period structures being strongly excited by earthquake shaking more than 500 km from the fault. Thus, these types of motions need to be carefully investigated as well.

### 2.2.4 Importance of Aftershocks

Large events trigger large numbers of aftershocks. Better understanding of the number, severity and location of aftershocks is needed. The cumulative effects of multiple aftershocks following large events on engineered structures and soil deposits need to be better understood and perhaps accounted for in design of structures. While aftershocks are generally smaller than the main shock, they can be located substantially closer to a particular structure or community than the main shock, and cause substantially more damage, as observed in Christchurch, New Zealand. Even small accumulations of damage during aftershocks raise questions about the safety of rescue workers and construction crews working on repairs or demolition. There is evidence in Japan (and elsewhere) that aftershocks re-damaged structures where damage from the main shock had already been repaired.

### 2.2.5 *Large Tsunami Accompany Large Earthquakes*

The tsunami resulting from earthquakes in Japan and Chile caused severe widespread damage in those countries as well as in other countries around the Pacific Ocean. Tsunami waves, with heights measuring up to 40 m in Japan, were responsible for the majority of earthquake casualties there and a large portion of the physical damage and loss. There is thus a need and opportunity to benchmark and improve abilities to predict tsunami waves, their interaction with costal geometries and structures, and the regions of expected flooding. The effectiveness of early warning systems and evacuation procedures should be studied. Similarly, there is a need to understand better and improve the behavior of engineered structures to tsunami wave action, impact by debris and scouring. In particular, the effectiveness of evacuation methods and the design of structures as vertical evacuation shelters should be investigated.

### 2.2.6 *Liquefaction- or Settlement-Related Damage to Structures*

Extensive liquefaction has been observed Chile, New Zealand and Japan in areas near rivers and coastlines. In Japan, permanent vertical and horizontal displacements of the soils supporting a structure's foundation were responsible for much of the damage to structures away from the tsunami-affected zone. Liquefaction and differential settlement also contributed heavily to damages observed in and near Christchurch, New Zealand. Such damages were seen over a large geographic region, and occurred for a wide variety of soil types and ground motion characteristics. As such, there is an opportunity and need to improve understanding of the triggering of liquefaction and/or lateral spreading, the deformations that occur, and the consequences of soil spreading and liquefaction on the behavior of supported structures. Methods for repair and restoration of structures damaged by differential settlement need additional study as well.

### 2.2.7 *Damage to Lifelines, Industrial Facilities, Such As Nuclear Power Plants, Can Exacerbate Disasters*

There are a multitude of engineering issues raised by the response of the Fukushima Daiichi Nuclear Power Plant to the initial shaking and aftershocks as well as to the tsunami. In addition, more than 20 other fossil fuel and nuclear power plants were taken off line immediately following the earthquake in Japan. Many of these remained off line a year following the event. These outages have had a critical effect on Japanese businesses and overall quality of life. Special issues related the behavior of nuclear power plants (and other critical and hazardous industrial facilities) to earthquakes and tsunami, the effect of the inoperability of critical electric power facilities on a community, region and nation, and the special issues related to radioactive or possible chemical/biological contamination, are high priority topics for further investigation.

### *2.2.8 Disruption of Business and Social Systems*

Wide spread economic and social disruption has resulted from the damage to housing, schools, hospitals, commercial structures, factories and infrastructure systems in Japan and New Zealand. In many cases, a facility's structural system may not have been substantially damaged, but damage to nonstructural elements and equipment as well as loss of lifelines (power, water, gas, transportation, communications, etc.) rendered it inoperable. Transportation was impaired since several national and local highways were closed due to ground shaking and landslides, and in the tsunami-affected region, many highway and railway bridges were completely destroyed. In some areas, manufacturing and other critical facilities in the tsunami-affected zone suffered little structural damage, but were inoperable due to water damage or the presence of debris. Thus, in addition to general economic, business and related studies, investigations on improving the seismic resistance of the nonstructural components and equipment and operationally critical lifelines are needed.

### *2.2.9 Effect of Earthquake Shaking on Engineered Facilities*

Modern buildings and other structures are not designed to be damage free during rare earthquakes. While there were substantial numbers of damaged structures in Japan, Chile and New Zealand, many damaged buildings tended to be older with known deficiencies. In Chile, newer tall reinforced concrete buildings suffered damage in part due to extending the use of systems that had behaved well in the past to structures having considerably greater height. In both Chile and New Zealand, newer taller buildings suffered structural damage and total or partial collapse due to discontinuous or irregular structural systems. In many buildings in heavily shaken areas, substantial damage to nonstructural elements caused expensive and disruptive delays in restoring use of a facility.

A particularly significant consequence of the vibration and damage to buildings in Japan is a lack of confidence in the safety of structures. The vibrations of tall buildings in particular, while not indicative of structural damage or danger, frightened many occupants so that they were reluctant to re-enter buildings and in some cases property values diminished greatly.

## 2.3 Resilient Structures

As noted above, there are many fruitful topics for study to improve the assessment and design of earthquake resistant structures. However, it appears that new issues related to the ability of a building to be repaired quickly following an earthquake, and for a community or region to recover rapidly, are becoming of more concern. As such, many researchers and practitioners worldwide are working to develop economical systems that can dependably permit engineered facilities and lifelines to continue functioning even following a large seismic event.

Such solutions are critical in the case of facilities such as hospitals, emergency command centers, and lifeline systems needed for emergency response and recovery. The types of facilities where post-earthquake functionality should be maintained can be much larger, if consideration is given to facilities needed to ensure rapid recovery of a community or organization. For example, the San Francisco Planning and Urban Research Association (SPUR 2009) has indicated that many schools, residential buildings, and even retail and commercial buildings are necessary to provide the housing and services that would enable a community to function following an earthquake. Similarly, the vitality and perhaps viability of a business depends on its ability to provide continuous service to its customers.

As such, there is increasing recognition that the traditional approach of designing "ductile" structures for collapse prevention, while necessary, may not be sufficient in a modern society. Engineering solutions that can with confidence preserve operations, or provide the promise of minimizing the disruption and cost of repairs following major earthquakes, are needed.

To minimize post-earthquake disruption, a structure or system must be able to limit the damage occurring in structural and nonstructural components, and should not exhibit permanent displacements. Structures could simply be made stronger so that they remain essentially elastic for the level of earthquake for which resilience is desired. However, this strength-based approach (1) requires more materials that add to the cost and carbon footprint of the structure, and (2) results in higher accelerations that necessitate special attention in the design of nonstructural components, equipment and contents. Moreover, such strength approaches are only as good as the estimates of future shaking.

While many types of structures that behave nonlinearly are being explored to enhance resilience, many share the following characteristics:

- They are designed with an explicit nonlinear deformation mechanism that has only modest strength in order to reduce cost as well as accelerations,
- They incorporate highly durable or easily replaceable energy dissipation devices to reduce forces and displacements in structural and nonstructural elements not associated with the nonlinear deformation mechanism, and
- They exhibit self-centering characteristics that minimize permanent lateral and vertical displacements of the structure.

Systems considered include ones with beams, columns, walls, braces and ones that exhibit origin-oriented hysteretic loops (Fig. 2.1a); this behavior is often

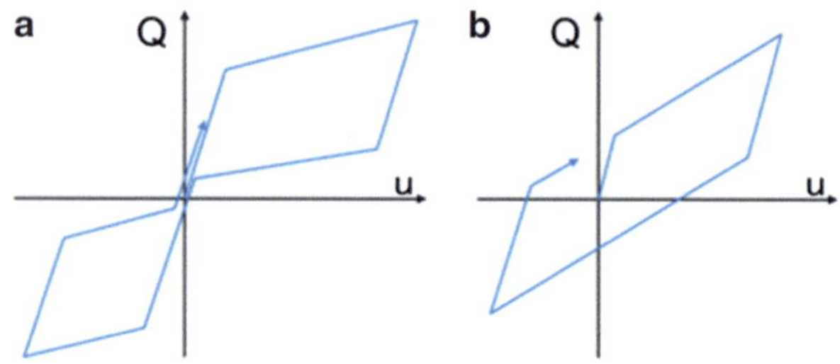


**Fig. 2.1** Hysteretic shapes for (**a**) Self-centering elements and (**b**) Traditional isolation devices

achieved by use of a combination of yielding or friction to provide hysteresis and use of gravity or post-tensioning to provide self-centering capabilities. Special materials, such as shape memory alloys, can provide these characteristics.

Another approach is to install viscous or similar damping devices. When installed in flexible systems that remain essentially elastic during the response, viscous dampers can be quite effective in reducing forces, accelerations and drifts in the structure. However, traditional design approaches in the US do not focus on limiting drifts and accelerations to levels necessary to achieve resilience.

Viscous dampers and elements having self-centering characteristics are typically installed pervasively throughout a structure. Where large nonlinear deformations are expected during the dynamic response, such systems may limit forces and accelerations experienced by structural and nonstructural elements, but they may not be able to limit displacement-related damages.

An alternative approach is seismic isolation. While seismic isolation has been in use for decades, design codes and guidelines are often formulated to provide seismic performance comparable to fixed base buildings. While it is presumed that isolated buildings perform better than traditional structures, generally building codes permit substantial damage in the event of major earthquakes.

### *2.3.1 Isolated Facilities*

For resilience, it is convenient to consider a limit on peak story drifts and residual displacements. The drift limit should prevent significant yielding of the structural elements that might result in damage requiring inspection and repair, and permanent residual displacements. Drift should also be limited to prevent damage to nonstructural elements that are sensitive to relative displacements between stories. It can be shown on the basis of first principles that steel braced frames will begin to buckle or yield bracing elements at a story drift ratio of about 0.25–0.4 % drift, where as moment frames in steel or concrete will yield at a drift ratio around 0.9–1.25 %. Many types of cladding and partitions begin to display damage at a drift ration of about 0.3 % with significant damage occurring by a drift ratio of 1 %. Thus, in setting design criteria it is necessary to consider drift limits considering likely damage to both structural and nonstructural elements. From the above example, it is technically feasible, and perhaps economical, to isolate a moment resisting frame, but it is likely that substantial damage will occur in partitions, cladding, elevators and so on, before the structural system yields. Thus, depending on the type of nonstructural elements utilized, braced frames (or shear walls or other stiff systems) may be more consistent with the requirement of limiting damage to nonstructural components. However, it may not be necessary to preclude all damage to nonstructural elements during the resilience targeted event, so long as continued functionality is preserved and the occupants are able to "shelter-in-place" during repairs.

Many types of nonstructural elements, equipment and contents are sensitive to in-structure accelerations. To prevent acceleration related damages, it is necessary to consider a limit of floor level accelerations, or on the characteristics of pseudo-acceleration spectrum at each floor. For simplicity, we might consider a design wherein the floor level accelerations are limited to 0.4 g, though other values may be appropriate for certain types of elements and equipment.

While near or completely elastic behavior in the portions of the facility away from the isolation plane will eliminate the possibility of residual displacement in those locations, some isolation systems may tend to have residual displacements. The acceptability of residual displacement in the isolation system following an intense earthquake should be carefully considered. If accounted for in the initial design with regards to accessibility and functionality of utilities, residual displacements in the isolators may not impair post-earthquake functionality, and thus be acceptable. Re-centering an isolated building following an earthquake may be economical in certain cases, if needed. In other cases, isolators can be designed to minimize residual displacements in the isolation plane.

For resilient designs, it is necessary to identify a level of shaking at which resilience is desired. This may be expressed as a fraction of the maximum considered earthquake, or in terms of the confidence with which resilience can be achieved over the life of the structure (i.e., 50 % confidence that the facility will remain operational following an earthquake having a 10 % probability of exceedance in 50 years, or a 90 % confidence that the structure will be inoperable due to earthquakes for less than 3 days). It is expected that different facilities will have different criteria, depending on their use and importance to the community or owner. For example, some structures would be expected to be operable immediately, while others might be acceptable in terms of resiliency if their function is restored within 90 days. In all cases, a high confidence of collapse resistance is assumed. For the purposes of this paper, resiliency for the maximum considered event is targeted, consistent with important or critical facilities.

### *2.3.2 Isolation Systems*

It has been noted that the design of isolated structures can be problematic due to the apparent over- or under-conservativeness of some design provisions, and also because of challenges encountered when designing isolated systems to resist intense

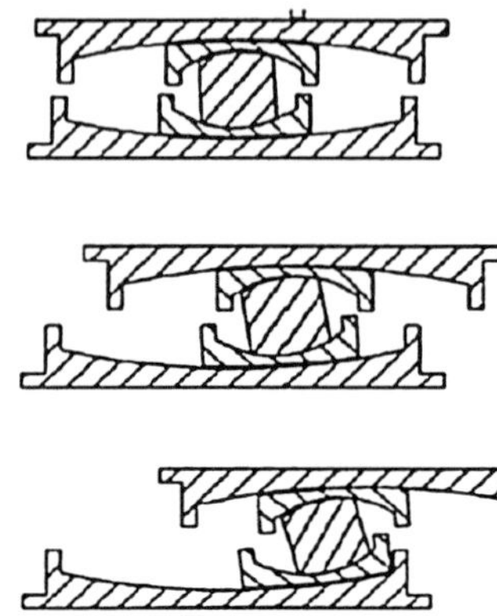

**Fig. 2.2** Triple pendulum bearing

near-field ground motions. To resist large near-fault ground shaking, relatively large and strong isolators are often necessitated to control displacements and maintain isolator stability. In these cases, isolators may not act effectively, especially for small events, and require relatively large design forces and trigger excessive accelerations in the superstructure (Morgan and Mahin 2007). Similarly, traditional designs have relied on isolation systems that have essentially bilinear hysteretic loop shapes with effective periods in the 2–3 s. range (Fig. 2.1b). This type of isolator may develop large forces in the supported structure due to the large lateral isolator forces that may develop in large earthquakes. Thus, efforts are underway to improve isolation for small events, limit drifts for larger events, and reduce floor accelerations.

Several approaches are being pursued at Berkeley and elsewhere to overcome these impediments to resilience, including novel combinations of elastomeric isolators acting in series with sliding bearings, and elastomeric bearings acting in parallel with nonlinear viscous dampers.

Another promising approach is the Triple Pendulum (TP) Slider (Earthquake Protection Systems 2007). This device has three independent pendulum mechanisms (Fig. 2.2). Strategically selecting friction coefficients and the effective radius for each pendulum mechanism can optimize hysteretic characteristics optimized for occasional, rare and very rare events. Figure 2.3 shows hysteretic loops for different combinations of friction coefficients and radii.

### *2.3.3 Isolated Building Tests and Analyses*

Tests (Fig. 2.4) and analyses (Morgan and Mahin 2011) demonstrate that TP devices can achieve about the same isolator displacements for a large event, but with smaller drifts and accelerations in the superstructure and with a far greater degree of isolation during smaller events (Fig. 2.3). Numerical simulations were conducted of a three-story structure similar to that shown in Fig. 2.4 subjected to the suite of ground motions corresponding to a 2 % probability of exceedance at a

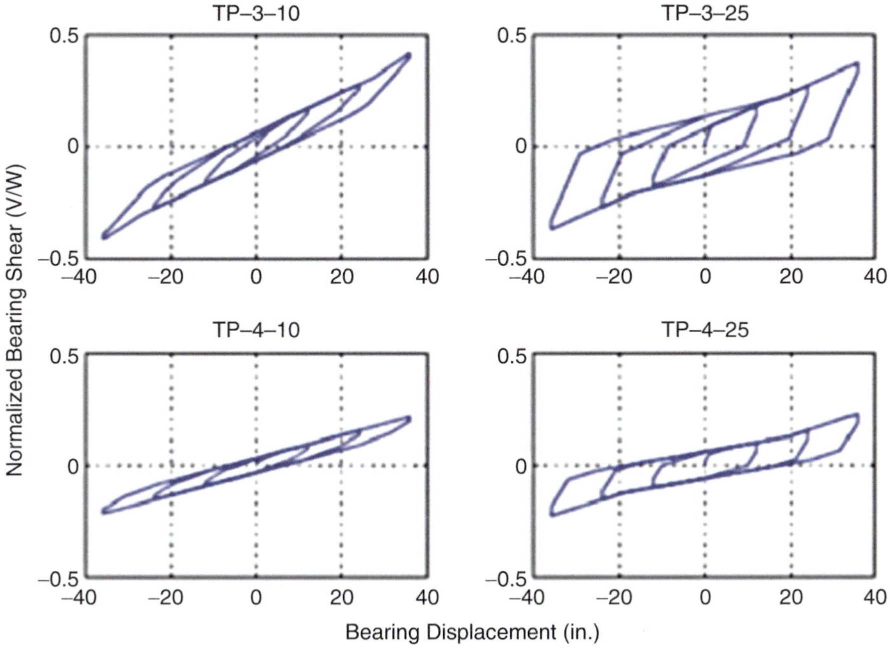


**Fig. 2.3** Representative hysteretic loops

**Fig. 2.4** Test of isolated 3-story building

**Fig. 2.5** Test of 3-story isolated superstructure

site in metropolitan Los Angeles. For TP bearings with an effective period of 4 s and damping of 10 % (see TP-4-10 plot in Fig. 2.3) the computed median isolator displacement is 770 mm, but the median peak story drift is less than 0.4 % and the median peak floor acceleration is less than 20 %g.

A current test program setup is shown in Fig. 2.5. This shows a simple shaking table capable of one-dimensional motion. In this case, the test specimen is designed with replaceable plastic hinges, made from clevises and replaceable steel rods. The rods are machined and located to produce a desired stiffness and moment capacity for the plastic hinge. In this case, various amplitudes and distributions of strength are provided, based on US building codes as well as other alternatives. These studies show that current US design provisions can result in significant yielding of the superstructure prior to the isolators reaching their ultimate displacement capacity and that the typical distribution of strength used for design in the US results in weak story behavior. Superior behavior is achieved where the strength of the structure is increased to remain elastic until the ultimate displacement capacity is reached and the distribution of story strength is consistent with that existing when the bearing displacement capacity is reached.

A number of challenges remain in the design of seismically isolated structures. One of these is the development of structural systems and non-structural components that are complementary in the attainment of resilience. Some structural systems are too flexible and result in significant damage to nonstructural systems. Similarly, many nonstructural elements are so fragile that they suffer substantial damage even if drifts are small. Moreover, current seismic isolation systems only provide protection for motions in the horizontal direction. Damage to nonstructural elements can occur in such cases.

## 2.4 Resilient Businesses

A business or service center may be located in a structure that is able to protect the occupants, nonstructural components and structural system, but still not be able to continue operations following an earthquake. As noted by the SPUR Resilient Communities initiative (SPUR 2009), businesses need staffing, raw materials, a means of distributing products or services, and various utilities and services in order to operate. As such, a growing number of consultants are providing services related to post-disaster emergency management and business continuity.

For staff to return to work (once the facility is restored to operability), they need housing, schools for their children, food, transportation, health care, and minimal utilities needed to live. The business will also need utilities such as water, power, sewerage, telecommunications, Internet services, and so on. They need access to banking and other business related services. Raw materials need to be shipped and delivered, and final products or services need to be sent to customers. In many cases, raw materials, suppliers, packaging, and shipping services will be located in the same region damaged by the earthquake and also not functioning. Thus, the operation of modern industrial societies is highly interdependent and the success of a region in carrying out business and industrial operations can be rapidly eroded by the failure of a few key services or lifelines.

Similarly, businesses and services well outside of the earthquake shaken area may be vulnerable to business interruption due to closure of businesses located in regions shaken by the earthquake. Thus, careful attention to supply chains in business operations is needed.

## 2.5 Resilient Communities

For a business to operate, it needs to be in a community that is functional following a large earthquake. For small and moderate earthquakes, local neighborhoods and communities can come to the aid of those who suffered damage, and traffic can be rerouted and other lifelines rapidly restored. However, as the extent of damage increases, it is difficult for neighboring areas to provide adequate relief. If housing is damaged moderately or severely so that many cannot stay in their homes, they will need to find temporary housing or move to other areas. If a factory or business is damaged, it may shift staff and operations to other locations. If workers cannot get food, or other services, they cannot work in their normal jobs or help with repairs. Thus, a critical balance is needed to maintain adequate housing, lifeline services, and business for a community to operate and recover rapidly from a large earthquake. While the balance of resilient facilities, lifelines, and services is a matter of public policy, realization of this balance and attaining a resilient society is a matter of engineering.

Lifelines such as transportation, power, water, sewers and so on are also needed. To this end, it is necessary to assess and design such infrastructure in terms of its criticality to post-disaster functionality of a community. Thus, hardening of water and electrical distribution systems may be needed. Nodes such as bridges, power plants, telecommunication centers, server farms, and so on may need to be designed with resilience as a design criteria.

Bridges are a particularly vulnerable aspect of transportation systems. Improved methods of designing bridges to withstand large earthquakes without significant damage or additional cost are the subject of considerable research. For example, various types of re-centering bridge columns have been suggested, along with bridge piers that are allowed to uplift during earthquakes. As with buildings, several strategies for improving the resilience of bridges are being explored associated with advanced materials that delay the onset of damage like cracking or spalling, supplemental energy dissipation devices, and seismic isolation.

## 2.6 Conclusions

The technology and design approaches described in this paper provide several alternative ways to achieve durable building systems that increase post-earthquake serviceability and reduce the need for repair. By permitting structures to undergo significant inelastic deformations during seismic events, yet suffer little damage that would require post-earthquake repairs and impair operability, structural engineers can achieve designs that are durable, dependable, and economical in terms of initial construction cost and the potential losses that might occur in the event of a damaging earthquake. As such, these approaches address the basic principles articulated by sustainable design and seismic resiliency.

Additional research is needed to refine these concepts, especially with regards to (1) reducing cost and increasing speed of construction, (2) reducing the societal and ecological impacts of construction, (3) and refining these concepts and developing more generally applicable design guidelines for resilient structures.

**Acknowledgements** The author appreciates the superb assistance of his graduate students and post-docs including Tracy Becker, Troy Morgan, Brian Olsen and Andreas Schellenberg. This paper includes work supported by the National Science Foundation under Grant No. CMMI-0724208. Opinions, findings, and conclusions are those of the authors and do not necessarily reflect the views of the sponsors.

## References

Earthquake Protection Systems (2007) Product description. Mare Island. Triple Pendulum ™ Bearing. http://www.earthquakeprotection.com/triple_pendulum_bearing.html. Accessed 2014

Morgan T, Mahin S (2007) Enhancing the performance capabilities of seismically isolated buildings using multi-stage friction pendulum sliding bearings. In: Proceedings, world forum on Smart Materials and Smart Structures Technology (SMSST'07), Chongqing & Nanjing

Morgan T, Mahin S (2011) The use of innovative base isolation systems to achieve complex seismic performance objectives. PEER report, Pacific Earthquake Engineering Research Center, University of California, Berkeley

SPUR (2009) The Resilient City: a new framework for thinking about disaster planning in San Francisco. Urbanist, San Francisco Planning and Urban Research Association, Feb 2009

# Chapter 3
# Towards the Bled Workshop in Future

**Hiroshi Akiyama**

**Abstract** The author sincerely expects further development of the Bled workshop and he would like to refer to some basic items important to keep continuation. One of the goals of the performance-based design is to control structural damage under earthquakes within a certain limit. Energy concepts in earthquake engineering can provide plausible way to achieve this goal. In other words, the performance-based design must be eloquently spoken by a technical language and the balance of energy can be a key concept to establish its grammatical construction. The mixed structure develops the most preferable structural mechanism under earthquakes due to collaboration between the elastic and elastic-plastic element. The concept is applied to different types of structures – base-isolated structures, seismically controlled structures, reinforced concrete shear walls and diagonal bracing systems in steel structures.

**Keywords** Energy concept • Energy balance • Cumulative plastic deformation • Mixed structure • Base-isolated structure • Seismically controlled structure • Shear wall • Diagonal bracing system

## 3.1 Introduction – The Bled Workshop

The Bled workshop started under the initiative of Prof. Peter Fajfar and Prof. Helmut Krawinkler in 1992 and through 20 years of activity, obtained great results on seismic engineering especially in the field of performance-based design.

H. Akiyama (✉)
Professor Emeritus, Faculty of Engineering, University of Tokyo, Tokyo 113-0032, Japan
e-mail: ktmu@fa2.so-net.ne.jp

M. Fischinger (ed.), *Performance-Based Seismic Engineering: Vision for an Earthquake Resilient Society*, Geotechnical, Geological and Earthquake Engineering 32,
DOI 10.1007/978-94-017-8875-5_3, 

This time of the Bled workshop (June 24–27, 2011) shows a turning point under the new leadership by Prof. Matej Fischinger and it is very important. Earthquakes happen incessantly somewhere in the world and hazard potentials seem to increase as civilization develops, since expansion of human activities always exceeds the technological development against natural hazards. Even the latest Great East Japan Earthquake again taught us the seriousness of Tsunami hazard. Seismic problems have not been solved yet. Therefore, continuation in our research activities is most important. I sincerely expect further development of the Bled workshop and I would like to refer to some basic items important to keep continuation herein.

## 3.2 Balance of Energy

Structural behavior under seismic ground motions are precisely described by the equation of motion which expresses the balance of forces between structural resistant forces and seismic forces (Akiyama 1985, 1999, 2003). By multiplying the deformation increment on both sides of the motion equation and integrating over the duration of ground motion, the equation of energy balance is obtained as follows:

$$W_e + W_p + W_h = E = \frac{MV_E^2(T)}{2} \tag{3.1}$$

where:

$W_e \ldots$ elastic vibrational energy
$W_p \ldots$ cumulative inelastic strain energy (structural damage)
$W_h \ldots$ energy absorption by damping
$E \ldots$ total energy input exerted by an earthquake
$M \ldots$ total mass of a structure
$T \ldots$ fundamental natural period
$V_E = \sqrt{\frac{2E}{M}} \ldots$ equivalent velocity for $E$

$V_E - T$ relationship is termed to be the energy spectrum for a specific earthquake ground motion. Thus, the total energy input by an earthquake is uniquely determined by the total mass and the fundamental natural period irrespectively of structural parameters such as mass distribution, strength distribution and stiffness distribution. Equation (3.1) is a general expression of energy balance. However, it is rather an intermediated expression. Since, the elastic vibrational energy tends to be changed into $W_p$ or $W_h$.

Therefore, the following expression is a more direct one.

$$W_p + W_h = E \tag{3.2}$$

One of the goals of the performance-based design is to control structural damage under earthquakes within a certain limit, and to know the real contents of $W_p$ becomes essentially necessary. Neglecting $W_h$ in Eq. (3.2), the following expression makes it possible to get rough but sound estimate of $W_p$:

$$W_p = E \tag{3.3}$$

$W_p$ is decomposed into local damage $W_{pi}$ as follows:

$$W_p = \sum_i W_{pi} \tag{3.4}$$

$$W_{pi} = Q_{Yi}\delta_{pi} \tag{3.5}$$

where:

$Q_{Yi}$ ... yield strength of i-th element
$\delta_{pi}$ ... cumulative plastic deformation of i-th element

Together with cumulative plastic deformation, the maximum plastic deformation, $\delta_{pmi}$ is a target of control in design. $\delta_{pmi}$ can be related to $\delta_{pi}$ as follows:

$$a_d = \frac{\delta_{pi}}{\delta_{pmi}} \tag{3.6}$$

Information about Eqs. (3.4) and (3.6) is indispensably necessary to complete design procedure and those knowledge must be accumulated through practical design experiences. The performance-based design must be eloquently spoken by a technical language and the balance of energy can be a key concept to establish its grammatical construction.

## 3.3 P-δ Effect

Any structure is primarily subjected to the gravity loading. When a structure deforms laterally under the seismic forces, columns of the structure incline with the inclination angle, $\theta_i$ in i-th story. The i-th story supports the gravity load of $W_i$ also. Then, the i-th story receives the following shear force due to $P$–$\delta$ effect, $\Delta Q_i$ in addition to the seismic force.

$$\Delta Q_i = \theta_i W_i = \frac{W_i \delta_i}{H} \tag{3.7}$$

where:

$\theta_i = \delta_i / H$
$\delta_i \ldots$ lateral story deformation
$H \ldots$ height of story

With the inclination of $\theta_i$, the weight which lies on i-th story, $W_i$, moves downward by the following amount:

$$\Delta_i = \frac{\theta_i^2 H}{2} \tag{3.8}$$

This movement corresponds to the release of the potential energy of the structure, $\Delta W_{Pi}$ in the gravitational field by the following amount.

$$\Delta W_{Pi} = W_i \Delta_i = \frac{W_i \theta_i^2 H}{2} \tag{3.9}$$

$\Delta W_{Pi}$ signifies another source of energy input due to the $P$–$\delta$ effect in addition to the seismic energy input.

The $P$–$\delta$ effect expressed by Eqs. (3.7) and (3.9) can cause very serious results when its presence is not fully recognized.

The potential energy of a structure stored in the gravitation field is far greater than the seismic input energy, and the plastification of the structure under an earthquake can render the release of potential energy and the total collapse of the structure.

The total released potential energy can be calculated from Eq. (3.9), by assuming constancy of $\theta_i$ $(= \theta_0)$ and mass distribution in the multi-storied structure being constant, as follows:

$$\Delta W_P = \sum_1^N W_i \Delta_i = \frac{W_0 \theta_0^2 N H}{4} = \frac{W_0 N \delta_m^2}{4H} \tag{3.10}$$

where:

$W_0 = \sum_1^N W_i \ \ldots$ total weight of structure
$\theta_0 = \delta_m / H$
$\delta_m \ldots$ maximum story drift
$N \ldots$ number of stories

One of the promising methods to eliminate the $P$–$\delta$ effect in structural responses is to limit $\Delta W_P$ within a certain extent compared to the seismic energy input E. This condition is described as follows:

$$\frac{\Delta W_P}{E} \le K_{P\delta} \tag{3.11}$$

where $K_{P\delta}$ is the limit value of $\Delta W_P / E$.

Based on Eq. (3.11), $\delta_m$ to be applied to seismic design is obtained as follows:

$$\delta_m \leq \sqrt{\frac{2K_{P\delta}V_E^2 H}{gN}} \tag{3.12}$$

To find out an appropriate value of $K_{P\delta}$ is a remaining problem.

The $P$–$\delta$ effect can be totally cancelled, if elastic elements which meet to Eq. (3.7) can be installed. The seismic coefficient of the elastic element, $\Delta\alpha_i$ is defined as follows:

$$\Delta\alpha_i = \frac{\Delta Q_i}{W_i} = \frac{\delta_i}{H} \tag{3.13}$$

Putting the maximum story deformation, $\delta_{mi}$ into $\delta_i$, the practical design condition of the elastic element is obtained as follows:

$$\Delta\alpha_i \geq \frac{\delta_{mi}}{H}, \text{under } \delta_i = \delta_{mi} \tag{3.14}$$

## 3.4 Flexible-Stiff Mixed Structure

The flexible-stiff mixed structure expresses a most preferable structural type in earthquake-resistant structures. It is defined by a mixed structure composed of an elastic flexible elements and an elastic-plastic stiff element. The restoring force characteristics of the stiff element are assumed to be of elastic-perfectly type to make simple the matter considered. The flexible-stiff mixed structure is characterized by a shear force ratio, $r_Q$ defined as follows:

$$r_Q = \frac{{}_f Q_m}{{}_s Q_Y} \tag{3.15}$$

where:

${}_f Q_m \ldots$ maximum shear force developed in the flexible element
${}_s Q_Y \ldots$ yield strength of the stiff element

The mixed structure becomes effective under the following condition.

$$r_Q \geq 1.0 \tag{3.16}$$

Therefore, the elastic rigidity in the flexible element is far lower than that of the stiff element. The distinction between flexible and rigid is based on this difference.

The mixed structure develops the most preferable structural mechanism under earthquakes due to collaboration between two elements. The flexible element, which remains elastic supports vertical loads, and can store energy tentatively, has a self-centering effect and can cancel the $P$–$\delta$ effect by its elasticity. On the other hand, the stiff element can absorb energy steadily with its high elastic rigidity and plasticity. As a result, compared with non-mixed structures, the mixed structure has following advantages:

A high efficiency in energy absorption is attained, or, the maximum deformation is restrained.

When the efficiency is expressed by $a_d$-value in Eq. (3.6), for the mixed structure, $a_d = 8.0$–$20.0$ compared to 2.0–4.0 for the non-mixed structure.

Residual deformation, $\delta_{ri}$ is restrained.

For the mixed structure, $\delta_{ri}$ is almost nullified, compared to the case of non-mixed structure to be around $\delta_{ri} = 0.5 \cdot \delta_{pmi}$.

Another merit of flexible-stiff mixed structure is found in its usefulness in categorize the earthquake resistant structures. The structure is divided into two types. One is the non-mixed structure. This structure has no flexible element and is considered to be consisted of a single rigid element. Most of ordinary structures belong to this type. This type of structures must be subjected to the $P$–$\delta$ effect. Another type of structure is the flexible-stiff mixed structure in which the flexible elastic element is intentionally installed. This type of structure can enjoy aforementioned advantage or high performance under earthquakes, and is divided further into two as follows:

- energy-concentrated type (base-isolated structure)
- energy-dispersing type (seismically controlled structure)

As is stated by Eq. (3.4), in the seismic design, to know the distribution of damage is essentially important. However, it is very difficult to predict the exact distribution of damage in non -mixed structures. Only the two extreme cases become predictable in the mixed structure. One is the case of sheer concentration of the total energy into one story and another is the case of uniform distribution of the total input over every stories.

The former was realized by the install of energy concentration story at the bottom of a structure. The bottom story is formed by the mixed structure and the technical break-through was realized by the invention of the laminated rubber bearing as a flexible element in the last quarter of twentieth century and now buildings even higher than 50 stories became possible to construct by the base-isolated technique.

The later is realized in multi-storied structures in which the major skeleton works as a flexible element and is designed elastically and energy absorbing devices are distributed evenly. The presence of the flexible element hinders the severe damage concentration, thus high-performance of energy absorption being attained.

To realize a higher performance in the seismic resistant structures, the introduction of the concept of the mixed structure is indispensable; however, the ordinary structure occupies the majority. Therefore, also more intensive continuous efforts must be paid toward the enhancement of performance for ordinary type of structures.

## 3.5 Shear Walls

Shear walls express a concept which has been introduced to keep resistance against lateral loadings such as earthquakes and winds. Reinforced concrete shear walls and diagonal bracing systems in steel structures combined with moment frames have worked as excellent seismic resistant elements. The mixed structure consisting of moment frames and shear walls belongs to the ordinary structure categorized in Sect. 3.4. The target of structural performance is basically divided into the following two:

- ultimate seismic resistance (guard for human life)
- reparability of damage (guard for human comfort)

For the first target, the mixed structure behaves excellently, however for the second target it cannot meet well. Nevertheless, the first target has a primarily importance for the ordinary structure. The mixed structure is simple in its geometrical form; however the exact structural behavior in the inelastic range being highly complicated remains to be solved in the future.

(a) **Reinforced concrete shear wall**
The reinforced concrete shear wall is characterized by its high rigidity. Therefore, the mixed structure can be assumed a seriated structure consisting of the first short period structure and the second long period structure. Then, referring to Eq. (3.3) the basic design criterion is described as follows.

$$\frac{{}_sW_p}{E(T_s)} = \frac{{}_mW_p}{E(T_m)} \tag{3.17}$$

where:

${}_sW_p \ldots$ ultimate energy absorption made by the shear wall
${}_mW_p \ldots$ ultimate energy absorption made by the moment frame
$T_s \ldots$ period of the shear wall
$T_m \ldots$ period of the moment frame

(b) **Diagonal bracing system in steel structures**
A single member in a bracing system is subjected to alternating compressive-tensile forces and its non-linear behavior is very complicated. However, a highly developed computational technology made is possible to analyze exact behavior of the mixed structure. The period of the diagonal bracing system and that of the moment frame are not so different, and the mixed structure can be considered to be a parallel system. Then, the design criterion for the system is described as follows, referring to Eq. (3.3):

$${}_DW_p + {}_mW_p = E(T_m) \tag{3.18}$$

where:

${}_DW_p$ ... ultimate energy absorption made by the diagonal bracing system
$T_m$ ... period of the mixed structure

${}_DW_p$ and ${}_mW_p$ are decomposed as follows:

$$\begin{aligned} {}_DW_p &= {}_DQ\,{}_D\delta_p \\ {}_mW_p &= {}_mQ_m\delta_p \end{aligned} \tag{3.19}$$

where:

${}_DQ$ ... strength of the diagonal bracing system
${}_mQ$ ... strength of the moment frame
${}_D\delta_p$ ... ultimate deformation capacity of the diagonal bracing system
${}_m\delta_p$ ... ultimate deformation capacity of the moment frame

${}_DQ$ and ${}_mQ$ can be easily defined, them ${}_D\delta_p$ and ${}_m\delta_p$ must be clarified.

## References

Akiyama H (1985) Earthquake-resistant limit-state design for buildings. University of Tokyo Press, Tokyo

Akiyama H (1999) Earthquake-resistant design method for buildings based on energy balance, Gihodo Shuppan (in Japanese)

Akiyama H (2003) Metodología de Proyecto Sismorresistente de Edificios Basada Balance Energético, Editorial Reverté S.A. (in Spanish)

# Chapter 4
# Global Challenges and the Role of Civil Engineering

**Žiga Turk**

**Abstract** The world is going through some profound changes: automation and general improvement of productivity is resulting in the abundance of industrial products, the domination of the West in global economy and politics is challenged by the rise of the BRICS economies, climate change is requiring a reconsideration of the energy system, particularly in Europe demographic changes are resulting in an ageing society, and finally, the electronic communication revolution is changing the ways in which elements in a society are held together influencing all aspects of economy, research, learning, living, media etc. Civil engineers and their forerunners have been shaping the infrastructure of societies for millennia. This paper explores how the listed trends will affect the civil engineering work and where civil engineers will be able to contribute. While the relative contribution of the construction industry to the jobs creation and economic growth will continue to decline, there are substantial opportunities in comparison with some other engineering industries, in particular in the area of climate change and globalization. There are some lessons; in particular with earthquake engineering – the notion of resilience – then can be borrowed by economics and finance.

**Keywords** Global future • Civil engineering • Construction • Climate change • Demography • Information technology • Earthquake engineering

---

Ž. Turk (✉)
Faculty of Civil and Geodetic Engineering, University of Ljubljana, Jamova 2, SI 1000 Ljubljana, Slovenia
e-mail: ziga.turk@fgg.uni-lj.si

M. Fischinger (ed.), *Performance-Based Seismic Engineering: Vision for an Earthquake Resilient Society*, Geotechnical, Geological and Earthquake Engineering 32,
DOI 10.1007/978-94-017-8875-5_4, 

## 4.1 Introduction

Civil engineers have been providing the infrastructure of the societies since the very beginning of civilization. Although the term "civil" appeared as the opposite of the "military" engineers, the term "civil engineering" gives justice to the role our profession has in society. In the beginning of the twenty-first century a lot in this society is changing. The long term global trends will be described in Sect. 4.2. Section 4.3 will address some general features of these trends. Section 4.4 will explore the impact these trends will gave on construction in general and on earthquake engineering in particular as well as what solutions our profession has there to offer.

## 4.2 Global Trends

The world is going through a series of disruptive changes, which, when coupled with the economic crisis, create a significant discontinuity. The future will not be like the extrapolation of the present but in many ways "a grand transformation" (Reinhardt and Roos 2008).

The five grand transformations may be listed alphabetically as:

(a) Automation and abundance;
(b) BRICs: Brazil, Russia, India, China and globalisation;
(c) Climate change and energy;
(d) Demography and ageing;
(e) E-everything, information, technology, computers, everything electronic.

### *4.2.1 Abundance and Automation*

Increased efficiency of manufacturing processes, automation and robotics are enabling the industry to produce more and more with less and less work. Markets are saturated with products that consumers hardly need at all – especially the consumers in the rich societies with money to spend (Pink 2006). A significant shift in economy is taking place where the industrial jobs are being replaced by jobs in services; routine production jobs are replaced by jobs non-routinely creating or caring (Levy and Murnane 2007; OECD 2010). Increasing share of the purchase-value of a product is not in the material, energy or routine labor embedded in the product, but in the meaning that a product (or service) evokes in the consumer. Products and services of the future are not only performing the basic function. They are well designed, beautiful, of a known brand, with a positive ethical connotation – "home-made", "environmentally friendly", "trustworthy", "ethically produced", "green" ... (Stehr 2008; Turk 2008).

While growth in the developing economies and poorer segments of the developed ones is possible in the old way, the challenge of the developed societies is how to create new needs. Re-shaping those needs from material to symbolic is a great challenge for a society as a whole. It will not only grow new domestic industries but could also help save the planet.

The EU can either use its rich cultural and ethical capital or reorient its economy towards a creative economy, by combining science, innovation and creativity to create products and services that are more than what they are useful for; maintain its world class brands and create new ones that embed our culture and values. Or it can engage in an uphill struggle on price and function only.

### 4.2.2 BRICs and Globalization

Brazil, India, Russia, China and others are the changing of the tides in global economic, technological, scientific, political and military power. By 2020 the EU will not be #1 economy in the world any more, but #3. It will drop from about a quarter of global GDP to less than one fifth (NIC 2008).

Increased prosperity around the world is a good thing. This means billions more will raise out of poverty. It also means many more will have similar needs to ours. It means bigger markets like ours for products and services that we are offering already.

But it also means more competition. And not only competition to European worker, also the competition for the innovator, engineer, scientist and scholar. Since the Renaissance our capacity for cutting-edge science, innovation and creativity has been the source of our strength and wealth. The role that the EU and European countries play on the international scene is linked to the vitality of its science and innovation, strengths of its population, vibrancy of its economy and potency of its diplomacy and military. Europe's share in all those areas is likely to diminish in the future (Fig. 4.1).

### 4.2.3 Climate Change and Energy

Current ways of using the natural resources are unsustainable. In particular the use of the atmospheric capacity to absorb and process the greenhouse gasses is most likely leading to significant warming of the planet with dire consequences for the life on it (IPCC 2007; Stern 2006). Standard of living cannot be maintained without energy and developing world cannot be deprived of the same standard that we have. The world will simply have to find a way to achieve the same quality of life by using less energy and a way to create almost all energy from renewable resources. This is a revolution, because in about 40 years we will need to replace the power base and much of the infrastructure that it took centuries to build.

Price instability of fossil fuels and concentration of supply from a few rather volatile regions is another motive for a change in energy policy.

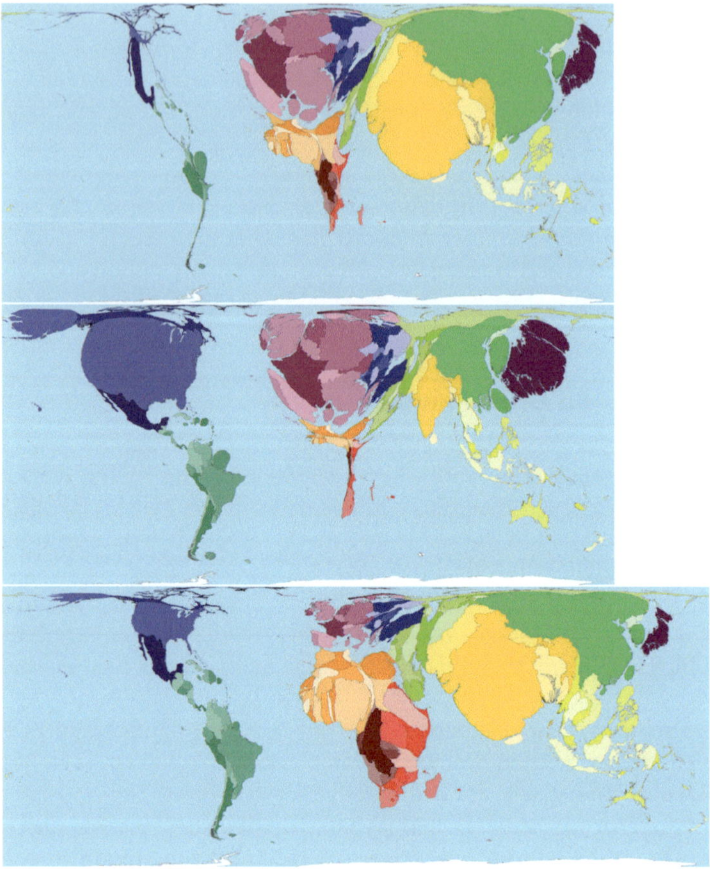

**Fig. 4.1** Global GDP (proportional to the area) in 1500 (*top*), 1960 (*middle*) and talent pool in 2050 (*bottom*) (Source: Worldmapper.org)

### *4.2.4 Demography and Ageing*

The EU is facing a major demographic challenge. The population is expected to peak at around 520 million in 2030 and then dropping back to 505 million by 2060 (Giannakouris 2008). This population will on average be much older today's, less optimistic and risk taking, while the EU will face demographically youthful societies globally and particularly on its southern and southeastern borders. By 2025 the EU, even with expected expansions, will be 6.5 % of the global population (EC 2009). With human talent becoming the main economic resource (Florida 2005), this also gives a perspective on the global share of the EU in the tomorrow's world.

The EU can either begin again to celebrate life, children and family, work longer and allow for immigration, or face severe worsening of the relation between the working and the supported population.

### *4.2.5 Electronic Communication*

Technological and scientific development is rapid in all areas. Amazing breakthroughs are happening in the fields of medicine, biotechnology, nanotechnology, genetics and other new sciences. But no other field has such a pervasive impact on all areas of life than information and communication technology. It is communicating among people that makes a family, enables collaboration in business, science, politics and governance. In the next decades the communication revolution around the internet will change how people collaborate, learn, work, it will change the mass media and will even have an impact on the kind of democracy and citizen participation is practiced (Turk 2010). It is an event of the scale of invention of affordable paper and print of half a millennium ago. The digital world is global. Countries and Unions do not exist on the net. There are no borders. Anyone is a few mouse clicks away. Everyone is a few clicks away.

## 4.3 Global, Historic and Immediate Nature of the Disruptions

**The trends are historic.** Abundance of industrial products is signaling the historic shift from industrial economy to a post industrial economy creative economy where information and meaning add most value (UN 2008). The rise of BRICS is signaling the end of the dominance of the West that lasted some 500 years. The climate change problem is, after 100 years, triggering an energy revolution, this time from below ground to above ground energy. Demographic trends are, for the first time in history, creating a pessimistic old society and opening the question if per capita growth is possible without demographic growth. The electronic communication revolution is comparable to a Gutenberg revolution 500 years ago.

**The trends are fueling the economic crisis in the short term.** Many of the disruptions played a part in the crisis. The rise of the BRICS after the end of the cold war created the global imbalances in trade, exchange rates and savings. The financial sector, that was the epicenter of the crisis, was responding to the desire of many for a safe old age. Because of the lack of social security system, some (like in China) were saving too much and because of its existence, some (like in Europe and the US) too little. Both stock and real estate bubbles were created by savings that did not go into real investment – into new factories, machines, infrastructures. Because the rich economies failed to introduce a new sustainable model of growth this is a crisis of the Western economies. Very high energy prices and fears of inflation stopped the availability of easy money. The crisis, when exploded, spread with the speed of the Internet and depth of panic only new media can create.

**Disruptions are global, not European.** Except for the demography where European problems are shared with Japan and China, the disruptions are global. Europe's problems are not global problems any more. The cold war was the last European war that became the global war. Europe's peace is not global peace. The next few years will define, if the crisis revitalized Europe or whether it will provide the date that will be used by historians to mark the beginning of the decline of a great European civilization.

As a general response to the challenges, the Reflection Group on the Future of Europe (Gonzalez et al. 2010) presented two options to reform or to decline.

## 4.4 Construction and ABCDE

The related challenges and decisions of the construction industry are not as dramatic, but nevertheless important. They are addressed in the next section.

### *4.4.1 Abundance*

The gains of productivity that have characterized some other industries did not take place in the construction industry. In fact some studies suggest that the productivity has been stagnating since the 1970s. Although the infrastructure and the housing is abundant in the developed world, much could be improved and a substantial portion of the infrastructure needs maintenance and upgrading.

On the down side, construction is fulfilling the objective functional requirements – countering the natural forces of gravity, earthquakes, floods, heat and cold. As we have shown in the previous section, however, much of the global economy is shifting attention from function to meaning. Function is a commodity. In the built environment the "meaning" part is created by the architects and the function by the various types of engineers. There is an increasing price difference between a square meter of a facility that simply provides the function of a sheltered space and the price of a square meter of a facility that the architect made desirable. The fact that architects are getting involved in the design of bridges is also a demonstrator of this trend. What used to be a rational engineering work that provides a function – provide a dry passage over a river for example – not has to tell a story, make sense, meaning and there seems to be an agreement that only an architect can do that.

One could simplify this with the 1:10:100:1,000 rule. If 1 is the cost of the structural system, 10 is the cost of facility, 100 is the cost of facility use and 1,000 is the cost of business in that facility. The figures are illustrative and not precise; however, they invite the rethinking of any savings in the construction phase, particularly the savings in the structural system.

### *4.4.2 BRICS, Globalization and Construction*

The real growth of the construction industry will take place in the BRIC economies. The real need and the growth are there. The effect of the globalization on the

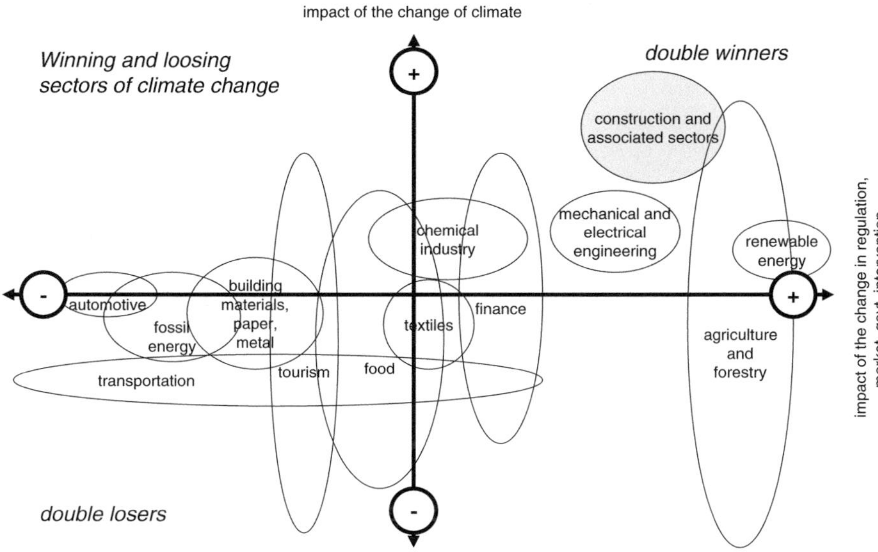


**Fig. 4.2** Construction industry as a double winner in climate change (Source: Deutsche Bank)

construction industry is different if we observe construction as a material industry or as an information industry. Naturally it would be classified as the first but the fact is that some construction processes are material (like actual building) but some are informational (like designing, planning). All information work is getting globalized and the construction related information processes and services will be even more globalized in the future. It makes much sense for the local contractors, queries and concrete plants to be involved in local construction projects but the designs could actually be done anywhere.

### *4.4.3 Climate Change, Energy and Construction*

Climate is getting warmer and the reduction of the $CO_2$ emission will remain one of the important policy goals in the future (Bakkes 2010; ECF 2010). Construction industry is a double winner in this context. On one hand new business opportunities will be generated by the changes of the climate – constructing dams for the high seas, irrigation infrastructure, retrofitting buildings for warmer climate etc. On the other, construction will benefit from the government policies that will stimulate the greening of the economy and society: reduction of energy use in buildings, investments in public transport, renewable energy power plants etc. Most of what are regarded as low hanging fruit of the improved energy efficiency will create new business opportunities for the construction industry (Fig. 4.2).

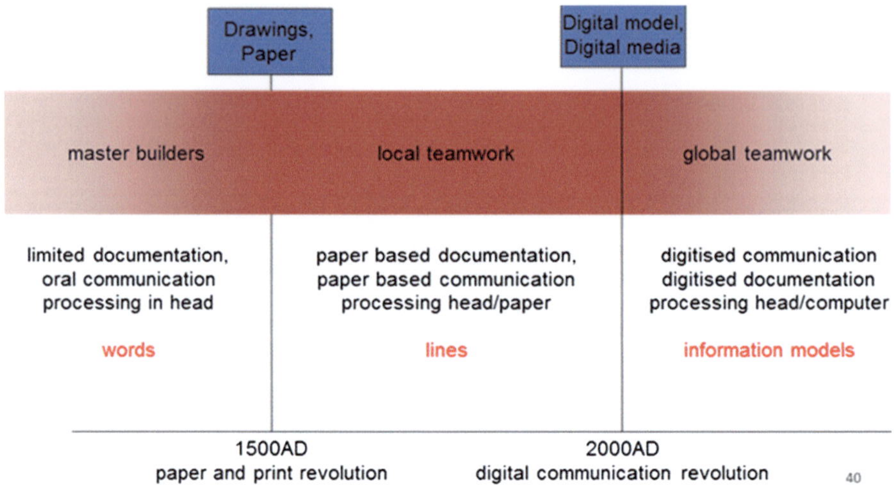


**Fig. 4.3** The three paradigms of construction organization, determined by information technology (Turk 2001)

## *4.4.4 Demography, Ageing and Construction*

If the human talent is the ultimate economic resource then economies should be involved in a competition for that talent. Florida claimed that talent is attracted by tolerance, other talent and technology. The technology not only means the IT infrastructure but the general living conditions to which a lion's share constitute the industries that shape the built environment: civil engineering, architecture and urban planning. He invented the concept of Creative City where of mix of these three professions provide a stimulating ecosystem for the creative class and new high tech as well as other innovative startups.

## *4.4.5 eTechnologies and Construction*

Ever since the introduction of systematic design documentation, information technology has been enabling the fragmentation and specialization of the profession. The author has studied the impact of communication revolutions on construction and claimed that (1) the communication revolution of the cheap paper that allowed for drawing and printing and (2) the digital communication revolution created three different organizational paradigms of construction (Turk 2001): the period of master builders before introduction of cheap paper, the period of local teamwork organized around the exchange of paper documentation and the period of global teamwork. The latter is organized around information sharing and collaboration on the internet. Internet will be enabling the globalization of the information processes is construction (Fig. 4.3).

## 4.5 Lessons for and from Earthquake Engineering, Discussion

The earthquake engineering aspect of construction addresses perhaps the most functional aspect of buildings – making sure the structure withstands the forces of nature. Somehow humanity is taking for granted that we can outsmart nature in any area. While challenges clearly remain, the investors in particular should bear in mind the 1:10:100 rule and the fact that the function is a commodity, that function is relatively a small part of today's products.

Globalization of the construction information processes may introduce new risks. The respect for earthquakes is an inherent part of the culture in areas where earthquakes happen. While designers from places where earthquakes have never been an issue may be knowledgeable about the theories and equipped with the right software the tacit and intuitive elements of their knowledge may be missing. On the bright side expertise from countries with a history of earthquakes can be more easily shared around the world.

The trends listed in Sect. 4.2 are, in spite of their disruptive nature, long term trends that are known. The unknowns that they introduce are the known unknowns. History, however, are not just trends, but also the so called Black Swan events that come as a surprise. Earthquake Engineers have been dealing with Black Swans forever and methods that have been used to abstract and model the unexpected in earthquake engineering could perhaps be reused in the sciences of forecasting of the future.

Many events in a society are exhibiting similar phenomena as in geology. For example, social scientists wonder why unrests and revolutions in a society happen apparently without a warning or why sudden crashes happen in a stock market. Studies that model pressures in a society or in financial markets could borrow from the models of pressures that are building along the geological fault lines before the energy is released by an earthquake. The potential strength of the earthquake and or the strength of the social revolution could be related to the length of the fault line – that in the earth crust or that in a society.

The financial systems could borrow the principles of the resilience of buildings against earthquakes for the building up of the resilience of the financial system. If let to itself even with good but unchanged regulation, the financial system would develop itself into a highly optimized organization that would minimize the costs and maximize the profits but would be extremely vulnerable to unexpected shocks. Much like an engineering product that is optimized for weight or speed is not very resilient. Earthquake engineers know how to build resilient, time proof structures. The designers of Formula 1 cars do not. Lessons from earthquake engineering could be used to build resilience into societal and economic systems.

ABCDE are societal trends. Engineers have a role to play. They are part of the solution as innovators. Engineering problem solving method is useful because there are similarities between an engineering and political work. Both professions are dealing with incomplete information and approximate models. Both must have a holistic perspective. Civil engineers in particular should not forget that, they are also the creators of civilization. They should keep in mind the statement by philosopher Martin Heidegger that "the essence of technology is nothing technical".

**Acknowledgement** The author has been the Secretary General of the Reflection Group on the Future of Europe that was given the task by the European Council to look into the issues that the Union may be facing the next twenty years and how it could respond to it. First part of this paper presents author's personal views and thoughts which may or may not overlap with the upcoming report of the Reflection Group.

## References

Bakkes J (2010) Getting into the right lane for 2050. Netherlands Environmental Assessment Agency (PBL), PBL publication number 500150001, Bilthoven, Oct 2009
EC (2009) THE WORLD IN 2025: rising Asia and socio-ecological transition. European Commission, Directorate-General for Research Socio-economic Sciences and Humanities, EUR 23921 EN, Brussels, Belgium
ECF (2010) Roadmap 2050 to a prosperous, zero-carbon Europe. European Climate Foundation Report, Brussels, Belgium
Florida R (2005) The rise of the creative class: and how it's transforming work, leisure, community and everyday life. Basic Books, New York, USA
Giannakouris K (2008) Ageing characterizes and the demographic perspectives of the European societies. Eurostat, Brussels
Gonzalez F et al (2010) Project Europe 2030: challenges and opportunities. A report to the European Council by the Reflection Group on the Future of the EU 2030, European Union, 2010 – RS 048/2010
IPCC (2007) Climate change 2007: the physical science basis – summary for policymakers. Intergovernmental Panel on Climate Change. http://www.ipcc.ch/SPM2feb07.pdf
Levy F, Murnane JR (2007) How computerised work and globalisation shape human skill demands. In: Suárez-orozco MM (ed) Learning in the global era: international perspectives on globalization and education. University of California Press, Berkeley
NIC (2008) The world in 2025. National Intelligence Council, Washington DC, USA
OECD (2010) PISA 2009 results: what students know and can do – student performance in reading, mathematics and science, vol 1. doi:10.1787/9789264091450-en
Pink DH (2006) A whole new mind: why right-brainers will rule the future. Penguin Books, New York
Reinhardt U, Roos T (2008) Future expectations for Europe. Primus Verlag, Darmstadt
Stehr N (2008) The moralization of the markets in Europe, Society. Springer, New York. ISSN 0147–2011 (Print), 1936–4725 (Online), 45(1)
Stern N (2006) Stern review on the economics of climate change (pre-publication edition). Executive summary. HM Treasury, London. Archived from the original on 31 Jan 2010. http://www.webcitation.org/5nCeyEYJr
Turk Ž (2001) Internet information and communication systems for civil engineering – a review. In: Topping BHV (ed) Civil and structural engineering computing. Saxe-Coburg Publications, Stirling, pp 1–26

Turk Ž (2008) Opening addressed. V: WALRAVENS, Arnold (ur.). Leadership scenarios for Europe's future: 6th international conference proceedings, Bled, June 5 and 6, 2008. European Leadership Centre, IEDC-Bled School of Management, Bled, pp 7–10
Turk Ž (2010) Europe 2030: the citizens, the sun and the Union. Eur View 9(1):79–92, ISSN 1781–6858, Springer
UN (2008) Creative economy report 2008. United Nations 2008, UNCTAD/DITC/2008/2. ISBN 978-0-9816619-0-2

# Chapter 5
# Earthquake-Resistant Bridges of the Future with Advanced Materials

**Saiid M. Saiidi, Ashkan Vosooghi, Carlos Cruz, Sarira Motaref, Chadi Ayoub, Fatemeh Kavianipour, and Melissa O'Brien**

**Abstract** This paper presents the highlights of several studies on seismic performance of bridges. The results showed that superelastic shape memory alloys, fiber-reinforced grouts, built-in elastomeric pads, concrete-filled fiber-reinforced polymer (FRP) columns, and FRP-wrapped segmental columns successfully resisted earthquake forces while substantially reducing apparent damage.

**Keywords** Concrete-filled FRP columns • Elastomeric pads • Fiber-reinforced concrete • FRP-wrapped columns • Segmental columns • Shape memory alloys

S.M. Saiidi (✉) • M. O'Brien
Department of Civil and Environmental Engineering, University of Nevada, MS 258, Reno, NV 89557, USA
e-mail: saiidi@unr.edu; lissakayob@hotmail.com

A. Vosooghi
AECOM, 2020 L Street, Suite 300, Sacramento, CA 95811, USA
e-mail: Ashkan.Vosooghi@aecom.com

C. Cruz
Department of Civil and Environmental Engineering, University of Alberta, Edmonton, AB T6G 2R3, Canada
e-mail: cruznogu@ualberta.ca

S. Motaref
Department of Civil and Environmental Engineering, University of Connecticut, 261 Glenbrook Road, Unit 3037, Storrs, CT 06269, USA
e-mail: motaref@engr.uconn.edu

C. Ayoub
C. Engineering Inc., 3122 Ashford Bend Dr, Houston, TX 77082, USA
e-mail: chadi.ayoub@hotmail.com

F. Kavianipour
TTG Corporation, 300 N. Lake Ave., FL. 14, Pasadena, CA 91101, USA
e-mail: fkavianipour@ttgcorp.com

M. Fischinger (ed.), *Performance-Based Seismic Engineering: Vision for an Earthquake Resilient Society*, Geotechnical, Geological and Earthquake Engineering 32,
DOI 10.1007/978-94-017-8875-5_5, © Springer Science+Business Media Dordrecht 2014

## 5.1 Introduction

The driving force behind innovation may be merely curiosity or search for a solution to a known problem. Whereas innovation in science is often instigated by curiosity, innovation in engineering is linked to potentially practical solutions to problems that are either clearly or vaguely defined. The purpose of this paper is to briefly describe several innovative approaches related to the seismic performance of bridges. The paper presents the highlights of several research projects focused on detailing of highway bridges using novel concepts or advanced materials such as fiber-reinforced polymer (FRP) materials, elastomeric plastic hinges, shape memory alloys, and high-performance fiber-reinforced cementitious materials.

## 5.2 Shape Memory Alloy (SMA)

Shape memory alloys are able to fully recover deformations even after yielding through application of external heat or removal of stress. The latter group, known as superelastic materials, was the subject of the UNR studies. The focus of the study has been on Nickel-Titanium SMA or Nitinol (NiTi). The yield stress of NiTi can be approximately the same as that of 400 MPa steel, but its modulus of elasticity is approximately one-third of the steel modulus. A typical stress-strain relationship for superelastic NiTi is shown in Fig. 5.1. Upon yielding slight strain hardening is observed. At a strain of 6 % major strain hardening occurs. When the stress is released, the stress-strain curve relationship follows a path that leads to a flag-shaped response. The area within the curve presents the dissipated energy. SMA

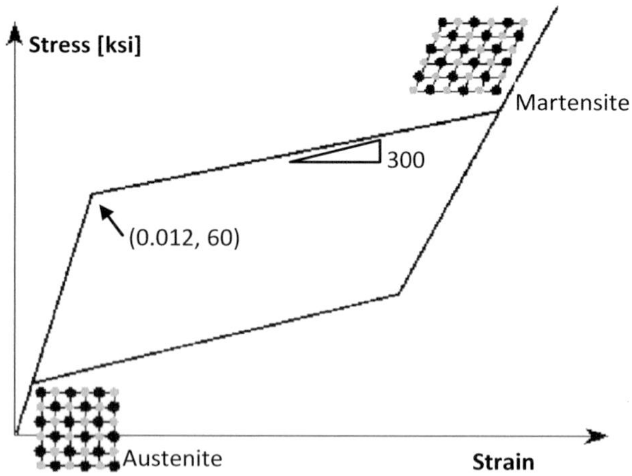


**Fig. 5.1** Superelastic behavior

is hence able to dissipate the earthquake energy, but with no residual strain once the stress is removed. The effectiveness of deformation recovery of SMA bars used as reinforcement in concrete members was investigated. Details of the studies are presented in Saiidi and Wang (2006), Saiidi et al. (2007, 2009), and Cruz and Saiidi (2012).

## 5.3 Engineered Cementitious Composites (ECC)

The low tensile strength of concrete is responsible for its cracking and spalling. Various fiber types have been added to concrete mixes to address this issue. Among these fibers is a polyvinyl fiber with special coating that allows for partial slippage and formation of microcracks but hardens and prevents widening of the crack. As a result new microcracks are formed. The relatively large number of microcracks enables the material to maintain tensile stresses up to strains of 3–5 %, which is substantially higher than the tensile strain capacity of conventional concrete. The tensile strength of ECC varies from 2.5 to 6.0 MPa (Kenser and Bilington 2001; Fischer and Li 2003). A cement-based grout incorporating polyvinyl fibers by 2 % volume, known as ECC, was studied in several bridge columns under earthquake loading to determine the effectiveness of ECC in reducing column damage under large deformations. Details of the study are presented in Saiidi and Wang (2006), Saiidi et al. (2009), Motaref et al. (2011), and Cruz and Saiidi (2012).

## 5.4 Elastomeric Plastic Hinges

Because of their low modulus of elasticity, elastomeric materials can undergo large deformations without damage. They have been used in seismic isolators for several decades, in which they help soften the structure through their shear deformation, thus elongating the vibration period and reducing seismic forces. A new application of elastomeric pads was explored in Kawashima and Watanabe (2006) by replacing part of the concrete in the plastic hinge of columns with rubber. Unlike seismic isolation, elastomeric pads acted mainly in flexure in this application. The column reinforcement was passed through the pad, and the pad was post-tensioned. The detail was successful in that there was no damage in the elastomeric pad under cyclic loads. However, at drift ratio of 3 % the longitudinal column bars began to buckle and failed under low-cycle fatigue. The pad did not provide sufficient lateral restraint against buckling of the longitudinal reinforcement. A modified version of the pad was developed at UNR by incorporating steel shims in the pad while increasing the thickness to approximately one-half the column diameter. To eliminate shear deformation an unbounded central steel pipe was included (Fig. 5.2). Studs were

**Fig. 5.2** UNR elastomeric plastic hinge

welded to the end plates of the pad to anchor it to the concrete above and below the pad. Similar to the pad in Kawashima and Watanabe (2006) the longitudinal column bars passed through the pad and the columns were post tensioned. More details about the design and properties of the pad are provided in Motaref et al. (2011) and Cruz and Saiidi (2012).

## 5.5 FRP Columns

One application of FRP is CFFT (concrete-filled FRP tubes) columns. Glass fibers aligned at ±55° were used to provide both flexural and shear strength for the column and confine the concrete. Studies of different types of FRP columns showed that glass CFFTs are more advantageous in seismic applications because of better hysteretic response and higher displacement ductility capacity. These GFRP (glass FRP) tubes enhance the displacement ductility capacity of the columns by increasing the plastic hinge length compared to conventional RC columns. The GFRP tubes can serve both as the formwork for the concrete and as structural components. Consequently, the construction cost and time is reduced.

Another form of FRP application is in segmental columns that are wrapped with carbon FRP sheets to satisfy the shear and confinement reinforcement requirements. These segments were connected to each other by a prestressed unbounded high strength threaded rod. The application of the FRP-wrapped segmental columns reduce the time and increase the quality of construction by using pre-fabricated segments. In addition, FRP wraps facilitate construction procedure by reducing steel reinforcement in the column segments.

## 5.6 Experimental and Analytical Studies

The initial studies of SMA-reinforced beams consisted of testing a series of simply-supported reinforced concrete beams subjected to symmetric two-point loading with SMA or steel reinforcement in the constant moment region (Saiidi et al. 2007). The objective of the study of the beams was to determine the deflection recovery characteristics of SMA-reinforced concrete members. Because of promising results the study was extended to cantilever reinforced concrete columns with longitudinal SMA reinforcement in the plastic hinge zones. Superelastic SMA bars were used to reduce permanent drift of the columns under lateral cyclic loads (Saiidi et al. 2009). To minimize earthquake damage, the effectiveness of ECC in the plastic hinge was also explored both in original and repaired column models. Some of the column models were subjected to slow cyclic loading and others were tested on a shake table under simulated earthquakes. The study of combined SMA/ECC columns was extended to the bottom plastic hinges in one of the piers of a 33-m long, four-span bridge model tested on shake tables (Fig. 5.3) (Cruz and Saiidi 2012). The bridge model was subjected to bi-directional motions simulating one of the 1994 Northridge earthquake records. The top plastic hinges in the same columns were constructed of conventional reinforced concrete.

**Fig. 5.3** Four-span bridge model

**Fig. 5.4** Elastomeric plastic hinge embedded in the column cage

The damage resistance of ECC to dynamic excitations was also evaluated in two other shake table studies, one involving segmental columns and the other a precast reinforced concrete column in which concrete in the plastic hinge was replaced by ECC (Motaref et al. 2011). Steel reinforcement was used in these models. In both cases similar models made with conventional concrete had been tested and hence benchmark data existed. The lateral reinforcement in all cases was made of steel and met current seismic code requirements.

The performance of elastomeric plastic hinges was investigated in two studies one on segmental columns and the other in the bottom plastic hinges of one of the piers of the aforementioned four-span bridge model. The details of the elastomeric element were the same in both cases. Figure 5.4 shows the elastomeric element embedded in one of the columns of the four-span bridge model.

The performance of FRP columns was investigated through shake table studies of a four-span bridge model. The bridge was a quarter scale model with three, two-column bents with different innovative details and construction methods. Side bents consisted of CFFT columns, one of them cast-in-placed (CIP) and the other with precast footing and columns. The middle bent was a double-curvature structure, with four segments which were wrapped with two layers of CFRP and connected with a posttensioning tendon. Ten biaxial motions with white noise in both directions were applied using three shake tables and two actuators for abutment motion input.

In all studies OpenSEES computer program was used for extensive analytical modeling and parametric studies of the test models prior to finalizing the design and loading of the test models and after the completion of the tests.

## 5.7 Results

A sample measured lateral load-drift hysteretic behavior of a column with SMA/ECC at the plastic hinge is shown in Fig. 5.5. It is evident that loading performance of the column is what is normally expected of conventional reinforced concrete columns. The initial parts of the unloading curves are also similar to those of steel-reinforced concrete columns. However, as the unloading curves approach the horizontal axis they bend towards the origin, thus leading to a very small permanent drift as they cross the axis. This feature is drastically different than that of conventional RC members, which exhibit relatively large drifts at zero force. The damage in the upper and lower plastic hinges of the columns with SMA/ECC at the bottom plastic hinges at the conclusion of testing the four-span bridge is shown in Fig. 5.6. It can be seen that there was substantial spalling in the conventional construction at the top of the column exposing longitudinal and transverse reinforcement. Furthermore a large number of other flexural cracks turning to shear cracks formed as can be seen in the left photo in Fig. 5.6. In contrast the damage in the SMA/ECC plastic hinge was minimal with minor cracking and spalling (the right photo in Fig. 5.6).

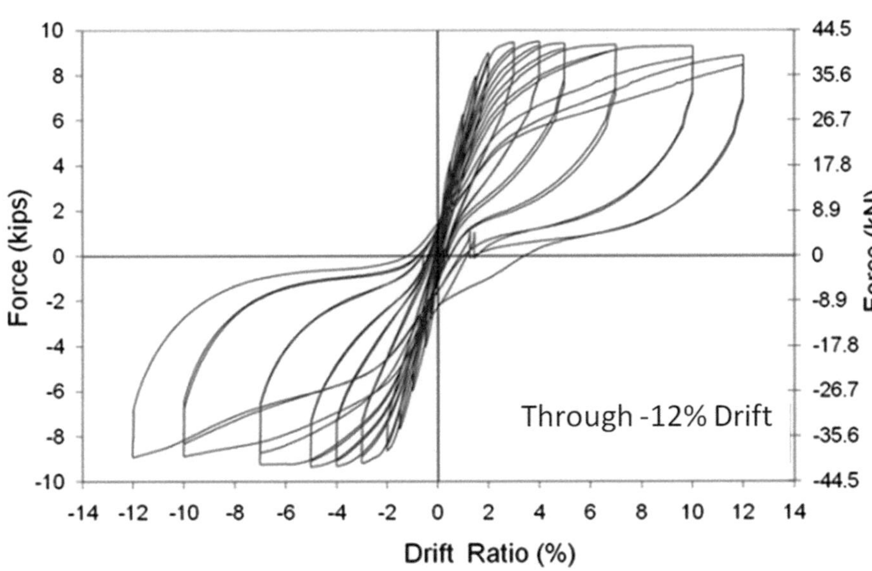


**Fig. 5.5** Performance of column with SMA/ECC plastic hinge

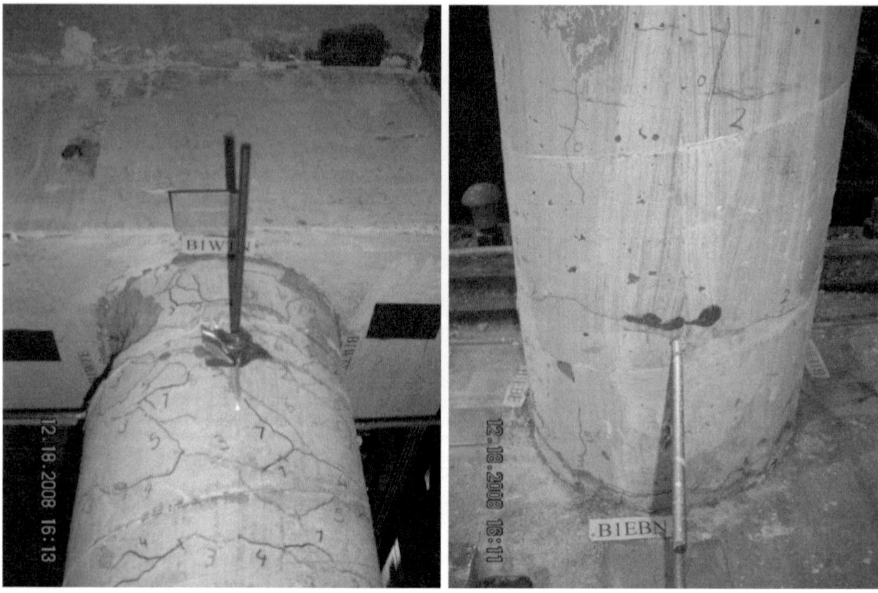


**Fig. 5.6** Damage at RC (*left*) and ECC (*right*) plastic hinges in 4-span bridge column

The contrast between the performance of ordinary concrete and ECC can also be seen in damage in Fig. 5.7 and in the envelopes of the measured force-displacement relationships in Fig. 5.8. The damage was initiated at the junction of the base segment and the adjacent segment due to the compressive failure of the concrete or ECC. It is clear in Fig. 5.7 that concrete damage was substantially more extensive in terms of the surface area and depth of damage than the damage in ECC. Figure 5.8 shows the degradation of the lateral load capacity in SC-2 (with concrete) starting at approximately 80 mm lateral displacement. It can be seen that SE-2 (with ECC) was able to maintain its strength up to 170 mm and degradation after this displacement was at a much lower rate. Cumulative force-displacement hysteresis curves for SC-2 and SE-2 are shown in Fig. 5.9. This figure indicates that the energy dissipation in both columns was comparable.

No damage was observed in elastomeric plastic hinges in testing of the 4-span bridge model or the cantilever segmental column. Testing of the 4-span bridge was terminated due fracture of steel reinforcement in the top plastic hinges constructed with conventional RC. The bottom plastic hinges that were constructed with built-in elastomeric pads experienced no damage in the rubber and only minor cracks in the adjacent concrete (Fig. 5.10).

The progression of damage in the four-span bridge model with FRP columns was monitored using visual inspection, video recordings, and instruments installed on key locations in the bridge. In the bents with CFFT columns, there was some spreading of cracks followed by concrete spalling in the surface of the footing

**Fig. 5.7** Damage at RC (*left*) and ECC (*right*) plastic hinges of segmental columns

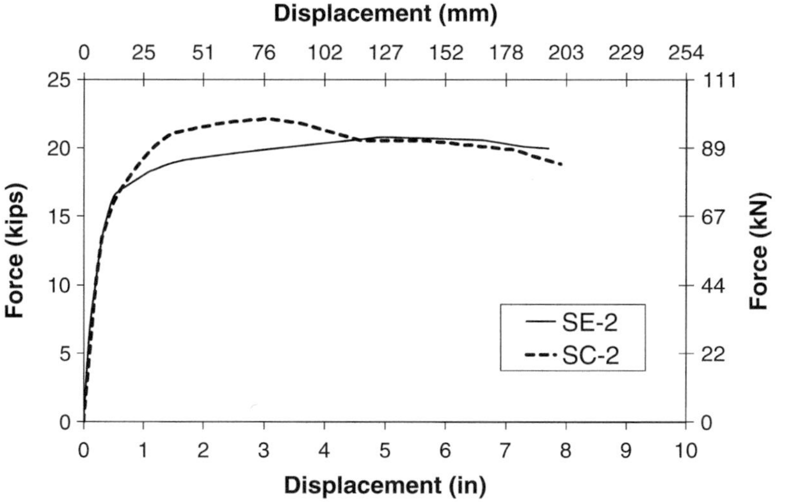


**Fig. 5.8** Measured response envelopes of RC (SC-2) and ECC (SE-2)

(Figs. 5.11 and 5.12). The performance of CFFT cast-in-place (CIP) and precast bents were similar and the FRP tubes remained intact in all columns. In the FRP-wrapped, segmental bent, no sign of gap opening between segments was observed during testing. In addition, no visible crack was noticed on the surface of FRP wrap

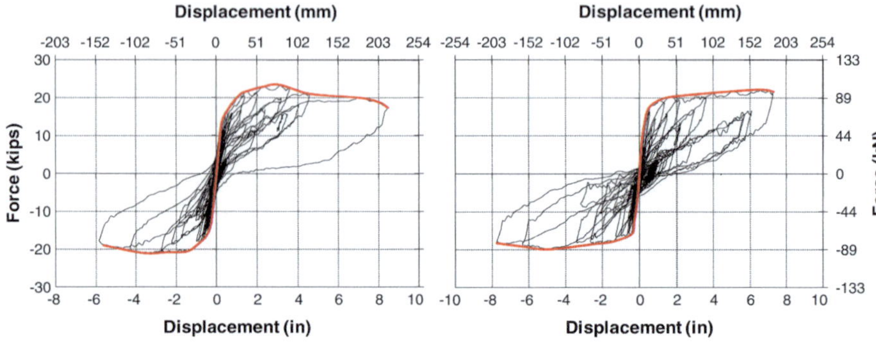


**Fig. 5.9** Cumulative force-displacement hysteresis curves for RC (*left*) and ECC (*right*)

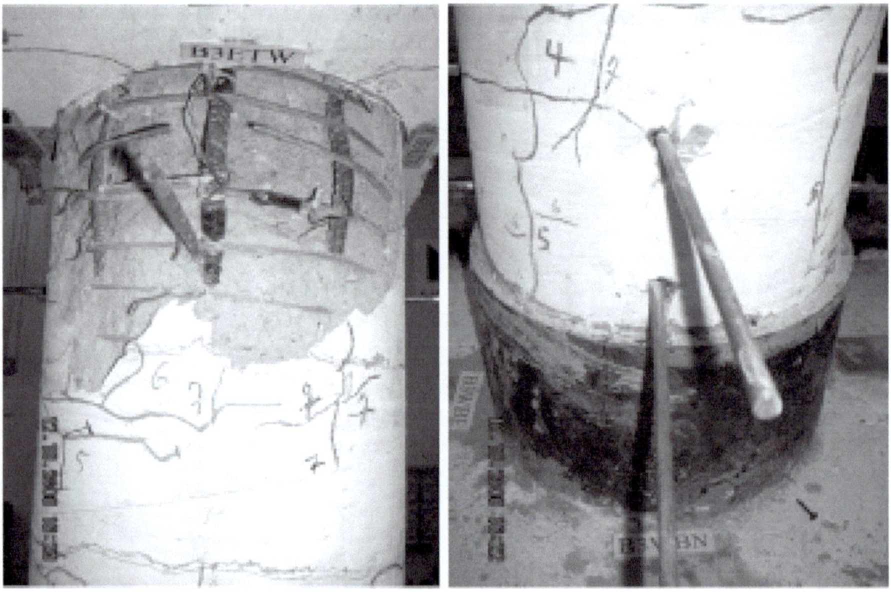

**Fig. 5.10** Damage at RC (*left*) and rubber (*right*) plastic hinges of 4-span bridge

(Fig. 5.13). After testing, the FRP was removed to examine the condition of concrete and longitudinal reinforcement. No sign of buckling or rupturing was evident.

Measured backbone curve and bilinear idealized force-displacement relationship in the transverse direction for the bent with elastomeric plastic hinges and cast-in-place CFFT columns are plotted in Figs. 5.14 and 5.15, respectively. These graphs indicate that the bents underwent a considerable plastic deformation with a significant post-yielding stiffness.

**Fig. 5.11** Precast column condition after final run

**Fig. 5.12** CIP column condition after final run

**Fig. 5.13** FRP-wrapped, segmental bent after shake table test

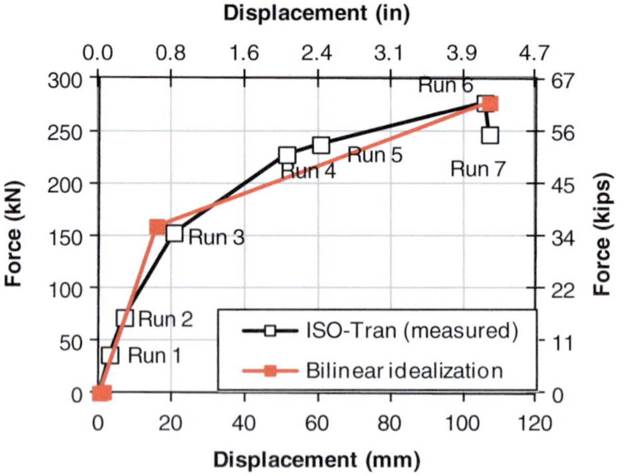


**Fig. 5.14** Measured and idealized backbone curves for bent with elastomeric plastic hinges

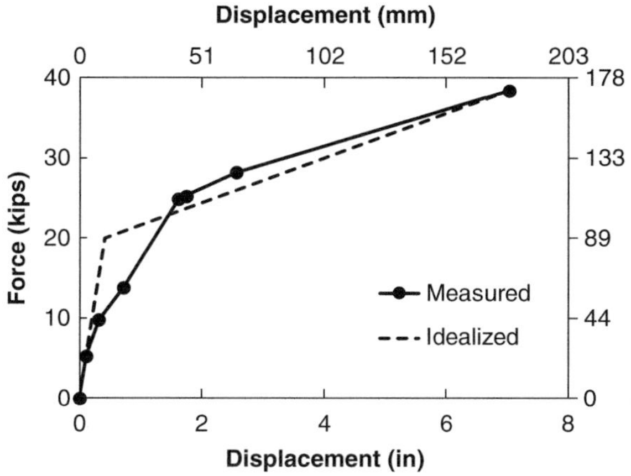


**Fig. 5.15** Measured and idealized backbone curves for bent with cast-in-place CFFT columns

## 5.8 Conclusions

The summary results presented in this article also demonstrated that a new paradigm of damage-free bridges is feasible in earthquake-resistant design of bridges by using advanced materials and details. Through innovation in detailing and materials it is possible to eliminate or substantially reduce earthquake damage while accomplishing satisfactory performance under strong earthquakes. A superelastic Nickel-Titanium alloy showed that it could replace reinforcement steel in plastic hinges and help the structure recover lateral deformation. The fiber-reinforced grout, or the ECC, experienced substantially less damage during demanding shake table tests discussed in this paper and its strength degradation was minor. Furthermore, built-in elastomeric pads in column plastic hinges were found to remain completely free from damage even under large rotational demand. Finally, concrete-filled FRP columns and FRP-wrapped segmental columns were damage-free during strong earthquake simulations. These columns can be used in seismic areas to facilitate accelerated bridge construction.

**Acknowledgement** The studies presented in this article were funded by the National Science Foundation (Grants CMS-0420347, CMMI-0650935, and CMS-0402490) and various grants from the California Department of Transportation.

## References

Cruz Noguez C, Saiidi M (2012) Shake-table studies of a four-span bridge model with advanced materials. J Struct Eng 138(2):183–192

Fischer G, Li VC (2003) Deformation behavior of FRP reinforced ECC flexural members under reversed cyclic loading conditioned. ACI Struct J 100(1):25–35

Kawashima K, Watanabe G (2006) Seismic performance of unbonded columns and isolator built-in columns based on cyclic loading tests. In: Proceedings of the international association of bridge management and safety, Porto, 2006
Kesner KE, Billington SL (2001) Investigation of ductile cement based composites for seismic strengthening and retrofit. In: Proceedings of the fourth international conference on fracture mechanics of concrete and concrete structures, Cachan, May-June 2001, pp 65–72
Motaref S, Saiidi M, Sanders D (2011) Seismic response of precast bridge columns with energy dissipating joints. Center for Civil Engineering Earthquake Research, Department of Civil and Environmental Engineering, Report no. CCEER-11-1, University of Nevada, Reno, May 2011, 760 p
Saiidi M, Wang H (2006) An exploratory study of seismic response of concrete columns with shape memory alloys reinforcement. Am Concr Inst ACI Struct J 103(3):436–443
Saiidi M, Zadeh M, Ayoub C, Itani A (2007) A pilot study of behavior of concrete beams reinforced with shape memory alloys. J Mater Civil Eng ASCE 19(6):454–461
Saiidi M, O'Brien M, Zadeh M (2009) Cyclic response of concrete bridge columns using superelastic nitinol and bendable concrete. Am Concr Inst ACI Struct J 106(1):69–77

# Chapter 6
# Inelastic Shear Response and Strengthening of RC Bridge Hollow Box Piers

**Tatjana Isaković and Matej Fischinger**

**Abstract** Several existing bridges located in Central Europe (in Slovenia) are supported by hollow box columns whose structural detailing is inadequate, by today's standards, for seismic regions. This chapter includes an overview of experimental and analytical studies of such columns. During the experiments, shear failure of the investigated columns was observed. For this reason it was necessary to increase their shear strength. Two techniques were analysed and compared: RC jacketing and CFRP wrapping. In both cases the minimum amount of strengthening effectively increased the shear strength. Although shear failure was prevented, other unfavourable failure types, induced by other deficiencies in the structural detailing, were activated. The analytical estimation of the shear strength and cyclic response of the as-built and the strengthened columns was quite challenging since the disagreement between different available methodologies was considerable. The most successful analytical procedures were identified.

**Keywords** Bridges • Shear strength • Strengthening • Shear strengthening • Cyclic response • Concrete jacketing • CFRP wrapping • Experiment • Eurocode • Numerical models

## 6.1 Introduction

Several bridges located in Central Europe (in Slovenia), built in the 1970s and 1980s, are supported by hollow box columns, whose structural detailing is substandard and are nowadays considered to be inadequate for seismic regions. The deficiencies in the detailing are mostly related to the transverse reinforcement in the

T. Isaković (✉) • M. Fischinger
Faculty of Civil and Geodetic Engineering, University of Ljubljana,
Jamova 2, SI 1000 Ljubljana, Slovenia
e-mail: tisak@ikpir.fgg.uni-lj.si; matej.fischinger@fgg.uni-lj.si

M. Fischinger (ed.), *Performance-Based Seismic Engineering: Vision for an Earthquake Resilient Society*, Geotechnical, Geological and Earthquake Engineering 32,
DOI 10.1007/978-94-017-8875-5_6, 

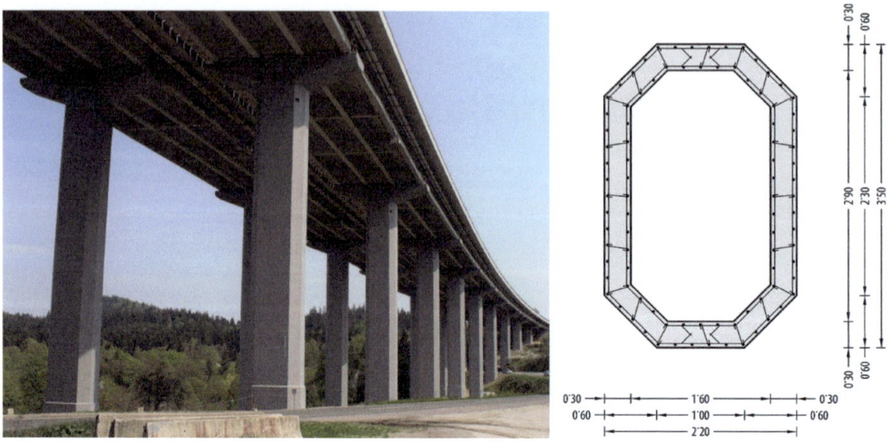

**Fig. 6.1** Typical European viaduct with hollow-box piers

columns, which typically cannot ensure adequate shear strength and confinement of the concrete core, nor prevent buckling of the longitudinal flexural bars.

An example of such a bridge is presented in Fig. 6.1. This bridge was constructed on the one of the Slovenia's main stretches of motorway. There have been several concerns regarding its seismic safety since its piers have many structural deficiencies such as the following: (a) the lap splices are formed in the region of potential plastic hinges, (b) the transverse reinforcement was placed on the inside of the longitudinal bars, (c) the amount of the transverse reinforcement was gradually reduced from the base to the top of the column, (d) plain bars were used for the longitudinal as well as for the transverse reinforcement. Such piers are obviously susceptible to shear failure.

The results of studies so far performed in connection with shear resistance and strengthening are mostly related to piers with solid cross-sections (e.g. Saiidi et al. 2001, Kawashima et al. 2001), so they are not directly applicable to hollow-box columns. Only a few investigations have been performed in the case of columns of this type (e.g. Calvi et al. 2005). However, most of these results are only partly applicable to the investigated columns, since many of them were obtained in the case of columns that are typical for bridges in Taiwan (Mo et al. 2003) and Japan (Kawashima et al. 1990). These bridges do not include exactly the same kind of structural detailing.

Since knowledge about the seismic response of the investigated columns was quite limited, it was difficult to identify reliable analytical tools, which could be used to estimate the most probable type of their failure. For this reason an experimental investigation was performed. The columns were tested cyclically on a scale of 1:4 (see Fig. 6.2). In this Chapter only the tests performed on short columns, which failed due to insufficient shear strength, are described (see Sect. 6.2). More details about the whole study can be found elsewhere (Isaković et al. 2008).

**Fig. 6.2** Analyses were performed on large-scale structures, which were tested experimentally

Taking into account the results of the experiments, the most adequate analytical procedures, which could be used to estimate the shear strength of the investigated columns, were identified (see Sect. 6.3).

Since the as-built short columns failed due to insufficient shear strength, shear strengthening was also performed. The effectiveness of two methods: RC jacketing and CFRP wrapping were investigated. Both methods were experimentally examined, since not much data was available about their possible efficiency when used to strengthened hollow box columns (Mo et al. 2004; Cheng et al. 2005). The performed experiments are described in Sect. 6.4. The strengthening of the investigated columns was particularly challenging due to the poor accessibility of their inner surfaces.

Prior to the experiments, analytical studies of the strengthened columns were performed. The same analytical procedures which were used to estimate the shear strength of the as-built columns were also employed to predict the shear strength of the strengthened columns. The results of this study are presented in Sect. 6.5.

## 6.2 The Cyclic Response of the As-Built Columns

A typical column, with an aspect ratio of 1.86, was chosen to be examined experimentally. The main properties of the 1:4 scale model are presented in Fig. 6.3.

The compressive strength of the concrete was 41.6 MPa. The yield stress of the steel was 324 and 240 MPa for the longitudinal and transverse reinforcement, respectively. The longitudinal reinforcement ratio at the base amounted to 1.5 % of the gross cross-sectional area. At the top of the columns this amount was reduced to 0.5 %. At the base of the column, the transverse reinforcing bars had a diameter of 4 mm (16 mm in the prototype), and were spaced at a distance of 5 cm (20 cm in the prototype). At the top, the diameter of the transverse reinforcement was reduced

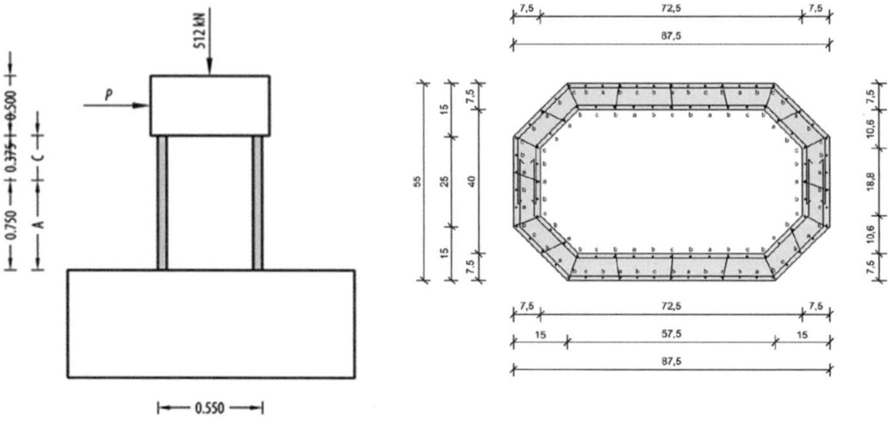


**Fig. 6.3** The 1:4 scale model of the tested column

to 2.5 mm (10 mm in the prototype), but the distance between the bars was kept the same as at the column base (5 cm; 20 cm in the prototype). The shear reinforcement was placed inside the longitudinal reinforcing bars (see the layout, which is shown in Fig. 6.3). The lap splices were constructed close to the column foundations.

The model was subjected to a horizontal cyclic load. The vertical load was applied to the top of the column, and kept constant during the whole experiment. The applied axial loading was typical for the columns of bridges that are built in central Europe. Thus, the normalized axial loading was relatively low (equalling about 7 % of the compressive strength of the concrete). The column was loaded up to failure.

The column failed due to its insufficient shear strength, after the yielding of the longitudinal bars (see Fig. 6.4). Considering the poor structural detailing, a relatively large displacement ductility capacity of 4 was obtained (see the cyclic response, presented in Fig. 6.5). It was provided by the favourable hollow box cross-section with a relatively large compression zone, by the low axial loading, and by the relatively high strength of the concrete.

More details about the response of the investigated column can be found in electronic form in the NEEShub-JEE database (Elnashai et al. 2011).

The cyclic response of the column was modelled using three different macro models: (1) A beam element with concentrated plasticity, where the cyclic behaviour of the hinges was modelled using Takeda hysteretic rules; the element which was included to the OpenSees program system (Mazzoni et al. 2009) by Japanese researchers (Takahashi 2009) was used, (2) A beam element with the hinges available in the OpenSees program system, and (3) a MVLEM model which is included in OpenSees at the University of Ljubljana (Fischinger et al. 2004).

All the selected types of elements were able to describe the cyclic response of the investigated column with reasonable accuracy (see Figs. 6.6, 6.7 and 6.8). In all cases a reduction of the shear stiffness proportional to the reduction of the flexural stiffness (as proposed by Calvi et al. 2005) was taken into account. More details about the models used in the study can be found elsewhere (Vidrih 2012).

Fig. 6.4 Shear failure of the as-built column

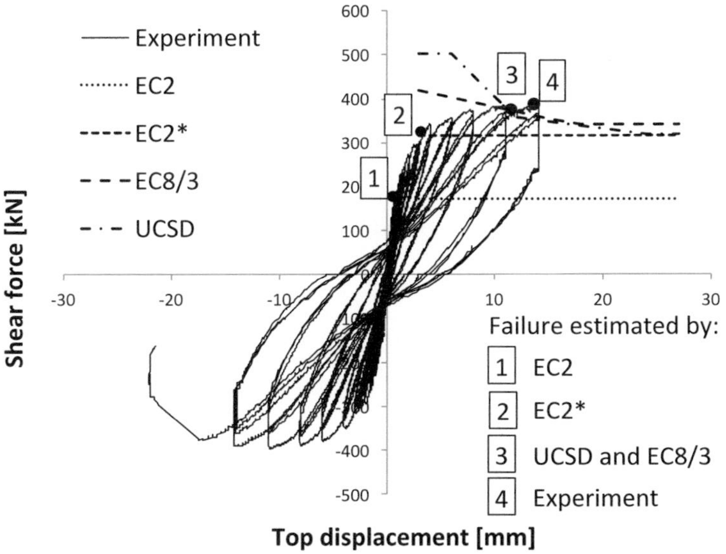


Fig. 6.5 Experimentally observed cyclic response of the as-built column; its shear strength, determined by different analytical procedures, is compared with the experimentally observed shear demand

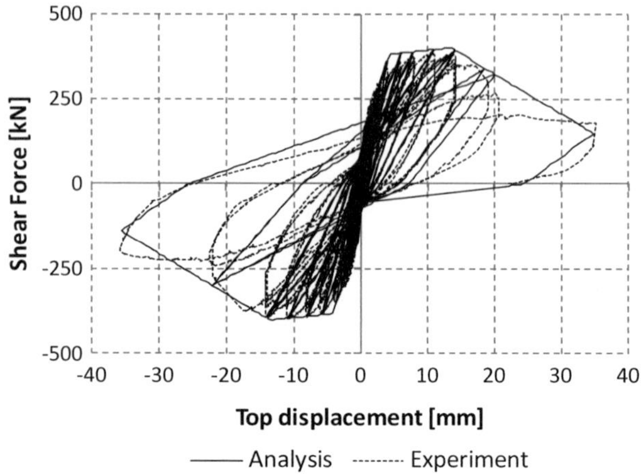


**Fig. 6.6** Comparison of the cyclic response obtained in the experiment, and in the case of a beam element with concentrated plasticity

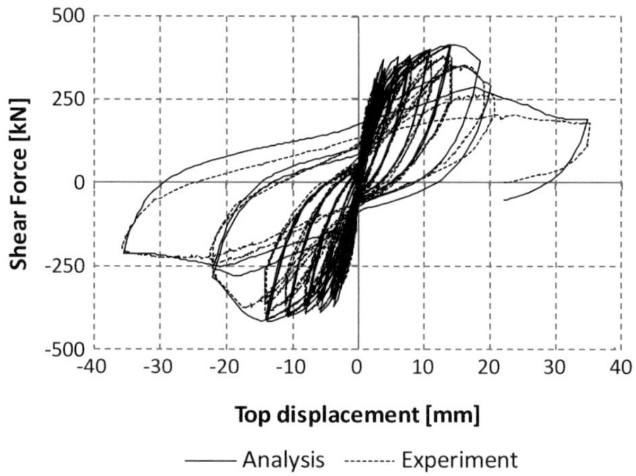


**Fig. 6.7** Comparison of the cyclic response obtained in the experiment, and in the case of the "beam with hinges" model

## 6.3 The Shear Strength of the As-Built Columns

The shear strength of the as-built columns was estimated using the procedures defined in the Eurocode 2 standard (CEN 2004), the Eurocode 8/3 standard (CEN 2005), and the procedure proposed at UCSD (Priestley et al. 1996). Since the actual strength of the column was investigated, all safety factors were excluded.

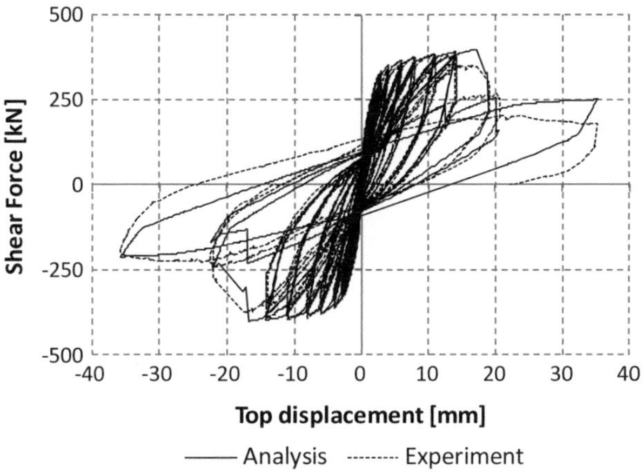


**Fig. 6.8** Comparison of the cyclic response obtained in the experiment, and in the case of the MVLEM model

**Table 6.1** The shear capacity of the as-built column, predicted using different methods

| Method | Predicted displacement ductility | $V_C$ | $V_N$ | $V_w$ | $V_R$ | $V_R/V_{exp}$ |
|---|---|---|---|---|---|---|
| EC2 | – | 93(0) | 54(0) | 171 | 318 (171) | 82 % (44 %) |
| EC8/3 | 3.9 | 117 | 110 | 146 | 373 | 96 % |
| UCSD | 3.9 | 83 | 110 | 171 | 364 | 93 % |

Note: The value of $V_R$, denoted in brackets, is obtained by neglecting the contributions of $V_c$ and $V_N$

In general, all the investigated procedures determine the shear strength of the columns in the same way. Three different mechanisms are taken into account in order to define the shear resistance ($V_R$), as is stated in Eq. (6.1).

$$V_R = V_c + V_N + V_w \tag{6.1}$$

where $V_c$ is the shear strength of an element without shear reinforcement (i.e. the contribution of the concrete to the shear capacity), $V_N$ is the contribution of the compressive stresses to the increase in shear strength, and $V_w$ is the contribution of the shear reinforcement.

However, the ways in which these mechanisms are considered, in different methods, are quite different, so that the estimated values of shear strength, and in particularly the contributions of the concrete, can differ significantly.

The shear strength of the investigated column was analysed at the bottom and at the top of the column, since the shear reinforcement was gradually reduced from the bottom to the top of the column. At the bottom part of the column the values of the shear strength, determined according to UCSD and Eurocode 8/3 (EC8/3), matched the experimental data quite well (see Table 6.1). This is also evident from Fig. 6.5, where the shear capacity, estimated by the considered procedures, is compared with the experimentally determined shear demand. The shear strength, as well as the corresponding displacement demand, both determined by these two procedures (points 3 and 4 in Fig. 6.5), was almost exactly the same as that observed during the experiment.

According to standard Eurocode 2 (EC2), the column would fail in shear, prior to yielding of the longitudinal bars (point 1 in Fig. 6.5). This unrealistic conclusion was obtained taking into account the significantly underestimated shear strength. This was too low because the contributions of the concrete shear strength and compressive stresses ($V_c + V_N$) were neglected, as is stated in the standard (EC2 neglects these two contributions when the demand exceeds the sum of their values). In the investigated case, this was evidently too conservative, since these two mechanisms contributed as much as approximately half of the total shear strength. If they had been taken into account, the shear strength (see point 2 in Fig. 6.5) would have been comparable with those estimated by the other two methods for large displacement demands, although the predicted type of failure would still have been incorrect (see Fig. 6.5 for more details).

In order to increase the shear strength of the investigated column, the angle of the concrete compression strut could be decreased to the minimum permissible value of 22° (as has been suggested by some researchers). This would consequently increase the contribution of the shear reinforcement $V_w$ by a factor of 2.5, so that the total shear strength of the column would be 428 kN. It is clear from Fig. 6.4 that a concrete strut angle of 45° was observed during the experiment. Thus, taking into account such a smaller value would be only an analytical bypass of the very conservative assumption related to the strength of the concrete in EC2.

Based on previous results and some other observations in the case of similar columns (Calvi et al. 2005), it can be concluded that the current EC2 requirements are not adequate for estimation of the shear strength of hollow box bridge columns and similar structural elements (e.g. RC walls).

Both of the standards EC8/3 and EC2 were less accurate when the shear strength was estimated at the top of the column. In the investigated case, the EC8/3 method was too conservative in the low displacement ductility demand range (note the significant difference between the shear strength corresponding to smaller displacements, predicted by EC8/3 and UCSD, in Fig. 6.5).

The UCSD method was the most accurate and was able to control well all the important parameters: the shear capacity of the columns, the type of failure, and even the location of the failure. However, it should be noted that the shear strength was estimated taking into account the experimentally observed angle between the concrete compression strut and the longitudinal column axis of 45°, instead of a value of 30°, which was originally proposed by the authors of the method.

More details about this study can be found elsewhere (Isaković et al. 2008).

## 6.4 The Cyclic Response of the Strengthened Columns

Based on the results obtained in the experiment, presented in the previous section, shear strengthening of the critical columns was performed. This strengthening was particularly challenging, since it was possible to construct the strengthening jacket only on the outer side of the column. Thus, the outer jacket had to provide sufficient shear strength, but yet not be too strong, since that could worsen the response of the inner parts of the column. This is because a very strong jacket could substantially increase the maximum possible compression deformation at the outer column edges, but at the same time it would also increase the deformation at the inner non-strengthened edges. This would cause spalling of the concrete cover, and buckling of the longitudinal bars located on the inner side of the column.

Since, in the investigated case, the shear demand was not much larger than the shear capacity of the as-built column, the minimum amount of jacketing was provided. Two types of strengthening were investigated: RC jacketing and CFRP wrapping. Both methods were examined experimentally. The tested specimens and their main properties are summarized in Figs. 6.9 and 6.10.

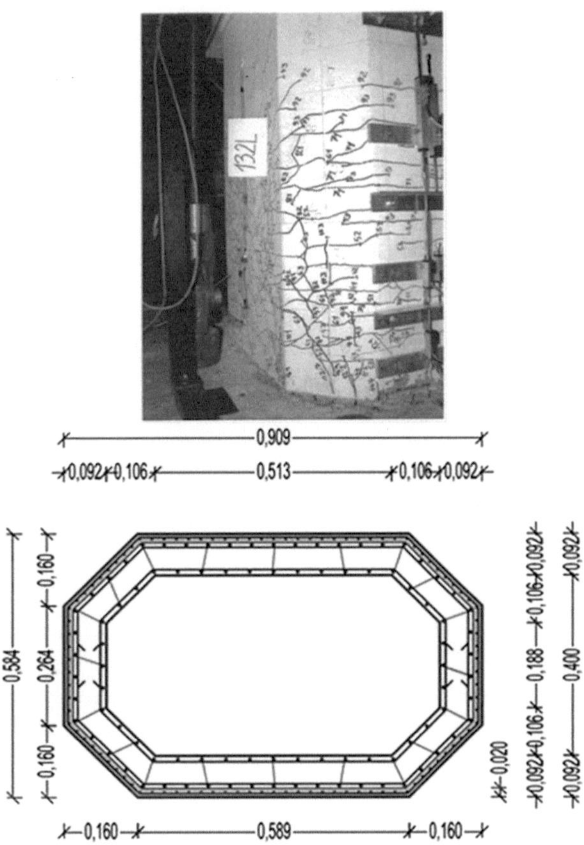


**Fig. 6.9** Strengthening of the column, using concrete jacketing – the test specimen

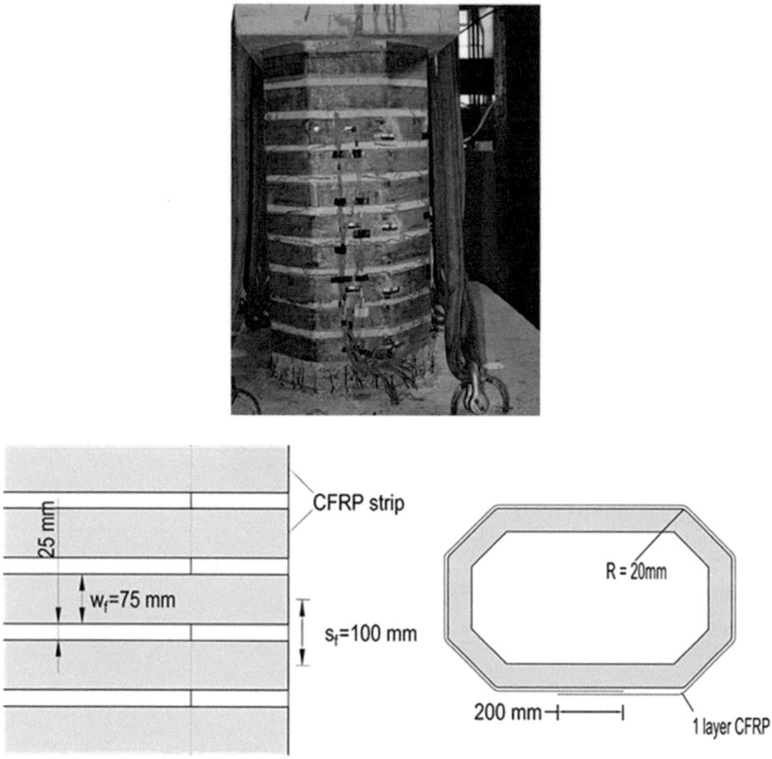


**Fig. 6.10** Strengthening of the column, using CFRP strips – the test specimen

In general, the response of both strengthened columns was similar. They effectively prevented shear failure of the column (see Figs. 6.11 and 6.12). The type of failure was completely different than that of the as-built column. Although shear failure was prevented, other unfavourable failure mechanisms induced by other deficiencies in the column were activated. Spalling of the concrete cover at the outer and inner edges was first observed. This was followed by the buckling and then the rupture of the longitudinal bars. The pullout of some of the longitudinal bars was also observed. A pronounced horizontal crack was observed at the bottom of the column near the footing. The observed rupture of the longitudinal bars was followed by substantial rocking of the column.

The maximum shear strength of column was 470 and 450 kN in the case of RC jacketing and CFRP wrapping, respectively. At the moment of failure (defined by a 20 % reduction in the maximum strength), the displacement was 28 and 24 mm in the case of RC jacketing and CFRP wrapping, respectively. Compared to the as-built column, the shear strength was increased by a factor of about 1.5 and 1.3, in the case of the RC jacketing and CFRP strips, respectively.

In the case of RC jacketing, the energy dissipation capacity of the column was improved. The CFRP wrapping was somewhat less effective. Compared to the

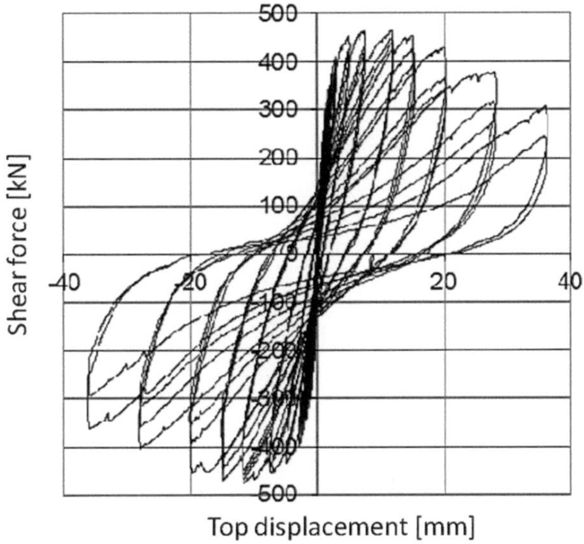


**Fig. 6.11** Experimentally observed cyclic response of the column strengthened by RC jacketing

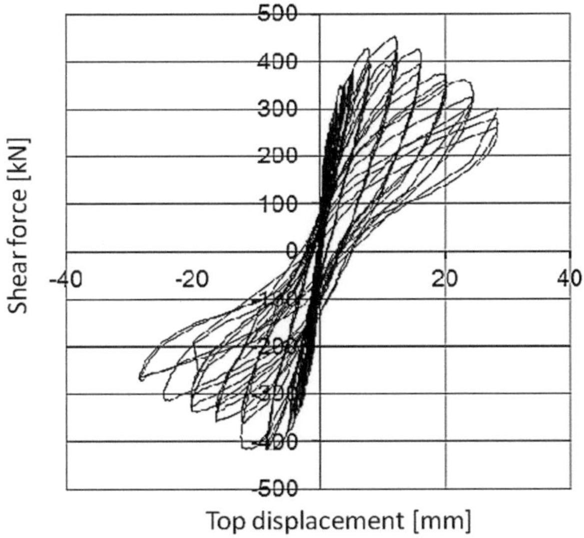


**Fig. 6.12** Experimentally observed cyclic response of the column strengthened by CFRP strips

concrete jacketing, there was no additional layer of concrete, which would prolong the spalling of the concrete around the longitudinal bars and their buckling. The deterioration of the column strength was more pronounced, and so the energy dissipation capacity was lower. However, it should be noted that the less favourable response of the column wrapped by CFRP strips was partly caused by the pre-corrosion of some of the longitudinal bars.

## 6.5 The Shear Strength of the Strengthened Columns

The shear strength of the strengthened columns was estimated analytically using the same procedures as for the as-built column. The contribution of the jackets to the shear strength was added to the other three mechanisms, defined in Eq. 6.1. The shear strength of the CFRP wrap was estimated according to CEN (2005) and Priestley et al. (1996). In the case of RC strengthening, the shear strength of jacket was determined taking into account its shear reinforcement in the same way as the quantity $V_w$ had been taken into account.

The estimated shear strength was compared with the shear demand, which was determined experimentally (see Fig. 6.13). All the methods, including the one using EC2 (with the contributions of $V_C$ and $V_N$ included), proved that the shear

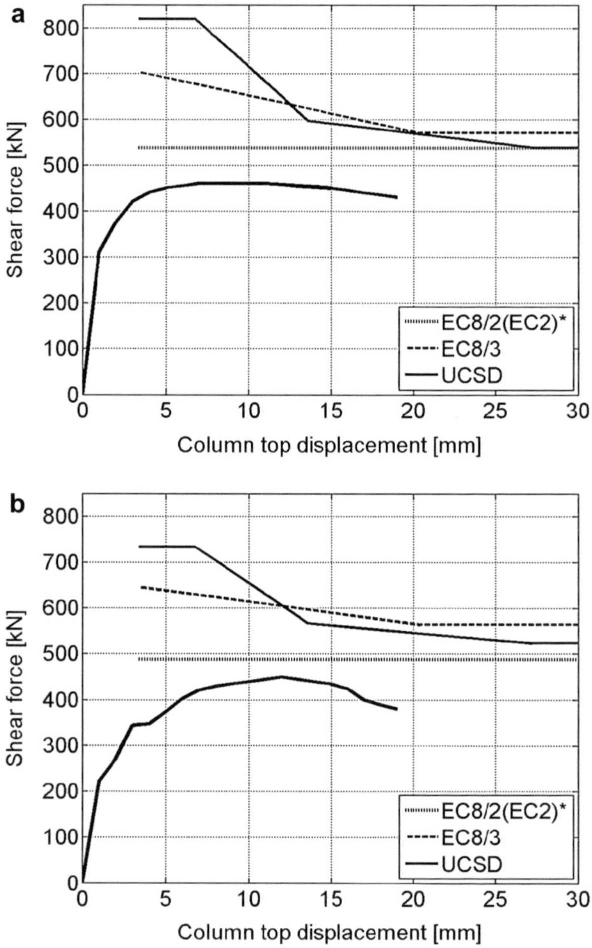


**Fig. 6.13** The shear capacity of the strengthened column: (**a**) RC jacketing, (**b**) CFRP wrapping

strength was reliably increased by both strengthening techniques, and that there was no danger of shear failure. This was later confirmed by the results of the experiment.

## 6.6 Conclusions

Experimental and analytical studies of the cyclic response of short reinforced concrete hollow box columns, which are typical for viaducts built in the 1970s and 1980s in central Europe, were performed. These columns have certain deficient construction details, which are nowadays considered inappropriate for seismic regions.

An experimental cyclic investigation of these columns demonstrated that, in spite of their poor structural detailing, their displacement ductility capacity was relatively good, and therefore still acceptable for moderate seismic demand regions (e.g. Central Europe). This ductility was provided by the favourable hollow box cross-section with its large compression zone, by the low axial forces, and by the relatively high strength of the concrete.

The shear strength of the investigated columns was estimated analytically by using standard models, which yielded quite different results. The location and type of failure, estimated using the procedure given in the EC2 standard, were unrealistic because of the significantly underestimated shear strength. This was too low since the contributions of the concrete's shear strength, and of the effect of the compressive stresses, which together contributed approximately half of the total shear strength, were neglected. If these effects were to be taken into account, the shear strength would be comparable with the values estimated by using the other two methods for large displacement demands. However, the predicted type of the failure would still be incorrect.

A more suitable estimation of the shear strength at the bottom part of the investigated columns was obtained by the procedure which is included in the standard EC8/3. However, the location of the failure was estimated incorrectly, since this standard underestimated the shear strength at the top of the column, where the ductility demand was low. Good estimation of the shear capacity, the type and the location of failure was achieved when the UCSD method was used.

The investigated column was strengthened using RC and CFRP jacketing. Both strengthening techniques successfully increased the shear strength with a minimum amount of wrapping. However, when sufficient shear strength was provided, other unfavourable types of failure induced by other construction deficiencies were activated. Particularly critical were the buckling and rupture of the longitudinal bars, as well as bar slip at the column base. The concrete strengthening was more efficient, considering the achieved enhancement of column ductility and energy dissipation capacity. In order to improve the ductility capacity of the investigated type of column using CFRP wrapping, a larger amount of CFRP strips would be needed (note that the absolute minimum amount of CFRP strips was used).

**Acknowledgments** The described experiments were performed in cooperation with the Slovenian National Building and Civil Engineering Institute (ZAG). The work was partly funded by the Company for Motorways in the Republic of Slovenia (DARS). The CFRP jacketing was provided by SIKA d.o.o., Slovenia. The research was also supported by the Ministry of Higher Education, Science and Technology of the Republic of Slovenia. Some of the presented results were obtained within the scope of work performed for the preparation of Ph.D. theses by Jaka Zevnik and Zlatko Vidrih.

## References

Calvi GM, Pavese A, Rasulo A, Bolognini D (2005) Experimental and numerical studies on the seismic response of R.C. hollow bridge piers. Bull Earthquake Eng 3:267–297

CEN (2004) Eurocode 2: design of concrete structures – part 1-1: general rules and rules for buildings. EN 1992-1-1, Comité Européen de Normalisation, Brussels

CEN (2005) Eurocode 8: design of structures for earthquake resistance. Part 3: strengthening and repair of buildings. EN 1998–3, European Committee for Standardization, Brussels

Cheng CT, Mo YL, Yeh YK (2005) Evaluation of as-built, retrofitted, and repaired shear-critical bridge columns under earthquake-type loading. J Bridge Eng ASCE 10(5):520–529

Elnashai A, Ambraseys NN, Dyke S (2011) NEEShub-JEE database. J Earthquake Eng Datab. http://nees.org/resources/3166

Fischinger M, Isakovic T, Kante P (2004) Implementation of a macro model to predict seismic response of RC structural walls. Comput Concrete 1(2):211–226

Isakovic T, Bevc L, Fischinger M (2008) Modelling the cyclic flexural and shear response of the R. C. hollow box columns of an existing viaduct. J Earthquake Eng 12(7):1120–1138

Kawashima K, Unjoh S, Lida H (1990) Seismic inspection and seismic strengthening of reinforced concrete bridge piers with termination of main reinforcement at midheight. In: Proceedings of the 1st US Japan Workshop on Seismic Retrofit of Bridges, Tsukuba, Japan 1990

Kawashima K, Hosotani M, Yoneda K (2001) Carbon fiber retrofit of reinforced concrete bridge bents. Toward new generation seismic design methodology of bridges. Tokyo Institute of Technology, Tokyo

Mazzoni S, Mckenna F, Scott MH, Fenves GL, Jeremic B (2009) Open system for earthquake engineering simulation (OpenSees) – command language manual. UCB, PEER, Berkeley

Mo YL, Wong DC, Maekawa K (2003) Seismic performance of hollow bridge columns. ACI Struct J 100(3):337–348

Mo YL, Yeh YK, Hsieh DM (2004) Seismic retrofit of hollow rectangular bridge columns. J Compos Constr ASCE 8(1):43–51

Priestley MJN, Seible F, Calvi GM (1996) Seismic design and retrofit of bridges. Wiley, New York

Saiidi MS, Wehbe NI, Sanders DH, Caywood CJ (2001) Shear retrofit of flared RC bridge columns subjected to earthquakes. ASCE Bridge J 6(3):189–197

Takahashi Y (2009) Code for the Giberson's element using Takeda's hysteretic rules included into OpenSees. http://opensees.dpri.kyoto-u.ac.jp/japanese/binaries.html. Accessed Sept 2009

Vidrih Z (2012) Seismic response of concrete bridges with deficient structural detailing. Ph.D. thesis (in the Slovenian language), University of Ljubljana

# Chapter 7
# Developments in Codifying Direct Displacement-Based Seismic Design

**Nigel Priestley**

**Abstract** Recent efforts to codify Direct Displacement-based Seismic Design (DDBD) as an alternative to current force-based (FBD) code approaches are discussed. First, the reasons why the change from a force-based to a displacement-based philosophy is necessary are presented, together with a brief summary of the DDBD procedure. Currently DDBD is generally utilized by compliance with codes that permit design verification by non-linear time history analysis (NTHA). While this is appropriate for important structures it is excessively demanding for routine structures. As a consequence, codification exercises have been undertaken to provide codified DDBD procedures for a number of seismic design codes. This work is on-going, and is briefly outlined in this chapter.

**Keywords** Design codes • Displacement • Drift limits • Force • Limit states • Seismic design • Strain limits

## 7.1 Introduction

Over the past 20 years a considerable effort has been put into developing Direct Displacement-Based Seismic Design (DDBD) into a seismic design approach with sufficient scope (in term of structural options) and detail (in terms of structural actions) to be considered as a viable and rational alternative to current force-based seismic design (FBD) approaches. The advantages of DDBD over FBD include:

---

N. Priestley (✉)
Jacobs School of Engineering, Department of Structural Engineering, University of California, San Diego, CA, USA

ROSE School European School for Advanced Studies in Reduction of Seismic Risk, Pavia, PV, Italy

Priestley Structural Engineering, P.O.B. 4, Diamond Harbour 8941, New Zealand
e-mail: nigelpriestley@xtra.co.nz

M. Fischinger (ed.), *Performance-Based Seismic Engineering: Vision for an Earthquake Resilient Society*, Geotechnical, Geological and Earthquake Engineering 32,
DOI 10.1007/978-94-017-8875-5_7, 

- Lack of dependency on elastic stiffness. In FBD an estimate is needed of the effective elastic stiffness (normally measured as secant stiffness to first yield) for calculation of structure periods, required for determining design base shear, and for distributing the base shear between different seismic-resisting elements of the structure. However, member stiffness is proportional to member strength, which is not known at the start of the design process, and distribution of seismic resistance based on member elastic stiffness often results in illogical distribution of strength for structures responding inelastically to seismic attack. DDBD relies on the yield characteristics of members and entire structures. These are independent of strength depending only on structural geometry and yield strain of flexural reinforcement, and are hence known at the start of the design process.
- DDBD allows structures to be designed to achieve a specified performance limit state under a given level of seismic excitation, based on limit state strains or non-structural drifts. At best, FBD provides an unreliable means for checking, at the end of the design process, whether or not a performance limit state has been exceeded.
- Different hysteretic characteristic of different structural systems can directly be accommodated in DDBD. FBD typically, and incorrectly assumes that hysteretic characteristics do not influence displacement response.
- Effects of enhanced displacement response from near field effects are directly incorporated into the design procedure with DDBD, whereas subjective amplification factors are needed for FBD.
- Effects of structural and accidental eccentricity can be rationally incorporated into the design process with DDBD, recognizing that stiffness eccentricity can be controlled by modifying the strength of different members within the constraints of defined section dimensions. The important parameter is the effective stiffness at design displacement response, not an artificial estimate of elastic stiffness eccentricity, as adopted in current FBD procedures.
- Emphasis on displacements in DDBD forces the designer to make proper consideration of displacement capacity of non-structural as well as structural elements of the structure, which is not always the case in FBD, where estimates of displacement response are often poor.

## 7.2 Fundamentals of Direct Displacement-Based Design

The DDBD principles have been presented in many previous documents (e.g. Priestley et al. 2007), and are only briefly summarized here with reference to Fig. 7.1. The structure to be designed (shown as a frame building in Fig. 7.1a, but the process is identical for all structures – structural walls, bridges, wharves etc.) is modelled by a single-degree-of-freedom (SDOF) representation based on the fundamental inelastic mode of response.

The characteristic structural stiffness is taken as the secant stiffness from the origin to the design displacement, rather than the elastic stiffness as in FBD, as

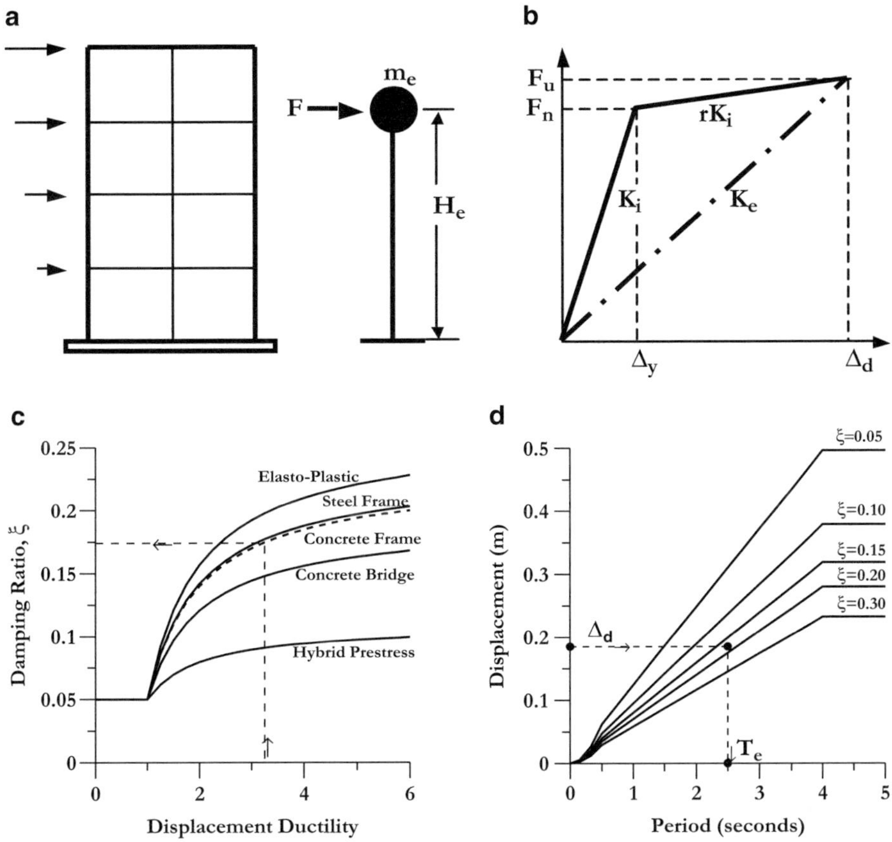


**Fig. 7.1** Fundamentals of direct displacement-based seismic design

shown in Fig. 7.1b. The design displacement depends on the limit state considered, and is found by integrating limit state strains or drifts. It should be noted that the design displacement can generally be determined without reference to the design strength, which is, of course, unknown at the start of the design process.

The equivalent viscous damping depends on the structural system considered (Fig. 7.1c), and the ductility level corresponding to the design displacement. Since the yield displacement is independent of the strength, the design ductility is also known at the start of the design process.

A displacement spectra set representing the seismicity of the limit state considered for different levels of equivalent viscous damping is used together with the known design displacement and known equivalent viscous damping to determine the effective period at the design displacement (Fig. 7.1d). The SDOF relationship between mass, stiffness and period is inverted to determine the effective stiffness. Multiplying this by the design displacement determines the design base shear.

Other aspects of the approach, including structural analysis to determine the design moments in potential plastic hinge regions and capacity design principles and details have been presented in detail elsewhere (Priestley et al. 2007).

## 7.3 Current Utilization of DDBD

Until recently, DDBD has largely been used for seismic design of major structures where inelastic deformation and hysteresis characteristics are such that force-based design is seen to be inappropriate – in particular where determination of behaviour (force-reduction) factors are not defined in existing design codes. Examples include marginal wharves, where piles of different free length result in different ductility demand on different structural elements, and structures where unbonded prestressing provides seismic resistance (the PRESSS system, (Priestley 1996)) and similar concepts in steel and timber). Direct displacement-based seismic design has been found to be simple to apply in such cases. Final designs have generally been verified by non-linear time-history analyses (NTHA) to ensure compliance with code drift and capacity design details. This approach (DDBD followed by NLTHA verification) has been used in many countries, including USA, New Zealand, Israel, Columbia and Italy.

## 7.4 Codification of DDBD

Recently a number of projects have been initiated to provide codification of DDBD, with the intention that regular and more conventional structures could be routinely designed in accordance with the principles of DDBD without the need for non-linear time history verification. A draft "straw man" code for DDBD of building structures was included as chapter 14 of Priestley et al. (2007). This was used as a starting point for a more ambitious coordinated research project in Italy (Sullivan et al. 2012), in which different Italian research institutes developed appropriate design information for structural types including buildings (frames, wall structures, mixed wall/frame structures) bridges, foundations, retaining walls and structures with isolation and added damping.

Preceding this, seismic design criteria compatible with DDBD principles were developed for the Port of Los Angeles (POLA) (POLA 2009) for seismic design of marginal container wharves. These principles have been used informally for the past 10 years by consultants working for POLA. Codification has formalized the approach. As with most DDBD codification efforts, the design recommendations have been prepared with parallel DDBD and FBD procedures, where the FBD procedures have been adjusted to be as compatible as possible with the DDBD approach. Limit-state strains are defined for the ductile piles. These are used as the starting point of the DDBD procedure. In the FBD procedure pile strains are

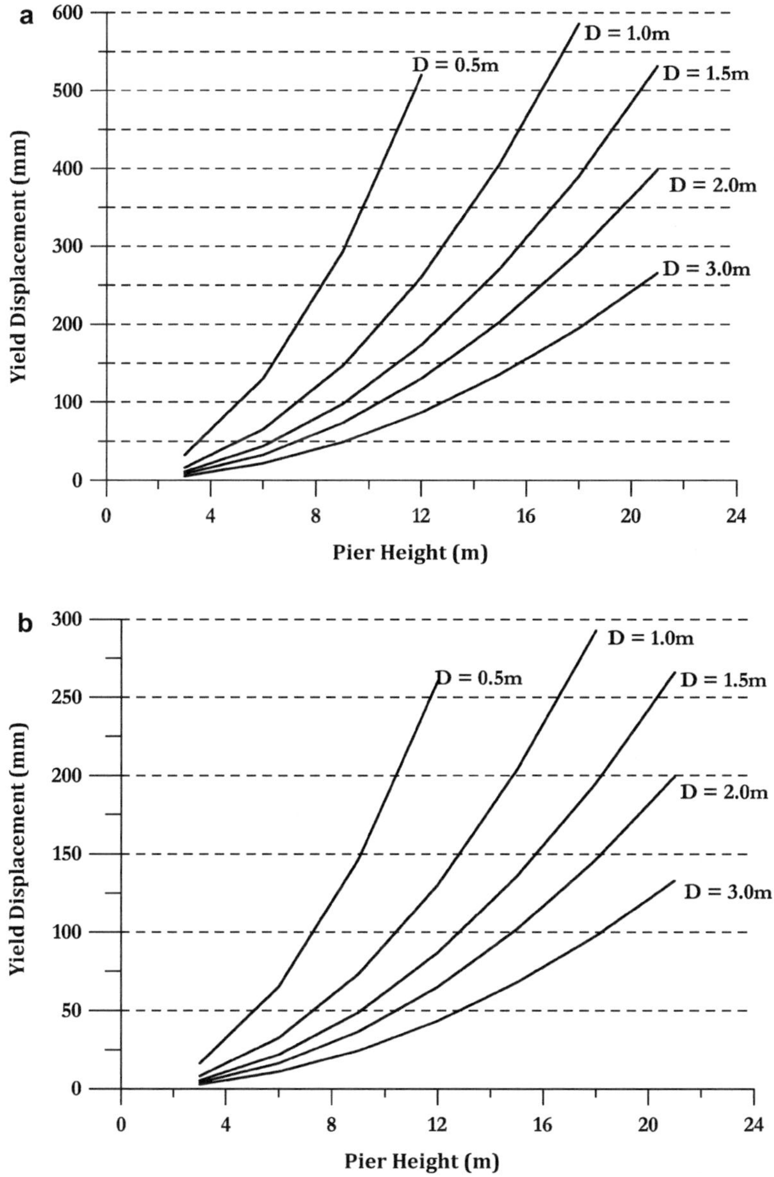


**Fig. 7.2** Bilinear yield displacement for rigid base bridge piers. (**a**) Cantilever Pier (Single Bending), (**b**) Pier in Double Bending

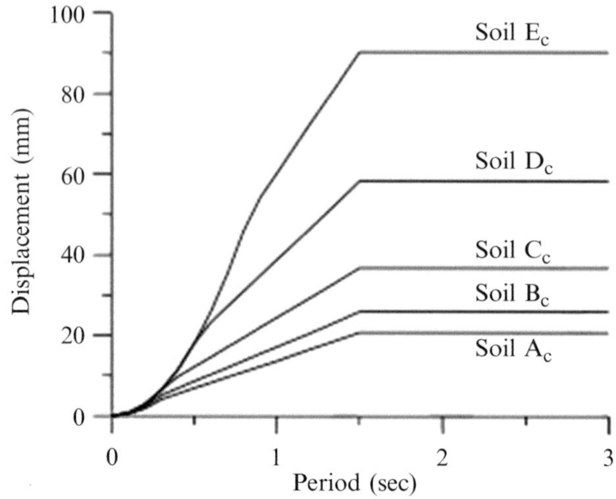


**Fig. 7.3** Displacement spectra for PGA = 0.08 g, (e.g. Sydney) and different soil conditions for Australia (500 year return period)

checked, by analysis, at the end of the design process to confirm code compliance. The Port of Long Beach (POLB) adjacent to POLA has recently adopted similar seismic design provisions (POLB 2009).

A draft DDBD code has been prepared for seismic design of bridges in Australia (Priestley et al. 2012). The low seismicity of Australia makes DDBD particularly attractive, as it allows a simple means for determining whether a bridge designed for gravity loads can be expected to respond elastically to the design seismicity level, and hence qualify for simplified detailing provisions. Displacement response spectra for Australia are based on a low corner period of 1.5 s. This, combined with typically low PGA values for most of Australia result in low displacement demand, particularly for long-period bridges. Effective yield displacements, modified from Fig. 7.2 to account for foundation flexibility may be compared with displacement spectra, such as shown in Fig. 7.3.

If the pier yield displacements exceed the plateau displacement for $T > 1.5$ s, then elastic response is assured. A second check, based on the calculated period, if this is less than 1.5 s, releases further bridge piers from specific seismic design requirements. Only if these two conditions are not met is specific DDBD required. As with the POLA and POLB codes, a force-based alternative is available to the designers. Seismic design forces for structures designed to DDBD provisions are significantly lower than for force-based designs to the same seismic intensity as a result of excessive conservatism in the latter case.

A similar codification exercise is underway for DDBD of New Zealand bridges.

A study group is about (Sept, 2012) to be convened in New Zealand with the aim of preparing a DDBD alternative to force-based design of building structures, and the author is aware of informal DDBD code documents that have been prepared, and used, in both Chile and Columbia.

## References

POLA (2009) Code for seismic design, upgrade and repair of container wharves. Port of Los Angeles, Los Angeles, California

POLB (2009) Wharf design criteria, version 2.0. Port of Long Beach, Los Angeles, California

Priestley MJN (1996) The PRESSS program – current status and proposed plans for phase III. PCI J 41(2):22–40

Priestley MJN, Calvi GM, Kowalsky MJ (2007) Displacement-based seismic design of structures. IUSS Press, Pavia, 721 p

Priestley N, Sedra S, Forster G, Bennett M (2012) New seismic design rules for Australian bridges. In: Proceedings of the Australian Bridge conference, Surfers Paradise, Queensland, Australia

Sullivan TJ, Priestley MJN, Calvi GM (2012) Model DDBD code, vol 2. IUSS Press, Pavia

# Part III
# New Vision After Recent Earthquakes

# Chapter 8
# A Lesson from the 2011 Tohoku Earthquake – The Necessity for Collaboration and Dialog Among Natural Scientists, Engineers, Social Scientists, Government Agencies, and the General Public

**Masayoshi Nakashima, Tracy C. Becker, Tomohiro Matsumiya, and Takuya Nagae**

**Abstract** The March 11th, 2011 Tohoku earthquake and subsequent tsunami caused great damage over a large region of North-Eastern portion of Japan. The magnitude of the event was not predicted and thus found Japan unprepared, especially for the effects of the tsunami. This article is a summary of observation of damage and disruption based primarily on the information available within 3 months after the disaster. Also presented are the lessons that the authors believe have been learned and should be shared within the international community of earthquake disaster mitigation researchers and practitioners. The major issues discussed are the ground motion, tsunami, building damage, and post-event response. Recent research efforts in response to the disaster are also touched upon briefly.

**Keywords** Tohoku earthquake • Tsunami • Liquefaction • Disaster response • Disaster mitigation

---

M. Nakashima (✉) • T.C. Becker
Disaster Prevention Research Institute (DPRI), Kyoto University, Gokasho, Uji, Kyoto, Japan
e-mail: nakashima@archi.kyoto-u.ac.jp; becker.tracy.75c@st.kyoto-u.ac.jp

T. Matsumiya
Department of Architecture and Building Engineering, Kinki University, Kowakae, Higashi-Osaka, Osaka, Japan
e-mail: matsumiya@arch.kindai.ac.jp

T. Nagae
Hyogo Earthquake Engineering Research Center, National Research Institute for Earth Science and Disaster Prevention, 1501-21 Nishikameya, Mitsuta, Shijimi, Miki, Hyogo 673-0515, Japan
e-mail: nagae@bosai.go.jp

M. Fischinger (ed.), *Performance-Based Seismic Engineering: Vision for an Earthquake Resilient Society*, Geotechnical, Geological and Earthquake Engineering 32,
DOI 10.1007/978-94-017-8875-5_8, 

## 8.1 Introduction

The March 11th, 2011 Tohoku earthquake, hit the North-Eastern portion of Japan, causing extremely serious damage. The damage and disruption, the majority caused from the tsunami, displays a range engineering and social organization deficiencies. Although historical records were available, an event of this magnitude was not predicted. The earthquake was large enough to seriously alter the Japanese topography and caused a major tsunami, affecting over 700 km of coastline. Although the majority of infrastructure was undamaged, previously identified problems, such as damage from landslides and shear failures in older building occurred. Major problems occurred due to the loss of utilities after the disaster.

The authors prepared the first draft of this article for the Bled4 Workshop – performance-based seismic engineering – a vision for an earthquake engineering society. The workshop was held 3 months after the Tohoku disaster; hence the materials included in the first draft were collected based on the information made available within the 3 months after the disaster. The authors read the draft 1 year after the event and confirmed that the materials presented remained valid in most portions. To maintain the atmosphere of "three months after the disaster", the authors had decided to present this article as faithfully as possible to the first draft. At this end of this article, recent, notable research efforts that had been initiated in compliance to the lessons learned from the disaster are introduced to fill the gap between the time when the Bled4 workshop was organized and the time of writing of this article.

In what follows, observations are presented on the ground motion, tsunami, building damage, and post-event response. In addition major lessons that the authors believe have been learned from the disaster are summarized in terms of the predictions, tsunami warnings and design, energy dependency, community and system based engineering, and post-disaster response.

## 8.2 Ground Motion

The Tohoku earthquake rupture zones extended over 400 km in length with a width of about 200 km. Figure 8.1 (Report of Japan Meteorological Agency 2011) shows the regions of rupture that triggered the earthquake and Fig. 8.2 (Report of Japan Meteorological Agency 2011) shows the intensities at the affected areas. Japan uses a scale named Shindo similar to the Modified Mercali Intensity scale. While the Shindo was developed based on visual inspection of damage it is now tied to numerical measurements of local acceleration intensity as shown in Fig. 8.2. Shindo six plus (6+) to seven (7) means very large and corresponds approximately to MM of XI to XII.

Although the Tohoku region (in the North-Eastern part of Japan) was hard hit this time, Japan has suffered from other damaging earthquakes in recent years, most

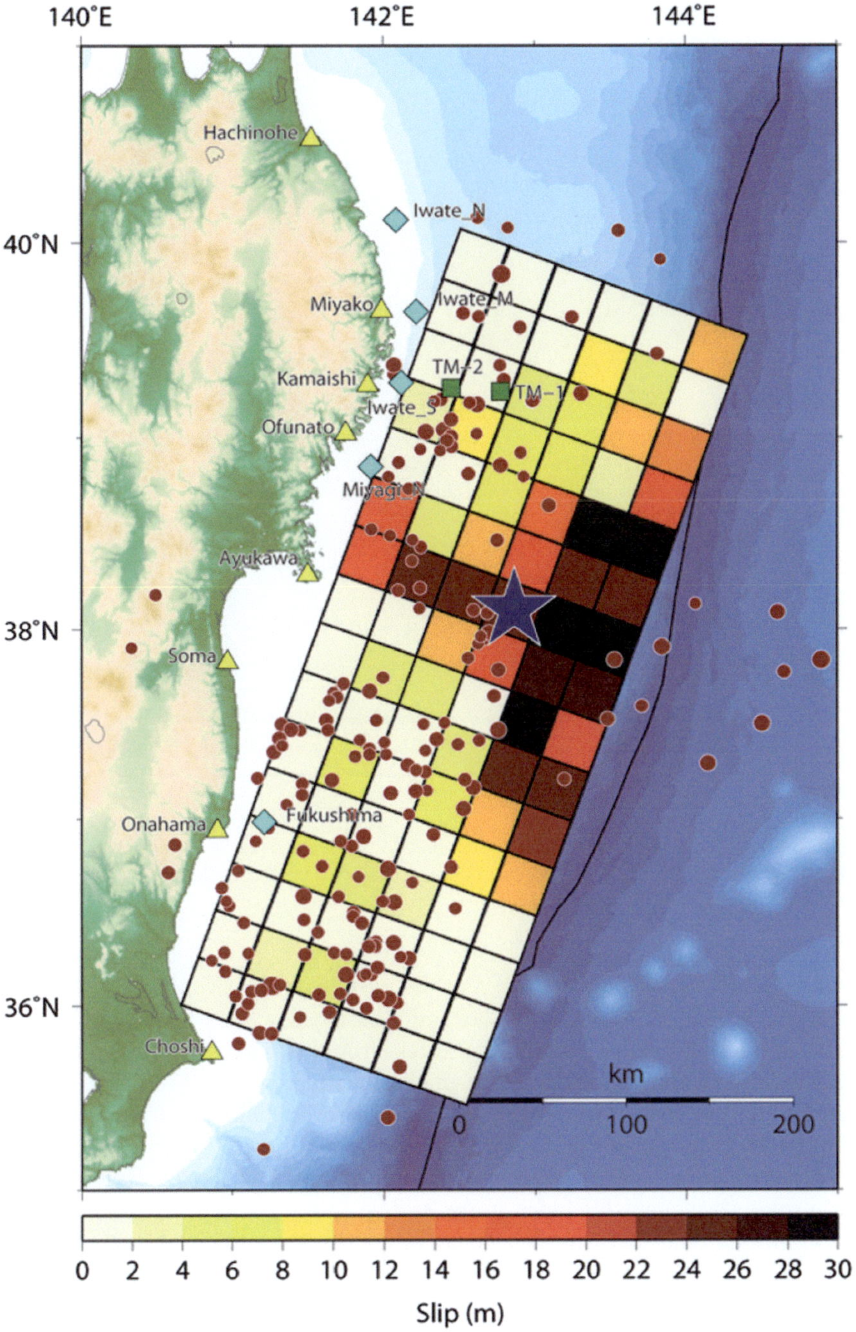


**Fig. 8.1** Map of fault slip in rupture region in the Tohoku earthquake, dots indicate epicentre locations of aftershocks occurring within 1 day of the main event (Report of Japan Meteorological Agency 2011)

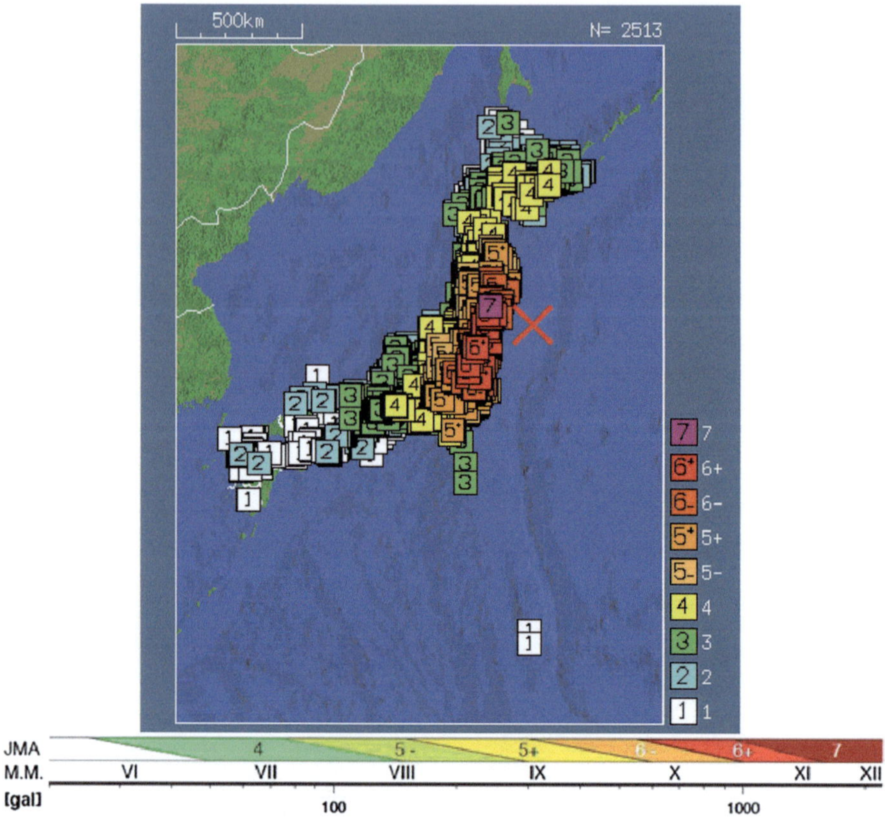


**Fig. 8.2** (*Top*) Earthquake intensities based on the Japanese Shindo scale (Report of Japan Meteorological Agency 2011) (*Bottom*) Comparison between the Japanese Shindo scale and Modified Mercalli Intensity scale

notably the 1995 Kobe earthquake. For the past 16 years between the 1995 Kobe and 2011 Tohoku earthquake, 16 damaging earthquakes had hit various parts of Japan. In addition, Japan periodically suffers from large ocean ridge quakes along the Nankai Trough that run deep off of the Pacific coast of Western Japan. The trough is divided into three regions named Tokai, Tonankai, and Nankai regions from the East, shown in Fig. 8.3. Historical records show that these three regions had ruptured periodically with average intervals of 100–150 years. The last major ocean ridge earthquake hit Japan in the middle of the last century. It follows that shaking larger than that experienced in Tohoku is expected in the not-so-distant future. This makes it all the more important to learn from the recent events and apply that knowledge to our engineering practice.

Unfortunately, in examining the data provided from the Tohoku earthquake, many future earthquake scenarios that have been explored for years were found invalid. Seismologists did not predict such large ruptures as shown possible in

**Fig. 8.3** Map of major subduction fault zones off of Japan

Tohoku earthquake. However, there are historical records of an earthquake roughly ten centuries ago that was believed to have had a rupture on the same magnitude as this recent event. Thus, earthquake prediction models for the coming Nankai Trough earthquake, mentioned above, have been considered to under-predict the potential hazards. It is of the upmost necessity for the seismological community in Japan to improve prediction modelling to adequately prepare for the next large event.

Besides major building shaking and the resulting tsunami, the earthquake caused subsidence, liquefaction and landslides. Serious subsidence can be observed along the coastline. According to GPS observations, some land subsided by nearly 2 m. Figure 8.4 (Report of Ministry of Land, Infrastructure, Transport and Tourism 2011) shows the change in elevation from before the earthquake. The area of regions at sea level increased dramatically, causing permanent flooding in some locations, exacerbating the tsunami damage and limiting the post-tsunami recovery.

Earthquake induced liquefaction occurred in many locations, especially in reclaimed lands in Tokyo and Chiba although these locations were far from the epicentre. Figure 8.5 shows examples of damage from liquefaction in these regions. Observed liquefaction was more significant than expected. Smaller excitation amplitudes with many more cycles than predicted and differences in behavior between natural soils and reclaimed soils, among others, are considered responsible for the discrepancy. This shows a need for the major re-examination of the stability of reclaimed land.

As seen in countless previous earthquakes, large fires (Fig. 8.6 (http://www.sankeibiz.jp)) occurred after the event. Fires after the Tohoku event were numerous including some at large oil tank farms, posing great danger to humans.

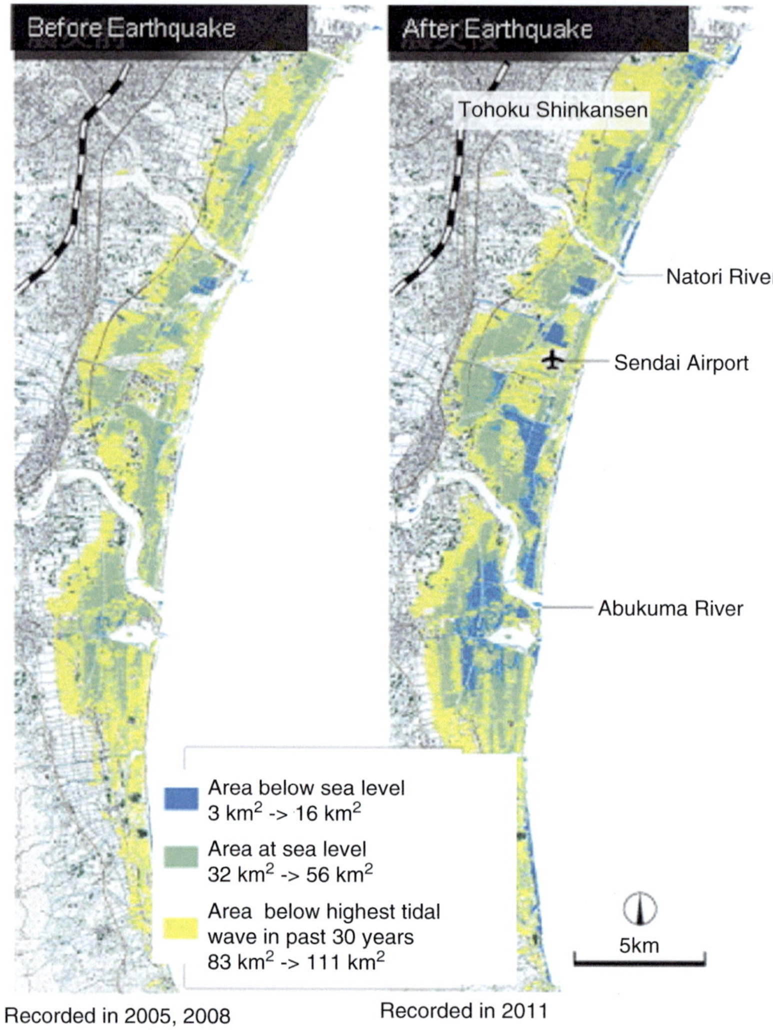


**Fig. 8.4** Map showing coastline elevation before (*left*) and after (*right*) of the Tohoku earthquake (Report of Ministry of Land, Infrastructure, Transport and Tourism, 2011)

## 8.3 Tsunami

The tsunami that occurred after the earthquake affected nearly 700 km of coastline. Figure 8.7 (Quick Report of the Field Survey and Research on "The 2011 off the Pacific coast of Tohoku Earthquake (the Great East Japan Earthquake)" 2011) shows the distribution of tsunami height over the affected land. The enormous tsunami caused complete devastation of many towns and villages and large loss

**Fig. 8.5** Pictures showing damage from liquefaction in Chiba and Tokyo Bay regions (**a**) Abiko (**b**) Shinkiba (**c**) Urayasu

**Fig. 8.6** Fires occurring after the Tohoku earthquake

of life. Damage and deaths from the tsunami were much greater than those from the earthquake shaking. Over 90 % of life lost was due to drowning. The elderly were the major victims, with those over 60 years of age comprising 65 % of casualties.

The tsunami was significantly beyond expected height. For example, a tsunami of 6 m was anticipated at the Fukushima power plant location, while the recorded tsunami height even reached 12 m. This discrepancy was caused by fault ruptures far larger than anticipated. In light of this earthquake, major work is needed to re-examining tsunami predictions due to the large troughs of off Japan's coasts.

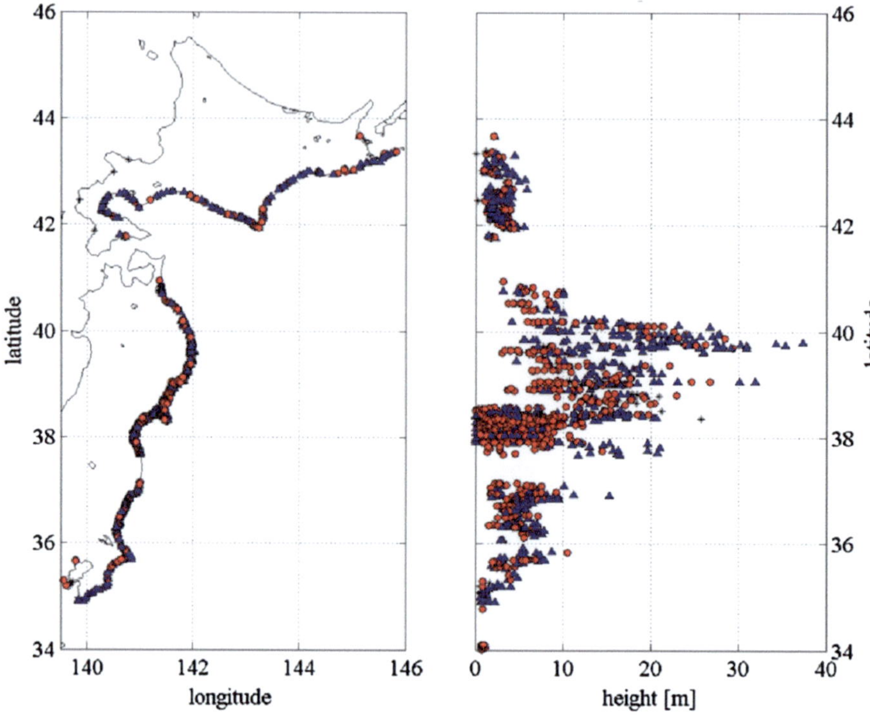


**Fig. 8.7** Tsunami height at affected regions, red circles denote inundation measurements and blue diamonds denote run-up measurements (Quick Report of the Field Survey and Research on "The 2011 off the Pacific coast of Tohoku Earthquake (the Great East Japan Earthquake)" 2011)

## 8.4 Building Damage from Earthquake

After the earthquake, a large investigation into the building performance was undertaken. The overall performance of buildings was found to be good for the size and magnitude of shaking. However, it was found that investigations of buildings without explicit damage were far more difficult than investigations of explicitly damaged ones. Building owners afraid of large monetary loss did not want their buildings to be examined. Thus, it was difficult to collect detailed data on the performance of undamaged buildings.

### *8.4.1 Sendai*

Sendai is a modern, large city with over one million inhabitants. It was the closest major city to the epicentre, and thus experienced strong shaking on the level of

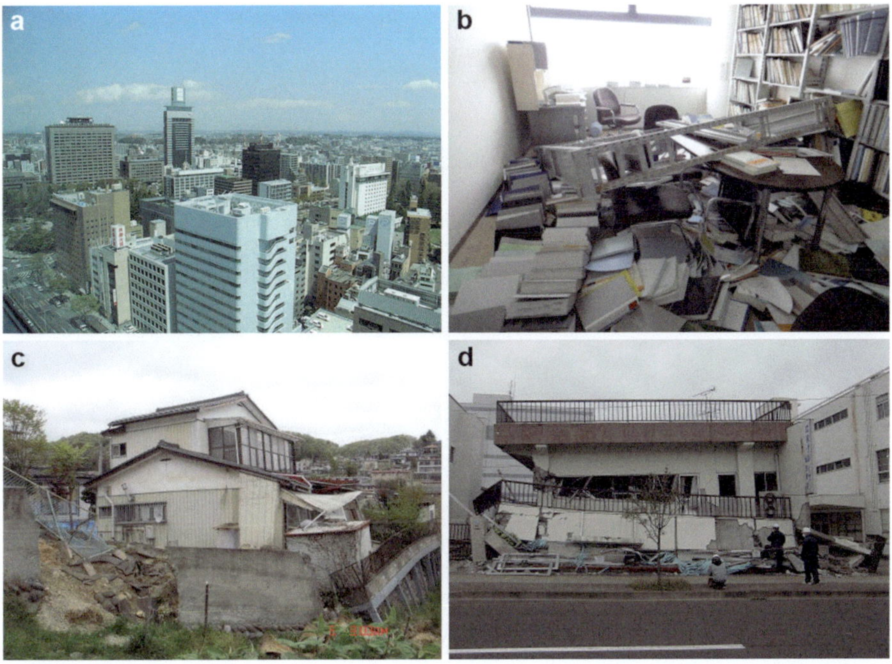

**Fig. 8.8** Response of buildings in Sendai area to the earthquake (**a**) Downtown Sendai after event (**b**) Nonstructural damage (**c**) Landslide damage (**d**) Shear failure and collapse of two-story RC building

Shindo 6 to 6+ (MM XI). An overview of the damage to buildings is shown in Fig. 8.8. Damage such as shear failures to RC columns and soft story collapses occurred in older buildings. Although shaking was intense, the damage to recent buildings remained limited. This demonstrates that current seismic design may have worked reasonably well. However, significant damage to nonstructural elements and building contents occurred.

Landslides were the major cause of damage to residential wood houses, especially in hillier areas. Insufficient soil compaction during land levelling is a likely cause. Many large aftershocks as well as heavy rain aggravated the damage to land.

Sendai has more than a dozen high-rises. Although they experienced strong excitation, no serious structural damage was reported, and all high-rises buildings maintained continuous occupancy. People in a 31 story SRC building built in 1998 during the shaking reported: difficulty in standing, partitions overturning, and books thrown from shelves. People outside the building said it looked as if the building might break in the middle. However, no people were injured and people evacuated orderly using stairs.

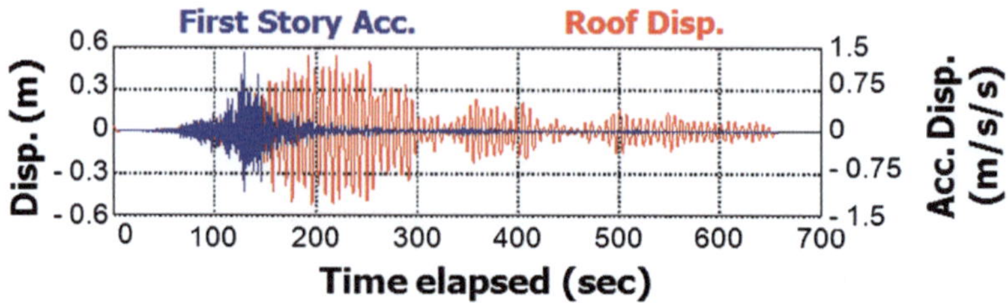


**Fig. 8.9** Input acceleration and roof displacement time histories for a 55 story building in Tokyo

### *8.4.2 Tokyo*

The Tokyo metropolitan area experienced shaking on the Shindo 5 level (out of 7), which is medium for the Japanese standard. Unlike many other shakings, the motion contained significant long-period components, which tend to promote the responses of high-rises and base-isolated buildings. There are nearly 1,500 high-rises and over 1,000 base-isolated buildings in the Tokyo metropolitan area. The Performance of hundreds of high-rises and base-isolated buildings was satisfactory. No reports of serious damage were given. Available data will be documented to study for future earthquakes. However, data from instrumented buildings are often privately owned and difficult to obtain.

An example of common observed behaviour comes from one high-rise with available data, a 55 story steel building, built in 1975 and retrofitted with viscous dampers not long before the earthquake. The ground input was relatively low, with a maximum acceleration of 0.35 g. The peak roof drift of 0.5 m, is within the expected range for a building of this height. The most significant finding is that, although input acceleration and maximum displacement were not large, the duration of building motion, shown in Fig. 8.9 continued for over 10 min. These long responses have the potential to cause low cycle fatigue in buildings. In addition, this behavior can cause discomfort and unease to the building inhabitants.

## 8.5 Building Damage from Tsunami

The tsunami was not only responsible for the majority of deaths but also caused the majority of damage to infrastructure. Wood buildings were swept away, steel buildings were reduced to warped frames, and many concrete buildings were overturned with damage only to the foundations. However, the behavior of buildings under the tsunami loads was unpredictable. For example, Fig. 8.10 shows two RC buildings, one that survived, while the other one was overturned.

**Fig. 8.10** Overturned versus standing RC buildings after the tsunami

## 8.6 Post-event Response

### *8.6.1 Evacuation*

There were several hundred thousand refugees from the earthquake and subsequent tsunami and nuclear crisis. Post-earthquake responses of the central and local governments were seriously tested. Issues arose regarding where to evacuate people to, what services to provide, how and where to build temporary shelters, and how long the people should stay at the evacuation centers and temporary housing. Japan had experience with these issues immediately after the 1995 Kobe earthquake, and many of the lessons learned then were applied and worked effectively. However, in the Tohoku earthquake and tsunami, the area of the affected regions was significantly larger, which made the application of many response measures impracticable.

### *8.6.2 Disruption of Utilities*

Although structural damage remained relatively limited after the earthquake, there was a large disruption to utilities. Loss of electricity was a major problem. In the Tohoku area, over 50 % of households (nearly 4.5 million) were without electricity. Even after 1 month, the recovery was not complete with nearly 150,000 households without electricity. Water supply was cut to over one million households to locations as far away as Chiba, with a nearly 100,000 without water over a month after the disaster. In addition, gas supplies were cut for over 400,000 homes, primarily in Miyagi prefecture where the city of Sendai is located.

In the Tokyo metropolitan area, about 15 % of households suffered from blackouts. Blackouts in the city caused widespread disruption (Fig. 8.11). Blackouts stopped trains, resulting in large traffic jams. Many people were forced to stay in

**Fig. 8.11** Consequences of power shortages and mass panic in Tokyo (**a**) Major traffic jam (**b**) Line for the train (**c**) People sleeping in the train station (**d**) Grocery store without food

train stations overnight. The next day, under limited train service, people had to wait for in lines for hours to ride on a train. Rumors were abound, and a couple of days later, basic food and commodities disappeared from the supermarket shelves.

The large loss of power and effects thereof were not anticipated. After the earthquake, the entire Kanto region, in which the Tokyo metropolitan is included, was affected by the subsequent electricity shortage with 15 % mandatory electricity cuts for large users and assigned rotating blackouts. The loss of utilities was a problem repeated from the 1995 Kobe earthquake, indicating that it is an issue of upmost importance in disaster mitigation planning.

## 8.7 Lessons for the Future and Necessary Actions

### *8.7.1 Predictions*

It is well agreed that our ability to predict coming events will greatly influence our ability to prepare and responds to them. Thus, it is of upmost importance to have the best possible predictions for ground motions and consequent tsunamis. The Tohoku earthquake and tsunami proved current prediction methods for large off shore trench earthquakes to be insufficient. Fault sizes were significantly underestimated leading to underestimation of the tsunami size and inundation. This has inspired a significant research effort and there is much for seismologists, tsunami researchers and structural engineers alike to learn and apply.

### *8.7.2 Tsunami Warnings and Design*

As mentioned above better, predictions for tsunami generation, propagation, and inundation are of upmost importance. However, much can be done to enhance current tsunami warning systems. Current methods based on quake motion records are too indirect. Accurate and reliable technology, such as sensors (like GPS-based ocean wave meters, seabed pressure gauges) that can directly measure generated tsunami height, must be deployed.

While some concrete buildings survived the tsunami, others were washed away. What had made the difference in performance may include soil conditions, foundation designs, superstructure designs, and local tsunami forces. The specific reasons must be identified so that an "anti-tsunami" design methodology can be established. This is particularly important for critical infrastructure such as emergency shelters, fire stations, and hospitals. With the next Nankai Trough earthquake coming in the foreseeable future, unless drastic measures are taken for anti-tsunami design, many coastal regions in the central and western parts of Japan will suffer from similar damage.

### *8.7.3 Energy Dependency*

Technical and social response to nuclear accidents is the largest issue for Japan resulting from the Tohoku disaster. The Fukushima power plant problem continues, and now how to safely shut down the reactors and eventually demolish the plant is the central issue. Many people have been semi-permanently displaced from their homes with the nuclear exclusion zone and the contamination has created large food scares and a depression on the agricultural and fishing industries in the surrounding areas. There has been a social push away from the use of nuclear energy and future energy policies are under constant debate.

Business disruption due to shortage of electricity has had an enormous impact on Japanese industry. This is a unique opportunity to reconsider our contemporary life, which heavily depends on electricity. As shown in Fig. 8.12, Japan relied on nuclear for about 30 % of its electricity before the Tohoku disaster. The other major sources are coal and liquefied natural gas (LNG). Currently, 54 nuclear power plants are located throughout Japan, with the vast majority of them shut down for maintenance.

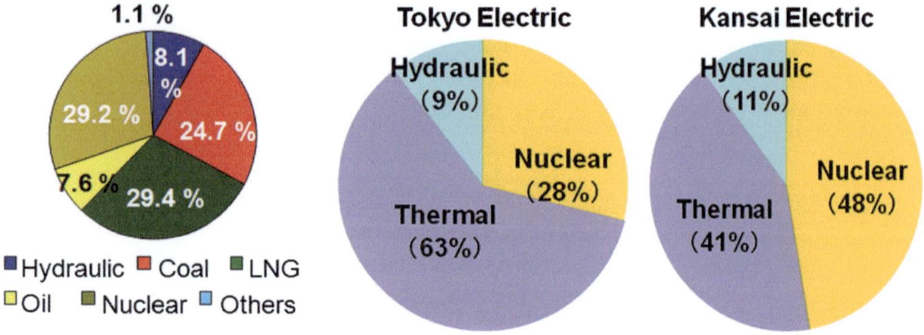


**Fig. 8.12** Energy source dependency of Japan (*far left*) and two major metropolitan regions of Japan (Statistics of before the Tohoku earthquake)

The dependency on nuclear energy differs from region to region (see Fig. 8.12). In Tokyo, which the Fukushima power plant provided electricity for, about 30 % of electricity is generated by nuclear plants, while the Kansai area which includes Osaka, Kyoto and Kobe relies heavily on nuclear, with nearly 50 % of power coming from nuclear plants.

Thus, how to save energy (or to reduce electricity consumption) has become a central subject of research and practice for coming years. For both political and social reasons, this subject is deemed even more critical than measures to mitigate future earthquake and tsunami disasters.

### *8.7.4 Community and System Based Engineering*

The Tohoku disaster has shown the need for emphasis on the full picture, from fault modelling to the ability to mobilize disaster response teams. In general focus must be put on community based engineering. This earthquake revealed that our metropolitan areas may be much weaker against earthquakes than what was previously thought. This ranges from building performance to energy dependency.

While the concept of performance-based design/engineering is good, we must understand that the performance of an individual structure is not governed by its own performance but by the interaction with the performance of other entities within the same society. Thus, focus on the performance-based design on individual structures may decrease the overall performance of a community. Objective, quantitative examinations are needed to assess the damage to Sendai and its vicinities, including damage to lifelines and utilities. Such information can be used for the careful characterization of community damage and appropriate measures must be identified in preparation for the next large ocean-ridge earthquake.

### 8.7.5 *Post-disaster Response*

New mechanisms for post-quake responses are absolutely needed, in which interaction and mutual assistance among local municipalities are to be sought. Multiple prefectures experienced serious damage simultaneously, and emergency assistance from the central government was necessary. Measures must be taken to secure communication and cooperation amongst local agencies and ensure rapid response during and after the disaster. If immediate response can be made, effects can be mitigated, in some cases such as fires, the spread may be limited.

Recovery and rehabilitation of seacoast towns and villages is another important issue. A collective (rather than local) effort is needed to prepare for tsunamis of a minimum 50 year return period. Practical solutions should be offered as to the mechanism to achieve this goal, which must be both socially and environmentally friendly as well as secure for life safety.

## 8.8 Conclusions

The March 11th, 2011 Tohoku earthquake and tsunami caused an overwhelming amount of destruction and subsequent disruption in Japan. The lessons learned from the Tohoku earthquake encompass a full range of disciplines. The disaster emphasized the need for improvements for earthquake and tsunami prediction models, tsunami warning systems, landslide and liquefaction mitigation, emergency planning and response and energy conservation.

"Resiliency" is currently a popular keyword to describe earthquake engineering, but it is lacking true quantification. Here, "resiliency" is defined as the ability to recover to normal conditions as quickly as possible. True resiliency cannot be obtained by focusing on individual components separately. Only when there exists full cooperation and exchange between all disciplines can true resiliency be achieved. Currently, the closest codified approximation of this approach is seen in "importance factors" placed on the design level of critical facilities. However, as long as building performance is investigated on only an individual basis, a full picture of the community performance cannot be obtained.

Last, and perhaps most important, the accumulation and spread of knowledge derived from disasters such as this must be promoted not just locally but globally. Many countries can learn from the Japanese experience and have paid serious attention to the Japanese response. It is the responsibility of Japan to disseminate the knowledge gained from the Tohoku disaster internationally.

This article was based on the observations 3 months after the Tohoku earthquake. Since that time, extensive efforts by numerous organizations and individuals had been made towards detailed investigations into multiple damage mechanisms and damage recovery efforts. Two examples of relevant reconnaissance reports are available (Quick Report of the Field Survey and Research on "The 2011 off the Pacific coast of Tohoku Earthquake (the Great East Japan Earthquake)" 2011; Architectural Institute of Japan (AIJ) 2011a), among others. The central and local governments, major political parties, research and professional societies, universities and other research institutions, and many other agencies have released "lessons learned" and "actions to make" in accordance with their missions and expertise. For instance, the Architectural Institute of Japan (AIJ), a learned society of about 36,000 members with expertise in all areas associated with buildings and inhabitants, had published a six-page proposal, entitled "Returning to Origin of Architecture – Restoration and Renovation of Our Living Environment in Response to the 2011 Tohoku Earthquake Disaster (First-Stage Summary and Recommendations)", in September 2011 (Architectural Institute of Japan (AIJ) 2011b).

A few very large research projects have been launched since 2011, with the Ministry of Education, Culture, Sports, Science and Technology (MEXT) as the funding body. One is a project in which many ocean-bottom seismometers are deployed along the subduction fault zones off of Japan to better estimate the tsunami generation and promote the earthquake prediction research. Another is a project in which drastic mitigation is sought regarding the damage and disruption that large cities might sustain in the future large earthquake events (Ministry of Education, Culture, Sports, Science and Technology (MEXT) 2012). According to the Headquarters of Earthquake Research, established in MEXT, a national budget that is equivalent to about 450 million US dollars has been appropriated in the year of fiscal 2012 to the investigation and research of earthquake prediction and mitigation and distributed to various government agencies. The budget was tripled from that appropriated in 2010.

## References

Architectural Institute of Japan (AIJ) (2011a) Preliminary reconnaissance report of the 2011 Thoku-Chiho Taiheiyo-Oki Earthquake, July 2011 (in Japanese)
Architectural Institute of Japan (AIJ) (2011b) Returning to origin of architecture – restoration and renovation of our living environment in response to the 2011 Tohoku Earthquake Disaster (First-Stage Summary and Recommendations), Building Science, No. 126:59–64 (in Japanese)
http://www.sankeibiz.jp/econome/photos/120712/ecc1207121231002-p1.htm
Ministry of Education, Culture, Sports, Science and Technology (MEXT) (2012) Special project for mitigating urban vulnerability for mega earthquake disasters
Quick Report of the Field Survey and Research on "The 2011 off the Pacific coast of Tohoku Earthquake (the Great East Japan Earthquake)" (2011) Technical Note, National Institute for Land and Infrastructure Management No. 636 May 2011, Building Research Data No. 132 May 2011 (in Japanese)
Report of Japan Meteorological Agency (2011) (in Japanese) http://www.jma.go.jp/jma/kishou/books/saigaiji/saigaiji_201101/saigaiji_201101.html
Report of Ministry of Land, Infrastructure, Transport and Tourism (2011) (in Japanese) http://www.mlit.go.jp/report/press/river03_hh_000327.html

# Chapter 9
# Lessons Learned from the 2010 Haiti Earthquake for Performance-Based Design

**Eduardo Miranda**

**Abstract** The January 12, 2010 Haiti earthquake caused more than 300,000 deaths and left more than one million people homeless. This earthquake is now considered one of the worst natural hazard disasters in history. Although it is clear that the Haitian people and its built environment were unprepared for this event, there are many other lessons that the earthquake community must take from this event. After a brief background on the country and on the seismological aspects of this event, a number of reflections on this earthquake are presented. In particular, several aspects that make this earthquake different to almost any other earthquake event are presented. It is argued that many of the factors that contributed to this catastrophe are the result of combination of a complicated socio-political history of the country coupled with being located in a multi-hazard setting. The earthquake led to perhaps the most complicated and challenging post-earthquake disaster management faced to date that overwhelmed the world's humanitarian aid infrastructure. Challenges to improve earthquake resilience in developing countries are discussed.

**Keywords** Economic loss • Developing countries • Poverty • Construction practices • Seismicity • Earthquake resilience • Performance based design • Multi-hazard • Socio-economic factors • Housing • Shelter • Displaced population • Gross domestic product • Topographical effects • Recovery • Risk transfer

---

E. Miranda (✉)
Department of Civil and Environment Engineering, Stanford University, Stanford, CA 94305-4020, USA
e-mail: emiranda@stanford.edu

M. Fischinger (ed.), *Performance-Based Seismic Engineering: Vision for an Earthquake Resilient Society*, Geotechnical, Geological and Earthquake Engineering 32,
DOI 10.1007/978-94-017-8875-5_9, 

## 9.1 Introduction

On February 12, 2010 a Mw 7.0 earthquake struck the island of Hispañola at 4:53 pm local time. The epicentre of the event was located approximately 25 km south west of Haiti's capital, Port-au-Prince on a previously unmapped fault now known as Léogâne fault near the Enriquillo-Plantain Garden fault (Calais et al. 2010; Hayes 2010). Estimate on the number of deaths produced by the earthquake range from 250,000 to 316,000, 300,000 injured and more than 1.3 million homeless (GORH 2010) making it the most destructive earthquake that any country has experienced when measured in terms of the number of people killed relative to its population (Cavallo et al. 2010). In particular, the estimated number of deaths exceeds those of the 1976 Tangshan, China and 2004 Sumatra, Indonesia earthquakes.

This earthquake demonstrated, like any other, the impacts that strong earthquakes can produce when occurring near large densely populated urban areas in developing countries. Aware of the earthquake risk in the island of Hispañola the author has participated for a number of years now in multiple events and courses in the Dominican Republic as part of the activities of the Instituto Dominicano de Ingeniería Superior y Desastres Naturales, Vitelmo Bertero (Dominican Institute of Advanced Engineering and Natural Disasters, Vitelmo Bertero). In particular, I participated in a symposium in November 2009, in the city of Santiago of los Caballeros, which is a city of approximately two million people whose metropolitan area is located less than 10 km south of a major strike-slip fault (the Septentrional fault) and that has been destroyed multiple times by earthquakes in 1564, 1783, 1842, 1887, and 1897. In that symposium I stressed to the attendees of the significant seismic risk in the island and the importance of seismic risk mitigation. I never imagined that, less than three months later, I would return to the island to witness one of the worst natural hazard disasters in modern times.

The purpose of this work is to summarize some of the main factors that, in the author's opinion, contributed to this catastrophe and to draw some lessons for performance based seismic design and seismic resilience in developing countries. Most observations are based on two trips that the author made to Haiti in February and June 2010 as part of an earthquake reconnaissance mission and a data gathering RAPID trip. For more details on the earthquake and its impacts the reader is referred to the special issue of Earthquake Spectra (Comerio and DesRoches 2011) and other reconnaissance reports (e.g. Eberhard et al. 2010).

## 9.2 Factors That Contributed to the Catastrophe

Although the main reason for the large impact of this natural hazard is of course the occurrence of the earthquake itself, seismic events of this magnitude or larger occur on average approximately 15 times per year in different parts the world,

**Fig. 9.1** Collapse of the National Palace in Port-au-Prince, Haiti

however the impacts are significantly smaller than those observed in this event. The main question is then why this magnitude earthquake had such a large impact. This section will summarize several aspects that to the author's opinion played a major role in the enormous and long-lasting impact of this earthquake (Fig. 9.1).

### *9.2.1 Seismicity of the Island*

The island of the Hispañola is located in the boundary of the Caribbean and North American plates. This boundary experiences relative motions of approximately 20 mm/year of east northeast motion. The tectonics of the island are characterized by two major strike-slip faults, the Septentrional fault on the north and the Enriquillo-Plantain Garden Fault on the south and two subductions zones on the northern and southern coasts (Prentice et al. 1993, 2010; McCann 2006). One of the main factors that played a major role in this earthquake was the lack of major earthquakes in the southern zone of Haiti in recent history. Although major earthquake are known to have caused damage in Port-au-Prince in 1701, 1751, 1770 and 1860, *no major earthquakes occurred in the southern zone of Haiti in the last 150 years*. This played a major role in this event because inhabitants, their parents or grandparents had not experienced a major earthquake leading to a lack of knowledge and relevance of seismic risks. In contrast, people in California, Mexico, Chile, Japan, New Zealand, Greece, Italy and most countries located in seismic regions have a recent history of destructive earthquakes that helps promote earthquake preparedness. For example, the development of seismic provisions and current state of earthquake

**Fig. 9.2** Photographs contrasting levels of observed damage in bidonvilles (*shanty towns*) just outside of Petion-Ville (*left*) and on a ridge (*right*) in the Port-au-Prince metropolitan area

resistant design practice in Mexico was strongly influenced by the occurrence of destructive earthquakes such as the 1957, 1979 and 1985 earthquakes. Similarly, the development of seismic provisions and current state of practice in California was strong influenced by earthquakes that occurred in 1906, 1933, 1952, 1971, 1989 and 1994 earthquakes.

### *9.2.2 Topographic and Geotechnical Conditions*

In addition to Port-au-Prince, the earthquake cause major destruction in the cities of Léogâne, Carrefour, Delmas, Pétion-Ville, Tabarre and Croix-des-Bouquets. The regions where these cities are located are characterized by Miocene deposits, Pliocene fan deposits and quaternary deposits. The patterns of damage in these areas were variable, ranging from some areas with limited damage to others with almost complete destruction. Although characteristics of the construction changes in these areas, there is evidence that geotechnical and topographic effects played also a major role in the observed damage patterns (Rathje et al. 2011). The largest amount of damage was observed in areas underlain by Holocene alluvium and artificial fills. Of particular interest are topographical effects that caused damage concentrations at the top and sides of ridges. As an example Fig. 9.2 shows *bidonvilles* (shanty towns) with similar types of deficient construction with completely different levels of damage. Although topographical effects have been identified for many years (e.g., Sánchez-Sesma and Campillo 1991) and have been incorporated for many years into the French code and the Eurocode 8 more recently (AFNOR 1999; CEN 2004) most seismic codes in the Americas (including seismic provisions in the United States) and many other seismic regions do not incorporate these effects.

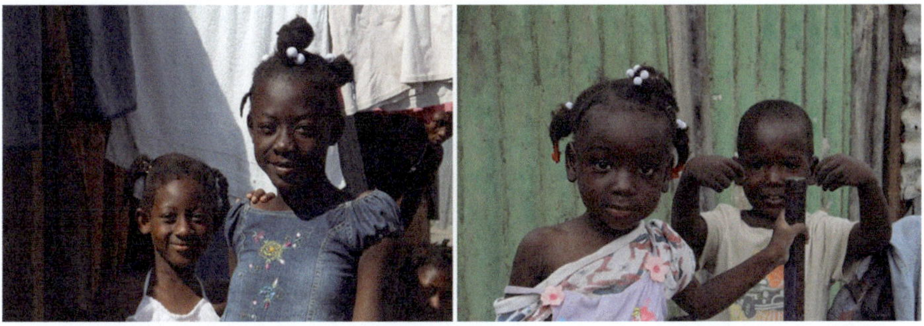

**Fig. 9.3** Haitian children in temporary housing in Port-au-Prince

### *9.2.3 Socio-economic Factors*

Haiti has a complex history characterized in the last century by a U.S. military occupation from 1915 to 1934 that was succeeded with ruling by a series of dictators including Duvalier and his son who were in power from 1957 to 1986. The first free election was held in 1990 won by Aristide but who was soon overthrown in 1991 and returned to office in 1994 with the support first of the U.S. in 1994 and by election in 2000. Préval was president first from 1996 to 2001 and from 2006 to 2011. The United Nations has had a Stabilization Mission in Haiti since 2004.

Haiti is the poorest country in the Western Hemisphere with a population of more than nine million with approximately a third of the population concentrated in the Port-au-Prince metropolitan area. The urban population in the country is 52 %. 36 % of the population is less than 14 years old (Fig. 9.3). The gross domestic product per capita in 2009 was $625 USD, which is less than two dollars a day (UNSD 2010).

In addition to its poverty a factor that also played an important role is the lack of professionals in the construction industry. In 2007 the university population of Haiti was approximately 40,000 students of which 28,000 (70 %) were in public universities and 12,000 (30 %) in private ones (INURED 2010). This represents approximately only 0.4 % of the Haiti population. To put this number in perspective the total undergraduate and graduate enrolment in degree-granting institutions in the same year in the U.S. was 18.2 million according to the National Center for Education Statistic of the U.S. Department of Commerce, which represents approximately 6 % of the U.S. population. Not only the university population was extremely small prior to the earthquake but it has been estimated that 80 % of the university graduates leave the country (Comfort et al. 2011).

**Fig. 9.4** Example of heavy slab built primarily for resisting gravity loads and hurricanes

## *9.2.4 Location Within a Hurricane-Prone Region*

Haiti is located in the Caribbean region, which it is hit by more than 20 hurricanes every year. In particular, the 2008 season was one of the worst in its history with hurricanes Fay, Gustav, Hanna and Ike causing major flooding that resulted in approximately 800 deaths, more than 20,000 homes destroyed and more than 80,000 damaged and causing an economic loss of approximately 5 % of its gross domestic product (GDP). In addition to erosion and weakening of foundations and weak structures of flood rampages worsened by severe deforestation this means that the primary natural hazard in the mind of the Haitian population are Hurricanes leading in many cases to excessively heavy constructions which leads to increasing inertia forces during earthquakes (see Fig. 9.4 for a representative example).

## *9.2.5 Inadequate Construction Practice*

Several earthquake reconnaissance reports have pointed out the many inadequacies of earthquake resistant practice in Haiti. For example, several reports that the author has had access to have pointed out the lack of adequate detailing of reinforcing steel, the lack or a proper confined masonry construction, use of inadequate construction materials, etc (Comerio et al. 2011). However, something that I believe is perhaps more revealing is the lack of adequate construction practices even for gravity loads or wind loads which are primary design actions. Even if there is no formal design, my experience, having grown in a developing country and having travelled extensively and observed construction in many poor countries, is that in most other countries even when dealing with self-construction there is usually some basic

**Fig. 9.5** Examples of poor construction practice even for resisting gravity loads

knowledge of construction practices that is passed from generation to generation and that is shared among family relatives, neighbours, friends, etc. Basic construction knowledge in Haiti even for resisting gravity loading is appalling. For example, the author noticed in many construction sites, that even some involving rich owners in Petion-ville or in some cases even on sites owned/operated by international religious organizations that were able to afford hiring a construction crews, there was evidence of lack of elementary construction practices that highlights deeper problems beyond an inadequate earthquake resistant construction detailing. Some examples are illustrated in Fig. 9.5 that show situations that could lead to eminent collapses putting in danger their own life even during construction.

## 9.3 Lessons for Developing Countries in Seismic Regions

The Haiti earthquake is one of the worst, if not the worst, earthquake catastrophe in modern history. It is important to reflect upon this tragedy and to learn from it in order to avoid to the extent possible a similar situation. This section includes some reflections by the author regarding this event.

Inherent in the concept of earthquake resilience is the ability of the society to recover from an earthquake event. Resilience then encompasses on one hand a measure of the impact of earthquake on society and on the other the capacity to recover from the disaster. This means that two cities with the same ability to recover from an earthquake will have different times of recovery depending on the size of the impacts and similarly for a given size of impacts two cities may have very different recovery periods depending of their capacity to recover from disasters.

The impact of earthquakes and other natural hazards is commonly measured in terms of the economic impact it had. In developing countries this is typically

**Table 9.1** Economic losses in a selected number of earthquakes

| Earthquake | Country | Year | Loss (in Bn. USD) | Loss as % of GDP |
|---|---|---|---|---|
| Hyogo-ken-Nambu | Japan | 1995 | 80 | 2.8 |
| Northridge | United States | 1994 | 40 | 0.4 |
| Maule | Chile | 2010 | 30 | 15 |
| Izmit | Turkey | 1999 | 20 | 10 |
| Loma Prieta | United States | 1989 | 8 | 0.2 |
| **Port au Prince** | **Haiti** | **2010** | **7.8** | **120** |
| Guatemala | Guatemala | 1976 | 6.1 | 18 |
| Michoacan | Mexico | 1985 | 5 | 3 |
| Managua | Nicaragua | 1972 | 2 | 40 |
| San Salvador | El Salvador | 1986 | 1.5 | 31 |

quantified by the gross dollar loss and by the insured losses. The difference between the two provides valuable information on the portion of the risk that is transferred to insurance and reinsurance companies. However, although commonly used measures, they are absolute measures of the loss and provide no information of how large or small the losses are relative to the economic activity of the city/region or country. A much better measure is to normalize the losses by for example the gross domestic product of the country. Table 9.1 lists economic losses in ten earthquakes that occurred in the last 40 years sorted by economic loss. As shown in this table the largest economic loss prior to the 2011 Tohoku earthquake in Japan was the Hyogo-ken-Nambu (Kobe) earthquake with $80 billion (in short scale) USD followed by economic losses in the 1994 Northridge earthquake which were the results of earthquake in urban area in the second and first largest economies at the time.

If these losses are, however, normalized by the gross domestic product a very different perspective becomes apparent, that is that even though these losses were very large in absolute terms, relative to the size of these economies the losses were relatively small compared to those experienced by developing countries, where it is clear that the Haiti earthquake was not only the deadliest earthquake in terms of human deaths relative to its population but also the costliest relative the size of its economy (Fig. 9.6). Estimates of the economic loss of the 1972 Managua earthquake relative to its GDP vary greatly in the literature ranging from 30 % to close to 100 % of its GDP. Although typically not widely reported, an even better measure would be the economic loss that is retained by the country (i.e., that is not transferred to foreign countries through insurance or reinsurance) normalized by GDP. This is important because in some cases, as for example in the case of Chile in 2010 approximately one third of the losses were insured and 95 % of that insured loss was ceded via reinsurance that would reduce the normalized loss from 15 % to approximately 10 %.

But even more important than economic losses are the human losses, human suffering and enormous social disruption caused by the event. Two years after the earthquake there are more than 300,000 people are still living in tents or makeshift shelters made with thin metal sheets and tarps with unhealthy and unsecure settings.

**Fig. 9.6** Complete collapsed building in downtown Port-au-Prince

One of the most important lessons from the Haiti earthquake was the importance of the survival of the local government to deal with the post-earthquake disaster management. In this event the National Palace, the Parliament building and most ministerial and public administration buildings were destroyed with many public servants killed in these buildings. International aid organizations in case of disasters typically rely on local governments in assisting in distribution of aid to the population in the aftermath of the disaster. Direct impact on the public infrastructure and the death and injuries of civil servant severely hindered the ability to assist the population.

The Haitian government was aware of the risks from natural disasters in the country. For example, through the Association of the Caribbean States there was a model code that included a model building code for earthquakes created in 2003 through financial support of the InterAmerican Development Bank and the Italian government. Furthermore, Haiti hosted in 2007 a high level conference on disaster reduction of the Association of Caribbean States (ACS). The final document of the conference included the five priorities for action stemming from the Hyogo Framework for Action, adopted by the World Conference on Disaster Reduction, held in 2005: (1) Ensure that disaster risk reduction is a national priority with a strong institutional basis for implementation; (2) Identify, assess and monitor disaster risks and enhance early warning; (3) Use knowledge, innovation and education to build a culture of safety and resilience at all levels; (4) Reduce the underlying risk factors; (5) Strengthen disaster preparedness for effective response.

The challenges in earthquake risk reduction are well exemplified by this earthquake, which illustrates that challenges do not lie only within the responsibilities of developing countries. The 2007 Haiti conference on disaster reduction was co-financed by the United Nations International Strategy for Disaster Reduction. It is sad, but at the same time ironic, that even the United Nations which is one of the most important sponsors and supporter of efforts on risk reduction from natural hazards had not dealt adequately with evaluation of some of their own facilities in the country and experienced the collapse of the its main UN headquarters in the country that resulted in the death of 101 of its personnel including its mission chief and top three ranking officials in the country leading to the largest loss of life experienced in the history of the United Nations.

In closing I would like to highlight another important lesson from this event, which is the role of the seismic performance of residential housing in disasters. This earthquake illustrates, perhaps better than any other, the consequences of experiencing large percentages of collapse and heavy damage in the residential stock in densely-populated urban areas in developing countries. As several studies have shown, damage to housing residents leads to essentially instantaneously-created large number of displaced households that, even in wealthy developed countries, poses enormous challenges that in the case of developing countries often leads to practically unmanageable situations that greatly exacerbate social and economic recovery. In California the city of San Francisco recently launched an initiative whose one of its main goals is to ensure that after a major earthquake most of its residents can "shelter in place" meaning they will only sustain damage that will enable them to stay in their homes while they are being repaired. This initiative, which is being developed and implemented in one of the richest cities of the state with the largest economy in the nation in the country with the largest economy in the world, is perhaps even more important in developing countries.

**Acknowledgements** This work in dedicated to the memory of my colleague and friend Prof. Helmut Krawinkler who dedicated his life to reduce the loss of life and effects of earthquakes. Among its many long lasting contributions to the field of Earthquake Engineering is in collaboration with Prof. Peter Fajfar having brought together specialists from all over the world to workshops in a beautiful and quiet place in Slovenia to exchange and discuss new trends in Earthquake Resistant Design. Special thanks are also extended to Profs. Matej Fischinger and Bozidar Stojadinovic, organizers of the Bled 4.

## References

Association Française de Normalisation (AFNOR) (1999) PS-92 Règles de construction parasismique: Règles PS applicables aux bâtiments. Normes NF P 06–013, Troisième Tirage; 1999

Calais E et al (2010) Transpressional rupture of an unmapped fault during the 2010 Haiti earthquake. Nat Geosci 3:794–799. doi:10.1038/ngeo992

Cavallo EA, Powell A, Becerra O (2010) Estimating the direct economic damage of the earthquake in Haiti, IDB working paper series no. IBD-WP-163, Inter-American Development Bank, Washington, DC

CEN (2004) Eurocode 8: design of structures for earthquake resistance. Part 1: General rules, seismic action and rules for buildings. EN 1998–1, Euro Commit for Stand, Brussels, December 2004

Comerio M, DesRoches R (eds) (2011) Special issue on the 2010 Haiti Earthquake, Earthquake Spectra, Earthquake Engineering Research Institute, Volume 27, Number S1, 507 pages

Comerio M, DesRoches R, Eberhard M, Mooney W, Rix G (2011) Overview of the 2010 Haiti earthquake. Earthq Spectra 27(S1):S1–S21. doi:10.1193/1.3630129

Comfort L et al (2011) Transition from response to recovery: a knowledge commons to support decision making following the 12 January 2010 Haiti Earthquake. Earthq Spectra 27(S1):S411–S430. doi:10.1193/1.3633342

Eberhard MO, Baldridge S, Marshall J, Mooney W, Rix GJ (2010) The $M_W$ 7.0 Haiti earthquake of January 12, 2010: USGS/EERI Advance reconnaissance team report, USGS Open File Report 2010–1048, U.S. Geological Survey, Reston, VA, 58 pp

Government of the Republic of Haiti (GORH) (2010) Action Plan for National Recovery and Development of Haiti: immediate key initiatives for the future. http://www.haiticonference.org/Haiti_Action_Plan_ENG.pdf. Accessed 20 June 2011

Hayes GR (2010) Complex rupture during the 12 January 2010 Haiti earthquake. Nat Geosci 3:800–805. doi:10.1038/ngeo977

Interuniversity Institute for Research and Development, INURED (2010) The challenge for Haitian higher education: a post-earthquake assessment of higher education institutions in the Port-au-Prince Metropolitan area, Port-au-Prince http://webarchive.ssrc.org/challenge-haiti-report.pdf. Accessed 12 June 2012

McCann WR (2006) Estimating the threat of tsunamigenic earthquakes and earthquake induced-landslide tsunamis in the Caribbean. In: Mercado A, Liu P (eds) Caribbean tsunami hazard. World Scientific, Singapore, pp 43–65

Prentice C et al (1993) Paleoseismicity of the North American-Caribbean plate boundary (Septentrional fault), Dominican Republic. Geology 21(1):49–52

Prentice CS et al (2010) Seismic hazard of the Enriquillo Plantain Garden fault in Haiti inferred from palaeoseismology. Nat Geosci 3:789–793. doi:10.1038/ngeo991

Rathje EM et al (2011) Damage patterns in Port-au-Prince during the 2010 Haiti Earthquake. Earthq Spectra 27(S1):S117–S136. doi:10.1193/1.3637056

Sánchez-Sesma FJ, Campillo M (1991) Diffraction of P, SV, and Rayleigh waves by topographical features: a boundary integral formulation. Bull Seism Soc Am 81:2234–2253

United Nations Statistics Division (2010) 2010 World statistics Pocketbook Country profile. United Nations http://data.un.org/CountryProfile.aspx?crName=Haiti. Accessed 12 June 2012

# Chapter 10
# L'Aquila Earthquake: A Wake-Up Call for European Research and Codes

**Iunio Iervolino, Gaetano Manfredi, Maria Polese, Andrea Prota, and Gerardo M. Verderame**

**Abstract** From the L'Aquila 2009 earthquake three issues, among others, strongly emerged to be addressed for the engineered structures, at least in Europe. They are related to near-source effects, non-structural damage, and reparability. Although they are well known since quite long time, still regulations seem giving little, if any, practice-ready tools to account for them. In the chapter, evidences from the event and scientific needs are briefly reviewed and discussed. The modest aim of the paper is to stimulate debate and research in the light of next generation of seismic codes.

**Keywords** Near-source • Directivity • Pulse-like records • Seismic hazard • Inelastic displacement ratio • Response spectrum • Non-structural elements • Infills • Reinforced concrete • Seismic assessment • Seismic design • Capacity • Codes • Damage • Collapse • Reparability • Substandard structures • Shear failure • Soft-storey • Residual drift • Non-linear structural analysis

## 10.1 Introduction

The April 6, 2009 L'Aquila earthquake ($M_W$ 6.2) caused about three hundreds of fatalities, more than a thousand injuries, and extensive and severe damage to buildings and other structures. About 66,000 residents were temporarily evacuated, and more than 25,000 were medium-term homeless.

The area of interest is known to be seismically active since a long time. Several events comparable in magnitude to this last earthquake, are reported by the

I. Iervolino (✉) • G. Manfredi • M. Polese • A. Prota • G.M. Verderame
Dipartimento di Ingegneria Strutturale, Università degli Studi di Napoli Federico II, Via Claudio 21, 80125 Naples, Italy
e-mail: iunio.iervolino@unina.it; gaetano.manfredi@unina.it; mapolese@unina.it; andrea.prota@unina.it; verderam@unina.it

M. Fischinger (ed.), *Performance-Based Seismic Engineering: Vision for an Earthquake Resilient Society*, Geotechnical, Geological and Earthquake Engineering 32,
DOI 10.1007/978-94-017-8875-5_10, 

national seismic catalogue. The main documented events, considering an estimated magnitude larger than 6.5, date to 1315, 1349, 1461, 1703 and 1915. In fact, the first modern-era seismic classification of L'Aquila refers to 1915, after the catastrophic Avezzano earthquake. The subsequent estimations of seismic hazard lead to the current value of expected peak ground acceleration, or PGA, (for a return period of 475 years) equal to about 0.26 g on rock. In fact, it is one of the largest seismic hazard sites according to the national hazard map (http://esse1.mi.ingv.it/).

Analysis of buildings stock in the city of L'Aquila shows a percentage of reinforced concrete buildings equal to 24 %, while the masonry buildings are 68 %; a percentage of 8 % refers to buildings of different typology. For what concerns the age of construction, 55 % of buildings was built after 1945. Therefore, the most of existing buildings was built using some seismic provisions; nevertheless with non-modern seismic standards. Moreover, seismic demand during the 2009 earthquake was, locally, much larger than the design one. This seems also due to near-source directivity effects. Generally, the seismic performance of buildings stock in L'Aquila was considered unsatisfactory. In fact, in the 6 months after the mainshock, the national department of civil protection organized a global survey of all the buildings in the area affected by earthquake. About 80,000 field surveys were performed by specialized teams. The results show that about 20,000 building suffered of large structural damage, while about 10,000 buildings suffered of light structural damage and/or non-structural damage. Lacks and deficiencies of seismic design are believed to be responsible for such a bad performance, which is going to have a large reconstruction cost for the country.

Starting from L'Aquila experiences, some remarks on possible improvements of next generation of codes are discussed in the following. Of the many facets of seismic risk, which a number of researchers studied after the earthquake, by far the best documented event in Italy, three are those briefly discussed in this chapter: (i) directivity-related near-source effects of engineering interest and their predictability; (ii) seismic behaviour and structural dynamics' influence of infills in reinforced concrete structures; (iii) reparability and the possibility to explicitly include this limit-state in design.

The reader may argue these are well known earthquake engineering topics since quite some time, yet European codes, at least, are somewhat lacking in their respect, while level of scientific knowledge is such they may be considered in the next generation of seismic regulations.

## 10.2 Near-Source Pulse-Like Engineering Issues

Near-source (NS), or directivity, effects in ground motion depend on the relative position of the site with respect to the fault rupture, and typically appear by means of large velocity pulses concentrating energy in the starting phase of the fault-normal, or FN (i.e., normal to the rupture's strike), component. This results in waveforms different from *ordinary* ground motion recorded in the far field, or in geometrical

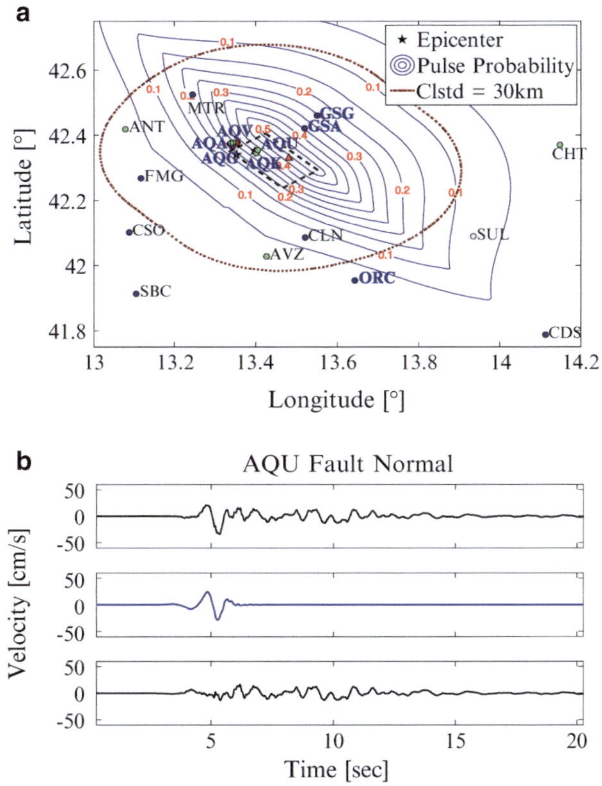


**Fig. 10.1** Near-source stations in L'Aquila 2009 earthquake and pulse occurrence probability contours (according to the model of Iervolino and Cornell 2008) together with actual identified (*in bold*) velocity pulses (**a**); impulsive signal of AQU station in the same earthquake; from *top to bottom*: velocity time history, extracted pulse and residual velocity (**b**)

conditions not favorable with respect to directivity. Near-source pulse-like traces were found in the records of L'Aquila earthquake (Chioccarelli and Iervolino 2010).

In Fig. 10.1a the seismic stations, which have recorded the event close to the source, are shown and superimposed to a probability model for occurrence of pulse-like records (Iervolino and Cornell 2008). In bold there are the stations where the records, in which pulses supposed to be originated by directivity, were found; consistency may be observed. In Fig. 10.1b, for one accelerometric station (AQU), the original velocity signal is shown together with the extracted pulse and the residual ground motion once the pulse is removed (according to the algorithm in Baker 2007).

In Chioccarelli and Iervolino (2010), analyzing the NGA database (http://peer.berkeley.edu/nga/), it was found that the three main characteristics of pulse-like records of earthquake engineering interest are:

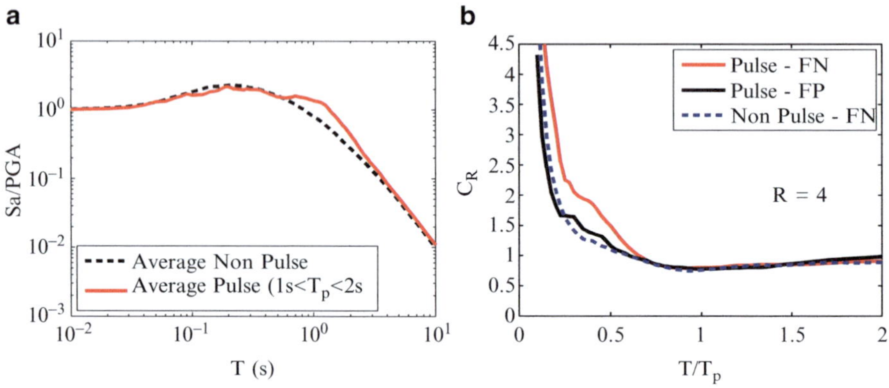


**Fig. 10.2** Elastic 5 % damped spectra for FN pulse-like with 1 s < $T_P$ < 2 s and ordinary records (**a**); empirical $C_R$ for FN pulse-like records, for their fault-parallel (FP) components, and for ordinary records, for a strength reduction factor R = 4 (**b**) (Adapted from Iervolino et al. 2012)

1. pulse-like signals are characterized by fault normal records generally stronger than both fault parallel components and non-pulse-like ground motions;
2. fault-normal pulse-like records are characterized by a non-standard spectral shape with an increment of spectral ordinates in a range around the pulse period (Tp), Fig. 10.2a;
3. inelastic-to-elastic seismic spectral displacement ratio ($C_R$) for pulse-like records can be 20–70 % higher than that of ordinary motions depending on the non-linearity level; such increments are concentrated in a range of period between 30 and 50 % of pulse period of each record, Fig. 10.2b.

These points show that near-source directivity is of interest to earthquake resistant design, which has to be adjusted for near-source because: (i) traditional hazard assessment may be unable to predict (1) and (2), that is, the elastic peculiar features of pulse-like records; (ii) point (3) shows that current static design, based on *equal displacement rule*, may fail when pulse-like records are concerned; (iii) it is not implicit in current design practice that safety margins are consistent between ordinary and pulse-like records.

Regarding (i): in near-source conditions the probabilistic seismic hazard analysis (NS-PSHA), expressed in terms of mean annual frequency of exceeding a spectral acceleration (Sa) level, or $\lambda_{Sa}$, is a linear combination of two terms, which account for the occurrence or the absence of the pulse, $\lambda_{Sa,NoPulse}$ and $\lambda_{Sa,Pulse}$, weighted by the pulse occurrence probability, Eq. (10.1). Moreover, NS-PSHA requires dealing with two more tasks, which are not faced in traditional hazard analysis: pulse period prediction; and pulse amplitude prediction.

$$\lambda_{Sa}(x) = \lambda_{Sa,NoPulse}(x) + \lambda_{Sa,Pulse}(x) \tag{10.1}$$

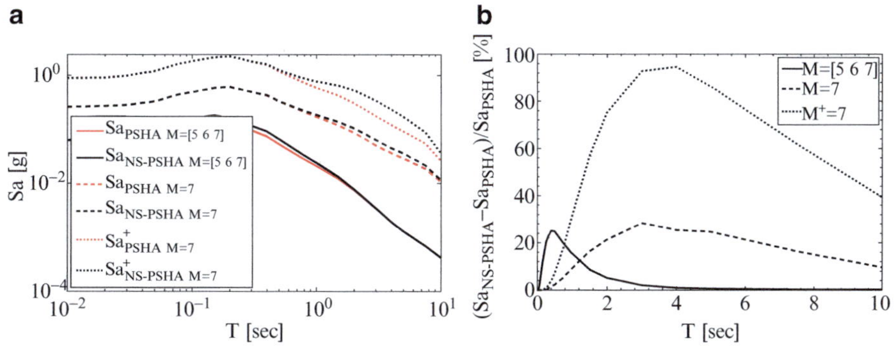

**Fig. 10.3** 475 years UHSs with modified and classical PSHA (**a**); increments due to directivity effects (**b**) (Adapted from Chioccarelli and Iervolino 2013)

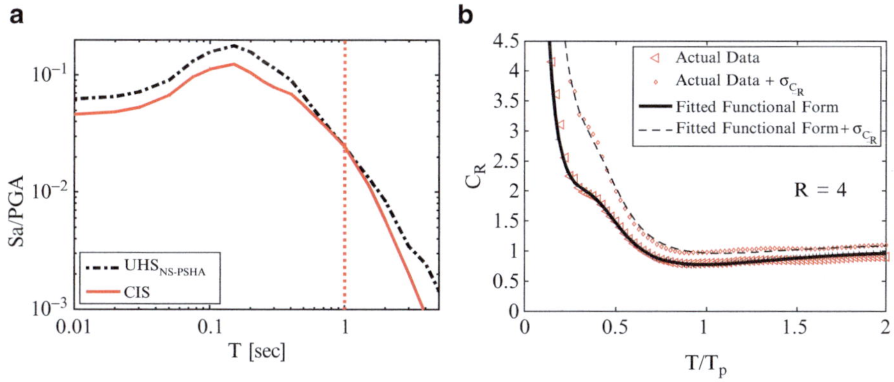

**Fig. 10.4** (**a**) Near-source possible design scenarios (Chioccarelli and Iervolino 2013); (**b**) inelastic displacement ratio functional form for pulse-like records fitted to actual data (Iervolino et al. 2013)

As an example, in Fig. 10.3a near-source uniform hazard spectra (UHS, 10 % in 50 years), are compared to UHS from traditional PSHA for different cases of magnitude distribution and source-to-site-geometry for a strike-slip fault (Chioccarelli and Iervolino 2013). In Fig. 10.3b increments of NS-PSHA with respect to PSHA are given. It is to observe that these may be significant. It also appears the UHS is only able to capture the NS amplitude, not the peculiar spectral shape of Fig. 10.2a. Therefore, it seems necessary to find alternate solutions to get design seismic actions.

One possible approach is that proposed in Chioccarelli and Iervolino (2013) and named *close-impulsive spectrum* (CIS), conceptually consistent with the conditional mean spectrum Baker (2011). CIS is derived from disaggregation of NS-PSHA in terms of magnitude, distance from the source, and pulse period. In Fig. 10.4a, starting from the hazard for a 0.5 s Sa, the shape of CIS is reported for a case in which the design scenario is 0.7 s, 5, and 10 km, in terms of Tp, magnitude, and source-to-site distance, respectively. The figure also shows the UHS for the same example, to appreciate differences and how CIS reproduces a 'bump' around the design pulse period.

Regarding point (ii), it was found in Iervolino et al. (2012) that the inelastic displacement ratio of near-source pulse like records, requires a specific functional form to fit observed data as a function of T/Tp, which is substantially different from ordinary records. In fact, the functional form in Eq. (10.2), which consists of adding two opposite bumps in two different spectral regions to the traditional hyperbolic format of $C_R$ in codes (e.g., FEMA 2005), may be suitable. In Fig 10.4b this functional form is plotted (also plus one standard deviation) against data of Fig 10.1b once the θ coefficients are fitted.

$$C_R = 1 + \theta_1 \cdot \left(T_p/T\right)^2 \cdot (R-1) + \theta_2 \cdot \left(T_p/T\right) \cdot \exp\left\{\theta_3 \cdot \left[\ln\left(T/T_p - 0.08\right)\right]^2\right\} + \theta_4 \cdot \left(T_p/T\right) \cdot \exp\left\{\theta_5 \cdot \left[\ln\left(T/T_p + 0.5 + 0.02 \cdot R\right)\right]^2\right\} \quad (10.2)$$

## 10.3 Non-structural Elements and Their Influence on Structural Assessment/Design

Infill walls are usually employed in reinforced concrete (RC) buildings for partition use and for thermal/acoustic insulation. Hence, they are considered as non-structural elements; nevertheless, post-earthquake damage observation, experimental and numerical research, showed that their influence on seismic behaviour of RC buildings can be not negligible at all.

Modern seismic codes (e.g., FEMA 2000; CEN 2004; DM 2008) prescribe to account for the possible influence of infills on seismic behaviour of RC frames, both at local and global level. In particular, according to Eurocode 8 (CEN 2004), if walls take at least 50 % of the base shear from a linear analysis, the interaction of the structure with the masonry infills may be neglected. This may be taken to imply that it is allowed then to disregard the infills in the structural model. However, this is not always a safe assumption. An asymmetric layout of the infills in plan may cause torsional response to the translational horizontal components of the seismic action; so, according to Eurocode 8 part 1, infills with strongly asymmetric or irregular layout in plan should be included in a 3D structural model and a sensitivity analysis of the effect of the stiffness and position of the infills should carried out. Even though the awareness about this issue in earthquake engineering is not very recent, it is likely to state that practically no existing RC building was designed accounting for the presence of these elements.

This latter observation becomes a key issue in countries, such as Italy, in which most of the building stock was realized before 1990s. It has to be noted that not even limit state approach was provided in the Italian regulations until 1996,

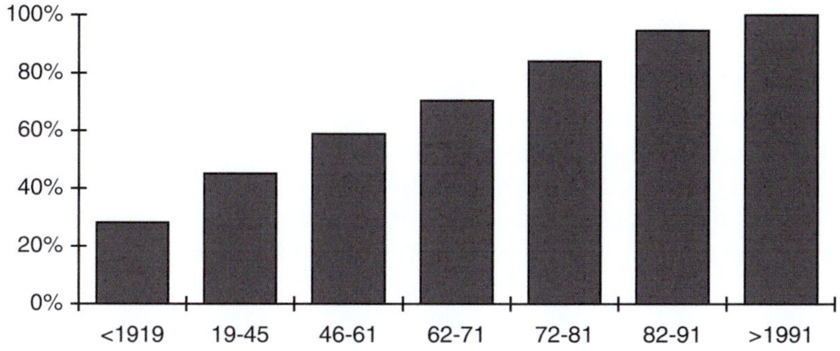


**Fig. 10.5** 2001 census ISTAT data for L'Aquila. The construction age intervals on the abscissa identify the percentage of buildings to add to the preceding bar to get the bar they correspond to, that is, to obtain the cumulative distribution (Ricci et al. 2011a)

**Fig. 10.6** (**a**) Shear failure of column adjacent to partial infilling panels; (**b**) diagonal cracking failure in concrete joint panel (Ricci et al. 2011a)

see Ricci et al. (2011a) for details. Figure 10.5 shows the empirical cumulative distribution of the age of construction of L'Aquila buildings that can be approximately considered representative of most of the Italian region in terms of urban development.

Observation of damage on RC structures after L'Aquila event emphasized the importance of taking into account the contribution of non-structural elements such as infills. Most of the observed damage was localized in infill panels, and their stiffness and strength contribution in some cases preserved the RC buildings from structural damage. On the other hand, it is to note that infill irregular distribution in plan and elevation can be addressed as one of the main causes of the structural collapses observed during in-field campaigns, together with brittle failure mode of columns and beam-column joints (see Fig. 10.6).

The lack of capacity design and shear-flexure hierarchy can lead to brittle failures in primary elements (De Luca and Verderame 2013), and interaction with masonry infills can favor the occurrence of them. Effect of infills on the whole structural behavior depends on different factors that are briefly recalled in the following.

First, the evaluated fundamental period of the infilled structure (Ricci et al. 2011b) changes significantly if compared to the period of the bare frame model, since the structure is significantly stiffer. Furthermore, infill presence can change regularity characteristics of the structure in plan and in elevation and consequently can affect the mode of vibration of the building. Second, infills are characterized by a brittle behavior and the high contribution in strength they provide to a RC building suddenly decreases for low values of drift. On the other hand, it is worth to note that infill impact on the structural performances of a building becomes a critical issue when RC primary elements are designed according to obsolete criteria and the structure is characterized by an insufficient global and local ductility. The latter can be observed when seismic performances of contemporary and existing infilled frame structures are compared (Dolsek and Fajfar 2005). If infill distribution is irregular in plan or elevation, their contribution introduces a source of irregularity (e.g. *Pilotis effect*) and the possibility to register a soft-storey mechanism is dramatically increased, especially when no capacity design criterion has been employed in the design of the bare frame.

Mechanical properties of the infills represent another critical factor that can vary the effects on the performances of the whole structure, since they are considered non-structural elements and their properties, not systematically checked, can vary significantly because of the local building practice. Regarding this latter issue, it is important to stress the relative weight that infills have with respect to the mechanical properties of the bare frame. Because of all the variables considered (seismic intensity level, old or contemporary design approach, distribution and mechanical properties of the infills) it is tough to say if structural contribution of these "non-structural" elements increases or decreases the overall seismic capacity of the building (Ricci et al. 2012).

In (Verderame et al. 2011) one of the few collapsed buildings after L'Aquila earthquake is assumed as case study; the building collapsed as a result of a soft storey mechanism at the first level (see Fig. 10.7). Observed damage points to collapse as a result of a brittle failure mechanism. Given the likely scenario collapse inferred by observed damage, an analytical model of the building was built taking into account nonlinear behavior of the infills; local interaction with columns was also considered by means of a three strut macro-model. Two parametric hypotheses based on Italian code prescriptions were assumed for infill mechanical properties. Time history analyses were carried out assuming as seismic input the three components of the real registered signals during the mainshock in the vicinity of the case study structure. Given the brittle failure highlighted by damage, beside capacity models suggested by codes, other shear failure mechanisms (*sliding shear failure*) not typical for columns were considered.

Numerical results seem to confirm the collapse scenario inferred by damage observation; the lack of proper structural and executive details was addressed as

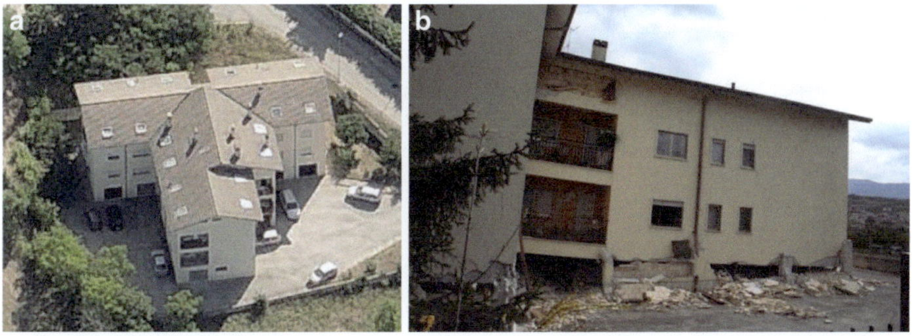


**Fig. 10.7** (**a**) Pre-event view of case study building, placed in Pettino (L'Aquila) (from Virtual Earth); (**b**). Collapsed building (Verderame et al. 2011)

the main cause that made a critical issue the local interaction between columns and infills, other than the strong vertical component registered.

The case-study structure, even if characterized by structural peculiarities, represents in itself a lesson for future code provisions and, above all, it emphasizes an effect typically disregarded in conventional assessment procedures.

## 10.4 Is a Reparability Limit State Needed and Feasible?

One of the most controversial problems in the aftermath of damaging earthquakes is the lack of agreed and transparent policies for acceptable levels of safety, as well as of advanced technical standards for repair and/or strengthening of damaged buildings. In fact, the technical difficulties for the assessment of the safety loss of damaged buildings and for the choice of the most appropriate method for repair and/or strengthening of damaged elements, avoided the development of sound and agreed re-occupancy criteria meeting engineering consensus on the different aspects to be considered.

The criteria for reconstruction funds assignments after L'Aquila earthquake distinguish between buildings classified as B/C (generally having slight damages) by the damage survey forms used by the Italian Civil Protection (Baggio et al. 2007), with respect to those classified as E (heavy structural damage). However, if the usual tagging procedures, based on an expert assessment of damage level and extent by a team of experienced practitioners, are deemed acceptable in an emergency phase in order to establish building usability, they cannot be considered as the only tool for choosing intervention categories and assigning grantable funds or insurance refunds for damaged buildings. A proper approach should rely on a performance based policy framework, as suggested in FEMA 308 (FEMA 1998b), where building's performance index and performance loss due to earthquake damage are the parameters considered to drive decisions among the alternatives of simply accepting the building as it is (insignificant damages), repairing or upgrading it.

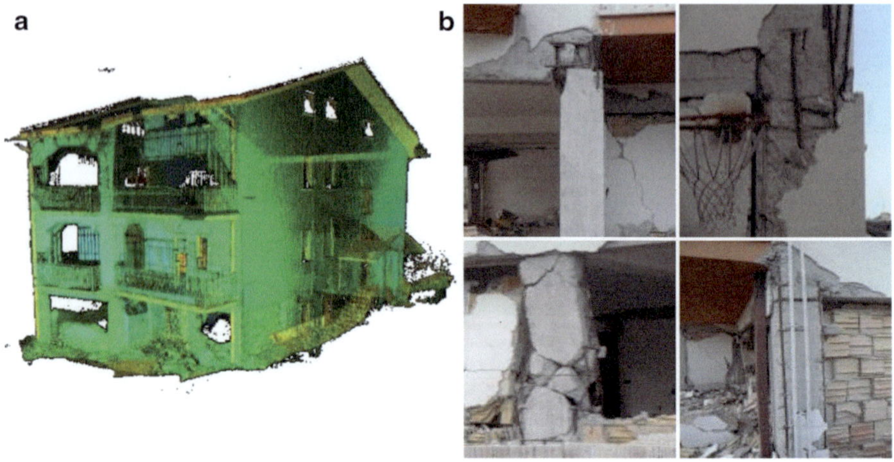


**Fig. 10.8** (**a**) Dense digital elevation model of the building in Pianola (AQ) obtained with laser-scanner 3D. (**b**) Examples of damages on the elements at the first storey (Polese et al. 2011)

Only partly in line with this approach, post L'Aquila regulations implicitly associate E buildings with medium-high performance loss, and the repairing and upgrading of buildings becomes mandatory when the performance index (ratio of the PGA corresponding to building collapse versus the design PGA in the nominal building life), evaluated for the intact building, is lower than 60 %. Moreover, with the purpose of avoiding length and costly investigations and analyses for heavily damaged buildings, an evaluation policy was introduced considering the amount of residual drifts as discriminative parameter for reparability. In particular, the code regulation OPCM 3881 (2010) establishes that it is possible to avoid demonstration of the economic convenience of demolishing and re-building if there are permanent drifts $\geq$1.5 % for at least 50 % of the columns at the same storey. Such value is somewhat greater than current literature values; for example, in ATC 58-1 (2011) for a residual drift of 1 % it is hypothesized that major structural realignment is required to restore safety margin for lateral stability, and the required realignment and repair of the structure may not be economically and practically feasible (i.e., the structure might be at total economic loss). However, considering the typical shear sliding mechanism at the beam-column interface of existing under-designed Italian buildings, due to masonry infill interactions with columns (Verderame et al. 2011), residual drift is often related to a local damage of some (typically external) columns, and larger permanent drifts may be expected, as reported in Polese et al. (2011). In fact, maximum residual drifts measured for two buildings in L'Aquila (Fig. 10.8 and Table 10.1), that were severely damaged at the first storey, is well beyond 1.5 % in both cases. However, the real measures for these two buildings show that it is very difficult to comply with the requirements of OPCM 3881 (2010) (at least 50 % of columns in a storey with residual drifts $>$ 1.5 %) even for very severely damaged buildings.

**Table 10.1** Permanent drifts ($\theta_A$ and $\theta_B$ on orthogonal sides of columns) evaluated at the first storey for two damaged buildings in Pettino (AQ) and in Pianola (AQ)

| Pettino (AQ) | | | Pianola (AQ) | | |
|---|---|---|---|---|---|
| Col. ID | $\theta_B$ (%) | $\theta_A$ (%) | Col. ID | $\theta_B$ (%) | $\theta_A$ (%) |
| 01 | +0.74 | – | 01 | +1.04 | −0.56 |
| 02 | +0.13 | – | 05 | +0.25 | +0.64 |
| 04 | +0.96 | – | 13 | −1.79 | +0.60 |
| 06 | +0.51 | – | 19 | +0.77 | +0.35 |
| 07 | +2.52 | +0.40 | 17 | +1.89 | −0.54 |
| 32 | −0.93 | +0.19 | 14 | +0.29 | 0.00 |

Only perimeter columns, where it was possible to measure residual deformations, are indicated (Polese et al. 2011)

**Fig. 10.9** Application of IN2 method for determining REsidual Capacity in the intact (REC) and damaged state ($REC_k$) (Polese et al. 2013)

$$REC_{Sa} = C_b \cdot \mu_{cap} \quad \text{for } T_{eq} \geq T_c$$

$$REC_{Sa} = C_b \cdot (\mu_{cap} - 1) \cdot \frac{T_{eq}}{T_c} + 1 \quad \text{for } T_{eq} < T_c$$

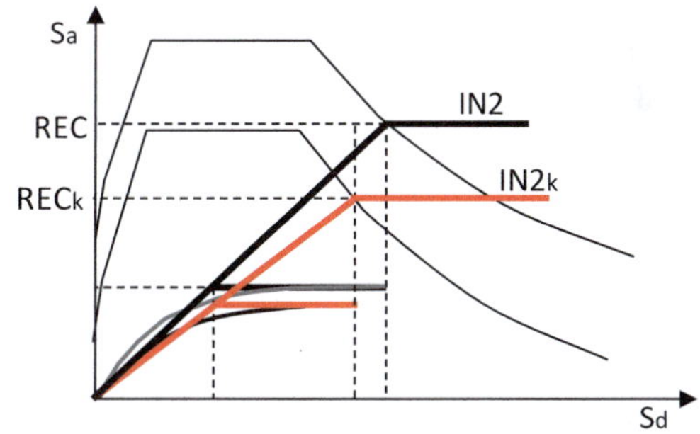


Obviously, there is a need to further investigate on the relationship between residual drifts, damages and performance loss in a performance-based policy framework. If varied vulnerability is to be considered in a consistent quantitative assessment framework, analytical modeling of building performance loss, accounting for building damage and residual drifts, is preferable. Building performance in the pre-event, damaged and eventually restored/upgraded state may be investigated with nonlinear static analysis considering proper variation of element's force-deformation relationships, as outlined in FEMA (1998a).

The seismic behavior of damaged buildings, and the relative seismic safety, may be adequately represented by their seismic capacity modified due to damage, the so called REsidual Capacity (REC). In the framework of a mechanical based assessment of seismic vulnerability, REC may be evaluated based on pushover curves obtained for the structure in different (initial) damage state configurations and accounting for possible residual drifts. In particular, residual capacity $REC_{Sa}$ is defined, for each global damage state $D_i$, as the minimum spectral acceleration (at the period $T_{eq}$ of the equivalent single degree of freedom system) corresponding to building collapse, and can be determined, among other approaches, with the IN2 method (Dolsek and Fajfar 2004), see Fig. 10.9.

A preliminary application Polese et al. (2013) shows that the adoption of nonlinear static analyses for damaged building, with explicit consideration of the damage and residual drifts in the main resisting elements, allows a consistent assessment of the building safety factor and performance loss with respect to intact structure. In this application, suitable modification factors for moment rotation plastic hinges of the columns have been calibrated based on a number of cyclic tests performed by the authors on non-conforming columns. However, models to predict the modification factors based on a wider number of experimental tests available in the literature (on columns representative of existing RC buildings of European Mediterranean regions) are currently under investigation (Di Ludovico et al. 2012).

Referring to the case study, an interesting conclusion is that while for structures that have been slightly or moderately damaged the ultimate deformation capacity does not significantly change with respect to the undamaged structure, at the same time these structures are more deformable (having higher $T_{eq}$) and have a lower ductility capacity, $\mu_{cap}$, with a consequent lower residual capacity. This circumstance is reflected in the increasing performance loss, that reaches values nearly of 10 % for slight damage, and up to 20 % for a structure that has reached a moderate damage state due to an hypothetical mainshock.

A key issue for a proper assessment of performance loss, to be further developed, is the experiment-based characterization of typical existing RC columns performance in the intact and damaged states, distinguishing their possible behavior modes (e.g. flexural, flexure/shear, sliding shear etc.) and significant damage levels. Moreover, in order to verify the effect of alternative retrofit solutions, also the influence of local retrofit interventions on such elements should be evaluated.

## 10.5 Conclusion

A few research opportunities and codes' needs were highlighted starting from the evidence of L'Aquila 2009 earthquake. In fact, analysis of records and damages to the built environment shows that the earthquake engineering research appears close to be capable of a practice-ready implementation on some issues, while other still require investigations, experimental tests, and competency building.

Three topics were found especially significant and briefly reviewed. In particular, near-source pulse-like effects, infills of reinforced concrete structures, and reparability, were analysed with respect to design and assessment.

On the near-source seismic demand side, it appears that there is a need for accounting for the peculiar pulse-like features in the elastic (i.e., seismic hazard) and inelastic range; state-of-the-art tools are available, while current provisions may be unable to capture those.

The influence of infills on the structural behaviour of engineering structures has to be considered in the seismic design for two main reasons: (1) infills can dramatically change the seismic behaviour determining unpredictable collapse modes with respect to the bare structure; (2) in a consequence-based framework, a reduction of infills damage can increase the immediate occupancy conditions and reduce the economic losses due to non-structural damage.

Finally, the definition of a reparability limit state determines potential improvements in seismic response; e.g., the definition of a clear condition of repair/upgrade convenience in the post-event situation, and new criteria in the choice of alternative retrofit solutions in a multi-criteria decision making method approach.

**Acknowledgements** The studies presented in this paper were developed within the activities of *Rete dei Laboratori Universitari di Ingegneria Sismica* (ReLUIS) for the research program funded by the *Dipartimento della Protezione Civile* (2010–2013).

## References

ATC 58-1 75% Draft (2011) Seismic performance assessment of buildings volume 1 – methodology. Applied Technology Council, Redwood City, California

Baggio C, Bernardini A, Colozza R, Di Pasquale G, Dolce M, Goretti A, Martinelli A, Orsini G, Papa F, Zuccaro G, Pinto AV, Taucer F (2007) Field manual for post-earthquake damage and safety assessment and short term countermeasures (AeDES), EUR 22868 EN – 2007. Joint Research Center, Ispra

Baker JW (2007) Quantitative classification of near-fault ground motions using wavelet analysis. Bull Seismol Soc Am 97(5):1486–1501

Baker JW (2011) Conditional mean spectrum: tool for ground motion selection. J Struct Eng 137(3):322–331

Chioccarelli E, Iervolino I (2010) Near-source seismic demand and pulse-like records: a discussion for L'Aquila earthquake. Earthq Eng Struct Dyn 39(9):1039–1062

Chioccarelli E, Iervolino I (2013) Near-source seismic hazard and design scenarios. Earthq Eng Struct Dyn 42(4):603–622

Comité Européen de Normalisation (CEN) (2004) Eurocode 8 – design of structures for earthquake resistance – Part 1: General rules, seismic actions and rules for buildings. EN 1998-1, CEN, Brussels

De Luca F, Verderame GM (2013) A practice-oriented approach for the assessment of brittle failures in existing RC elements. Eng Struct 48:378–388

Decreto Ministeriale (DM) del 14 gennaio (2008) Approvazione delle nuove norme tecniche per le costruzioni. G.U. n. 29 del 4/2/2008 (in Italian)

Di Ludovico M, Polese M, Gaetani D'Aragona M, Prota A (2012) Plastic hinges modification factors for damaged columns. In: 15th world conference on earthquake engineering – 15th WCEE, Lisbon, Portugal, 24–28 Sept 2012

Dolšek M, Fajfar P (2004) IN2- a simple alternative for IDA. In: 13th world conference on earthquake engineering. Vancouver, BC, Canada, 2004; Paper no. 3353

Dolšek M, Fajfar P (2005) Simplified non-linear seismic analysis of infilled reinforced concrete frames. Earthq Eng Struct Dyn 34:49–66

Federal Emergency Management Agency (1998a) FEMA 306: evaluation of earthquake damaged concrete and masonry wall buildings – basic procedures manual. Federal Emergency Management Agency, Washington, DC

Federal Emergency Management Agency (1998b) FEMA 308: repair of earthquake damaged concrete and masonry wall buildings. Federal Emergency Management Agency, Washington, DC

Federal Emergency Management Agency (2000) FEMA 356: NEHRP guidelines for the seismic rehabilitation of buildings. Federal Emergency Management Agency, Washington, DC

Federal Emergency Management Agency (2005) Improvement of nonlinear static seismic analysis procedures. Report FEMA 440. Washington DC, US

Iervolino I, Cornell CA (2008) Probability of occurrence of velocity pulses in near-source ground motions. Bull Seismol Soc Am 98(5):2262–2277

Iervolino I, Chioccarelli E, Baltzopoulos G (2012) Inelastic displacement ratio of near-source pulse-like ground motions. Earthq Eng Struct Dyn 41(15):2351–2357

OPCM 3881 (2010) Ulteriori interventi urgenti diretti a fronteggiare gli eventi sismici verificatisi nella regione Abruzzo il giorno 6 aprile 2009 (in Italian)

Polese M, Prota A, Manfredi G, Dolce M (2011) Spostamenti residui in edifici in c.a. danneggiati dal sisma. XIV Convegno Anidis, Bari, Italy (in Italian)

Polese M, Di Ludovico M, Prota A, Manfredi G (2013) Damage-dependent vulnerability curves for existing buildings. Earthq Eng Struct Dyn 42(6):853–870

Ricci P, De Luca F, Verderame GM (2011a) 6th April 2009 L'Aquila earthquake, Italy – reinforced concrete building performance. Bull Earthq Eng 9(1):285–305

Ricci P, Verderame GM, Manfredi G (2011b) Analytical investigation of elastic period of infilled RC buildings. Eng Struct 33(2):308–319

Ricci P, De Risi MT, Verderame GM, Manfredi G (2012) Influence of infill presence and design typology on seismic capacity of RC buildings: sensitivity analysis. In: 15th world conference of earthquake engineering, Lisbon, Portugal

Verderame GM, De Luca F, Ricci P, Manfredi G (2011) Preliminary analysis of a soft-storey mechanism after the 2009 L'Aquila earthquake. Earthq Eng Struct Dyn 40(8):925–944

# Chapter 11
# Lessons from the 2010 Chile Earthquake for Performance Based Design and Code Development

**Rubén Boroschek, Patricio Bonelli, José I. Restrepo, Rodrigo Retamales, and Víctor Contreras**

**Abstract** The February 27, 2010 $M_w$ 8.8 Maule earthquake in Chile generated MM VII intensity or higher, and PGA = 0.3 g or higher on a 100 km wide by 600 km long corridor, as well as a tsunami. Notwithstanding the large area affected by strong shaking, where about eight million people live, there were only 521 casualties. Approximately a third of the casualties were due to the tsunami alone. Nearly the rest were due to the collapse of non-engineered low-rise dwellings. Eight people only died in modern buildings. The number of severely damaged tall buildings, most likely requiring demolition or heavy structural intervention, has been estimated at around 50 out of 2,000 buildings. A large proportion of the structural damage in tall buildings concentrated in buildings 10 or less years old supported on intermediate or soft soils. Taking into account the total building stock exposure and its damage, and the total population exposure and its losses, this earthquake showed that the local engineering practices are effective at preventing

R. Boroschek (✉)
Department of Structural Engineering, University of Chile, Santiago, Chile
e-mail: rborosch@ing.uchile.cl

P. Bonelli
Department of Structural Engineering, Universidad Técnica Federico Santa María, Valparaíso, Chile
e-mail: patricio.bonelli@usm.cl

J.I. Restrepo
Department of Structural Engineering, University of California at San Diego, San Diego, CA, USA
e-mail: jrestrepo@ucsd.edu

R. Retamales
Rubén Boroschek and Associates, Santiago, Chile
e-mail: rodrigo.retamales@rbasoc.cl

V. Contreras
Rubén Boroschek and Associates, Santiago, Chile
e-mail: victor.contreras@rbasoc.cl

M. Fischinger (ed.), *Performance-Based Seismic Engineering: Vision for an Earthquake Resilient Society*, Geotechnical, Geological and Earthquake Engineering 32,
DOI 10.1007/978-94-017-8875-5_11, © Springer Science+Business Media Dordrecht 2014

loss of life. However, the disproportionate concentration of structural damage in newly built buildings, the collapse of three buildings, and widespread damage to nonstructural components and systems prompted the government to revise current design practice, in part because current societal expectations are different from expected performance tacitly or explicitly stated in the local design codes.

**Keywords** 2010 Chile earthquake • Seismic codes • Ground motion • Response spectra • Design spectra • Code development • Earthquake damage • Nonstructural elements • Reinforced concrete buildings • Reinforced concrete walls • Shear failure • Compression-tension failure • Lap-splice failure • Wall buckling • Special shear walls • Reinforcement detailing • Wall thickness • Shear force • Ductility • Confinement

## 11.1 Introduction

The February 27, 2010 $M_w$ 8.8 Maule earthquake in Chile generated MM VII intensity or higher, and PGA = 0.3 g or higher on a 100 km wide by 600 km long corridor, as well as a tsunami. About eight million people live in this corridor. The central government, and a large percentage of the domestic and export industries are located within the affected areas. The direct cost of damage may exceed US $25 billion, or about 16 % of GDP.

The earthquake occurred over the known Concepción-Constitución seismic gap in central Chile, where several prior studies had concluded the most likely near future occurrence of an earthquake of the characteristics observed (Boroschek and Domb 2006; Ruegg et al. 2009). The rupture models of the mainshock processed by the Delouis et al. (2010) present two or three asperities with high slips and energy release.

Several strong motion records were obtained by the University of Chile during this earthquake and are available to the scientific community (at http://terremotos.ing.uchile.cl). Peak ground accelerations, some of them in the order of 0.9 g, are discussed in Boroschek et al. (2010). Figures 11.1 and 11.2 show the ground motions recorded at Vina del Mar Marga-Marga and downtown Concepción stations, at epicentral distances of 390 and 60 km, respectively. These records are long, which typifies this kind of large magnitude subduction earthquakes. For example, the Vina del Mar Marga-Marga record presents strong motion duration based on 90 % of Arias Intensity of 30 s and an envelope with three periods of large amplitude vibrations.

For the case of downtown Concepción, Fig. 11.2a, the record shows duration of 85 s and includes more than 10 cycles of high accelerations with amplitudes greater than 0.10 g. This characteristic is strongly related with geotechnical conditions of the site (Class D with average $V_s$ of 220 m/s). Narrow band-pass filtering of this record at T = 1.9 s shows nine sinusoidal cycles of increased acceleration resulting from the response of the deep alluvial soil column where the station is located, see Fig. 11.2b. A more detailed description of the sites can be found in Arango et al. (2011).

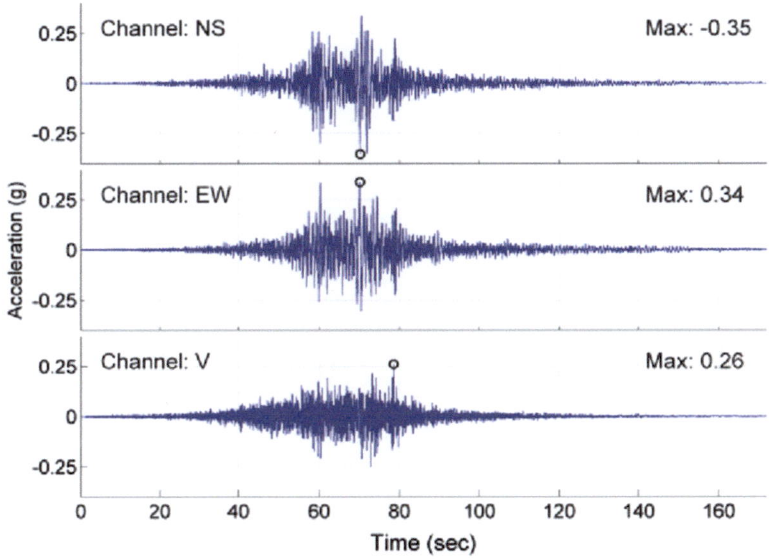


**Fig. 11.1** Corrected acceleration time histories. Station: Vina del Mar-Puente Marga Marga. Filter band [0.05 90] Hz

## 11.2 Seismic Design Codes and Standards

In Chile there are several seismic design codes which apply to structures depending on their occupancy. The main loading codes are: NCh433 (2009) for residential and office buildings; NCh2369 (2003) for industrial facilities; NCh2745 (2003) for base isolated buildings; and the Highway Manual (MOP 2012) for bridges. Additionally, special standards are used for important industries, such as ETG-1.015 (1987) and ETGI-1.020 (1997) for electrical facilities; and the Structural Seismic Design Criteria (2011) for the Government mining industry (CODELCO). The design codes and standards deal with seismic demands in different ways, which will be explained below. There are different performance objectives in each of the codes: Building and residential, shall not experience damage during a frequent event, and structural and nonstructural damage is acceptable under moderate events. No collapse shall occur during design level event.

Design codes follow, in most of the cases, those in the U.S. For example, NCh430 (2008), used for designing reinforced concrete buildings, that are predominant in Chile, is largely based on ACI 318-05 (2005). Steel structures are mainly used in Chile in heavy engineering industrial buildings. The design of these structures largely follows AISC-ASD (1989). In general, the seismic demand is represented by seismic zoning maps, a reference design zero period acceleration (ZPA) and an elastic design response spectrum. ZPA and response spectra are not strictly derived from a probabilistic or deterministic hazard study. The design spectrum shape has

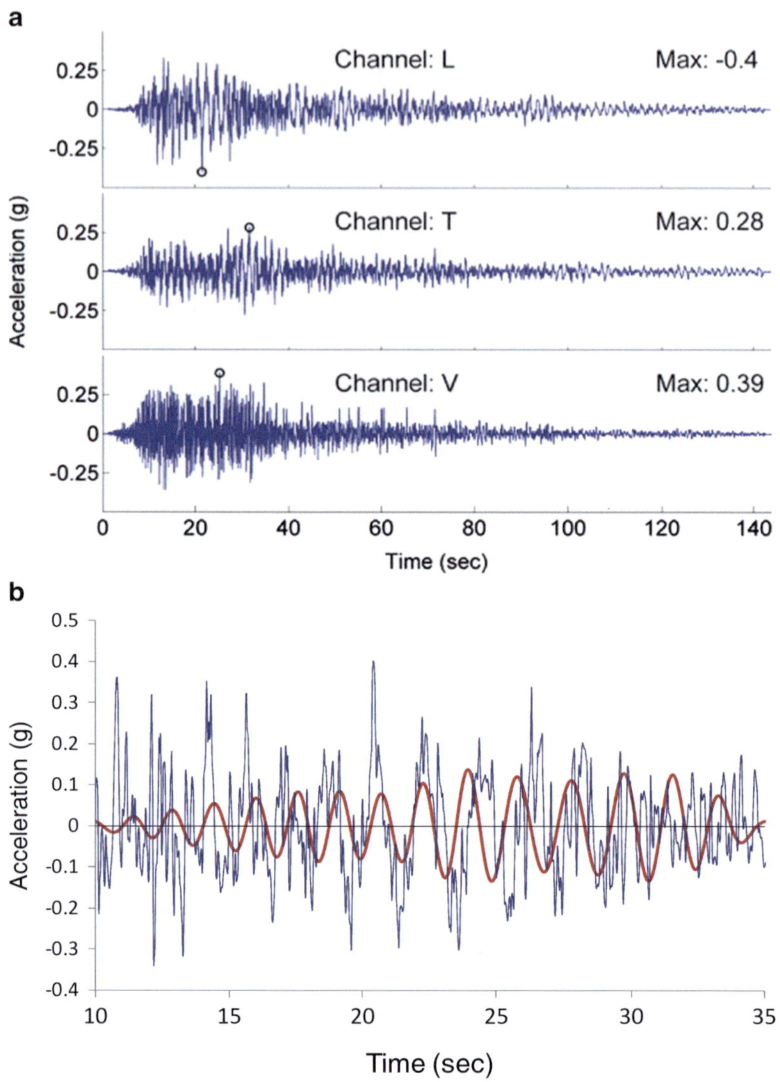


**Fig. 11.2** Corrected acceleration time histories. Station: Concepcion-Colegio Inmaculada Concepcion. (**a**) Acceleration signal, filter: [0.2 80] Hz. (**b**) Horizontal acceleration signal – S63°W component, filters: [0.2 80] Hz and [0.5 0.67] Hz

been calculated from the shape of average acceleration response spectra normalized by PGA for the database of historical ground motions occurred in Chile. This shape has been incorporated into the NCh433 and NCh2369 design codes. With such spectra, tall buildings that have become common in Chile since 1985, are subjected to unrealistically small displacement demands, see Fig. 11.3.

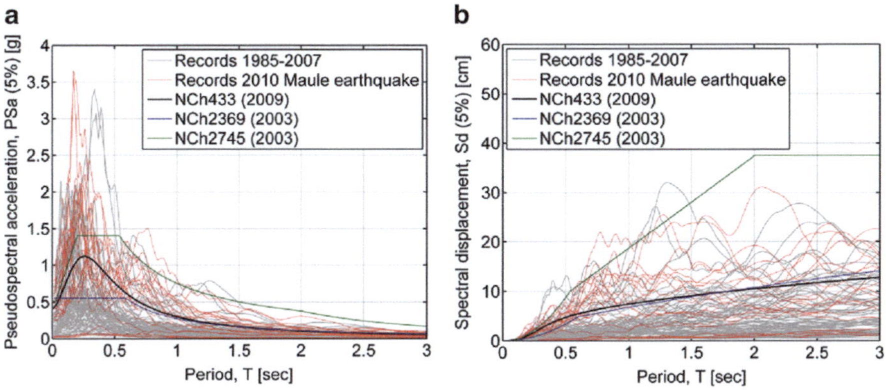


**Fig. 11.3** Response spectra at 5 % damping for records obtained in zone 3 and soil type II according to Chilean seismic design codes. Elastic demands of NCh433, NCh2369 and NCh2745 are shown. (**a**) Acceleration spectra. (**b**) Displacement spectra

On the other hand, Newmark type acceleration response spectra have been incorporated into the ETG-1.015 and NCh2745 loading codes. In NCh2745 the combined effective acceleration and spectral shape is scaled, cut or amplified based on the results of analysis on several typical reference structures. Then design spectrum in any Chilean code does not correspond to elastic seismic demands derived from local recording by large factors. As it can be seen in Fig. 11.3, the criteria for the shape and scale of each code design spectra are quite different. The NCh433 was calibrated so it has the average shape of the response spectra compatible with that obtained from historical records, but values are scaled so residential buildings with periods between 0.5 and 1.5 s have a base shear coefficient consistent with pre-1985 design standards. In NCh2369 the spectral shape contained in NCh433 is used as a reference. However, in the industrial code the spectral values are limited by a maximum value generating an extremely low equivalent plateau. This plateau was selected in order to obtain low period and rigid structures, equipment and systems designed with seismic demands comparable with those used prior to the development of the code.

Strong emphasis to displacement demands is given in the spectrum incorporated into NCh2745 for buildings with periods greater than 1.5 s. This is because it was deemed that the design of base isolated buildings should result in large displacement demands. Moreover, the ZPA value in this code does reflect the results of a probabilistic study; and the plateau in the design spectra represents a typical amplification factor for the acceleration controlled spectral branch observed in Chilean records.

Lateral forces obtained from design codes ground response spectra are modified by the importance of the structure, and by the inherent ductility and damping of the seismic resistant structural system. Minimum and maximum base shear limits may control these forces. Limits are also considered to control maximum interstory drifts.

All design codes in Chile only require linear elastic analyses. A traditional modal response spectrum method is the preferred choice. Nonlinear pushover analyses are not mandatory but such analyses have been performed to support the design of some tall residential buildings as part of the peer review process. Nonlinear time history analyses are rarely used, and have been conducted to support the design of some base isolated buildings and to verify the seismic response of buildings that incorporate supplementary damping devices. Soil-structure interaction is commonly not considered for design.

The NCh430 code did not have requirements for transverse reinforcement in walls boundary elements until 2008, which is a deviation from ACI 318. As it was described above, the detailing of steel structures is carried out in accordance with ASD89, with some additional requirements imposed by NCh2369. Low ductility demands in this case are expected because this type of structures are inherently strong and nonlinear incursions are deemed limited.

The electrical standard did not relay on ductility, rather it requires functionality for a specific response spectrum. Functionality shall be demonstrated by testing, or by analysis if authorized by the owner.

The industrial building code requires low loss of investment and functionality protection. The electrical standards require continuation of operation. The CODELCO standards require protection of investment and an extremely rapid recovery of production. This standard is the only one that explicitly calls and enforces performance check.

## 11.3 Effect of the Earthquake in Performance Based Design

### *11.3.1 Performance of Nonstructural Components*

The February 27, 2010 Maule earthquake evidenced the lack of seismic design of most nonstructural components and systems in Chile. Although most of structures behaved as expected, the performance of nonstructural elements was substandard, and this was in part because of a lack of seismic design and deficient installation of these elements.

Nonstructural damage affected the continuity of operation of more than 15 hospitals and an unknown number of office and commercial buildings located as far as 300 miles from the epicenter.

Although Chilean codes specified a procedure for the seismic design of nonstructural systems, partially based on ATC-3 (ATC 1978) provisions, its application has been scarcely enforced by the authorities. In general, structural engineers normally neither participate in the design process for nonstructural systems nor perform nonstructural design. Furthermore, the Chilean law that rules the peer review process explicitly excludes from the review process the nonstructural components. The most widespread damage observed in nonstructural components consisted of total or partial collapse of acoustical suspended ceiling systems. The damage occurred mainly due to the lack of lateral bracing and compression struts, and the use of weak runner sections. Many cases of floods caused by fire sprinkler breakage were reported, affecting the serviceability of malls, casinos and airports.

Similarly, severe and widespread damage was observed in partition wall systems. Although the Chilean code establishes that partition walls shall be able to accommodate (or resist) the building interstory drifts.

Additionally, severe damage and loss of operation was reported in practically 50 % of the elevators. The typical damage included counterweight and passenger cabin derailment, fall of counterweight blocks, damage due to impact between cabin and counterweights, shifting of driving machines and pulleys without adequate anchors, and damage to control panels.

Other frequently observed damage consisted of collapse of unreinforced masonry parapets, collapse of unbraced HVAC ductwork, severe damage to HVAC equipment, damage to unanchored water boilers, and damage to glazing systems. Minor damage was observed in curtain wall systems due to compliance to code requirements. Figure 11.4 highlights some of the nonstructural damage observed.

Immediately after the earthquake, the Chilean engineering community addressed the need for developing a new code for the seismic design of nonstructural components and systems, aiming at making compatible structural and nonstructural seismic performances. By the time of this writing, the new code is in its final review process. The new code, partially based on the seismic design provisions of Chap. 13 of ASCE/SEI 7-2010 (ASCE/SEI 2010), was adapted to the local seismic hazard and local seismic design practices. First of all, each project's specialist will have to provide information on the seismic design of the items (equipment and distributed lines) included in his project.

The new code requests prequalification of all nonstructural components by analysis, experiment, experience, or a combination of these three methods. The code makes explicit reference to recognized international standards for testing, designing, detailing and installing nonstructural components and systems due to the absence of specific local code.

A quality assurance program and an installation checklist shall be included as part of each nonstructural design. As a consequence of the new requirements, all projects will be fully defined since their design stage, avoiding improvisation during construction.

Other codes that are also under development by the time of this writing are the code for construction in tsunami areas, the code for design of strategic and communitarian facilities, the code for the seismic design of elevator systems, the code for the seismic design of earth retaining walls, and the code for retrofitting historical adobe construction, among many others.

In addition to the development of improved design codes, the use of passive seismic protection devices such as base isolation and passive energy dissipation systems, as strategy for structural and nonstructural protection, is becoming popular among government agencies, investors, stakeholders and the public.

**Fig. 11.4** Examples of nonstructural damage: (**a**) Damage to floating partition walls, (**b**) Damage to HVAC equipment, (**c**) Damage to driving system, (**d**) Damage to rigidly connected partition walls, and (**e**) Damage to suspended ceilings

### *11.3.2 Performance of Tall Reinforced Concrete Buildings*

The overall performance of buildings in Chile during the Maule earthquake was rather acceptable, in terms of the amount of damage observed. However, it is evident that newer high-rise buildings suffered a disproportionate percentage of the damage observed. Concentration of damage in high-rise buildings was observed in softer soils conditions. The ground motion obtained from Concepción has shown sensitivity to strength degradation in the period band corresponding to the fundamental period of damaged high-rise buildings built in the area.

The application of performance-based design in Chile is difficult due to the typology of tall bearing wall buildings. In Chile, these buildings have numerous walls and many of these walls have openings and distinct discontinuities, poor detailing of lap-splices, and little if any confinement of the concrete. Besides, walls are so closely spaced that the slabs provide significant coupling. In addition,

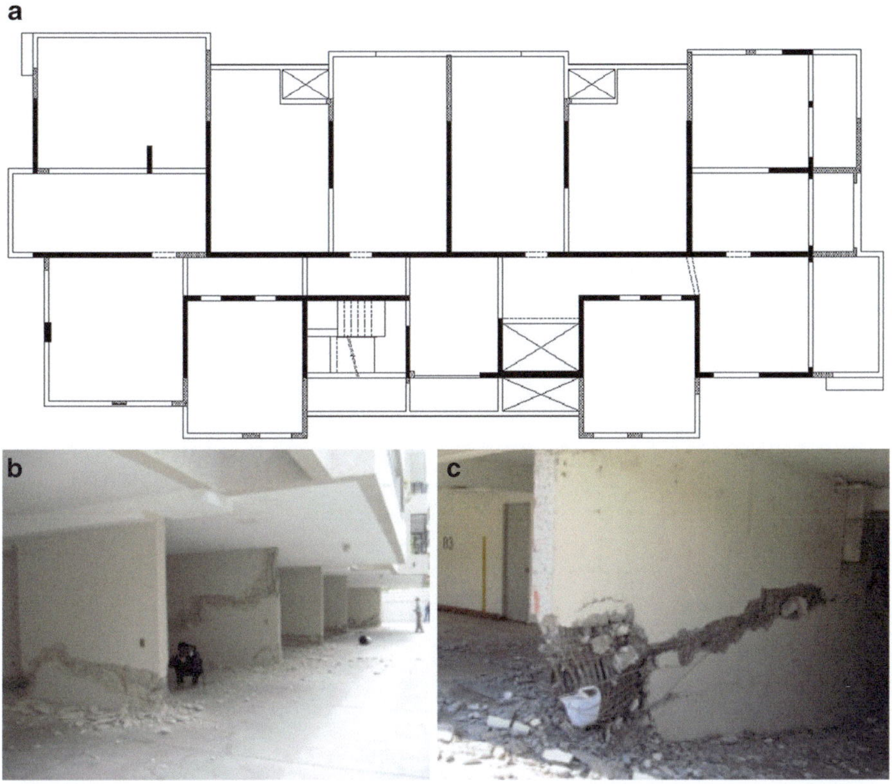

**Fig. 11.5** (**a**) Typical plant of R/C wall building in Chile, (**b**) Typical failures in R/C walls, localization caused by buckling of vertical reinforcement, and (**c**) Buckling of bars in a boundary element

soil-structure interaction is deemed to play an important role in the response of tall buildings founded on soft soils. In Chile, the soil-structure interaction is not considered in the design. All of this makes the prediction of the development of most damage states extremely challenging with most analytical tools at reach in design offices, and even with more specialized tools.

Most of the damaged buildings are similar to that one shown in Fig. 11.5a. Typical damage in R/C walls are shown in Figs. 11.5b and 11.6. Several R/C buildings had damage in wall boundaries as shown in Fig. 11.6. Buckling of vertical reinforcement produced localization of damage, hampering the spread of plasticity. Figure 11.6a shows a shear failure example in a new 25 story building. A compression-tension failure in elements supporting walls is shown in Fig. 11.6b. Endemic fracture of the longitudinal reinforcement after buckling was also observed, see Fig. 11.6c. Figure 11.6d depicts a lap-splice failure.

Through damage observations, the need of providing limits to the wall thickness was recognized. To avoid steel congestion in boundary elements and ameliorate the

Fig. 11.6 (a) Shear Failure Example in a new 25 story building, (b) Compression-tension failure in elements supporting walls. (c) Fractured of longitudinal reinforcement after buckling. (d) Lap-splice failure

likelihood of developing premature lap splice failures, a minimum wall thickness for walls and limits thickness to bar diameter and wall thickness ratio are necessary. Furthermore, the current detailing of the transverse reinforcement in walls in Chile essentially results in little if any confinement of the concrete core, and even if such reinforcement is present it is barely efficient. As a result, thin walls in Chile can attain very small ultimate compressive strains. For this reason, it is the view of the authors that an effort should be made in design to avoid compression-controlled walls sections.

### *11.3.3 Proposed Modifications to ACI318-08 to Design R/C Special Walls in Chile*

In an attempt to improve the seismic performance of tall wall bearing buildings in Chile, a set of complementary prescriptions were recommended superseding or

**Fig. 11.7** The whole flange width of a flanged section must be considered (T, L, C, or other cross sectional shapes)

supplementing those in ACI 318-08 for the design of special walls. The objective was to ensure designs that can reach inelastic deformations and maintain their lateral force and gravity load capacity when subjected to cyclic loading, while ensuring the cost of construction would be minimally impacted.

Urgent modifications were proposed to design special R/C walls in Chile, after the February 27th, 2010 earthquake. A law has been promulgated on February 14th, 2011 adopting ACI 318S-08 to R/C building design, but complemented by the additional requirements presented in this paper. The complementary provisions apply only to R/C special structural walls, defined in §21.1 in ACI 318-08.

Since many engineers had designed walls determining the amount of reinforcement for separate rectangular sections forming the whole section, §21.9.5.2 in ACI318-08 was changed to:

1. The whole flange width of a flanged section must be considered (T, L, C, or other cross sectional shapes), see Fig. 11.7.
2. The total amount of longitudinal reinforcement in the section must be considered when assessing the flexural strength due to combined flexural and axial loads.

The law was modified again on December 13, 2011, permitting engineers to apply directly ACI318-08 provisions in this mater.

To avoid the use of very large bar sizes in thin walls, the longitudinal reinforcement bar diameter was limited to be less or equal than one-ninth of the least dimension of the boundary element. In addition, the transverse reinforcement bar diameter shall be greater or equal than one-third of the diameter of the longitudinal bar being tied, and this reinforcement shall be anchored to the extreme longitudinal bars in a wall.

The use of 135 or 180° hooks, with a hook extension of $6d_b$ or 75 mm whichever was greater, was recommended to anchor the transverse reinforcement, see Fig. 11.8. This recommendation stemmed from the endemic poor performance of transverse reinforcement anchored with 90° hooks. The 75 mm was recommended because shorter hook extensions could be accepted when accounting for tolerances during bending of the reinforcement. We note that ACI 117-2006 (2006) permits a tolerance of 25 mm in the length of the hook extensions. Then, hoops made out of 8, 10 and 12 mm diameter could be accepted with a length of extension equal to 23, 35 and 47 mm, respectively. Under severe ground motions the anchorage of

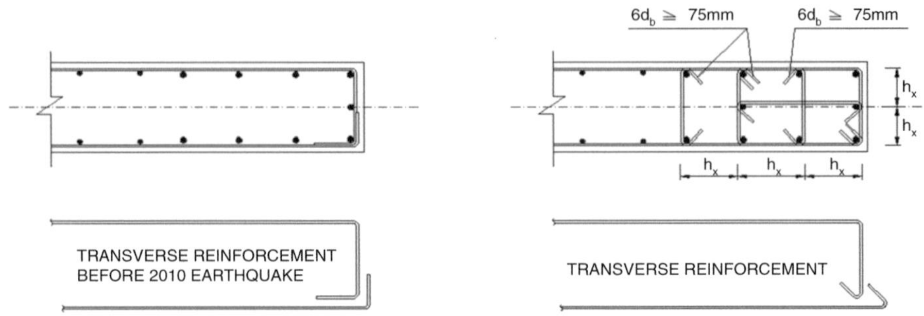


**Fig. 11.8** Transverse reinforcement in walls, before and after the 2010 earthquake

the transverse reinforcement with such short extensions could be lost progressively under cyclic loads, rendering ineffective the role of bracing of the longitudinal bars that the transverse reinforcement plays.

Many of the failures observed in bearing walls occurred in walls that were too thin, used in Chile since the 1980s. To avoid brittle failures under bending and axial forces, limitations to the depth of the neutral axis depth were proposed. A nominal strain in concrete equal to 0.003 has been adopted by ACI318-08 to define a nominal strength under axial load and bending. Under large deformations, compression strain in concrete is usually greater than this conventional value at critical zones. The tensile strain in steel when compression strain in concrete is equal to 0.004 in the opposite fiber is used as an index to identify the type of failure. Failures controlled by compression were not permitted in walls in the proposed document. Sections controlled by tension were recommended for walls, with transverse reinforcement to avoid premature buckling of longitudinal bars in extreme fibers or in boundary elements. Similar to provisions for beams in §10.3.5 in ACI318-08, tensile strain in steel $\boldsymbol{\varepsilon t}$ at nominal strength in special walls must be greater or equal than 0.004. Large lateral displacement must not be reached with large compression strains at confined concrete, longitudinal steel in opposite fiber must yield before the ultimate strength capacity is reached in concrete under compression. For effective confinement, walls need to be either thicker, or the confinement reinforcement should be spaced very closely, making construction difficult.

Then, maximum factored axial load permitted in walls, called as $\boldsymbol{P_4}$, is associated to a tensile strain in steel equal to 0.004 when compression strain in concrete reaches 0.003.

Vertical equilibrium of forces gives, $\boldsymbol{P_4 = C - T_f}$ where $\boldsymbol{P_4}$ is the factored applied axial force on the wall, $C$ is the compression force acting on boundary under compression and $\boldsymbol{T_f}$ is the tension force due to longitudinal steel in flange of a T section, or boundary element in compression in a rectangular section.

Figure 11.9 shows that the neutral axis depth increases proportionally with the axial load, and when the amount of steel in flange $\boldsymbol{A_{sf}}$, increases, and decreases when amount of compressed vertical steel in the boundary $\boldsymbol{A_{sw}}$, increases. Then, to decrease neutral axis depth longitudinal steel in boundary under compression can

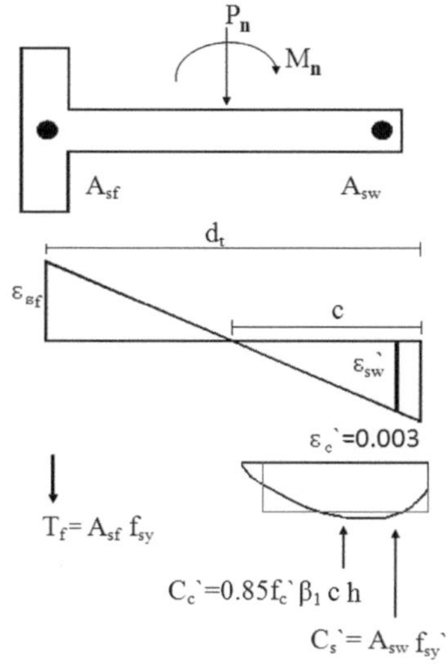


Strain compatibility:

$c_4 = 3/7 d_t a_4 = \beta_1 c_4$

Equilibrium of vertical forces:

$C = P_4 + T_f = P_4 + A_{sf} f_{sy}$

$P_4 = 0.85\, h\, \beta_1\, 3/7 d_t f_c' + A_{sw} f_{sy}' - A_{sf} f_{sy}$

Only if $f_{sy}' = f_{sy}$

$P_4 = 0.85\, h\, \beta_1\, 3/7\, d_t f_c' + f_{sy}\,(A_{sw} - A_{sf})$

And if $A_{sw} = A_{sf}$ assuming $\beta_1 = 0.85$,
$P_4 = 0.3 f_c' A_g$

Note:
$c_4 = (P_4 + A_{sf} f_{sy} - A_{sw} f_{sy}')\,/0.85\, h\, \beta_1 f_c'$

**Fig. 11.9** Strain and stress distributions

be added, $\boldsymbol{A_{sw}}$, thickness of boundary element can be increased $\boldsymbol{h}$, or external axial load could be decreased. Upper bound to axial load defining the transition zone can be increased adding longitudinal steel in the boundary element $\boldsymbol{A_{sw}}$, or increasing the wall thickness $\boldsymbol{h}$.

Then, to avoid compression failures in structural walls a special provision was added to ACI318-08 establishing that "The net tensile strain in the extreme tension steel, εt, must be equal or greater than 0.004 when the concrete in compression reaches its assumed strain limit of 0.003."

A year later this disposition was changed demanding a limit equal to 0.008 for the compressive strain in concrete calculated for the design displacement.

To avoid global wall buckling, the emergency code recommended that the transverse dimension of special structural walls had to be greater or equal to one-sixteenth of the lateral unsupported member length under compression, $\boldsymbol{l_u/16}$. One year later this provision was changed to *"If transverse dimension of special structural walls is less,* $\boldsymbol{l_u/16}$ *then global buckling must be checked"*.

ACI318-08 permits that capacity design rules be applied to determined shear due to seismic action, see § 9.3.4. Alternatively to the Capacity Design rules, seismic shear design can be obtained amplifying seismic shear from linear analysis as specified in ACI 318-08 for intermediate frames in § 21.3.3 (b), read R21.3 in ACI318-08.

Tests (Panagiotou and Restrepo 2010), analysis (Kawashima et al. 2010; Panagiotou et al. 2010; Rejec et al. 2012), and the damage observed, suggest that shear demands in walls can be considerably greater than shear force calculated from equivalent static methods or from modal analysis using reduction factors. The shear force demand in a wall is also affected by the coupling effect of beams and slabs. A shear force amplification of 2 was recommended in the emergency code but the last revision to the emergency code eliminated such recommendation.

A clear transfer mechanism from forces from one lap splice bar to other must be established since concrete cannot resist well tension stresses. An equivalent truss can be developed if struts and ties are present, if transverse reinforcement is there, tensile strength in concrete is not so important. Amount of transverse reinforcement has been obtained from a strut and tie model assuming 45 grade inclination for compressed struts and neglecting hardening of longitudinal steel. If the potential plane of failure must be crossed over, reinforcement in two orthogonal directions must be prescribed since lap spliced bars can be at any position, following the direction of the wall central axis or the orthogonal one.

## 11.4 Conclusions

The Mw 8.8 February 27, 2010 Maule earthquake tested many of the design procedures for buildings and other structures in Chile. The main lessons learned were the effect that long-duration motions in soft soils can have on the performance of tall bearing wall R/C buildings and, more general, on the performance of nonstructural components. Although the general performance of tall buildings in Chile is acceptable from the engineering perspective, it was not from the societal perspective. Many buildings experienced unnecessary and costly damage, which could had been prevented with appropriate detailing of the reinforcement with minimum economic impact. The detailing of the reinforcement in tall buildings in Chile have been lax. Bearing walls had reached very small widths in the last few years, resulting in walls that were compression-dominated with practically no detailing to ensure ductility.

From the point of view of performance-based seismic design, bearing wall buildings in Chile incorporate many discontinuities. With current methods of linear and nonlinear analyses used in design offices such discontinuities makes the prediction of damage states very difficult. Likewise, in a country that has had little in the seismic design and installation of nonstructural elements, performance based design of these elements to given damage states can also prove to be difficult.

## References

ACI (2005) ACI 318-05, building code requirements for structural concrete. American Concrete Institute, Farmington Hills, Michigan

ACI (2006) ACI 117-2006, specifications for tolerances for concrete construction and materials and commentary. American Concrete Institute, Farmington Hills, Michigan

AISC (1989) Specifications for structural steel buildings: allowable stress design (ASD). American Institute of Steel Construction, Chicago, Illinois

Arango MC, Strasser FO, Bommer J, Boroschek R, Comte D, Tavera H (2011) A strong motion database from the Peru-Chile subduction zone. J Seismol 15:19–41
ASCE/SEI (2010) ASCE/SEI 7-2010: minimum design loads for buildings and other structures. American Society of Civil Engineers, New York
ATC (1978) ATC-3: tentative provisions for the development of seismic regulations for buildings. National Bureau of Standards Special Publication, Washington, DC
Boroschek R, Domb F (2006) Estudio Nacional de Chile. Project IDB-CEPAL Programa de información para la gestión de desastres, CEPAL, Santiago
Boroschek R, Soto P, León R (2010) Registros del Terremoto del Maule, Mw = 8.8, 27 de febrero de 2010, RENADIC 10/05. https://terremotos.ing.uchile.cl. Oct 2010
CODELCO (2011) Criterio de Diseno Estructural Sísmico, SGP-GFIP-ES-CRT-002 Rev. 1, Vicepresidencia de Proyectos, Gerencia Funcional de Ingeniería y Procesos, CODELCO-Chile, 10 de noviembre de 2011 (in Spanish)
Delouis B, Nocquet J-M, Vallée M (2010) Slip distribution of the February 27, 2010 Mw = 8.8 Maule Earthquake, central Chile, from static and high-rate GPS, InSAR, and broadband teleseismic data. Geophys Res Lett 37(17)
INGENDESA (1987) ETG-1.015, Especificaciones Técnicas Generales-Diseno Sísmico, Mayo 1987 (in Spanish)
INGENDESA (1997) ETGI-1.020, Especificaciones Técnicas Generales-Requisitos de Diseno Sísmico para Equipo Eléctrico, Noviembre 1997 (in Spanish)
INN (2003a) NCh2369Of.2003, Diseno sísmico de estructuras e instalaciones industriales. Instituto Nacional de Normalización, Santiago (in Spanish)
INN (2003b) NCh2745, Análisis y diseno de edificios con aislación sísmica. Instituto Nacional de Normalización, Santiago (in Spanish)
INN (2008) NCh430Of.2008, Hormigón armado - Requisitos de diseno y cálculo. Instituto Nacional de Normalización, Santiago (in Spanish)
INN (2009) NCh433Of.1996 Modificada en 2009, Diseno Sísmico de Edificios, Norma Chilena Oficial. Instituto Nacional de Normalización, Santiago (in Spanish)
Kawashima K, Sasaki T, Kajiwara K (2010) Experimental study on the seismic response of bridge columns using E-Defense (2010). http://www.pwri.go.jp/eng/ujnr/tc/g/pdf/25/1-4.pdf. Accessed 18 Aug 2010
MOP (2012) Manual de Carreteras, MOP-DGOP-Dirección de Vialidad, 2012 (in Spanish)
Panagiotou M, Restrepo JI (2010) A displacement-based method of analysis for regular reinforced concrete wall buildings: application to a full-scale 7-story building slice tested at UC San Diego. J Struct Eng 137(6):677–690
Panagiotou M, Restrepo JI, Conte JP (2010) Shake table test of a full-scale 7-story building slice phase I: rectangular wall. J Struct Eng 137(6):691–704
Rejec K, Isaković T, Fischinger M (2012) Seismic shear force magnification in RC cantilever structural walls, designed according to Eurocode 8. Bull Earthq Eng 10(2):567–586
Ruegg JC et al (2009) Interseismic strain accumulation measured by GPS in the seismic gap between Constitución and Concepción in Chile. Phys Earth Planet Interiors 175(1):78–85

# Chapter 12
# Performance-Based Issues from the 22 February 2011 Christchurch Earthquake

**Kenneth J. Elwood, Stefano Pampanin, Weng Yuen Kam, and Nigel Priestley**

**Abstract** At 12:51 pm local time on 22 February 2011, a Mw 6.2 aftershock of the September 4, 2010, Darfield Earthquake shook the city of Christchurch, New Zealand. The aftershock occurred on an unmapped fault less than 8 km from the city center resulting in the collapse of two reinforced concrete office buildings and one concrete parking garage, and severe damage to numerous others. The region has continued to suffer from aftershocks and further damage to building structures throughout the year following the February earthquake. This paper summarizes the observed damage to buildings in the Central Business District (CBD), with a specific focus on identifying future research to support the development of performance-based design procedures.

K.J. Elwood (✉)
Civil and Environmental Engineering Department, University of Auckland, Auckland, New Zealand
e-mail: k.elwood@auckland.ac.nz

S. Pampanin
College of Engineering, Civil and Natural Resources Engineering, University of Canterbury, Christchurch, New Zealand
e-mail: stefano.pampanin@canterbury.ac.nz

W.Y. Kam
Beca, Auckland, New Zealand
e-mail: Kam.WengYuen@beca.com

N. Priestley
Jacobs School of Engineering, Department of Structural Engineering, University of California, San Diego, CA, USA

ROSE School European School for Advanced Studies in Reduction of Seismic Risk, Pavia, PV, Italy

Priestley Structural Engineering, Diamond Harbour 8941, New Zealand
e-mail: nigelpriestley@xtra.co.nz

M. Fischinger (ed.), *Performance-Based Seismic Engineering: Vision for an Earthquake Resilient Society*, Geotechnical, Geological and Earthquake Engineering 32,
DOI 10.1007/978-94-017-8875-5_12, 

**Keywords** Christchurch earthquake • Ground motions • Concrete buildings • Collapse • Code changes • Shear walls • Precast moment frames • Repair costs

## 12.1 Introduction

Six months after the 4 September 2010 $M_w$ 7.1 Darfield (Canterbury) earthquake, the $M_w$ 6.2 Christchurch earthquake struck Christchurch, New Zealand on the 22 February 2011. The $M_w$ 6.2 earthquake occurred on a previously unknown fault less than 8 km south-east of the Christchurch central business district (CBD), initiating at a shallow depth of 5 km. Unlike the 4 Sept event, when limited-to-moderate damage was observed in engineered reinforced concrete (RC) buildings (Kam et al. 2010), after the 22 February event about 16 % out of 833 RC buildings in the Christchurch CBD were severely damaged. Whilst there was no fatality in 4 September earthquake (also due to the time of occurrence i.e. at 4.35 am), there were 182 fatalities in the 22 February earthquake (occurring at 12.51 pm), 135 of which were the unfortunate consequences of the complete collapse of two mid-rise RC buildings.

This paper highlights important observations of damage to RC buildings in the 22 February 2011 Christchurch earthquake. Focus will be on damage to modern concrete frame and wall building and the identification of future research directions necessary to develop appropriate concrete code provisions to address the observations in Christchurch. Detailed description of damage to concrete buildings of all vintages and types can be found in Kam et al. (2011).

## 12.2 Ground Motions

Recorded ground shaking in the CBD matched or exceeded the 2,500 year return period design spectrum from the NZ Standard (NZS1170.5 2004), particularly in the east-west direction (Fig. 12.1). The large displacement demands shown at 1.5 s are a particular concern for the numerous 10–15 story buildings in the CBD. Using NZS 1170.5, normal buildings are designed for the 500 year return period design spectrum using ultimate limit states design principles, while post-disaster buildings are designed using the 2,500 year design spectrum shown in Fig. 12.1.

From disaggregation of the seismic hazard, a $M_w$ 6.2 earthquake at a distance of 10 km did not contribute significantly to the probabilistic seismic hazard model used for Christchurch (Stirling et al. 2002). With this in mind it raises the question if considering relatively large area sources in probabilistic seismic hazard assessment to account for earthquakes on unknown faults is sufficient to capture the risk unknown faults may pose to our urban regions. Should we consider deterministic scenarios (i.e., shaking from a specific magnitude at a specific distance) for seismically active regions where unknown faults may dominate the earthquake risk? Research is required on the most appropriate means of accounting for seismic hazard posed by earthquakes on unknown faults.

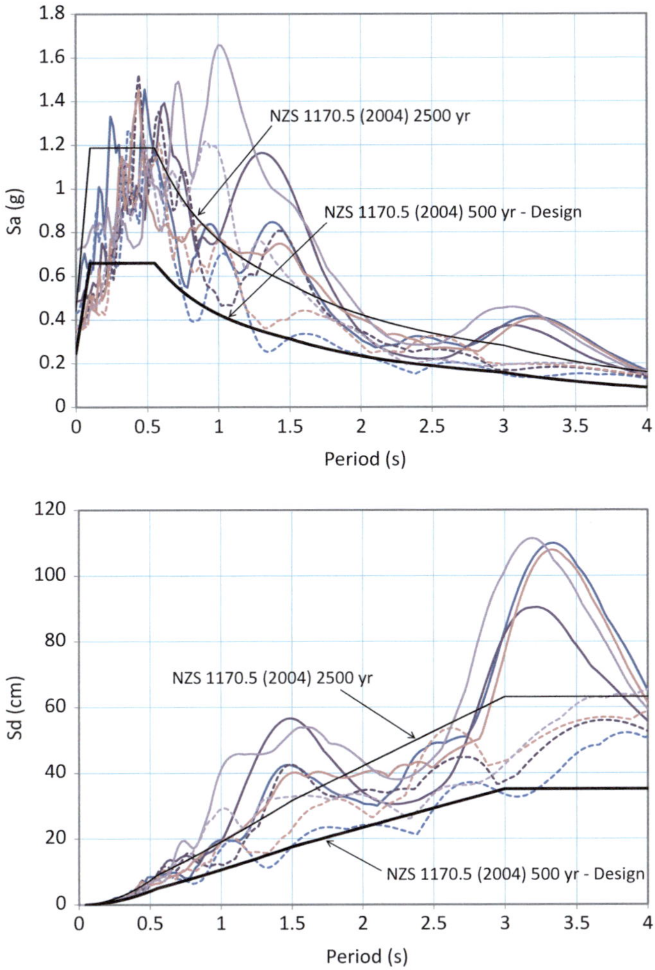


**Fig. 12.1** Elastic horizontal acceleration and displacement response spectra (5%-damped) recorded in Christchurch CBD and the NZS1170.5 design spectra (NZS soil class D, R = 20 km). *Dashed lines* are N-S demands, *solid lines* are E-W demands

One unique aspect of the 22 February earthquake was the very high vertical ground accelerations (as high as 2.2 g), frequently exceeding the peak horizontal accelerations at recording station within approximately 10 km of the epicentre. Similar to past earthquakes, the vertical accelerations were characterised by very high frequency content and peak values were only attained for a very short duration. Research is needed to define the importance of vertical excitations on building performance considering the high frequency content and phasing with lateral demands. Rapid attenuation of high frequency vertical ground motion should also be considered as this may only be critical for near-fault events.

## 12.3 Concrete Building Types and Damage Statistics

With a population of approximately 390,000, Christchurch is the second largest city in New Zealand and the economic centre of the South Island. As a relatively centralized city, approximately 25 % of the total employment in the city was located in the Christchurch CBD, leading to a concentration of buildings over five storeys in the city centre. In the CBD there were 127 buildings with at least six stories, with the tallest RC building being 22-storey (86 m). RC frames and RC walls are the most common multi-storey construction types. Out of 183 buildings with more than 5-storeys, 49 % are RC frame buildings, 22 % are RC wall buildings, 7.7 % are reinforced concrete masonry and 5.5 % are RC frame with infills. Only nine steel structures with more than 5-storeys were observed in the CBD.

Buildings constructed prior to the introduction of modern seismic codes in mid-1970s are still prevalent in the Christchurch CBD. Approximately 45 % of the total CBD building stock was built prior to the 1970s. Of this, 13.8 % or 188 pre-1970s buildings have 3-storey and more.

Precast concrete floor systems have been used for multi-storey RC buildings in New Zealand since the mid-1960s. From 1980s to present, the majority of multi-storey RC buildings use precast concrete floors or concrete composite steel deck systems. Ductile precast concrete frames, designed with wet connections to emulate cast-in-place construction, were introduced in early the 1980s and soon became the most popular form of construction for RC frames.

Two concrete office buildings collapsed as a result of the strong ground shaking from the 22 February earthquake (Fig. 12.2). Response of these buildings during the earthquake has been extensively studied and reported in Beca (2011) and DBH (2012) and summarized in Kam et al. (2011). This paper will focus on the performance of modern concrete buildings and implications for the development of performance-based design methodologies.

Kam et al. (2011) provide statistics for building safety evaluation placarding of buildings of different vintages and structural systems in the CBD as of June

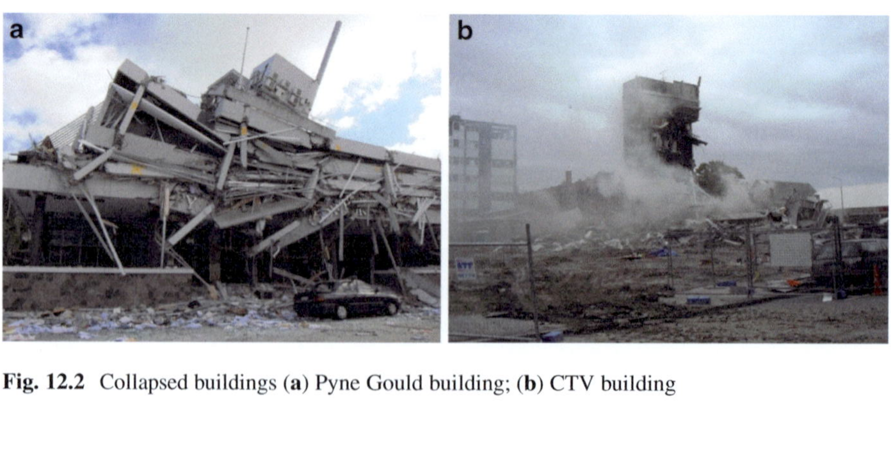


**Fig. 12.2** Collapsed buildings (**a**) Pyne Gould building; (**b**) CTV building

12, 2011 (i.e. prior to June 13 aftershocks). While not being a refined measure of damage, coloured placarding, following procedures similar to ATC-20 (NZSEE 2009), provides a general indicator of the distribution of damage to building types and vintages. As expected, unreinforced masonry buildings exhibited very severe damage with approximately 70 % of all URM buildings in the CBD receiving a red tag. Statistics for concrete frame buildings indicate a relatively consistent level of performance for concrete frames constructed prior to 1990, with a slight improvement in performance after 1990. In contrast, for concrete wall buildings the percentage of red placards was markedly higher for post-1990 buildings compared with buildings constructed prior to 1990. Furthermore, the percentage of green placards remained reasonably consistent for all vintage of wall buildings. The relatively high level of damage observed in concrete wall buildings will be discussed with reference to specific observations later in this paper.

When considering the structural performance implied by the statistics in the placarding statistics, it is important to recall the level of ground shaking relative to the earthquake demands assumed in design (Fig. 12.1). Normal buildings designed to NZS 1170.5 and the concrete design code (NZS3101 2006) are expected to have a "small margin against collapse" for the 2,500 year design spectrum, assumed to be 1.8 times the 500 year design spectrum (King et al. 2003). Considering the high spectral demands (Fig. 12.1) and that the two concrete office buildings that collapsed (Fig. 12.2) were designed in 1960s and 1980s, it might be concluded from the point of view of collapse and life safety performance that concrete buildings designed according to recent building codes performed well or as expected during the Christchurch earthquake.

While potentially satisfying the objectives of the code, the level of damage observed in many concrete buildings was severe, leading to a large percentage of buildings currently considered uneconomical to repair. Many of the concrete highrise buildings in the CBD are expected to be demolished as a result of damage from the February and subsequent earthquakes. Figure 12.3 shows beam hinging observed in a 22-storey moment frame building. The damage was consistent with the prevalent capacity-design philosophy, protecting the columns from damage and concentrating nonlinear response in the beam hinges. However, wide residual cracks in the beams have raised concerns regarding low-cycle fatigue of the reinforcement and repairability of the building. Moving forward, the financial risk and damage acceptance of ductile RC systems may require further considerations. It is not clear if the performance of this building is acceptable to society or if society is willing to pay more for better performance in future earthquakes. However, to enable the selection of different performance levels in the future, the engineering community should use the impacts of the Christchurch earthquake to promote the further development and implementation of performance-based seismic design approaches (e.g. ATC-58 2011). High insurance coverage by CBD building owners has also contributed to the decisions to demolish rather than repair damaged buildings.

**Fig. 12.3** Beam cracking in a 22-story RC moment frame (precast emulating cast-in-place) scheduled to be demolished

## 12.4 Examples of Damage to Concrete Wall Buildings

Perhaps some of the most important lessons for modern construction from the Christchurch earthquake relate to the performance of reinforced concrete wall buildings. Most shear walls in CBD buildings were tall slender walls where, after the 1982 Concrete Code (NZS3101 1982), capacity design concepts were applied to ensure flexural yielding at the base of the wall limited the shear demands and sufficient horizontal reinforcement was provided to avoid shear failure in the plastic hinge zone. While this design approach appeared to protect against shear failures in modern wall buildings, unexpected flexural compression and tension failures in numerous shear walls in Christchurch indicate the need to modify shear wall design provisions to improve the flexural ductility of slender walls. The following provides a brief summary of some examples of failure modes observed after the 22 February earthquake including web buckling, boundary zone and web crushing, and boundary zone steel fracture.

### *12.4.1 Wall Web Buckling*

Figure 12.4 shows the overall buckling of one outstanding leg of a V-shaped (or L-shaped) shear wall in a 7-story building. The width of the buckled web was 300 mm, with an unsupported wall height of 2.66 m, resulting in a height-to-thickness

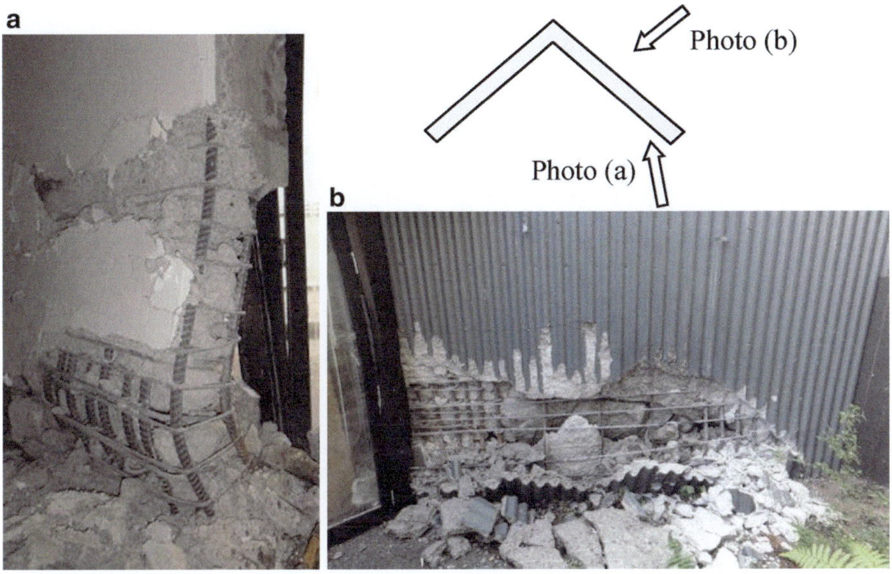


**Fig. 12.4** Seven-storey 1980s building with compression failure of the V-shaped RC shear wall

(slenderness) ratio of 8.9. The boundary zone extended approximately 1.2 m into the 4 m long web. The boundary steel at the damaged end of the wall consisted of 16–24 mm deformed bars confined by 10 mm smooth bars at 120 mm, with a 180° hook on every other longitudinal bar.

The wall buckled over a height of approximately 1 m and crushing extended over 3 m into the web. Horizontal cracks (approximately 1–1.5 mm width) were visible at the buckled end of the web, while inclined cracks in both directions at approximately 45° were apparent in the middle of the web over the first story height. Well distributed, primarily horizontal, cracking with widths less than 1.5 mm, were observed in the lower half of the first story of the generally undamaged flange. The damage pattern suggests that the web may have initially experienced flexural tension yielding of the boundary steel, followed by buckling of the unsupported web over the relatively short plastic hinge length. The L-shaped cross-section would have resulted in a deep compression zone with high compression strains at the damaged end of the web wall. Stability of the compression zone may have been compromised by a reduction in the web out-of-plane bending stiffness due to open flexural tension cracks from previous cycles. High vertical accelerations would also have contributed to the compression and tension demands on the wall.

Buckling in a wall with a height to thickness ratio of less than 10, a limit used in several international codes to avoid out-of-plane instability, suggests that a limit on *h*/*t* may not be enough to prevent wall buckling. Building on the work by Paulay and Priestley (1993), further research is needed to determine a relationship between wall buckling, length of plastic hinge and axial stiffness of the compression zone after bar yielding.

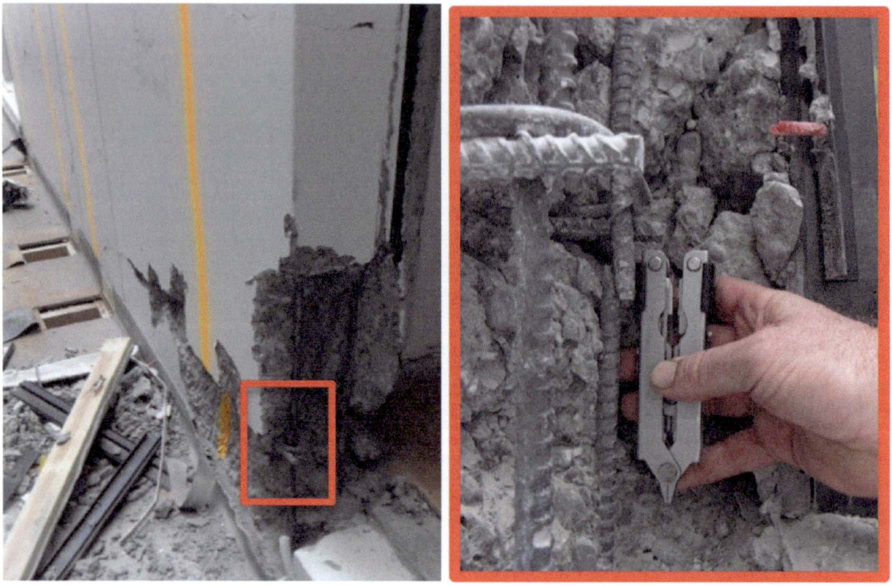

**Fig. 12.5** Fractured bars in lightly reinforced slender RC shear wall (Photo credit: D. Bull)

### *12.4.2 Boundary Bar Fracture*

Fracture of very light longitudinal reinforcement was also noted in modern high-rise buildings. In some lightly reinforced shear walls only exhibiting symptoms of flexural cracking (e.g. Fig. 12.5), bar fracture of multiple longitudinal bars was detected only after removal of cover concrete. The low vertical reinforcement ratios (in some walls as low as 0.002 with no concentration of steel in boundaries) and high in-situ concrete strengths resulted in concrete tension strength higher than the tensile strength of the vertical reinforcement, thus condensing the plastic hinge length. With the small diameter of rebar used, the strain-penetration length was low, leading to fracture at very low displacement ductility values. This has significant implications for both design of lightly reinforced walls but also inspection of similar walls after future earthquakes. Engineers performing post-earthquake assessments need to be cautious when assessing the extent of damage to lightly reinforced shear walls.

Fracture of boundary reinforcement was also observed in the 200 mm thick wall shown in Fig. 12.6. This 7-m long wall (coupled with a 2-m wall) was the primary E-W lateral force resisting system for an 8-story plus basement condominium. For the bottom four stories the wall was reinforced with 12 mm deformed bars at 100 mm in both directions, each face. The boundaries, extending 980 mm from each end, were confined with 6 mm smooth hoops at 60 mm, supporting at least every other longitudinal bar.

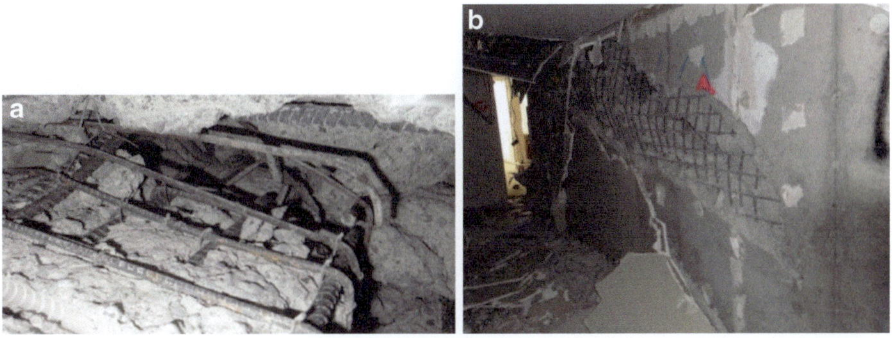

**Fig. 12.6** Boundary bar fracture (**a**) and web crushing (**b**) in modern 8-story apartment building

As shown in Fig. 12.6a, fracture of at least four of the 12 mm end bars occurred at the top of the ground floor. Core concrete generally remained intact in the confined boundary (except where fracture of bars occurred); however, crushing of the core extended into the unconfined web for approximately 3 m from the end of the confined region. The crushing in the web exposed spliced transverse bars, which could not contain the core concrete once the cover was spalled. The damage in the web extended diagonally downward from the fractured boundary (Fig. 12.6b), suggesting that high shear stresses may have also contributed to the observed damage.

In terms of future code development, the concrete crushing within the web of the wall in Fig. 12.6 suggests that cross ties may be required outside the confined end zones and splices should be avoided in transverse bars. Additionally, bar fracture of the lightly reinforced walls shown in Figs. 12.5 and 12.6 suggests that the minimum reinforcement provisions for boundary zones of shear walls should be reviewed.

## *12.4.3 Wall Crushing*

Figure 12.7 shows severe damage to a shear wall in the 22-story Hotel Grand Chancellor, designed and constructed in mid 1980s. Shortening of the ground-floor wall in the hotel lobby by approximately 800 mm led to a visible lean of the building and restricted access to the potential fall zone around the building. Significant structural layout irregularities influenced the seismic response of the building; most notably the east side of the building was cantilevered over an access lane. Furthermore, the seismic force resisting system for the lower 14 storeys consisted of shear walls, while perimeter moment frames were used for the upper stories. (Description below focuses on the performance of the damaged wall; further details about the building and the complex damage pattern observed can be found in Dunning Thornton (2011) or Kam et al. (2011).)

**Fig. 12.7** Crushing and out-of-plane movement of shear wall in 22-story Hotel Grand Chancellor: (**a**) Lobby wall; (**b**) close up of failure at splice; (**c**) out-of-plane movement of wall; (**d**) reinforcement details (Photos (**b**) and (**c**) from Dunning and Thornton 2011)

Damage shown in Fig. 12.7 indicates that the wall displaced downward along a diagonal failure plane through the thickness of the wall. The failure plane, extending the full length of the wall, appeared to initiate at the top of the lap splice in the web vertical reinforcement (Fig. 12.7b). The limited hoops in the boundary appeared to have opened allowing the boundary longitudinal bars to deform with the shortening of the wall. Crushing of concrete was also noted at the top of the lobby wall (Fig. 12.7a), likely to accommodate the out-of-plane movement of the wall as it slid down the diagonal failure plane.

The wall shown in Fig. 12.7 was likely supporting very high axial loads from several sources. First, the wall supported a disproportionately high tributary area since it acted as a prop for the cantilevered bay on the east side of the building. Secondly, the corner column of the upper tower perimeter moment frame would have imparted high axial loads to the wall due to overturning moments, particularly with any bi-directional movement to the south-east. Thirdly, vertical excitation of the cantilever structure could have exacerbated the axial load on the wall. Finally, the structural wall would have also attracted in-plane loads due to N-S earthquake excitation, leading to flexural compression stresses on one end of the wall. Considering the potential for simultaneous compression from all sources of axial loads described above, it is expected that the combined axial load and bending in the wall likely exceeded the concrete compression strain capacity given the limited tie reinforcement provided at the base of the wall.

Some out-of-plane drift of the wall during the earthquake excitation and the plane of weakness created at the end of the splice of the web vertical reinforcement at the base of the wall, may have further contributed to the location of failure observed in Fig. 12.7. Future research is needed on the influence of the out-of-plane movement of shear walls when combined with high axial loads.

Observed wall crushing failures, including that shown in Fig. 12.7, suggest that it is best to avoid compression-controlled walls. It should be recognized, however, that the true axial loads on walls are not well known, in part due to growth of a wall during shaking and the outrigger effect from gravity columns. Codified limits on wall axial loads are being considered by the New Zealand Department of Building and Housing in response to the damage observed in Christchurch (DBH 2012).

## 12.5 Displacement Compatibility

Similar to past earthquakes, the Christchurch earthquake demonstrated the need to carefully consider displacement compatibility in the design of concrete buildings. The following sections highlight three specific issues related to displacement compatibility observed in Christchurch; namely, collapse of precast concrete stairs, near-unseating state of precast floors, and severe damage of "gravity system" not detailed for adequate ductility capacity.

### *12.5.1 Precast Stairs*

Precast stair units collapsed in at least four multi-story buildings, and were severely damaged in several other cases, as a result of the 22 February 2011 Christchurch Earthquake. Figure 12.8 shows collapsed stair units from the 18-storey Forsyth Barr Building (described in detail by Beca (2011)) and the Hotel Grand Chancellor. Stair damage was particularly prevalent in buildings relying on moment-frames for lateral load resistance, although stair damage was also noted in some shear wall buildings. Collapse of stairs not only pose an immediate life safety hazard during the earthquake, but can result in death or injury after the earthquake as occupants

**Fig. 12.8** Collapse of precast stairs: (**a**) Forsyth Barr building; (**b**) hotel Grand Chancellor

attempt to evacuate the building in dark conditions without prior knowledge of the stair collapses. Loss of egress routes as a result of the stair collapses necessitated the evacuation of occupants from windows of high-rise buildings following the Christchurch earthquake.

Stairs are typically designed not to resist building deformations during earthquakes. Movement of the building relative to the stairs is generally accommodated by seismic gaps and seating provided at one end of the stair unit. The seismic gap and seating support must be sized for the expected drift demands during an earthquake. The drift demands from the 22 February earthquake exceeded the 500-year design drift demands required by the New Zealand loading standard (NZS 1170.5 2004) for most building periods. Review of stair collapses in the Forsyth Barr building (Beca 2011) indicates that the seismic gap provided was not large enough to avoid closure of the seismic gap and development of compression forces in the precast stair units. Debris or construction imperfections may have further decreased the seismic gap, increasing the likelihood of compression in the stair units. Expected compression loads could have resulted in yielding at the landing and shortening of the stair units (Beca 2011). Upon reversal of drift demands on the building, the shortened stairs were particularly vulnerable to unseating and collapse. Limited seating support would have also increased the likelihood of progressive collapse once collapse was initiated at one story.

Although damage to stairs was noted in some buildings with shear walls, damage observed in Christchurch suggests that buildings with moment frames may be particularly vulnerable to stair collapse. Hinging of beams in moment frames leads to shear distortions in a building frame bay resulting in lengthening of bay diagonal, and hence, more movement at the stair support. Furthermore, hinging in beams can result in beam elongation, if not sufficiently restrained by the diaphragm, leading to further displacement demands at the stair support. Stair collapses in the Clarendon

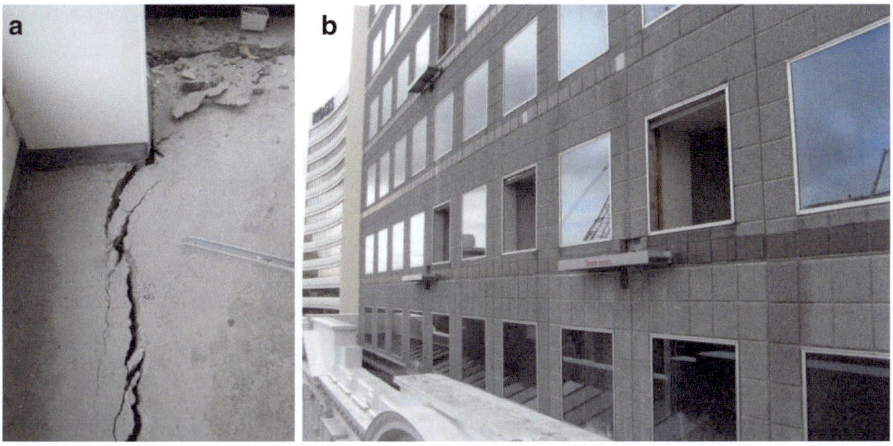

**Fig. 12.9** (**a**) Separation of precast floor slabs from supporting perimeter moment frame in 18-storey office building. Photo (**b**) (Courtesy of D. Bull) shows temporary supports for columns after earthquake

Towers, where precast diaphragms detached from the exterior frames (see Fig. 12.9), have been, in part, attributed to the additional deformation demands from beam elongation (Bull 2011).

The poor performance of stairs in the Christchurch earthquake raises several important considerations for the design and evaluation of buildings internationally. Considering the need to provide egress for occupants of damaged buildings after a major earthquake, it is important that the drift demand used to size the gap and seating be reflective of that expected in the maximum considered earthquake. In light of the uncertainty in the ground motions, differences in the linear and nonlinear displacement profile of the building, and the lack of redundancy when seating support is lost, a detail insensitive to construction tolerances or obstructions and allowing for larger than design displacements should be adopted. Ideally to avoid unintended compression in the stair unit, seismic gaps should be avoided by allowing the stair unit to slide on the top of the slab surface (Beca 2011).

### 12.5.2 *Precast Diaphragm and Beam Elongation*

Figure 12.9 illustrates an extreme example in which extensive floor diaphragm damage with near loss of precast flooring unit supports was accompanied by beam elongation in a perimeter moment frame. The building shown in Fig. 12.9 is a 17-storey building with ductile perimeter moment frames, internal gravity frames and flange-hung supported precast double-tee flooring. 60 mm topping with cold-drawn wire mesh reinforcement is used. The perimeter frames have typical 500 × 850 mm deep precast beams with 600 mm square and 800 mm square columns. The beam spans are typically 2,900 mm in the East-west direction and 5,800–6,500 mm in the North-south direction.

Ductile beam hinging mechanism in the North-south perimeter frames was observed (and repaired) after the 4 September 2010 earthquake. In addition, the mesh reinforcement in the precast floor topping had fractured at several levels where the precast floors butted together, with residual crack widths as small as 2 mm. In the 22 February event, the beams in the East-west perimeter frames experienced hinging. However, as the North-south perimeter frames were previously hinged and softened, the torsional resistance expected from the overall system would have decreased. Consequently, the building may have exhibited a moderate level of torsional response, which amplified the demand on the Northern East-west perimeter frame.

Due to the high beam depth-to-span ratio (850/2900), the beam elongation effects (geometrical elongation and plastic cyclic cracking) were significantly more pronounced in the East-west perimeter frames. As expected, the elongation of beams created tension in the connection between the precast floors and supporting perimeter beams. The largest horizontal crack parallel to the double-tee flange support was approximately 20–40 mm wide (see Fig. 12.9a). Slab mesh fracture was observed in floor topping close to the beam plastic hinges. In several locations at the Northern bays, the precast floors dropped vertically by approximately 10–15 mm, indicative of loss of precast floor seating support.

While floor collapse did not occur, the separation of perimeter frames from the diaphragms at multiple levels in the Clarendon Towers (Fig. 12.9) raises concerns of possible much greater consequences due to column buckling in aftershocks. To address this risk of collapse in future aftershocks, pairs of 25 mm rods spanning the full width of the building and attached to spreader beams were installed at each vulnerable column at four heavily damaged levels (see Fig. 12.9b). This building is now slated for demolition. Where similar precast floor systems are used, it is important to ensure there is strong connection between the diaphragm and the frames through a well reinforced topping slab in order to ensure separation between the frames and precast components is minimized. Welded-wire mesh should be avoided in topping for precast diaphragms given their limited strain capacity. Only ductile welded-wire mesh has been allowed in New Zealand since 2005 (DBH 2005).

### *12.5.3 Gravity Columns*

The Christchurch earthquake also reinforced the need to consider deformation compatibility of the gravity system with the seismic force resisting system, particularly for so-called “gravity columns” not assumed to resist lateral loads but essential for gravity load support. Figure 12.10 shows varying levels of damage to gravity columns in three different buildings. Figure 12.10c shows one of the critical

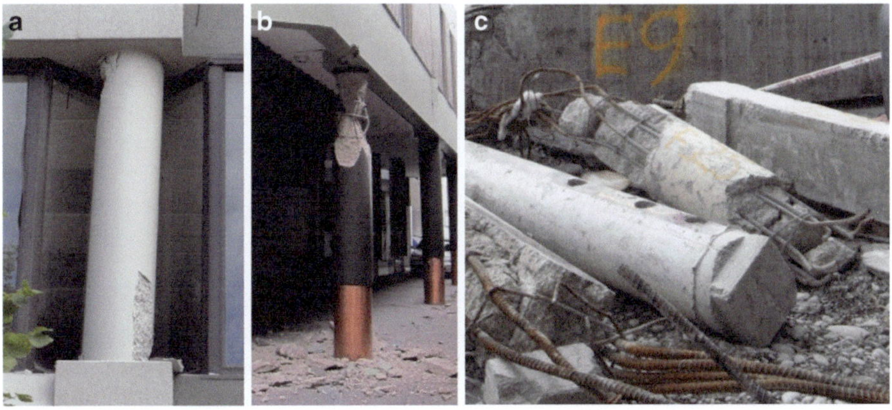


**Fig. 12.10** Damage to gravity columns: (**a**) Spalting and residual draft; (**b**) crushing; (**c**) failure of columns in CTV building

weaknesses of the collapsed CTV Building, 400 mm diameter columns with 6 mm ties at 250 mm (for further details on the CTV building see Kam et al. (2011) and DBH (2012)).

The gravity system must be able to accommodate the deformations imposed by the seismic force resisting system during strong ground shaking. After the axial load failure of several gravity columns during the 1994 Northridge Earthquake, ACI 318 introduced provisions for "components not part of the lateral force resisting system". For columns deformation compatibility must be explicitly checked or confinement must be provided to ensure adequate deformation capacity. Similar provisions have been adopted in other international codes (including New Zealand); however, continued research is required given the challenges of evaluating both deformation demands and capacity accurately.

## 12.6 Lessons and Conclusions

Codes controlling the seismic design of buildings evolve over time, with the greatest advances often accompanying damage observations from severe earthquakes. Observations from Christchurch are expected to ultimately impact codes in New Zealand (DBH 2012) and internationally. This paper has provided some examples of observed damage to concrete buildings that relate directly to future research needs and potential changes for seismic design practice and codes.

Some of the important observations and recommendations are summarized below:

- The recorded ground motions in the Christchurch CBD exceed the 2,500 year design spectrum from the NZ Standard, NZS 1170, for many period ranges of interest.

- Despite the excessively strong shaking demand, performance of most modern concrete buildings generally exceeded the life safety objectives of the code.
- While many buildings met code objectives, an increasing number of concrete buildings are now being considered uneconomical to repair, once again raising the question if collapse prevention is the appropriate target performance level for the building code.
- Research is needed to determine the effects of high frequency vertical ground motion on building structures and how this shaking should be accounted for in design.
- Research is needed to better understand the brittle damage and failure mechanisms observed in walls. Typical wall damage indicates confinement may be required in regions of distributed web reinforcement and over a height exceeding the assumed plastic hinge length. Similarly provisions to limit axial load ratio and slenderness ratio should be evaluated.
- In several occasions bar fracture at critical sections was accompanied by limited flexural cracking. Research is needed to understand this failure mode and develop relevant code provisions.
- Displacement compatibility was shown to be a critical issue in (a) collapse of precast concrete stairs, (b) close-to-unseating state of precast floors due to beam elongation in the moment resisting frames, and (c) the severe damage of "gravity columns" not detailed for adequate ductility capacity. Refinement of code provisions is needed to ensure all systems are able to withstand expected displacement demands.

## References


ATC-58 (2011) 75 % Draft guidelines for the seismic performance assessment of buildings: ATC-58-1. Applied Technology Council, Redwood City

Beca (2011) Investigation into the collapse of the Forsyth Barr building stairs on 22nd. A report to the Department of Building and Housing (DBH). Beca Carter Hollings & Ferner Ltd, Auckland

Bull DK (2011) Stairs and access ramps between floors in multi-storey buildings. A report o the Canterbury earthquakes royal commission. Holmes Consulting Group, Christchurch

DBH (2005) Beware of limitations: cold-worked wire mesh. Department of Building and Housing (DBH), Wellington

DBH (2012) Technical investigation into the structural performance of buildings in Christchurch – final report. Department of Building and Housing (DBH), Wellington, January 2012

Dunning Thornton (2011) Report on the structural performance of the hotel grand chancellor in the earthquake on 22 February 2011. A report to the Department of Building and Housing (DBH). Dunning Thornton Consultants Ltd, Wellington

Kam WY, Pampanin S, Dhakal RP, Gavin H, Roeder CW (2010) Seismic performance of reinforced concrete buildings in the September 2010 Darfield (Canterbury) earthquakes. Bull N Z Soc Earthq Eng 43(4):340–350

Kam WY, Pampanin S, Elwood K (2011) Seismic performance of reinforced concrete buildings in the 22 February Christchurch (Lyttelton) earthquake. Bull N Z Soc Earthq Eng 44(4):239–279

King AB, Bull DK, McVerry, GH, Jury RD (2003) The Australia/New Zealand earthquake loadings standard, AS/NZS 1170.4. In: Proceedings of the 2003 Pacific conference on earthquake engineering, New Zealand Society for Earthquake Engineering, Christchurch, New Zealand. Paper number 138, 13–15 Feb 2003

NZS1170 (2004) NZS 1170:2004 structural design actions. Standards New Zealand, Wellington

NZS3101 (1982) NZS 3101:1982 concrete structures standard. Standards New Zealand, Wellington

NZS3101 (2006) NZS 3101:2006 concrete structures standard. Standards New Zealand, Wellington

NZSEE (2009) Building safety evaluation during a state of emergency. Guidelines for territorial authorities, 2nd edn. New Zealand Society for Earthquake Engineering (NZSEE), Wellington

Paulay T, Priestley MJN (1993) Stability of ductile structural walls. ACI Struct J 90(4). American Concrete Institute, pp. 385–392

Stirling MW, McVerry GH, Berryman KR (2002) A new seismic hazard model for New Zealand. Bull Seismol Soc Am 92(5):1878–1903

# Part IV
# Vision in Japan and Asia

# Chapter 13
# Seismic Performance of a Bridge Column Based on E-Defense Shake-Table Excitations

**Kazuhiko Kawashima, Richelle G. Zafra, Tomohiro Sasaki, Koichi Kajiwara, and Manabu Nakayama**

**Abstract** As part of a major study on the seismic response of bridges by the National Research Institute for Earth Science and Disaster Prevention (NIED), Japan, a full-scale column incorporating an advanced material – polypropylene fiber reinforced cement composites (PFRC) at the plastic hinge region and part of the footing was recently tested on the E-Defense shake-table of NIED. The column was subjected to three components of the near-field ground motion recorded at the JR Takatori station during the 1995 Kobe, Japan earthquake. Excitations were repeated under increased mass and increased intensity of ground motion. After six times of excitation, experimental results showed that use of PFRC substantially mitigated cover concrete damage and local buckling of longitudinal bars. Measured strains of tie reinforcements and cross-ties at the plastic hinge were also smaller. Moreover, there was no visible damage in the core concrete at the plastic hinge after the series of excitations. The damage sustained by the column using PFRC was much less than the damage of regular reinforced concrete columns.

**Keywords** Bridges • Polypropylene fiber reinforced cement composites • E-Defense • Shake-table experiment • Seismic design • Seismic response

---

K. Kawashima (✉)
Department of Civil Engineering, Tokyo Institute of Technology, O-Okayama, Meguro, Tokyo, Japan
e-mail: kawashima@kba.biglobe.ne.jp

R.G. Zafra
Department of Civil Engineering, University of the Philippines Los Baños, Laguna 4031, Philippines
e-mail: rgzafra@gmail.com

T. Sasaki • K. Kajiwara • M. Nakayama
Hyogo Earthquake Engineering Research Center, National Research Institute for Earth Science and Disaster Prevention, 1501-21 Nishikameya, Mitsuta, Shijimicho, Miki, Hyogo 673-0515, Japan
e-mail: tomo_s@bosai.go.jp; kaji@bosai.go.jp; mnakaya@ba.kobegakuin.ac.jp

M. Fischinger (ed.), *Performance-Based Seismic Engineering: Vision for an Earthquake Resilient Society*, Geotechnical, Geological and Earthquake Engineering 32,
DOI 10.1007/978-94-017-8875-5_13, 

## 13.1 Introduction

Bridges are vital components of transportation facilities and they are vulnerable to seismic effects. Extensive damage of bridges occurred in past earthquakes such as the 1989 Loma Prieta, USA earthquake, 1994 Northridge, USA earthquake, 1995 Kobe, Japan earthquake, 1999 Chi Chi, Taiwan earthquake, 2008 Wenchuan, China earthquake, 2010 Maule, Chile earthquake and recently the 2011 East Japan earthquake.

In 2007–2010, a large-scale bridge experimental program was conducted by the National Research Institute for Earth Science and Disaster Prevention (NIED), Japan (Nakashima et al. 2008). In the program, shake table experiments were conducted for two typical reinforced concrete columns which failed during the 1995 Kobe, Japan earthquake (C1-1 and C1-2 experiments), a typical reinforced concrete column designed in accordance with the 2002 Japan design code (JRA 2002) (C1-5 experiment) and a new generation column using polypropylene fiber reinforced cement composites (PFRC) for enhancing the damage control and ductility (C1-6 experiment) (Kawashima et al. 2012).

Prior to the C1-6 experiment, a series of bilateral cyclic loading experiments were conducted on a 1.68 m high, 0.4 m by 0.4 m square cantilever reinforced concrete column and a column each using steel fiber reinforced concrete (SFRC) and PFRC at the plastic hinge region and the footing for deciding the material of C1-6 column (Kawashima et al. 2011). The column using PFRC had superior performance than the other columns due to the substantial mitigation of cover and core concrete damage, longitudinal bar buckling and deformation of tie bars at the plastic hinge region resulting from the high tensile strain capacity of PFRC which delays the propagation and widening of cracks and the high compression strain capacity of PFRC which avoids loss of integrity of cover concrete by crushing and spalling. PFRC is a type of engineered cementitious composites (ECC), belonging to the class of high performance fiber reinforced cement composites (HPFRCC). HPFRCCs exhibit multiple fine cracks upon loading in tension which leads to improvement in toughness, fatigue resistance and deformation capacity of structures (Matsumoto and Mihashi 2003; Parra-Montesinos 2005). As a result, C1-6 column was built using PFRC at the plastic hinge region and part of the footing. This paper presents the results of C1-6 column experiment.

## 13.2 E-Defense Shake-Table Excitations

### *13.2.1 Column Configuration and Properties*

Figure 13.1 shows C1-6 column which is a 7.5 m tall, 1.8 m by 1.8 m square, cantilever column. It was designed based on the 2002 Japan Specifications for

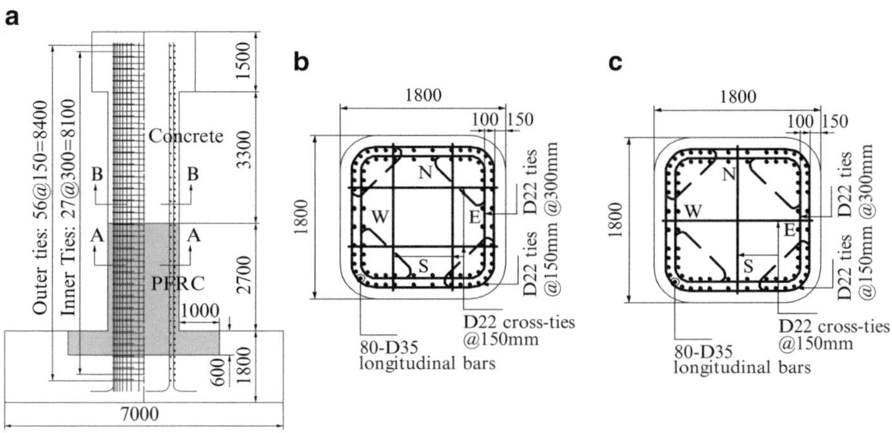


**Fig. 13.1** C1-6 column configuration and details (**a**) Reinforcement detail (**b**) Section A-A (**c**) Section B-B

Highway Bridges assuming moderate soil condition under the Type II design ground motion (near-field ground motion). PFRC was used at a part of the column with a depth of 2.7 m above the column base and a part of the footing with a depth of 0.6 m below the column base to minimize the volume of PFRC. The 2.7 m depth of PFRC is three times the estimated plastic hinge length of one-half the column width (JRA 2002) corresponding to 0.9 m to avoid failure at the PFRC-concrete interface. The 0.6 m depth of PFRC at the footing was decided to minimize damage. Regular concrete was used in the other parts of the column. The 2.7 m depth at the column and 0.6 m depth at the footing may be reduced after careful examination of damage at the PFRC-concrete interface.

The design compressive strength of PFRC was 40 MPa. PFRC was made by combining cement mortar, fine aggregates with maximum grain size of 0.30 mm, water and 3 % volume of polypropylene fibers. Fibrillated polypropylene fibers with diameter of 42.6 μm, length of 12 mm, tensile strength of 482 MPa, Young's modulus of 5 GPa and density of 0.91 kg/m$^3$ were used (Hirata et al. 2009). Superplasticizers were added to improve the workability of the mix.

The longitudinal and tie bars had nominal yield strength of 345 MPa (SD345). Eighty-35 mm diameter deformed longitudinal bars were provided in two layers corresponding to a reinforcement ratio $\rho_l$ of 2.47 %. Deformed 22 mm diameter ties with 135° bent hooks lap spliced with 40 times the bar diameter were provided. Outer ties were spaced at 150 mm and inner ties were spaced at 300 mm throughout the column height. Cross-ties with 180° hooks at 150 mm spacing were provided to increase confinement of the square ties. Volumetric tie reinforcement ratio $\rho_s$ within a height of 2.7 m from the column base was 1.72 %.

**Fig. 13.2** Experiment set-up using the E-Defense shake-table

### *13.2.2 Experiment Set-Up and Shake-Table Excitations*

Figure 13.2 shows the experiment set-up using the E-Defense shake table. Four mass blocks were set on the column through two simply supported decks. The decks were used to fix the mass blocks to the column but were not designed to idealize the stiffness and strength of real decks. Each deck was supported by the column on one side and by the steel end support on the other side.

A 78 ton (765 kN) mass block and a 45 ton (441 kN) mass block were fixed to each deck as close to the column as possible so that tributary weight in the transverse direction could be maximum. The total weight consisting of four mass blocks, two decks, two fixed bearings, two movable bearings, eight side sliders and 32 load cells was 307 ton (3,012 kN). Note that the tributary weight which generated the inertia force in the column in the transverse direction was 215 ton (2,109 kN), about 2/3 of the total weight. The total weight of the entire experiment set-up including the column and the two end-supports was 1,069 ton (10.5 MN), which was close to the payload of 1,200 ton (12 MN).

The column was excited using the near-field ground motion recorded at the JR Takatori Station during the 1995 Kobe earthquake. Although duration was short, it was one of the most destructive ground motions to structures with peak ground acceleration (PGA) of 0.62 g and peak ground velocity (PGV) of 1.19 m/s in the fault-normal direction (JRTRI 1999). Because the energy dissipation of a column anchored to a shake table is extremely less than the real energy dissipation of a column embedded in the ground (Sakai and Unjoh 2006), a ground motion with 80 % of the original intensity of the JR Takatori record was imposed as a command to the table in the experiment to take into account the effect of soil-structure

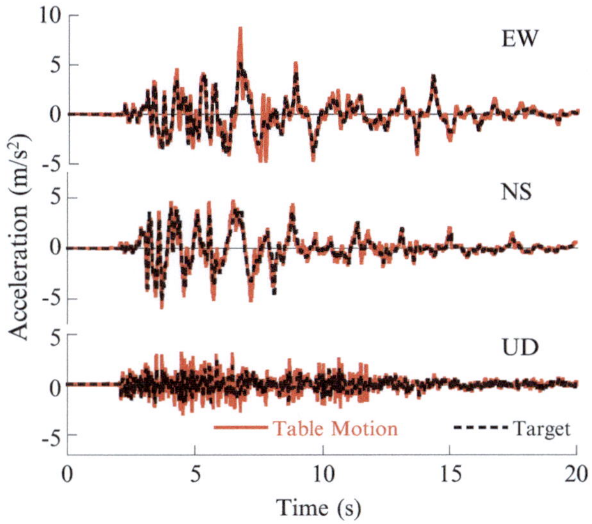


**Fig. 13.3** E-Takatori ground motion

interaction. This ground motion is called the 100 % E-Takatori ground motion. Figure 13.3 shows both the target and the recorded table acceleration during the first 100 % E-Takatori ground motion excitation. The EW, NS and UD components of the 100 % E-Takatori ground motion were applied in the longitudinal, transverse and vertical directions, respectively, of the bridge model.

Shake table excitations were conducted six times. Excitations were repeated to clarify column performance when subjected to much stronger and longer duration near-field ground motion. The column was excited twice with 100 % E-Takatori ground motion (1–100 %(1) and 1–100 %(2) excitations). After the mass in the longitudinal direction was increased by 21 % from 307 ton to 372 ton, excitations were conducted with 100 % E-Takatori ground motion once (2–100 % excitation) and 125 % E-Takatori ground motion three times (2–125 %(1), 2–125 %(2) and 2–125 %(3) excitations).

## 13.3 Column Seismic Performance

### *13.3.1 Damage Progress*

Figures 13.4, 13.5 and 13.6 show the damage progress within 1.2 m from the column base at the SW and NE corners during 1–100 %(1), 2–100 % and 2–125 %(3) excitations at the instance of peak response displacement where the SW corner was subjected to compression while the NE corner was subjected to tension. As shown in Fig. 13.4, during 1–100 %(1) excitation, only micro cracks were observed around the column. Although photograph during 1–100 %(2) excitation is not shown here,

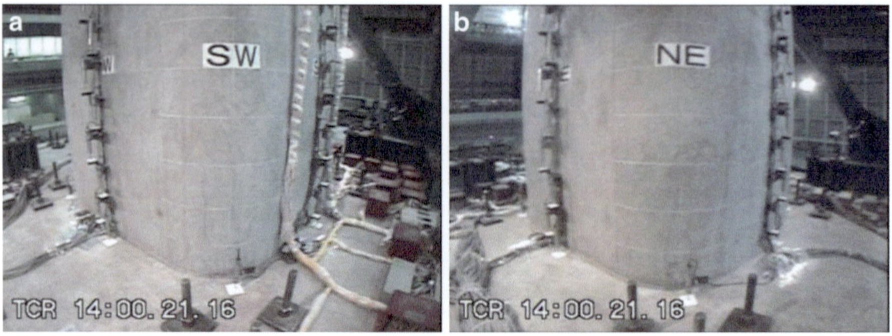


**Fig. 13.4** Damage of C1-6 column during 1–100 %(1) excitation (**a**) SW corner (**b**) NE corner

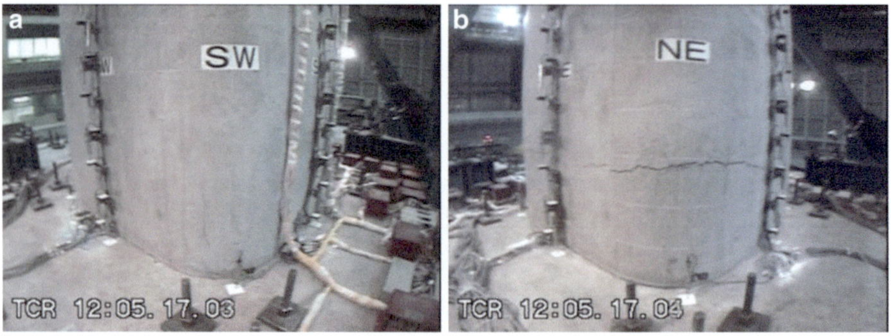


**Fig. 13.5** Damage of C1-6 column during 2–100 % excitation (**a**) SW corner (**b**) NE corner

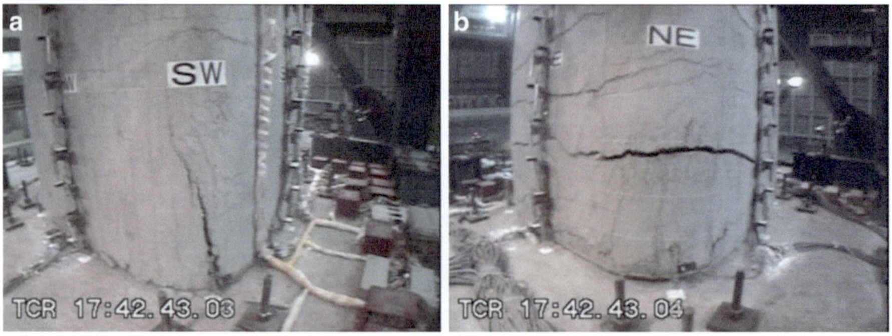


**Fig. 13.6** Damage of C1-6 column during 2–125 %(3) excitation (**a**) SW corner (**b**) NE corner

very thin flexural cracks as wide as 0.1–0.2 mm occurred within 1.6 m from the base all around the column.

During 2–100 % excitation, with the mass increased by 21 %, damage progressed as shown in Fig. 13.5. Flexural cracks propagated and a crack 0.6 m from the column base at the NE corner opened about 8 mm at the peak response displacement occurring at 6.78 s. After the excitation, the maximum residual crack at the above location was 1–2 mm wide. Although only flexural cracks occurred all around the column with the cover concrete remaining as a whole shell due to the bridging action of fibers, vertical hairline cracks started to occur at the NE and SW corners within 0.6 m from the column base due to the large strut action of cover concrete shell resulting from the footing reaction when the column was laterally displaced.

During 2–125 %(1) excitation, in which the seismic excitation intensity was increased by 25 %, at the peak response displacement which occurred at 6.97 s, the crack 0.6 m from the base opened to 14 mm at the NE corner which was subjected to tension while a vertical crack opened to 9 mm at the opposite SW corner subjected to compression. As the loading progressed, at the SW corner subjected to tension, a crack 1.2 m from the base opened to 9 mm and vertical cracks started to widen at the opposite NE corner.

Succeeding excitations resulted to further propagation of flexural cracks within 2 m from the base around the column and the widening of the vertical crack at the SW corner. As shown in Fig. 13.6, the damage progressed during 2–125 %(3) excitation wherein at the peak response displacement at 7.07 s, the crack 0.6 m from the base at the NE corner opened to 20 mm and the vertical crack at the SW corner opened to 15 mm. Note that at the NW corner, cover concrete spalled within 200 mm from the column base when it was subjected to compression while flexural cracks opened to 13 mm at the opposite SE corner subjected to tension. After the excitation, the cracks which opened to over 10 mm during the excitation almost closed with widths of only 5–8 mm in flexural cracks and 7–12 mm in vertical cracks. Moreover, majority of other small cracks closed to hairline cracks after the excitations due to the fiber bridging action of fibers. Cover concrete spalling was much restricted and there were no exposed longitudinal bars and ties in C1-6 column after 2–125 %(3) excitation.

To investigate how damage progressed in the core and the longitudinal bars after the last excitation, cover concrete was removed at the SW and NE corners using an electric drill and saw. Removal of cover concrete in the fiber mixed concrete was difficult due to the presence of fibers compared to that of regular reinforced concrete. Figure 13.7a shows the opened area at the NE corner after the outer ties were removed to facilitate inspection of the outer and inner longitudinal bars for local buckling. Three outer longitudinal bars buckled in between outer ties at 250 mm and 550 mm from the base. Note that ties at these locations have double tie area because of the 40 times bar lap splice and development of the 135° hook which increased the tie constraint, resulting to buckling of bars in between the 250 mm and 550 mm ties. The maximum lateral offset among the three longitudinal bars from their original vertical axis was 8 mm. On the other hand, the inner longitudinal bars did not buckle because they were constrained by the undamaged concrete between the outer and inner longitudinal bars.

Figure 13.7b shows that at the location where a crack opened to 20 mm, the crack occurred only in the cover concrete with a depth of 110 mm and did not propagate into the core concrete. Figure 13.7c shows the block of cover concrete that was removed at the bottom right portion of the NE corner where the presence of fibers

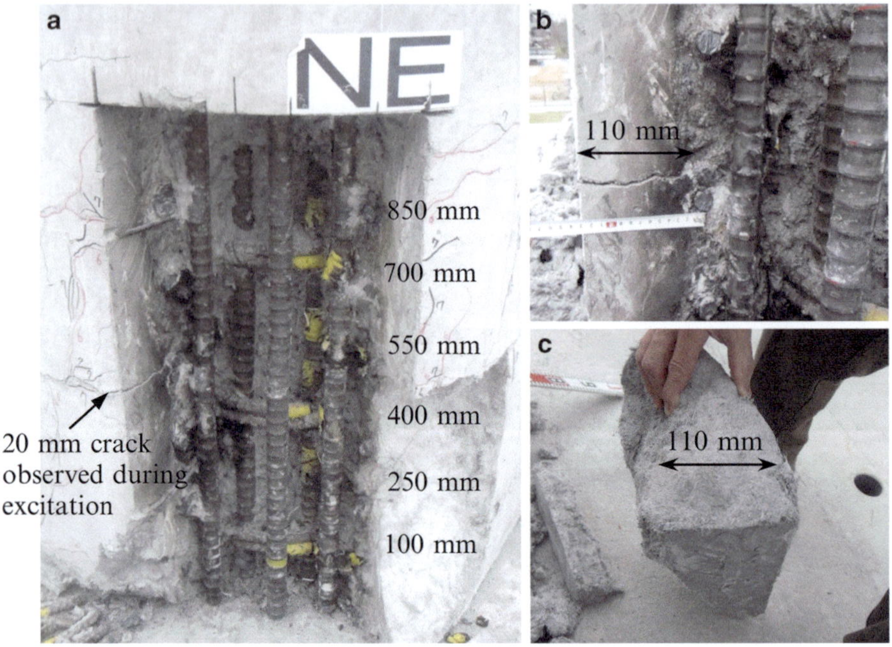


**Fig. 13.7** Longitudinal bar buckling and damage of PFRC cover concrete at the NE corner after 2–125 %(3) excitation (**a**) Opened area (**b**) Crack on cover concrete (**c**) Cover concrete block

held the cover concrete together preventing the disintegration of cover concrete. Hence, it is worthy to note that even after six times of excitation, the damage sustained by C1-6 column was much less than the damage of regular reinforced concrete columns.

### *13.3.2 Strains of Longitudinal and Tie Bars*

Figure 13.8 shows the axial strains of longitudinal and tie bars of C1-6 column at the plastic hinge zone (300–400 mm from the base) at the SW corner where the most extensive damage occurred. Only strains during 1–100 %(1), 2–100 %, 2–125 %(1) and 2–125 %(3) excitations are shown due to space limitation. Because longitudinal bars were set in two layers as shown in Fig. 13.1, strains of both the outer and inner longitudinal bars and tie bars are shown here. Noting that the yield strain of both longitudinal and tie bars was nearly 2,000 $\mu$, the longitudinal bars started to yield in tension during 1–100 %(1) while tie bars started to yield in tension during 2–125 %(1) excitation. The outer and inner longitudinal bars and tie bars exhibited similar response however the amplitude of strains were generally larger in the outer longitudinal and tie bars than the respective inner longitudinal

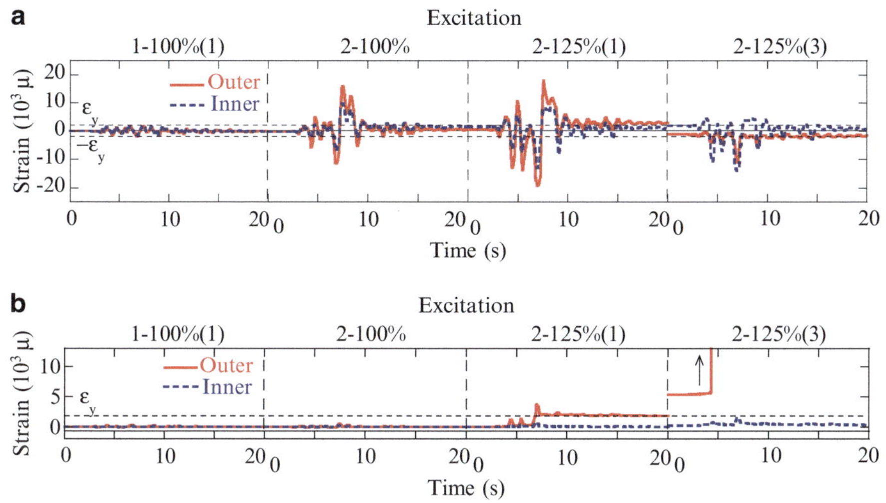


**Fig. 13.8** Axial strains of longitudinal bars and tie bars at the SW corner during 1–100 %(1), 2–100 %, 2–125 %(1) and 2–125 %(3) excitations (**a**) Longitudinal bars at 300 mm from the base (**b**) Tie bars at 400 mm from the base

and tie bars. The difference of strain amplitude between outer and inner tie bars is particularly large during and after 2–125 %(1) excitation resulting from local buckling of longitudinal bars, which will be described later.

An interesting point in Fig. 13.8 is that the compression strains of the outer and inner longitudinal bars became larger than tension strains during and after 2–125 %(1) excitation. For example, compression strain of the outer longitudinal bar reached 19,000 $\mu$ while tension strain reached 18,000 $\mu$ during 2–125 %(1) excitation. This obviously resulted from the low elastic modulus of PFRC. The large compression strain must have caused the longitudinal bar to buckle. However, in spite of the bar buckling, as described in Sect. 13.3.1, spalling of cover concrete did not occur indicating that the presence of fibers made the cover concrete remain as a whole shell.

On the other hand, the tie bar was still elastic during 1–100 %(1) until 2–100 % excitations. At 6.97 s when compression strain of the outer longitudinal bar sharply increased during 2–125 %(1) excitation, the outer tie strain started to increase to 3,700 $\mu$, indicating that the tie resisted the longitudinal bar buckling. Compression strain of the inner longitudinal bar also sharply increased at the same time, however, the inner tie strain did not increase indicating that the inner longitudinal bar did not buckle. This is because confinement for bar buckling was larger at the inner longitudinal bar than the outer longitudinal bar due to the resistance of core concrete between outer and inner ties which was still intact as shown in Fig. 13.7.

Figure 13.9 shows the interaction of a longitudinal bar with a tie bar for outer and inner bars. The tie strains during 2–125 %(3) excitation were larger than 5,000 $\mu$ and only reliable data are shown here. A sharp increase of the outer tie strain resulting

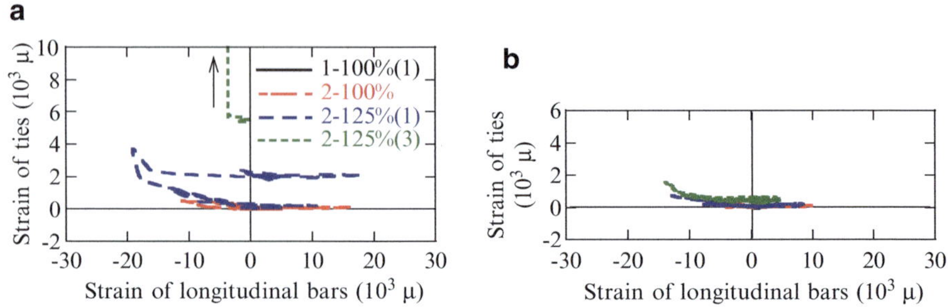


**Fig. 13.9** Strain of a tie at 400 mm from the base vs. strain of a longitudinal bar at 300 mm from the base at the SW corner (**a**) Outer bar (**b**) Inner bar

from restraining local buckling of the outer longitudinal bar under high compression strain is clearly seen during and after 2–125 %(1) excitation while the inner tie strain remained below 2,000 $\mu$ because inner longitudinal bars did not yet buckle.

### *13.3.3 Response Acceleration and Displacement*

Figure 13.10 shows the response acceleration and displacement at the top of column in the principal response direction and Table 13.1 summarizes the peak acceleration, displacement, residual displacement and moment at each excitation. The principal response direction is defined as the direction in which the response displacement was maximum. It is seen that the response acceleration has similar shape with the input ground acceleration at early excitations. However, in later excitations, the response acceleration tends to have almost uniform amplitude during the excitation, if several spikes with large amplitudes are eliminated, and this is due to the nonlinear response of the columns.

Due to the high acceleration pulse in the input ground motion, the column experienced high amplitude displacement during each excitation. The peak response displacement was equal to 0.078 m (1.0 % drift) during 1–100 %(1) excitation and progressed to 0.45 m (6.0 % drift) during 2–125 %(3) excitation. As the excitation progressed with increasing mass and intensity of ground motion, the response displacements increased due to column stiffness deterioration resulting from the damage. Residual displacement was only –0.004 m (0.05 % drift) after 2–100 % excitation, increased to –0.037 m (0.49 % drift) after 2–125 %(2) excitation then decreased to –0.013 m (0.13 % drift) after the last excitation. It is important to note that residual displacement not only increases but also decreases during seismic excitations. Because instantaneous stiffness vary, instantaneous period also vary which causes changes in the residual displacement (MacRae and Kawashima 1997). Since the allowable residual drift for a cantilever column based on the 2002 JRA code is 1 %, the residual drift of the column was still smaller than the allowable limit.

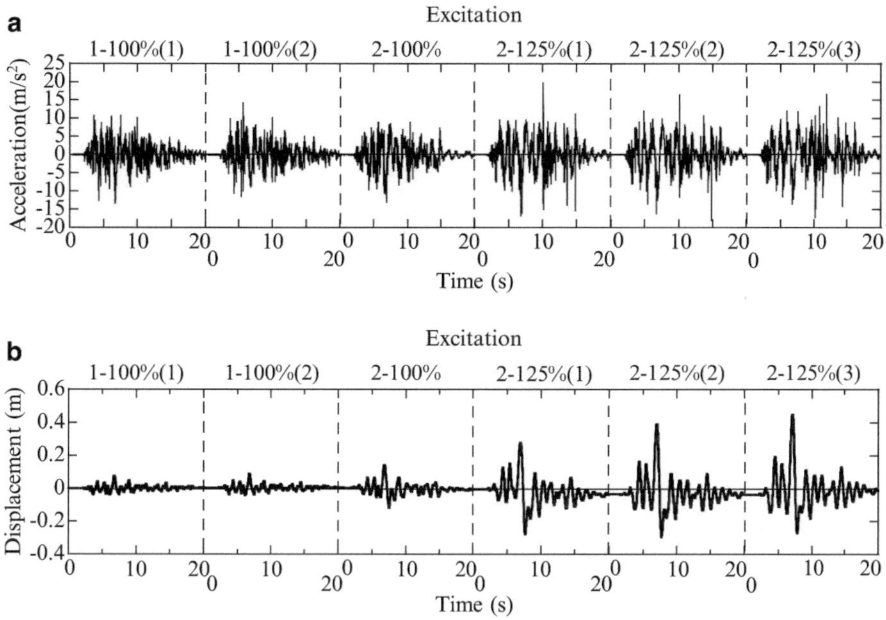


**Fig. 13.10** Response acceleration and displacement in the principal direction (**a**) Response acceleration (**b**) Response displacement

**Table 13.1** Column response in the principal direction

| Excitation | Response acceleration (m/s$^2$) | Response displacement (m) | Response displacement (% drift) | Residual displacement (m) | Moment (MNm) |
|---|---|---|---|---|---|
| 1–100 %(1) | −13.4 | 0.078 | 1.0 | 0.005 | 20.5 |
| 1–100 %(2) | 14.2 | 0.089 | 1.2 | 0.007 | 21.8 |
| 2–100 % | −13.0 | 0.144 | 1.9 | −0.004 | 24.0 |
| 2–125 %(1) | 19.9 | 0.280 | 3.7 | −0.035 | 24.3 |
| 2–125 %(2) | −17.9 | 0.392 | 5.2 | −0.037 | 25.3 |
| 2–125 %(3) | −17.1 | 0.450 | 6.0 | −0.013 | 24.9 |

### *13.3.4 Moment and Ductility Capacity*

Figure 13.11 shows the hysteresis of moment at the base vs. displacement at the top of the column in the principal response direction. The hysteresis during the entire six times of excitation is stable with sufficient energy dissipation. As summarized in Table 13.1, the peak moment gradually increased as the excitation progressed. A maximum capacity of 25.3 MNm at 5.2 % drift was developed during 2–125 %(2) excitation. During this excitation, flexural cracks further propagated all around the column and the vertical cracks at the SW corner widened as described in Sect. 13.3.1. During the subsequent 2–125 %(3) excitation, the peak drift increased

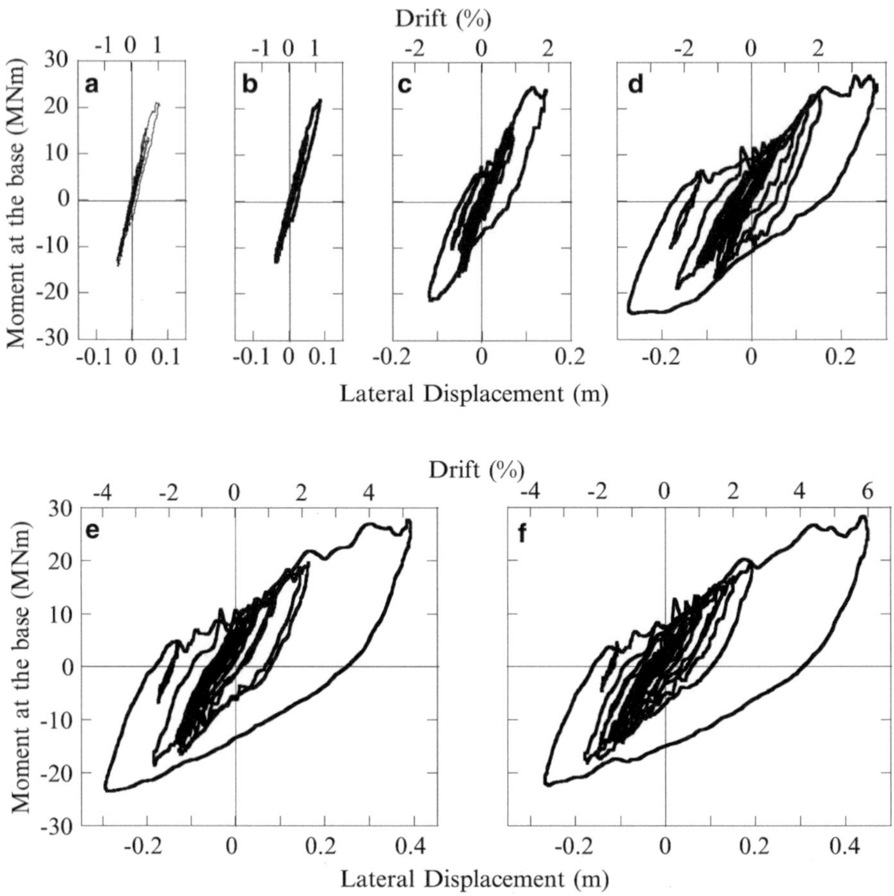


**Fig. 13.11** Hysteresis of moment at the base vs. displacement at the top of column in the principal direction (**a**) 1–100 %(1) (**b**) 1–100 %(2) (**c**) 2–100 % (**d**) 2–125 %(1) (**e**) 2–125 %(2) (**f**) 2–125 %(3)

to 6 % while the peak moment slightly deteriorated by 2 %. It should be noted that even during the 2–125 %(3) excitation, the moment vs. lateral displacement hysteresis was still very stable.

## 13.4 Conclusions

The seismic performance of a full-scale bridge column using polypropylene fiber reinforced cement composites (PFRC) subjected to near-field ground motions was investigated through shake table experiments. Based on the results presented, the following conclusions were obtained:

(a) Under a strong earthquake, the use of PFRC substantially reduced the apparent damage which can allow the bridge to be serviceable.
(b) PFRC did not have the brittle compression failure of regular reinforced concrete under repeated large inelastic deformation due to the bridging mechanism of fibers.
(c) As a consequence of (b), use of PFRC mitigated the buckling of outer longitudinal bars and the deformation of outer and inner tie bars. No visible buckling of inner longitudinal bars occurred due to the intact PFRC between outer and inner longitudinal bar layers.
(d) Because the PFRC cover concrete of C1-6 column did not spall in a brittle manner compared to standard reinforced concrete columns, the cover concrete resisted the compression from the footing reaction at the base due to strut action as shell component, although vertical cracks occurred on the cover concrete.
(e) As a result of the damage mitigation properties of PFRC, the column had a stable flexural capacity and enhanced ductility reaching until 6 % drift.

## References


Hirata T, Kawanishi T, Okano M, Watanabe S (2009) Study on material properties and structural performance of high-performance cement composites using polypropylene fiber. Proc Japan Concr Inst 31(1):295–300

Japan Railway Technical Research Institute (1999) West Japan Railway Company: JR Seismic Data. Kokubunji, Japan

Japan Road Association (2002) Specifications for Highway Bridges. Maruzen, Tokyo

Kawashima K, Zafra R, Sasaki T, Kajiwara K, Nakayama M (2011) Effect of polypropylene fiber reinforced cement composite and steel fiber reinforced concrete for enhancing the seismic performance of bridge columns. J Earthq Eng 15(8):1194–1211

Kawashima K, Zafra R, Sasaki T, Kajiwara K, Nakayama M, Unjoh S, Sakai J, Kosa K, Takahashi Y, Yabe M (2012) Seismic performance of a full-size polypropylene fiber reinforced cement composite bridge column based on E-Defense shake-table experiments. J Earthq Eng 16:463–495

MacRae G, Kawashima K (1997) Post-earthquake residual displacements of bilinear oscillators. Earthq Eng Struct Dyn 26:701–716

Matsumoto T, Mihashi H (2003) DFRCC terminology and application concepts. J Adv Concr Technol JCI 1(3):335–340

Nakashima M, Kawashima K, Ukon H, Kajiwara K (2008) Shake table experimental project on the seismic performance of bridges using E-Defense. In: 14th world conference on earthquake engineering, Beijing, China. Paper S17-02-010 (CD-ROM)

Parra-Montesinos G (2005) High-performance fiber-reinforced cement composites: an alternative for seismic design of structures. ACI Struct J 102(5):668–675

Sakai J, Unjoh S (2006) Earthquake simulation test of circular reinforced concrete bridge column under multidirectional seismic excitation. Earthq Eng Eng Vib 5(1):103–110

# Chapter 14
# Development of Building Monitoring System to Verify the Capacity Spectrum Method

**Koichi Kusunoki, Akira Tasai, and Masaomi Teshigawara**

**Abstract** Due to Kobe Earthquake (1995, M7.3), 6,434 people were killed, and 104,906 buildings were totally collapsed. After Kobe Earthquake, 17 large earthquakes occurred in Japan, which include Niigata Cyuetsu Earthquake (M7.2) in 2005 and the 2011 Off the Pacific Coast of Tohoku Earthquake (M9.0) in 2011. At the same time, a lot of earthquake ground motion data were measured by sensors. Some earthquakes, however, caused only slight damage to reinforced concrete structures although many earthquake records with large PGAs were measured. A monitoring building response is required to find out the reason of disagreement between analysis result and observation. In this paper, a building monitoring system with inexpensive sensors is proposed and the validity of the system is confirmed with an actual response of an instrumented building during the 2011 Off the pacific Coast of Tohoku Earthquake.

**Keywords** Health monitoring • Quick inspection • Aftershocks • Performance-based design • Building response • Tohoku earthquake

---

K. Kusunoki (✉)
Division of Disaster Mitigation Science, Earthquake Research Institute, The University of Tokyo 1-402, 1-1-1 Yayoi, Bunkyo-ku, Tokyo, #113-0032, Japan
e-mail: kusunoki@eri.u-tokyo.ac.jp

A. Tasai
Department of Architecture and Urban Culture, Yokohama National University, 79-5, Tokiwadai, Hodogaya Ku, Yokohama, Kanagawa, Japan
e-mail: tasai@ynu.ac.jp

M. Teshigawara
Department of Architecture, Yokohama National University, ES Building, Furo Cyo, Chikusa Ku, Nagoya, Aichi, Japan
e-mail: teshi@corot.nuac.nagoya-u.ac.jp

M. Fischinger (ed.), *Performance-Based Seismic Engineering: Vision for an Earthquake Resilient Society*, Geotechnical, Geological and Earthquake Engineering 32,
DOI 10.1007/978-94-017-8875-5_14, 

## 14.1 Introduction

Due to Kobe Earthquake (1995, M7.3), 6,434 people were killed, and 104,906 buildings were totally collapsed. The maximum PGA of 848 gal was measured at Kobe Station of Japan Meteorological Agency. After that, number of sensor was drastically increased. Nowadays, there are 4,400 seismic intensity observation points (150 before Kobe Earthquake) and Kyoshin Net (K-Net) of NIED has about 1,000 strong motion observation stations in Japan.

After Kobe Earthquake, 17 large earthquakes occurred in Japan, which include Niigata Cyuetsu Earthquake (M7.2) in 2005 and the 2011 Off the pacific Coast of Tohoku Earthquake (M9.0) in 2011. At the same time, a lot of earthquake ground motion data were stored measured by sensors. Some earthquakes caused severe damage of buildings and killed some people. Some earthquakes, however, caused only slight damage to reinforced concrete structures although many earthquake records with large PGAs were measured.

JCI (Japan Concrete Institute) organized 'a research committee for seismic performance evaluation of reinforced concrete under recent earthquakes' and a report was published in 2004. The comparison of the earthquake records and observed damages in structures were made in the report. It is mentioned in the conclusions that numerical analysis of records showed there must be but no buildings were observed severely damaged and monitoring building response is required to find out the reason of disagreement between analysis result and observation.

In order to promote building monitoring system, it must be attractive to owners. Therefore, a residual seismic capacity evaluation system with building monitoring system is proposed in this paper. Currently, buildings have to be investigated one by one by engineers or researchers after an earthquake. For example, 5,068 engineers and 19 days were needed to investigate 46,000 buildings on a damaged area at the Kobe earthquake (BCJ 1996). Nineteen days were too long and yet the number of investigated buildings was not enough. Moreover, many buildings were judged as "Caution" level, which needs detailed investigation by engineers. "Caution" judgment is a gray zone and it could not take away anxieties from inhabitants. Furthermore, the current quick investigation system presents a dilemma since buildings should be investigated by visual observation of engineers. Thus, judgment varies according to engineers' experience. In order to solve these problems, building monitoring system is proposed in this research.

## 14.2 Configuration of the System and Outline of the Evaluation

The system has basically one accelerometer on each floor and one judgement machine as shown in Fig. 14.1. The evaluation method is based on the performance design concept as shown in Fig. 14.2. The residual seismic capacity will be judged

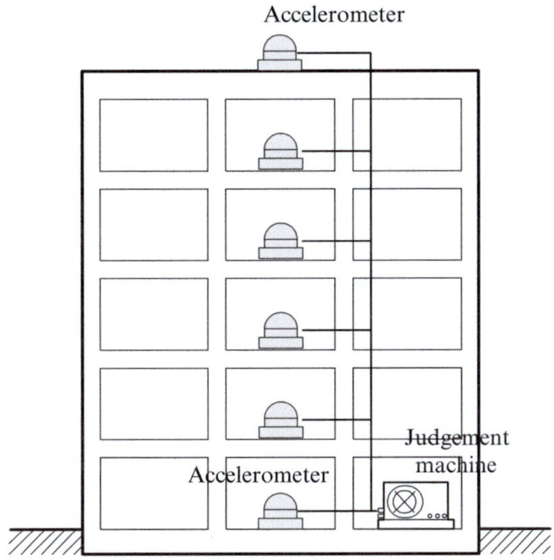


**Fig. 14.1** Configuration of the system

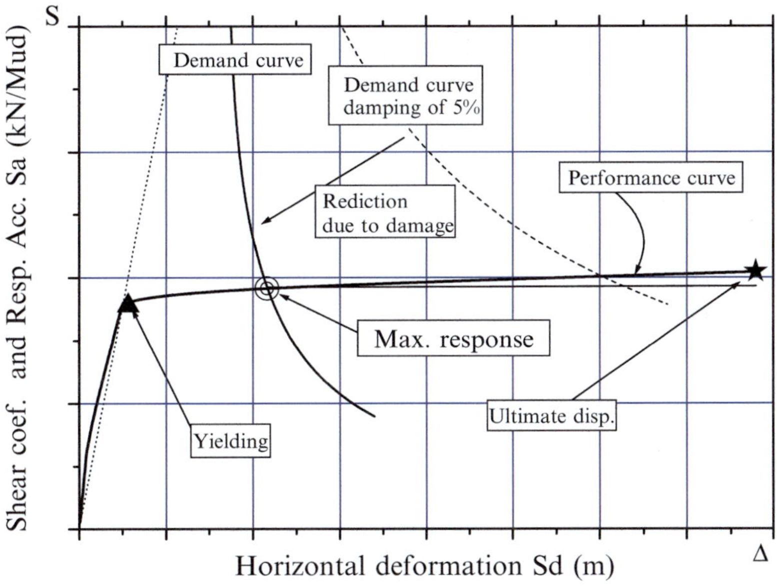


**Fig. 14.2** Outline of the evaluation based on the performance design concept

by comparing the measured performance curve of a structure and the measured demand curve (Kusunoki and Teshigawara 2003, 2004).

The performance curve is the relationship between the representative deformation $\Delta$ and the representative restoring force S, which shows the predominant response of a structure. The method to evaluate theses representative values is outlined below.

The calculated relative displacement vector at the basement $\{_M x\}$ from measured accelerations can be derived as (14.1) with the modal participation factor $_M\beta$, mode vector $\{_M u\}$, and the assumption that the $\{_M x\}$ is the unique vibration mode.

$$\{_M x\} = {}_M\beta \cdot \{_M u\} \cdot \Delta \tag{14.1}$$

The story shear (inertia force) of the first story $_M Q_B$ can be calculated using (14.2) with the measured absolute acceleration $\{_M\ddot{x} + \ddot{x}_0\}$ and mass $m_i$ of each floor.

$$_M Q_B = \sum m_i \cdot ({}_M\ddot{x}_i + \ddot{x}_0) \tag{14.2}$$

The equation of motion of a multi-degree-of-freedom system can be simplified down to a single- degree-of-freedom system as given in (14.3).

$$M \cdot \ddot{\Delta} + {}_M\tilde{C} \cdot \dot{\Delta} + {}_M\tilde{K} \cdot \Delta = -M \cdot \ddot{x}_0 \tag{14.3}$$

where, $M$ is the total mass of a structure, $_M\tilde{C}$ is the equivalent damping, $_M\tilde{K}$ is the equivalent stiffness, and $\ddot{x}_0$ is the ground acceleration, respectively.

The $_M Q_B$ can be calculated with (14.4). If the first mode is predominant enough, the calculated angular frequency, $_M\omega = \sqrt{\frac{_M\tilde{K}}{M}}$, can be the natural angular frequency of the first mode.

$$S = {}_M Q_B = M \cdot \ddot{\Delta} \tag{14.4}$$

Equation (14.5) can be derived from (14.1) by dividing both sides by $\Delta$. The inertia force acting on each floor $_M P_i$ can be derived as (14.6) by using (14.1) and (14.5).

$$_M\beta \cdot {}_M u_i = \frac{_M x_i}{\Delta} \tag{14.5}$$

$$_M P_i = m_i \cdot {}_M\beta \cdot {}_M u_i \cdot \ddot{\Delta} = m_i \cdot \ddot{\Delta} \cdot \frac{_M x_i}{\Delta} \tag{14.6}$$

The total mass $M$ can also be derived from (14.4) and (14.6), since the total mass M is the sum of each floor mass, i.e.;

$$M = \frac{_M Q_B}{\ddot{\Delta}} = \frac{\sum m_i \cdot \ddot{\Delta} \cdot \frac{_M x_i}{\Delta}}{\ddot{\Delta}} = \frac{\sum m_i \cdot {}_M x_i}{\Delta} = \sum m_i \tag{14.7}$$

Therefore, the representative displacement $\Delta$ can be derived as (14.8a).

$$\Delta = \frac{\sum m_i \cdot {}_M x_i}{\sum m_i} \tag{14.8a}$$

The representative acceleration, $\ddot{\Delta}$ is applied to the representative restoring force, S $\left(S = {}_M Q_S / \sum m_i = \ddot{\Delta}\right)$. If a system is elastic, the representative displacement, $\Delta$ and the representative acceleration, $\ddot{\Delta}$ can be calculated with (14.8b), i.e.;

$$\ddot{\Delta}^2 + 2 \cdot {}_M h \cdot {}_M \omega \cdot \dot{\Delta} + {}_M \omega^2 \cdot \Delta = -\ddot{x}_0 \tag{14.8b}$$

where, ${}_M h$ is the damping coefficient, and ${}_M \omega$ is the angular frequency, respectively.

As a result, the maximum representative displacement $\Delta_{max}$ and the absolute acceleration $\left(\ddot{\Delta} + \ddot{x}_0\right)_{max}$ correspond to the value from the response displacement and acceleration spectrum with a damping coefficient of ${}_M h$.

On the other hand, the demand curve is the relationship between the response acceleration (Sa) and displacement (Sd) spectrum. The intersection point of the demand curve and performance curve shows the maximum elastic response. However, the damage of a structure can dissipate some amount of an input energy, thus the damping effect can be increased. Therefore, the demand curve can be reduced according to the damage (Fig. 14.2). The intersection point of the reduced demand and performance curves shows the maximum inelastic response.

## 14.3 Performance Curve Decomposition with the Wavelet Transform Method

If a structure has more than one predominant vibration mode, the response of the structure cannot be estimated as a single-degree-of-freedom system, since the performance curve for the structure has more than one predominant slope, in other words, more than one predominant angular frequency. To overcome the problem with these kinds of structures, a method to decompose a calculated performance curve with the wavelet transform method (WTM) will be proposed in this section.

Recorded acceleration vector $\{{}_M\ddot{x} + \ddot{x}_0\}$ and integrated displacement from them $\{{}_M x\}$ are decomposed as (14.9) and (14.10) with the WTM.

$$\{{}_M x\} = \left\{ \sum_{i=1}^{N} g_{Disp,i} + f_{Disp,n} \right\} \tag{14.9}$$

$$\{{}_M\ddot{x} + \ddot{x}_0\} = \left\{ \sum_{i=1}^{N} g_{Accel,i} + f_{Accel,n} \right\} \tag{14.10}$$

where, $g_{Disp,i}$ and $g_{Accel,i}$ are components decomposed of rank $i$ of the displacement and acceleration, and $f_{Disp,i}$ and $f_{Accel,i}$ are eventual remaining of the displacement and acceleration, respectively.

$f_{Disp,i}$ and $f_{Accel,i}$ are generally error components, since they are single values and their periods are much longer than that of the structure. Thus, $f_{Disp,i}$ and $f_{Accel,i}$ can be ignored. The representative displacement and representative restoring force are decomposed as (14.11) and (14.12) using (14.9) and (14.10).

$$\Delta = \sum_r \frac{\sum_i m_i \cdot {}_i g_{disp,r}}{\sum_i m_i} \tag{14.11}$$

$$\ddot{\Delta} = \sum_r \frac{\sum_i m_i \cdot {}_i g_{Accel,r}}{\sum_i m_i} \tag{14.12}$$

Therefore, the representative displacement and representative restoring force for rank $r$, $\Delta_r$ and $\ddot{\Delta}_r$, which is a component decomposed by the WTM, are calculated by (14.13) and (14.14).

$$\Delta_r = \frac{\sum m_i \cdot {}_i g_{disp,r}}{\sum m_i} \tag{14.13}$$

$$\ddot{\Delta}_r = \frac{\sum m_i \cdot {}_i g_{Accel,r}}{\sum m_i} \tag{14.14}$$

As it is obvious from (14.13) and (14.14), the slope of the relationship between $\Delta_r$ and $\ddot{\Delta}_r$ is the square of the predominant angular frequency of rank $r$, ${}_r\omega^2$.

The number of rank decomposed by the WTM depends only on the number of data points, and the frequency range of each rank depends on the number of data points and sampling rate of the original signal. In other words, they are independent of the degree of freedom of the structure. Even if there are two modes in a rank decomposed by the WTM, it is impossible to separate these two modes numerically because of the uncertainty relation.

If no obvious correlation can be seen between $\Delta_r$ and $\ddot{\Delta}_r$, there is no predominant vibration mode in the frequency range of rank $r$. The maximum number of modes that the WTM is capable of decomposition is $n$, which is the number of ranks calculated by (14.15).

$$n = \log_2 N \tag{14.15}$$

The equivalent mass of rank $r$ ${}_rM$ is defined as (14.16). The representative displacement ${}_r\Delta'$ and the representative restoring force $-\left({}_r\ddot{\Delta}' + \ddot{x}_0\right)$ are calculated by (14.17) and (14.18) with the equivalent mass, respectively.

$$ {}_rM = \frac{\left(\sum m_i \cdot {}_ig_{disp,r}\right)^2}{\sum m_i \cdot {}_ig_{disp,r}^2} \tag{14.16} $$

$$ {}_r\Delta' = \frac{M}{{}_rM}\Delta_r \tag{14.17} $$

$$ \begin{aligned} {}_rQ &= -M \cdot g_{Accel,r} = -M \cdot \ddot{\Delta}_r \\ &= -{}_rM \cdot \left({}_r\ddot{\Delta}' + \ddot{x}_o\right) \\ \Longleftrightarrow {}_r\ddot{\Delta}' + \ddot{x}_0 &= \frac{M}{{}_rM}\ddot{\Delta}_r \end{aligned} \tag{14.18} $$

${}_r\Delta'$ and ${}_r\ddot{\Delta}'$ are calculated from the equation of motion with the predominant angular frequency of rank $r$, ${}_r\omega$ (14.19).

$$ {}_r\ddot{\Delta}' + 2 \cdot {}_rh \cdot {}_r\omega \cdot {}_r\dot{\Delta}' + {}_r\omega^2 \cdot {}_r\Delta' = -\ddot{x}_0 \tag{14.19} $$

Therefore, ${}_r\Delta'$ coincides with the values of the response displacement under the input motion of $\ddot{x}_0$ with the damping coefficient of ${}_rh$ and predominant angular frequency of ${}_r\omega$. Thus the comparison between ${}_r\Delta'$ and $-\left({}_r\ddot{\Delta}' + \ddot{x}_0\right)$ (performance curve) and demand curve calculated from the input motion gives the residual seismic capacity of the structure.

## 14.4 Health Monitoring of the Building in Yokohama National University

The proposed health monitoring system was installed into the building for the department of architecture of Yokohama National University in the beginning of the year of 2008. The building has eight stories and one underground floor. The height of the building is 30.8 m and its structural type is steel reinforced concrete. The building had been designed before 1981, when the Japanese building code was revised to confirm the ultimate strength of buildings. It was found that the building did not have enough ultimate strength, and then the building was retrofitted. The retrofitting construction was conducted from July 2008 to May 2009, and the sensors were removed at that time. The building before and after retrofitting is shown in Photo 14.1. EW direction is the longitudinal direction and NS direction is transverse direction.

Figure 14.3 shows the measured PGAs in horizontal and vertical directions during 1 year before and after the retrofitting construction. The maximum PGA was 37 cm/s$^2$. Figure 14.4 shows the skeleton curves of the measured performance curves in the NS direction before the retrofitting construction. It can be seen that the slopes of the skeleton curves were constant independent of the level of the earthquakes, which means that the measured natural periods are the same. Since the measured earthquakes were small, the building remained linear.

During the retrofitting construction, some walls were removed in the EW direction to reduce the weight of the building, and some walls were added in the NS direction to increase the strength. Figure 14.5 shows history of the natural periods in both EW and NS direction calculated from the slope of the skeleton curves. The natural period became longer in the EW direction and shorter in the NS direction due to the retrofitting construction, which coincides with the applied retrofitting method.

The health monitoring system worked well during the 2011 Off the pacific Coast of Tohoku Earthquake. Figure 14.6 shows the measured lateral accelerations on the basement and roof. The maximum acceleration was 91.5 cm/s$^2$ on the basement and 410 cm/s$^2$ on the roof. The predominant component of the acceleration lasted about 180 s.

The measured performance curve, skeleton curve from the performance curve, and the demand curve in the EW direction are shown in Fig. 14.7. The maximum representative displacement of 1.7 cm was measured in the positive direction. The equivalent period from the maximum displacement point in the positive direction was 0.48 s. The calculated viscous damping for the demand curve in order to get the same demand value for the period of 0.48 as the maximum response was 5.04 %, which is reasonable value.

Since the natural period in the EW direction before the earthquake was about 0.41 s as shown in Fig. 14.5, the equivalent period of 0.48 is longer than the period before the earthquake. Figure 14.8 shows the skeleton curve and the slopes for the periods of 0.41 and 0.48 s. It is clearly found that the stiffness degrading started at the representative acceleration of about 100 cm/s$^2$. The stiffness degraded down to 73 % according to the change of the period from 0.41 to 0.48 s.

Figure 14.9 shows the history of the lateral period in both EW and NS directions from the beginning of the monitoring, Feb. 10, 2008 to Sep. 15, 2011. As shown in Fig. 14.5, periods had changed due to the retrofitting construction, where period became longer in the EW direction and shorter in the NS direction. After the 2011 Off the pacific Coast of Tohoku Earthquake, the periods in the both direction became longer due to the damage to the building as shown in Fig. 14.9.

One of the advantages of the proposed method is that the representative force-displacement relationship can be obtained. If monitored response is evaluated only from the change of the period that is shown in Fig. 14.9, the ratio of the stiffness degrading is obtained that is 73 % in this case. However, it is difficult to judge if the degradation of 73 % means serious damage or not. On the other hand, the proposed method provides the force-displacement relationship as shown in Fig. 14.8. From the figure, it can be easily said that the stiffness degrading occurred due to cracking, not due to yielding, and the building still has more strength.

Figure 14.10 shows the relationship between the equivalent mass ratio (ratio of the equivalent mass calculated with (14.16) to the total mass) and representative displacement. It can be said that the calculated equivalent mass ratio is stable even from very small displacement, and constant at about 0.77. The stable equivalent mass ratio can be an evidence to evaluate the proposed method worked successfully.

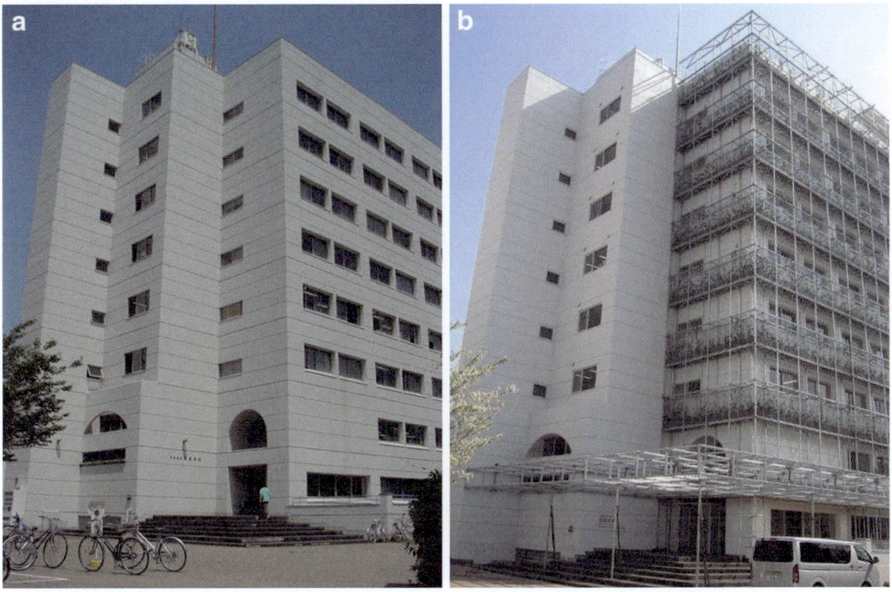


**Photo 14.1** Instrumented building (**a**) Before retrofitting (**b**) After retrofitting

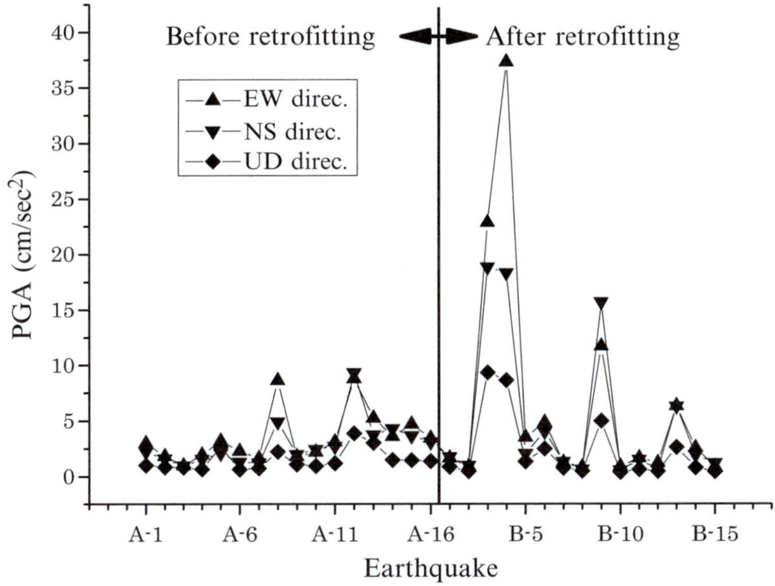


**Fig. 14.3** PGAs measured before and after retrofitting

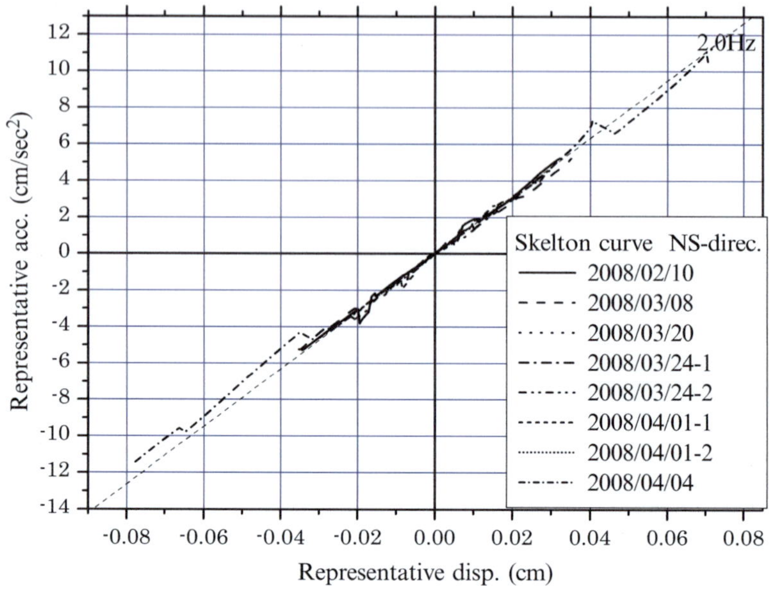


**Fig. 14.4** Skeleton curves in the transverse direction before retrofitting

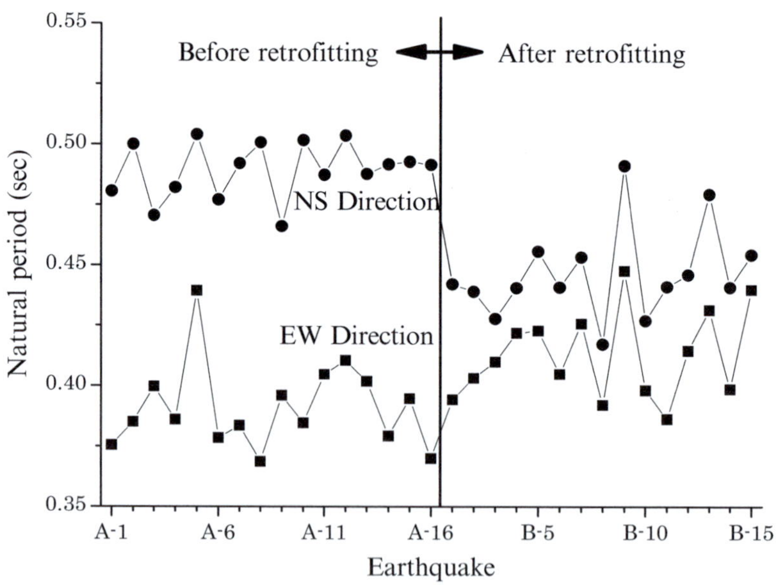


**Fig. 14.5** History of the natural period in the EW and NS direction due to retrofitting

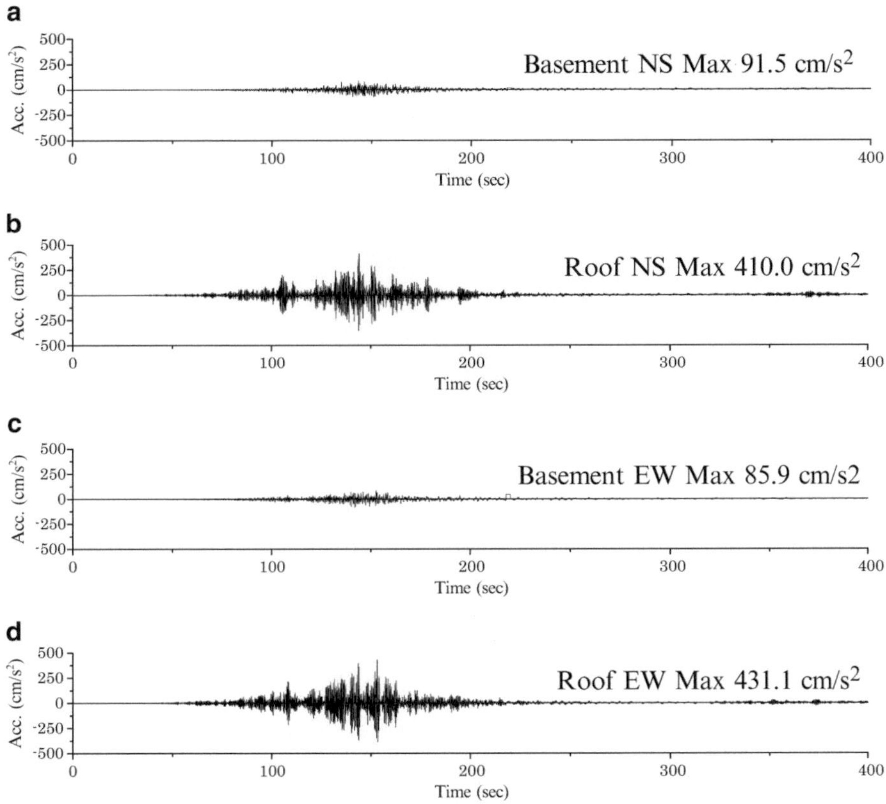


**Fig. 14.6** Measured earthquake during the 2011 Off the pacific Coast of Tohoku Earthquake (**a**) NS direction on the basement (**b**) NS direction on the roof (**c**) EW direction on the basement (**d**) EW direction on the roof

## 14.5 Concluding Remarks

A performance curve decomposition method using the Wavelet transform method was proposed to clear off the higher mode effects from a performance curve. The validity of the method was confirmed with the monitoring data of the building for the department of architecture of Yokohama National University. Results from the studies are as follows;

- A performance curve decomposition method using Wavelet transform method was proposed.
- The developed WTM can efficiently decompose the dynamic response into its primary response frequency bands.

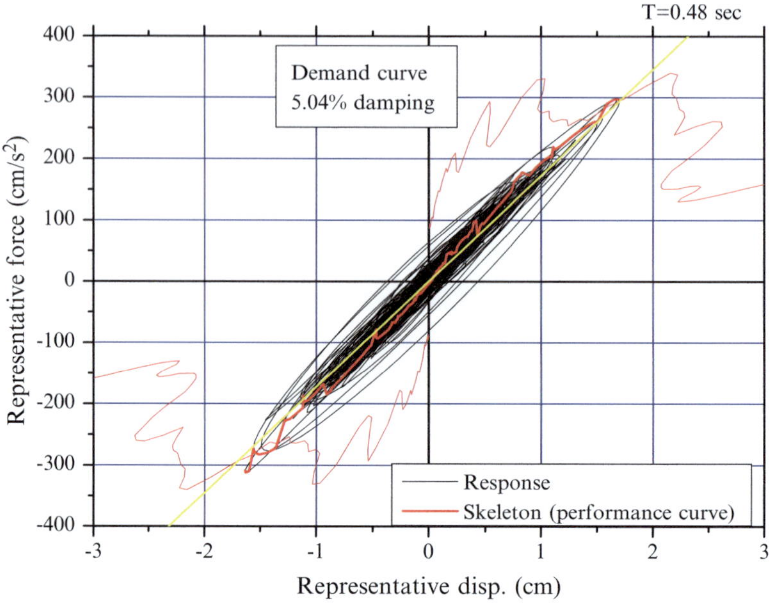


**Fig. 14.7** Measured performance and demand curves during the 2011 Off the pacific Coast of Tohoku Earthquake (EW direction)

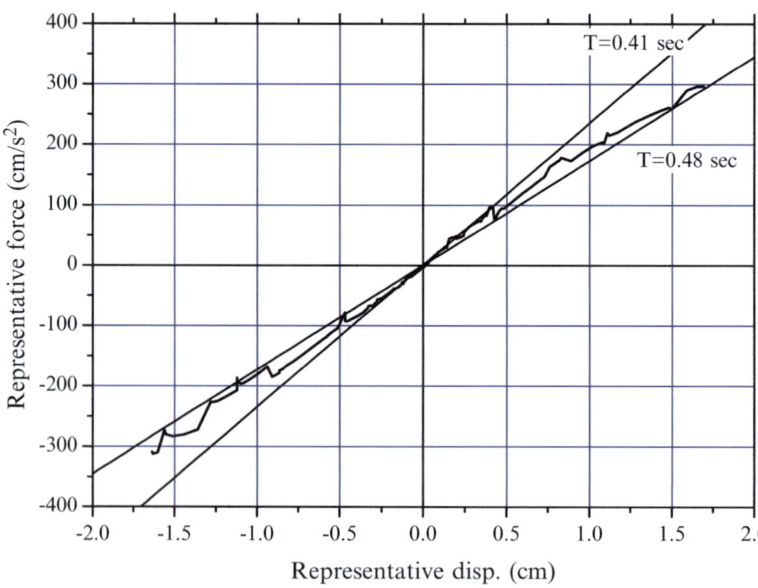


**Fig. 14.8** Skeleton curve of the measured performance during the 2011 Off the pacific Coast of Tohoku Earthquake (EW direction)

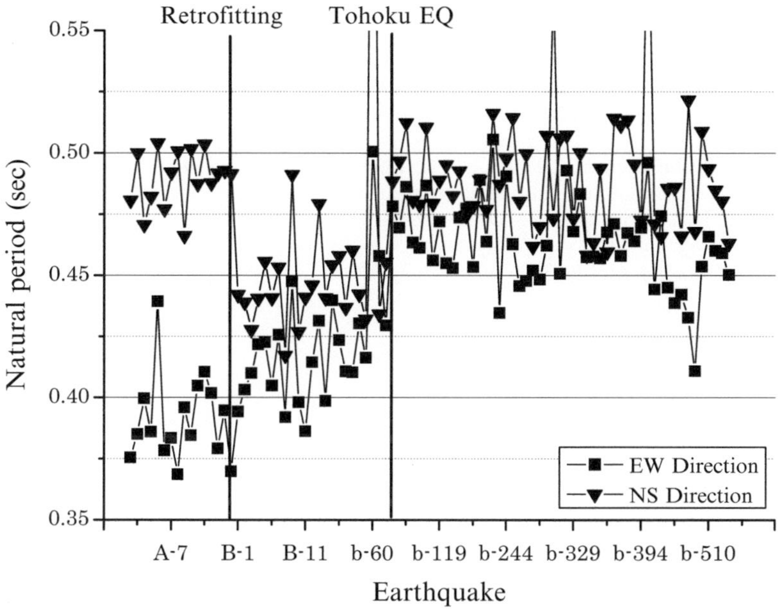


**Fig. 14.9** History of the lateral natural period

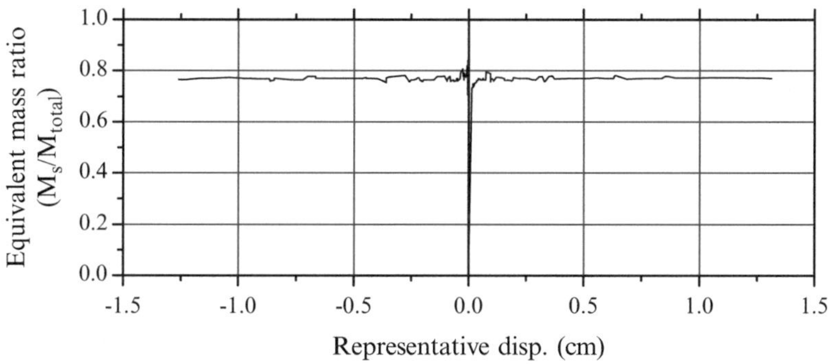


**Fig. 14.10** Equivalent mass ratio during the 2011 Off the pacific Coast of Tohoku Earthquake (EW direction)

- In the investigated cases, the predominant performance curves of the building for the department of architecture of Yokohama National University were successfully extracted.
- The building suffered cracks during the 2011 Off the Pacific Coast of Tohoku Earthquake and the degrading of the performance curve was successfully measured.

## References

Building Center of Japan (BCJ) (1996) Report on The Reconnaissance Committee on The Building Damages due to the Hanshin-Awaji Great Earthquake Disaster, 1995 – Summarization (in Japanese)

Kusunoki K, Teshigawara M (2003) A new acceleration integration method to develop a real-time residual seismic capacity evaluation system. J Struct Constr Eng 569:119–126 (in Japanese)

Kusunoki K, Teshigawara M (2004) Development of real-time residual seismic capacity evaluation system –integral method and shaking table test with plain steel frame. In: 13th world conference on earthquake engineering, CD-Rom, Beijing, China

# Chapter 15
# Evaluation on Flexural Deformability of Reinforced Concrete Columns with Wing Walls

**Toshimi Kabeyasawa, Yousok Kim, Toshikazu Kabeyasawa, and Hiroshi Fukuyama**

**Abstract** Tests on reinforced columns with wing walls were conducted in 2010 to investigate the flexural deformability following the shear tests in the previous years. The effects of the moment-to-shear ratios of loading, the reinforcement details and the width and length of the wing walls on the flexural deformability were investigated. The specimens with thin wing walls showed strength decay after the ultimate strength in flexure, due to the compression failure of concrete and buckling of the re-bars at the wall ends under the larger deformation amplitudes, while the specimens with thick wing walls showed much less strength decay generally. If the edge was well confined, the strength decay was much smaller. All specimens were ductile and stable in flexural failure mode up to the maximum loading drift level. The damage to the column could relatively be relieved owing to the inelastic energy dissipation by the wing walls. The ultimate strength and deformability are formulated for practical calculation based on a flexural theory and are compared with the test results, by which fair correlations are obtained. A simple formula on the deformability is derived based on the theory, by which the deformability rank of the member may be specified in the current code of seismic design.

**Keywords** Wing wall • Compressive failure • Confinement • Ultimate deformation • Flexural theory

T. Kabeyasawa (✉) • Y. Kim
Earthquake Research Institute, University of Tokyo, 1-1-1, Yayoi, Bunkyo-ku, Tokyo 113-0032, Japan
e-mail: kabe@eri.u-tokyo.ac.jp; yskim1220@yonsei.ac.kr

T. Kabeyasawa • H. Fukuyama
Building Research Institute, 1 Tachihara, Tsukuba-shi, Ibaraki-ken 305-0802, Japan
e-mail: tosikazu0911@gmail.com; fukuyama@mse.biglobe.ne.jp

M. Fischinger (ed.), *Performance-Based Seismic Engineering: Vision for an Earthquake Resilient Society*, Geotechnical, Geological and Earthquake Engineering 32,
DOI 10.1007/978-94-017-8875-5_15, 

## 15.1 Introduction

Full-scale shake table tests on reinforced concrete school buildings were conducted at E-Defense, the world largest three-dimensional shake table, in 2006 (Kabeyasawa et al. 2007a, b; Kabeyasawa and Kabeyasawa 2008). A dynamic collapse of the bare building occurred associated with shear and axial failure of short columns in case of fixed foundation, while an obvious input loss was observed in case of swaying foundation. Based on the test result, we have proposed "hyper-earthquake resistant system," which is based on a simple fail-safe design concept against extreme motions exceeding the design level, consisting of relatively strong superstructure and sway-slip foundation. The slip behavior would occur only under very high ground acceleration, so that the response of the superstructure could be controlled as minor as insensitive to the level and characteristics of possible extreme motion. Use of columns with wing walls is a simple but cost-effective design option to provide a superstructure with relatively higher capacity, up to required in the hyper-earthquake resistant system. The field investigations on the structural damage induced from past earthquakes and laboratory tests indicated that reinforced concrete columns with wing walls had relatively good seismic performance as earthquake-resistant members increasing column stiffness and strength in reinforced concrete buildings.

In recent seismic design practice of Japan, however, columns with wing walls have not been used very much as earthquake resistant elements. Instead, they have been mostly separated from structural members by installing seismic slits between the column and the wall. Since the columns with wing walls show different seismic behaviour from shear walls or independent columns, the evaluation methods of its strength and ductility, especially inelastic deformability, have not been clearly defined in the guidelines for design practice, which makes it difficult to design columns with wing walls. However, the wing walls attached to column undoubtedly increases the lateral strength of the column, therefore, installing seismic slits between the column and the wing walls might be inefficient in many cases of low-rise buildings, which could have been oriented to strength-based-design. Conventional evaluation methods for columns with wing walls may be found in Seismic Evaluation Standard for RC Buildings by The Japan Building Disaster Prevention Association (Building Center of Japan 2007) and also in Guidelines for Standard Requirements on Building Structures by The Building Center of Japan (Japan Building Disaster Prevention Association 2001). However, these conventional evaluation methods for ultimate strength and ductility are not rational in theoretical point of view. In addition, these methods are formulated by assuming columns with both-side wing walls, so the design equations are not basically applicable to columns with one-sided wing wall. Therefore, in the previous studies, shear tests on the specimens of the columns with wing walls were tested to investigate seismic performance of the members, or the shear strengths. A cumulative method of evaluating shear strength has been proposed as design formula in practice (Kabeyasawa and Kabeyasawa 2007), which gave better correlation with the observed failure strengths than the conventional design formula.

Following the series of tests on columns with wing walls, mostly failed in shear, in the previous years (Kabeyasawa et al. 2008, 2009, 2010; Tojo et al. 2008), flexural tests on six specimens were conducted in 2010 to investigate the ultimate strengths and deformability of the members. The method and results of the tests are reported in this paper. The specimens were six one-half scale reinforced concrete columns with wing walls on both sides in the loading direction. The shear span to depth ratio of loading was selected so that the calculated shear strength was to be higher than the shear at the flexural strength. The observed deformability was compared with calculation by assuming the compressive hinge region. The test results below have been reported (Kabeyasawa et al. 2011), to which the a simplified theoretical method of evaluating the deformability was added in the last section.

## 15.2 Test Specimens of Columns with Wing Walls

The sectional properties and reinforcement details of the six specimens SWF1 through SWF6 are listed in Table 15.1. The scale of specimens is half or two-thirds of typical sections of full-scale medium-rise buildings in Japan. The section and reinforcement details of each specimen were planned following the previous tests on columns with wing walls (Kabeyasawa and Kabeyasawa 2007; Kabeyasawa et al. 2008), as shown in Fig. 15.1. SWF1 and SWF2 have identical section and reinforcement details with a specimen in the series, with the column size of $400 \times 400$mm and the wall thickness of 100 mm, while the specimens SWF3, SWF4 and SWF5, the thickness of the wing walls was increase to 150 mm and instead, the column size was reduced to $400 \times 300$mm, depth of 400 mm and width of 300 mm in the loading direction, so that the total sectional area was made the same as that of SWF1 and SWF2. The amount of the shear reinforcing bars in the wing walls was increased keeping the same reinforcement ratio. The wall edge of the specimen SWF5 was specially confined with closed square hoop of D6 at the spacing of 50 mm. Only the moment-to-shear ratio of loading was changed also between SWF3 (SWF5) and SWF4. As for the specimen SWF6, the wall thickness was 100 mm but the wall length was made longer to 600 mm. The compressive strength of concrete was slightly varying from 26.1 through 28.5 MPa as shown in Table 15.1, measured from the material test conducted at the age of testing each specimen. The yielding strengths of D6, D10 and D16 were 349, 342 and 351 MPa as shown in the note of Table 15.1.

The shear strength under lower shear span ratio has been identified by the former shear test series and a method of estimating the shear strength, cumulative method has been proposed (Kabeyasawa and Kabeyasawa 2007; Kabeyasawa et al. 2008). Based on the calculation using the former equation, the shear span to depth ratio of loading was selected so that the calculated shear strengths would be higher by 1.1–1.2 times than the shear at the ultimate flexural strength. Then the flexural ultimate strength would be achieved and the ultimate deformability after yielding

**Table 15.1** List of test specimens(SWF1-SWF6)

| Name | Concrete strength (N/mm$^2$) | Column (mm) Width × Depth | Main bars | Hoop | Wing walls on both sides Length (mm) | Thick (mm) | End re-bars | Shear re-bars | Axial load (kN) | M/Q (mm) |
|---|---|---|---|---|---|---|---|---|---|---|
| SWF1 | 26.9 | 400 × 400 | 12-D16 | 2-D6@40 (0.0040 | 400 | 100 | 4-D13 | 2-D6@150 | 800 | 2,400 |
| SWF2 | 27.5 | | (0.0149) | or 0.0053) | | | | | | 1,800 |
| SWF3 | 26.1 | 300 × 400 | 10-D16 | | 400 | 150 | 6-D13 | 2-D6@100 | | 2,400 |
| SWF4 | 28.3 | 300 × 400 | (0.0166) | | | | 6-D13 | (0.0043) | | 1,800 |
| SWF5 | 28.5 | 300 × 400 | | | | | 6-D13[a] | | | 2,400 |
| SWF6 | 27.8 | 400 × 300 | | | 600 | 100 | 6-D13 | 2-D6@150 | | 3,000 |

[a]Special confinement detail for end region with closed hoops of D6@50
#1 Yield strengths of steel, D6, D13, D16: 349, 342 and 351(N/mm$^2$), respectively
#2 Tensile strengths of steel, D6, D13, D16: 489, 499 and 517(N/mm$^2$), respectively
#3 Nominal sectional area of reinforcing bars, D6, D13, D16: 32, 127 and 200(mm$^2$), respectively

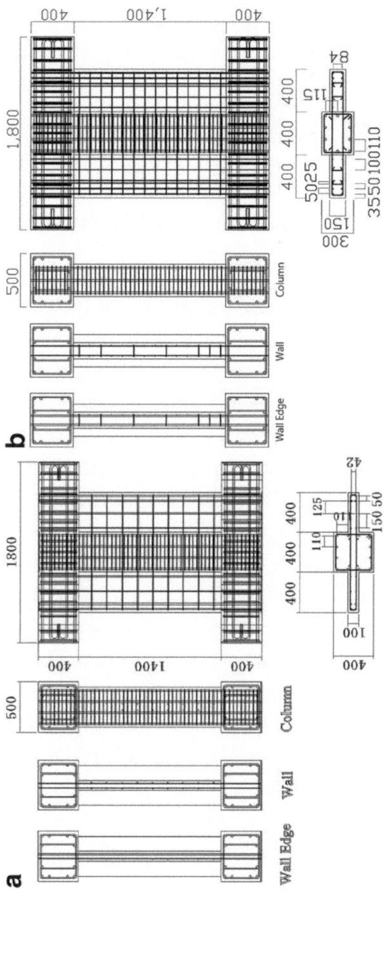


**Fig. 15.1** Test specimens of columns with wing walls **(a)** SWF1 and SWF2 **(b)** SWF3 and SWF4

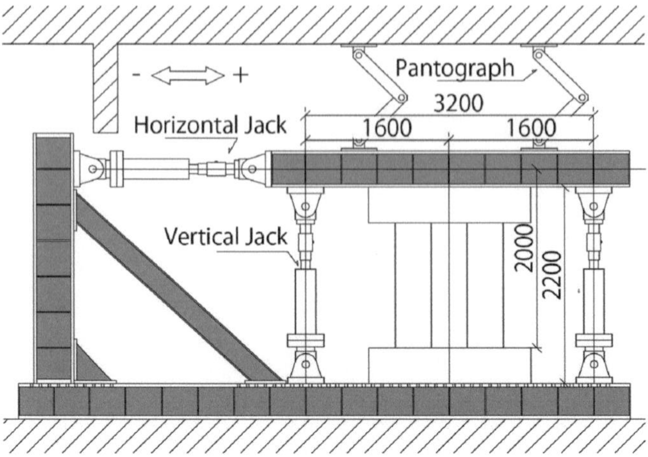


**Fig. 15.2** Test set-up at Earthquake Research Institute, University of Tokyo

would depend primarily on the compressive failure at the edge of wing walls. Also the effects of the loading would be comparable among the specimens. The shear span of loading for SWF1 and SWF2 were selected as 1,800 mm and 2,400 mm, where the shear span to depth ratio was 1.5 and 2.0, respectively. This was also the case for the two specimens SWF3 and SWF4. The loading condition of the specimen SWF5 was the same with SWF3 so as to investigate the effect of the confinement at the wall edge. As for the specimen SWF6 with total depth of 1,500 mm, the shear span to depth ratio was selected as 2.0, so that the shear span was 3,000 mm. The same constant axial load (N = 800 kN) was applied for all specimens. The axial load ratio for the column area was 0.2, N/(Ac*Fc) = 0.2, by using the column sectional area Ac of 400 × 400mm and Fc as compressive concrete strength of 25 Mpa, which is not exactly equal to the measured concrete strengths. The axial load ratio for the total area including the wing walls was 0.13.

The test set-up at ERI laboratory was used for the loading system as shown in Fig. 15.2. The constant axial load was applied with the two vertical oil jacks, each applying 400 kN. Then the lateral load Q was applied by the horizontal oil jack at the height of 2,000 mm, while the corresponding varying axial load ($\Delta$N) in proportion to the lateral resistance was also applied at the top of the specimen with the two vertical oil jacks to maintain the constant target moment-to-shear ratio or the shear span of loading $h_0$ as planned and shown in Table 15.1 ($h_0 = 2 + 3\Delta N/Q(m)$). The three jacks are controlled simultaneously by monitoring the loading conditions. The lateral loading was reversed at the peak drift ratios of ±1/400, ±1/300, ±1/200, ±1/150, ±1/100, ±1/75, ±1/50, ±1/37.5, ±1/25 and up to 1/12.5 rad. The maximum displacement was limited by the stroke capacity of the horizontal oil jack.

## 15.3 Observed Failure Modes and Hysteretic Relations

The six specimens generally failed in flexural mode following the similar process though the deformation levels were different for each specimen. Therefore, the detailed process of each specimen is not described here, but the behavior is described in common, while the ultimate failure states were shown in Fig. 15.3. Flexural bending and shear cracks occurred and progressed from the wing walls at the first loading cycle of 1/400 rad. Longitudinal reinforcement at the wall edge yielded then the maximum strengths were attained at 1/200 to 1/100 (1/50 in SWF5) in a flexural mode. Then apparent or slight strength decay was observed due to the compression failure of concrete and buckling and rupture of the re-bars at the wing wall ends under the deformation amplitudes greater than the maximum. Crushing of concrete was observed in 1/200 to 1/100 cycles of loading in the specimens SWF1, SWF2 and SWF6 with thin wing walls, while the slight crushing of cover concrete was observed in the same deformation levels in the specimens SWF3, SWF4 and SWF5 with thick wing walls, though the additional damage was much less at the larger deformation amplitudes. Although the strength decay was different, all specimens basically showed ductile and stable behavior in the flexural failure mode of column up to the maximum loading drift of 1/12, owing to the inelastic energy dissipation by the wing walls, by which the damage to the column might have been relatively

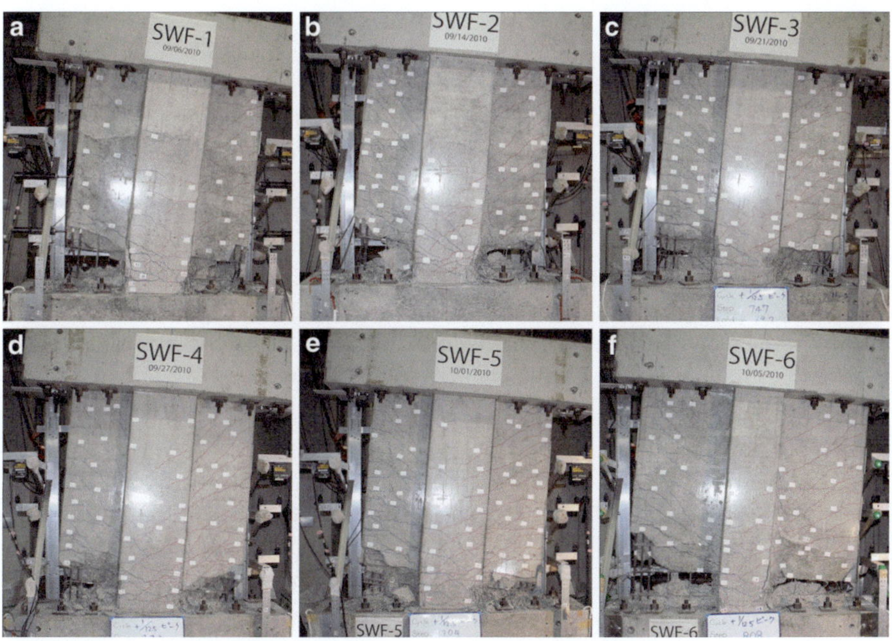


**Fig. 15.3** Observed failure modes at the maximum loading deformations. (**a**) SRF1 (**b**) SWF2 (**c**) SWF3 (**d**) SRF4 (**e**) SWF5 (**f**) SWF6

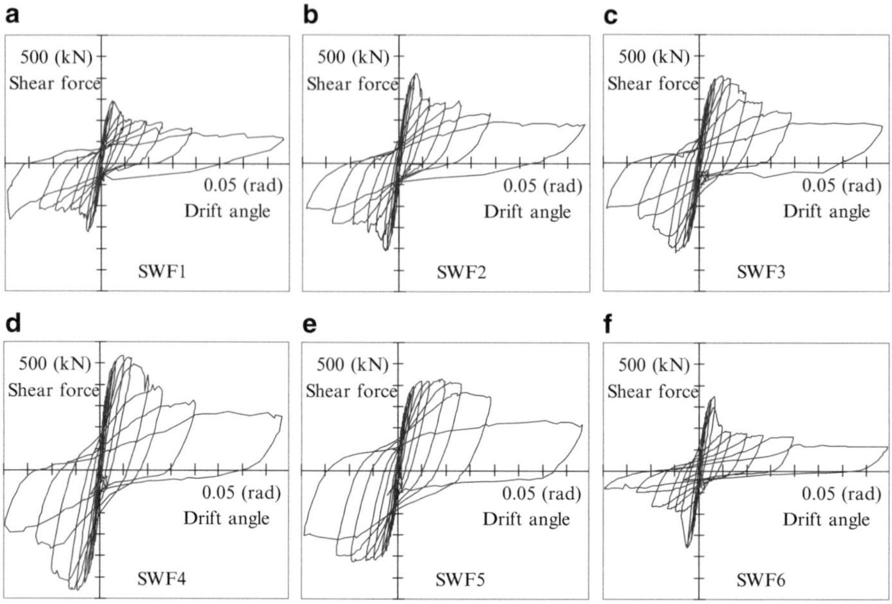


**Fig. 15.4** Observed hysteretic relations between the lateral load and drift rotation angles. (**a**) SWF1 (**b**) SWF2 (**c**) SWF3 (**d**) SWF4 (**e**) SWF5 (**f**) SWF6

relieved. After the lateral loading test, the axial compression test was carried out by which all the specimens could bear up to the axial load of 2000 kN, which was 2.5 times the constant axial load, the axial load ratio of 0.5 and the limit capacity of the oil jacks.

The observed hysteretic relations of the specimens are shown in Fig. 15.4, where the relationships are between the lateral shear forces and the lateral deformation at the upper loading beam level in terms of rotation angles. The measured peak values are summarized in Table 15.2, such as the maximum strengths in positive and negative directions with the drift rotations when the strengths were attained. The deformability was defined at the deformation with the strength decay to 80 % of the peak strengths. These characteristic values were analyzed in comparison with calculated values.

## 15.4 Calculation of Ultimate Strengths and Deformations

The ultimate flexural resistance of the specimens at the base section was calculated based on plastic flexural theory, using simple design equations derived and proposed for the columns with wing walls as in the forms of (15.1, 15.2, 15.3 and 15.4):

$$M_u = \sum \left(a_t \cdot \sigma_y \cdot j_t\right) \cdot +N \cdot j_N \tag{15.1}$$

**Table 15.2** Observed and calculated strengths and deformations

| Title | SWF1 | SWF2 | SWF3 | SWF4 | SWF5 | SWF6 |
|---|---|---|---|---|---|---|
| Maximum strength positive (kN) | 293 | 421 | 409 | 535 | 427 | 351 |
| Drift angle at max strength (rad) | 0.0052 | 0.0073 | 0.0098 | 0.0098 | 0.0189 | 0.0065 |
| Maximum strength negative(kN) | −323 | −409 | −422 | −564 | −427 | −360 |
| Drift angle at max strength (rad) | −0.0061 | −0.0056 | −0.0104 | −0.0102 | −0.0196 | −0.0052 |
| Calculated flexural strength (kN) | 284 | 379 | 347 | 462 | 347 | 322 |
| Maximum strength/ calculated | 1.14 | 1.11 | 1.21 | 1.21 | 1.22 | 1.12 |
| Observed deformability[a] (rad) | 0.0087 | 0.0096 | 0.0189 | 0.0189 | 0.0385 | 0.0069 |
| Observed deformability[a] (rad) | −0.0067 | −0.0090 | −0.0143 | −0.0204 | −0.0256 | −0.0063 |
| Calculated neutral axis (mm) | 293 | 421 | 409 | 535 | 427 | 351 |
| Calculated deformation $R_u$ (rad) | 0.0081 | 0.0081 | 0.0129 | 0.0129 | 0.0258 | 0.0058 |

[a]The ultimate deformation is defined as the deformation at 80 % of the peak strength reached after the maximum strength in each direction

where, $a_t$ and $\sigma_y$: area ($mm^2$) and yield strength ($N/mm^2$) of tensile longitudinal reinforcing bars, which includes all bars in the tensile regions, though the bars close to the neutral axis may be ignored, $j_t$: effective distance between the tensile longitudinal reinforcing bars and the center of compressive concrete stress block ($=d_t - L_{cc}$) (mm), $d_t$ : effective depth of the bars from compressive fiber(mm), $N$: constant axial load(N), $j_N$ : axial load ($=L/2 - L_{cc}$) (mm), $A_{cc}$: area of compressive concrete block($mm^2$), here given by Eq. (15.2), ignoring the compressive bars and reduction due to substitute to full-plastic concrete block, which would have compensating effects:

$$A_{cc} = \frac{\sum \left(a_t \cdot \sigma_y\right) + N}{F_c} \tag{15.2}$$

and, the centroid of compressive concrete stress block from the compression fiber (mm) $L_{cc}$ can be given by the following Eqs. (15.3) or (15.4):

$$L_{cc} = A_{cc} / \left(2t_w\right) \quad \text{if} \;\; A_{cc} \leq A_{w1} \tag{15.3}$$

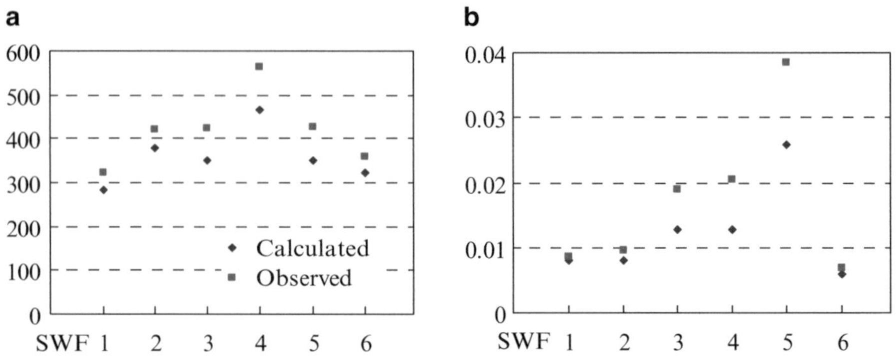


**Fig. 15.5** Observed and calculated ultimate strengths and deformation capacity (**a**) Ultimate strength (kN) (**b**) Ultimate deformation capacity (rad)

$$L_{cc} = \frac{A_{w1}}{A_{cc}} \cdot \frac{L_{w1}}{2} + \left(1 - \frac{A_{w1}}{A_{cc}}\right)\left(L_{w1} + \frac{A_{cc} - A_{w1}}{2B_c}\right) \quad \text{if} \quad A_{cc} > A_{w1} \tag{15.4}$$

where, $t_w$ and $L_{w1}$: thickness and length of the compressive wing wall (mm), $A_{w1} = L_{w1} * t_w$,: area of the wing wall, $B_c$: width of column(mm). The Eq. (15.4) is to formulate the centroid of T-shaped compressive region of the wing wall and the column in total area of $A_{cc}$.

The flexural strengths calculated from above formula for the six specimens are shown also in Table 15.2 and Fig. 15.5a, in comparison with the observed ultimate strengths, which is taken as either of the highest values of the strength in the positive and negative loading directions. The ratios of the measured maximum strength to the calculated strength were 1.14, 1.11 and 1.12 for the specimens SWF1, SWF2 and SWF6 with thin wing walls, while they are 1.21–1.22 for the specimens SWF3, SWF4 and SWF5 with thick wing walls. The ratios are generally higher in the latter case, probably because the deformation levels at the maximum strengths are larger owing to the stable behavior of the compressive concrete block in the cases of thick walls, so that the effects of strain hardening in the tensile bars were much higher.

The calculation on the ultimate deformability is proposed here and formulated based on a simple flexural model as follows. The ultimate deformation angle $R_u$ is assumed in proportion to the ultimate curvature $\phi_u$ and the compressive hinge zone length $l_h$ in the form as:

$$R_u = c \times l_h \times \phi_u \tag{15.5}$$

and, the ultimate curvature may be given using the neutral axis $x_n$, and the ultimate compressive strain of concrete $\varepsilon_{cu}$, in the form as:

$$\phi_u = \varepsilon_{cu} / x_n \tag{15.6}$$

The compressive hinge zone length is here assumed in proportion to, for example as observed in the test, taking the twice of the wall thickness as:

$$l_h = 2t_w \tag{15.7}$$

Then, the ultimate deformation may be written in the form as:

$$R_u = c \times 2t_w \times \varepsilon_{cu}/x_n \tag{15.8}$$

By taking the inner end of the compressive concrete in the same forms as the Eqs. (15.2), (15.3) and (15.4), the neutral axis $x_n$ can be derived as follows:

$$x_n = 2L_{cc} = A_{cc}/t_w \quad \text{if} \quad A_{cc} \le A_{w1} \tag{15.9}$$

$$x_n = L_{w1} + \frac{A_{cc} - A_{w1}}{B_c} \quad \text{if} \quad A_{cc} > A_{w1} \tag{15.10}$$

The ultimate strain is taken as $\varepsilon_{cu} = 0.003$ for the unconfined concrete, and $\varepsilon_{cu} = 0.006$ in case of the specimen SWF5 with the confined end region. Then, the constant factor $c$ is determined here empirically as $c = 6$. The factor reflects theoretically the effects of additional deformation components, such as (1) the elastic deformation besides the hinge region, (2) the elastic and inelastic shear deformation, and (3) the relatively conservative assumption on the ultimate strain. The ultimate deformability in the test is defined as the deformation at 80 % of the maximum strength in the skeleton of the hysteretic relations, and larger of positive or negative, and are compared with the calculated results are shown in Table 15.2 and Fig. 15.5b.

The calculation gives a fair and conservative estimation of the observed deformability varying with the test parameters, such as the wall thickness, the length and the confinement detail. The assumptions above are to be verified through other past test data.

To verify partially the assumptions made in above formula, local strains measured in the test on the columns with wing walls are examined. Local strains in concrete measured at compressive wall tip region for the specimens SWF1 through SWF6 are shown in Fig. 15.6. The strains are measured at the constant height of 300 mm from the base, which corresponded to the three times the wing wall thickness ($3t_w$) in case of SWF1, SWF2 and SWF6, and the twice ($2t_w$) in case of SWF3, SWF4 and SWF5. The measured compressive strains are shown in the figure in relation to the overall lateral deformation angle in positive and negative directions. The slid and dashed lines show the measured relations of peak deformations at the load reversals and the compressive strains for the six specimens. The circular and square marks show the strains measured at the ultimate deformations at the 80 % strength decay after the peak strength, which is selected as the definition on the ultimate deformability. The measured compressive strains at the ultimate

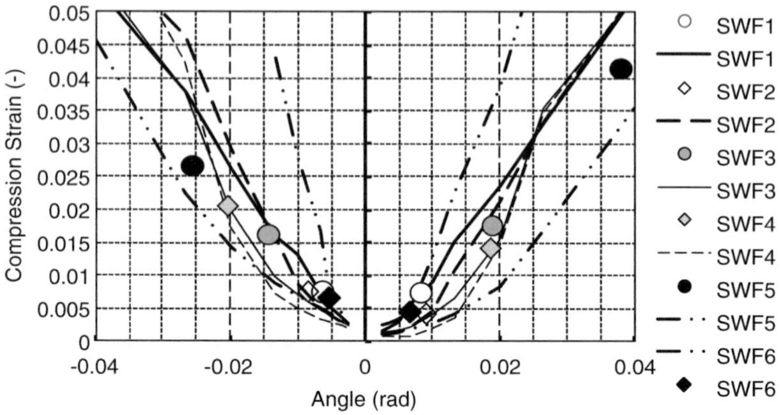


**Fig. 15.6** Local strains in concrete measured at compressive wall tip region (SWF1 through SWF6)

deformations were approximately (a) 0.005–0.007 for the specimens SWF1, SWF2, SWF6 with the wall thickness of 100 mm ($t_w = 100$ mm), (b) 0.018–0.02 for the specimens SWF3, SWF4 with the wall thickness of 150 mm ($t_w = 150$ mm), and (c) 0.04 for the specimen SWF5, where the wall thickness is 150 mm ($t_w = 150$) and well confined. To compare these measured strains are compared with the assumptions above, the measured strains are twice 0.003 in case of (a), and six times 0.003 in case of (b) and twelve times 0.003 (six times 0.006) in case of (c), which means that the measured strains was not constant through the different wall thickness as assumed in the formula but much more increasing with the thickness, though the effect of measured length shall be investigated further in detail.

In the Building Standard Law (BSL) of Japan, the ultimate lateral load-carrying capacities of structures are calculated by non-linear analysis, mostly now by pushover analysis, and are confirmed to be higher than the required capacities, which are specified based on the deformability of members, classified into four ranks, such as in cases of columns and beams, FA: high, FB: medium, FC: limited ductile in flexure, FD: brittle in shear. The capacity is required higher in case of consisting of less ductile members.

The formula on the ultimate deformability derived from the flexural theory as Eq. (15.8) can be transformed into the following form as Eq. (15.11) using also Eq. (15.2):

$$
\begin{aligned}
R_u &= c \times 2t_w \times \varepsilon_{cu}/x_n \\
&= c \times 2t_w \times \varepsilon_{cu} \times t_w/A_{cc} \\
&= c \times 2 \times (t_w)^2 \times \varepsilon_{cu} \times \frac{F_c}{\sum (a_t \cdot \sigma_y) + N}
\end{aligned} \tag{15.11}
$$

in case of Eq. (15.3), such as $A_{cc} \leq A_{w1}$, with relatively long wing walls so that the neutral axis is in the wall section. To simplify the calculation more, the total area of the tensile reinforcement are assumed as whole longitudinal bars in the column, while the wall tensile reinforcement are assumed to be equal to those in the compressive region as in the following:

$$p_g \sigma_y = \sum (a_t \cdot \sigma_y)/A_c \tag{15.12}$$

$$\rho_g = p_g \frac{\sigma_y}{F_c} \tag{15.13}$$

Then the deformability can be expressed as in the form:

$$\begin{aligned} R_u &= c \times 2 \times \frac{(t_w)^2}{A_c} \times \varepsilon_{cu} \times \frac{F_c}{p_g \cdot \sigma_y + \sigma_0} \\ &= c \times 2 \times \frac{(t_w)^2}{A_c} \times \varepsilon_{cu} \times \frac{1}{\rho_g + \eta_0} \end{aligned} \tag{15.14}$$

where,

$$\sigma_0 = N/A_c : \text{Stress by axial force}$$

$$\eta_0 = N/\left(A_c F_c\right) : \text{Axial force ratio}$$

Let us assume these ratios from normal design as constant upper values, such as the longitudinal reinforcement ratio $p_g = 0.02$, and the axial load ratio as $\eta_0 = 0.2$ and $\sigma_y/F_c = 15$ then $\rho_g = 0.02 \times 15 = 0.3$, and using the constants from the tests as $c = 6$ and $\varepsilon_{cu} = 0.003$ for unconfined detail, then the Eq. (15.14) for the ultimate deformability is expressed in a simple form as:

$$R_u = 0.072 \times \frac{(t_w)^2}{A_c} \tag{15.15}$$

In case of rectangular or square column section ($Ac = B \times D$, $B = D$), for example, the deformability may be calculated using the wall thickness ratio as:

$$\begin{aligned} & R_u = 0.008,\ \ 0.0045,\ \ and\ 0.003, \\ \text{if} \quad & \frac{t_w}{B} = \frac{t_w}{D} = \frac{1}{3},\ \frac{1}{4},\ and\ \frac{1}{5}, \quad \text{respectively.} \end{aligned} \tag{15.16}$$

The deformability ranks of the columns with wing walls in practical design (BCJ 2007; JBDPA 2001) may be specified simply by using the ratio of the wall thickness to the column section size referring to above simplified estimation. The ultimate strain may be increased considering confinement details in addition.

## 15.5 Conclusions

Six specimens of the column with wing walls were tested to flexural failure with different shear span ratio, wall thickness and length and confinement details. All specimens failed in flexure, yielding longitudinal reinforcement and then compression failure at the wall edge causing strength decay after the ultimate maximum strength in flexure. The specimens with thin wing walls showed strength decay, due to the compression failure of concrete and buckling of the re-bars at the wall ends under the larger deformation amplitudes. As for the specimens with thick wing walls, the strength decay was much less, and was very slight in case of well-confined details at the wall edges. The observed ultimate deformations at 20 % strength decay from the maximum were compared with the calculation, constantly factored from the simple theoretical flexural deformations assuming the ultimate strain of concrete and the length of compressive hinge region. The calculation gives a fair correlation with the observed variation with the test parameters, though the constant factor needs be investigated further. By assuming the constants in regular design, the equation on the deformability can be simplified using the wall thickness ratio to the column size, which may be used to specify the deformability ranks of the members in the design practice of Japan.

## References

Building Center of Japan (2007) Guidelines for standard requirements on building structures (in Japanese). BCJ

Japan Building Disaster Prevention Association (2001) Standard for seismic evaluation of existing reinforced concrete buildings (in Japanese). JBDPA

Kabeyasawa T, Kabeyasawa T (2007) Shear design equation in practice for columns with Wing walls (in Japanese). In: Proceedings of the 5th annual meeting, JAEE, pp 248–249

Kabeyasawa T, Kabeyasawa T (2008) Nonlinear soil-structure interaction theory for low-rise reinforced concrete buildings based on the full-scale shake table test at E-Defense. In: Proceedings of 14th world conference on earthquake engineering, Beijing, China

Kabeyasawa T, Kabeyasawa T, Matsumori T, Kabeyasawa T, Kim YS (2007a) 3-D collapse tests and analyses of the three-story reinforced concrete buildings with flexible foundation. In: Proceedings of the 2007 structures congress, Long Beach, 16–19 May

Kabeyasawa T, Matsumori T, Kabeyasawa T, Kabeyasawa T, Kim YS (2007b) Plan of 3-D dynamic collapse tests on three-story reinforced concrete buildings with flexible foundation. In: Proceedings of the 2007 structures congress, Long Beach, 16–19 May

Kabeyasawa T, Kabeyasawa T, Tojo Y, Kabeyasawa T (2008) Experimental study on columns with wing-walls failing in shear (in Japanese). Proc JCI 30(3):115–120

Kabeyasawa T, Kabeyasawa T, Kim Y, Kabeyasawa T, Bae K (2009) Tests on reinforced concrete columns with wing walls for hyper-earthquake resistant system. In: Proceedings of the 3rd international conference on advances in experimental structural engineering, San Francisco, USA, Oct 2009, 12 pp

Kabeyasawa T, Kabeyasawa T, Kim Y, Kabeyasawa T, Kunkuk B, Van Quang P (2010) Strength and deformability of reinforced concrete columns with Wing Walls. In: Proceedings of 9th US National/10th Canadian conference on earthquake engineering, Toronto, Ontario, Canada, 25–29 July, EERI, Paper 813, 10 pp

Kabeyasawa T, Kim Y, Sato M, Hyunseong H, Hosokawa Y (2011) Tests and analysis on flexural deformability of reinforced concrete columns with Wing Walls. In: Proceedings of the 2011 Pacific conference on earthquake engineering (PCEE2011), New Zealand Society for Earthquake Engineering, Paper 102, pp 1–8

Tojo Y, Kabeyasawa T, Kabeyasawa T, Kim YS (2008) Experimental study on column with wing-walls yielding in flexure. Proc Jpn Concr Inst 30(3):109–114 (in Japanese)

# Chapter 16
# Seismic Performance and Reinforcement of Japanese High-Rise Buildings Facing Subduction Earthquakes: E-Defense Shake Table Tests

**Takuya Nagae, Takahito Inoue, Koichi Kajiwara, and Masayoshi Nakashima**

**Abstract** The seismic capacity of high-rise steel buildings is a matter of concern, particularly when they are subjected to long-period ground motions. A series of large-scale shaking table tests conducted at E-Defense disclosed fractures in beam-to-column connections and represented the effects of retrofit for such high-rise steel buildings. Damage to office and residential rooms was also reproduced.

**Keywords** High-rise building • Long-period ground motion • Shake table test

## 16.1 Introduction

Periodical occurrences of large ocean-ridge earthquakes having a magnitude over eight along the subduction zones in the southwest part of Japan have been documented in historical materials. Figure 16.1a shows a map of Japan and an ocean ridge, called the Nankai trough, running deep along the three regions: Tokai, Tonankai and Nankai, from east to west. For many centuries, slips and ruptures along the three regions have been occurring at an interval of 100–150 years. Such earthquakes are known to generate long-period ground motions on land, especially in the basin areas where large cities such as Tokyo, Nagoya and Osaka are located (Kamae et al. 2004). Long-period ground motions tend to resonate high-rise

T. Nagae (✉) • T. Inoue • K. Kajiwara
Hyogo Earthquake Engineering Research Center, National Research Institute for Earth Science and Disaster Prevention, 1501-21 Nishikameya, Mitsuta, Shijimi, Miki, Hyogo 673-0515, Japan
e-mail: nagae@bosai.go.jp; dinoue@bosai.go.jp; kaji@bosai.go.jp

M. Nakashima
Disaster Prevention Research Institute (DPRI), Kyoto University, Gokasho, Uji, Kyoto, Japan
e-mail: nakashima@archi.kyoto-u.ac.jp

M. Fischinger (ed.), *Performance-Based Seismic Engineering: Vision for an Earthquake Resilient Society*, Geotechnical, Geological and Earthquake Engineering 32,
DOI 10.1007/978-94-017-8875-5_16, 

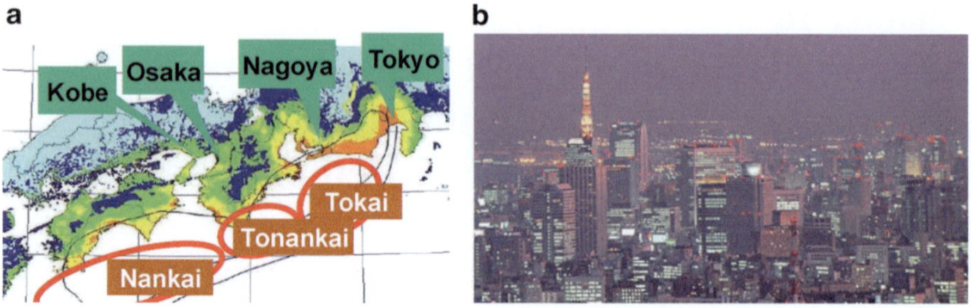


**Fig. 16.1** (**a**) Ocean-ridge earthquakes anticipated in Japan, (**b**) office buildings located in the Tokyo metropolitan area

buildings whose fundamental natural periods are several seconds. Because high-rise buildings perform very important roles in the Japanese economy (Fig. 16.1b), severe damage to them will cause extreme difficulties throughout Japan. In this study, a series of large-scale shaking table tests on high-rise steel buildings were conducted by using E-Defense. The tests on the seismic capacity of existing high-rise steel buildings were conducted in 2008, and again on the effects of retrofit for those buildings in 2009. In addition, damage due to large floor response and the effects of preparation were verified for office and residential rooms.

## 16.2 Test Method for High-Rise Buildings

The average height of Japanese high-rise buildings constructed in the past 30 years is about 80 m (BRI. 2005). This height was chosen in this study, and a 21 story building was adopted as the prototype. The height of the building is about four times what the E-Defense shaking table facility (Ogawa et al. 2001) can accommodate. In this study, a substructure test method was employed, and a test specimen that represented a 21 story high-rise building was shaken on the E-Defense shaking table (Chung et al. 2010). Figure 16.2 shows the procedure for constructing the modified model within a few degrees of freedom of the original model with 21 degrees of freedom. The responses of the three masses in the substitute layers were assumed to represent the responses of the ninth, fourteenth, and nineteenth floors of the original model.

Figure 16.3 shows the designed test specimen. The lower part is the test frame. The upper part consists of three substitute layers made of thick concrete slabs and rubber bearings. The test frame has a two and one span in the longitudinal and transverse directions. The member sizes were determined using the allowable stress design method, with the base shear ratio $C_b$ of 0.12. In the longitudinal direction, a built-up wide flange section of H $600 \times 200 \times 9 \times 19$ was arranged with the shop weld connection detail. In the transverse direction, a honeycomb section of H $800 \times 199 \times 10 \times 15$ was arranged with the field weld connection detail. The

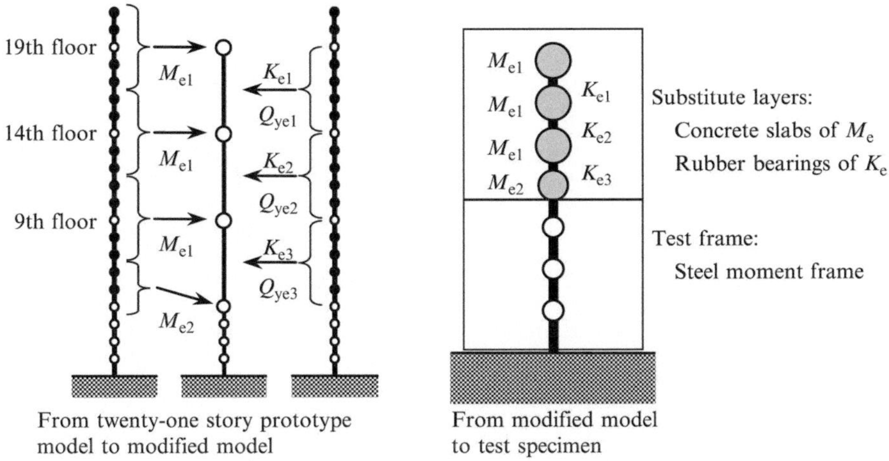


**Fig. 16.2** Development from 21 story prototype model to test specimen

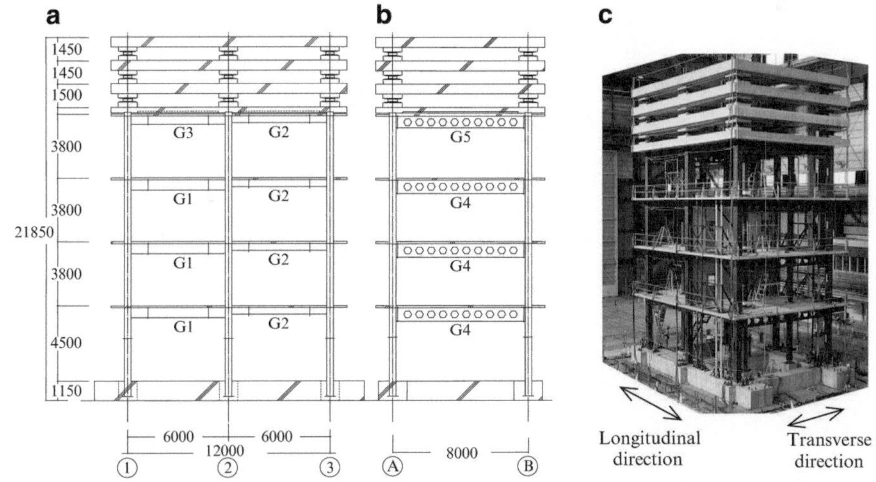


**Fig. 16.3** E-Defense test specimen (unit: millimeters): (**a**) elevation-A, (**b**) elevation-1, (**c**) overview of setup

details of beams and beam-to-column connections were chosen based on the early time design practice. WUF-B was adopted for the field weld connection detail. The columns were stronger than beams by a column-to-beam strength ratio of about 1.5. Reinforced concrete slabs with a thickness of 120 mm were cast at every floor. Rubber bearings and steel dampers were placed at each substitute layer. The steel dampers were used to mimic the nonlinearity of the upper part, and the stiffnesses and strengths of the devices were carefully adjusted.

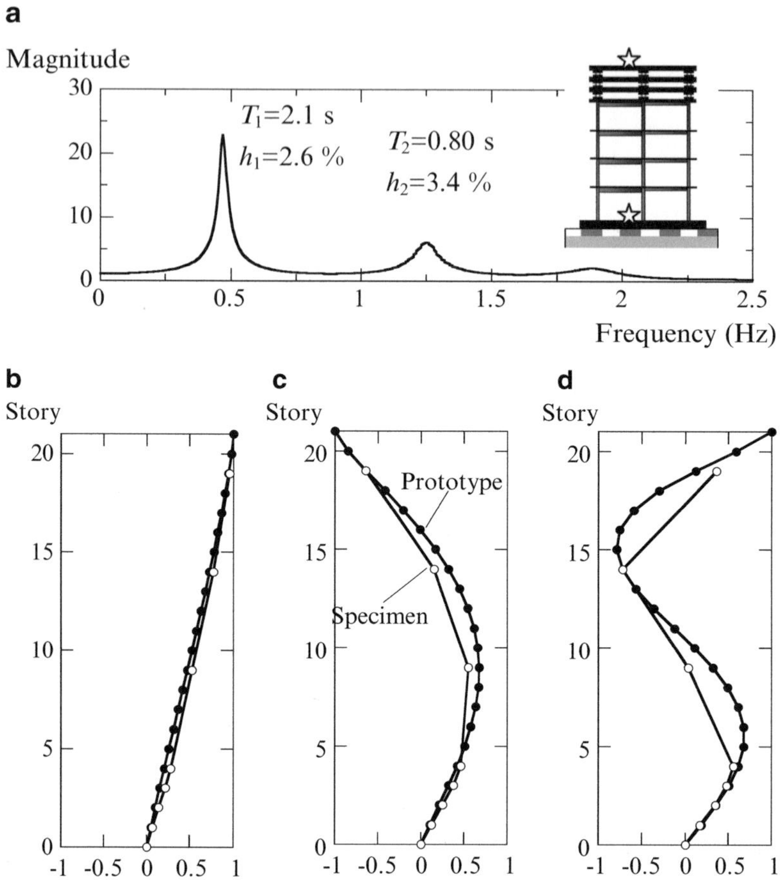


**Fig. 16.4** Test results from white noise vibrations: (**a**) transfer function curve, (**b**) first mode shape, (**c**) second mode shape, (**d**) third mode shape

Figure 16.4 shows the frequencies and damping ratios of the first three modes of the test specimen obtained from the white-noise test applied prior to each main test. The first-mode periods were 2.13 and 2.24 s respectively in the longitudinal and transverse directions. As for the first three mode shapes of the specimen, the first, second and third substitute layers are plotted at the equivalent heights of the prototype, that is, at the ninth, fourteenth, and nineteenth floors. A comparison of the two mode shapes is very reasonable for all three modes.

Figure 16.5 shows the time histories and velocity response spectra of the input waves. For long period ground motions, two synthesized waves were adopted. The HOG wave (PGV = 0.40 m/s) was predicted at a Kawasaki site, which is next to Tokyo, and rupture of the Tokai trough was supposed. The SAN wave (PGV = 0.51 m/s) was predicted at a Nagoya site, and simultaneous ruptures of

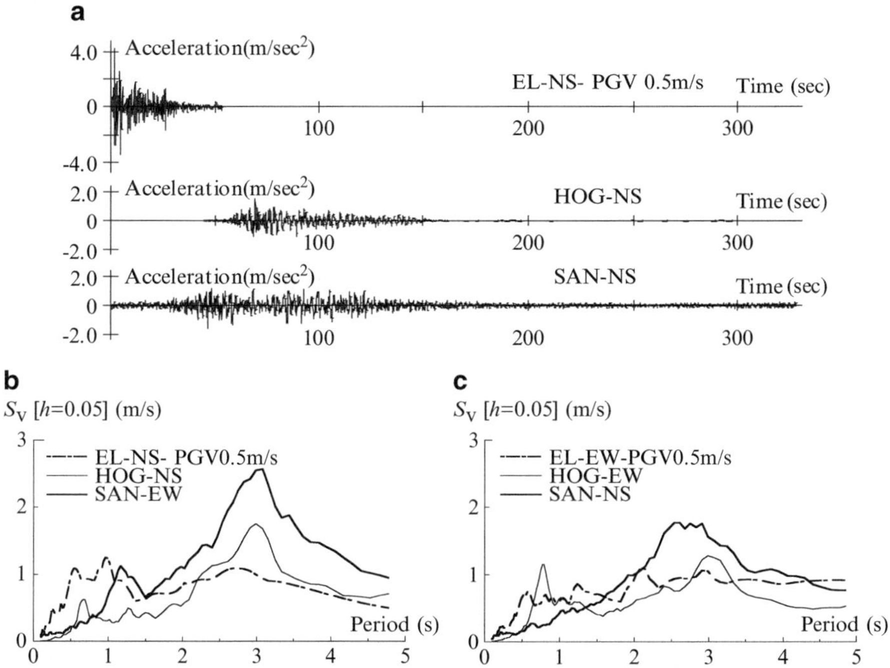


**Fig. 16.5** (**a**) Time histories of input waves, (**b**) and (**c**) pseudo velocity response spectra of input waves: longitudinal direction (**b**) and transverse direction (**c**)

the Tokai and Tonankai troughs were supposed. These two waves had predominant periods of about 3 s and durations of 200 and 320 s respectively. At the period of 2.4 s expected in the inelastic specimen, the HOG wave is about 1.2 times larger in amplitude than EL2, which is the level 2 El Centro wave scaled to 0.5 m/s in PGV and was used for the Japanese seismic design. The SAN wave is about two times larger than the EL2 wave. The EL2 wave, HOG wave and SAN wave were applied sequentially.

## 16.3 Seismic Performance of Existing High-Rise Steel Buildings

Figure 16.6 shows the time histories at the second story of the test frame. For the EL2 wave, the maximum inter-story drift was slightly smaller than the design limit of 0.01 rad. For the HOG wave and SAN wave, the maximum inter-story drifts were 0.011 rad and 0.017 rad. However, a number of inter-story drifts of more than 0.01 rad were repeated over and over in the long-period input waves. Figure 16.7

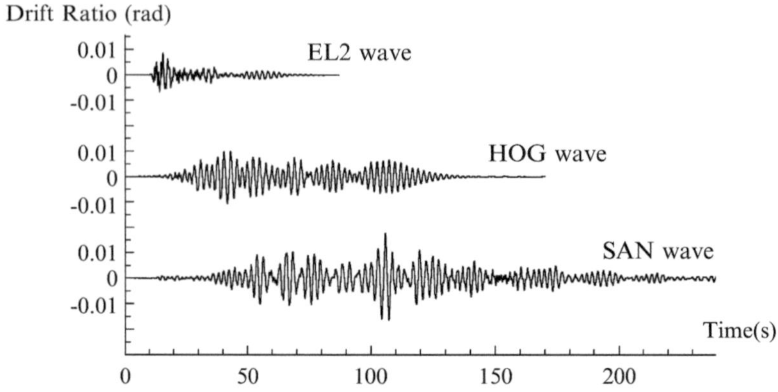


**Fig. 16.6** Inter-story drift in the second story of test frame (longitudinal direction)

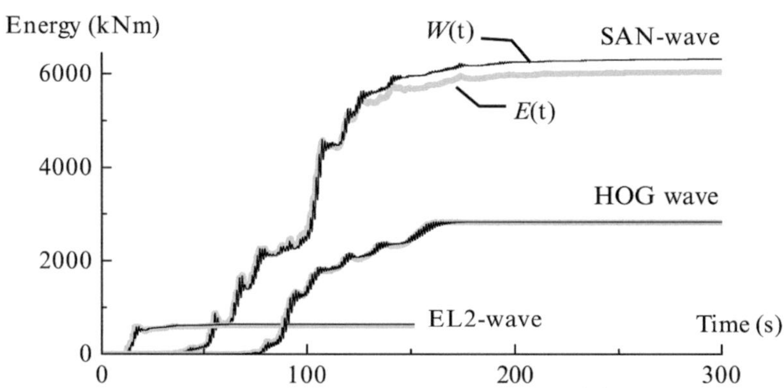


**Fig. 16.7** Total input energy of test specimen (longitudinal direction)

shows the input energy to the test specimen in each test. The test specimen when subjected to the HOG wave and SAN wave exhibits large input energy more than four times that of the EL2 wave in the durations of 200 and 320 s respectively.

The story ductility ratio, $\mu$, was defined as the maximum inter-story drift divided by the story yield drift. The story yield drift angle was estimated as 0.0055 and 0.0045 rad for the longitudinal and transverse directions, respectively. The story cumulative plastic deformation ratio, $\eta$, was defined as the total energy dissipation of each story divided by the product of the story yield drift and the corresponding story shear force (Akiyama 2002). Both $\mu$ and $\eta$ reached their largest values in the second story. The $\mu$ values observed in both the EL2 and HOG waves were about two, and that observed in the SAN wave was about three. On the other hand, the $\eta$ values of the HOG and SAN waves reached four times and fifteen times that observed in the EL2. Large values of $\eta$ relative to $\mu$ are very notable in the long period input waves.

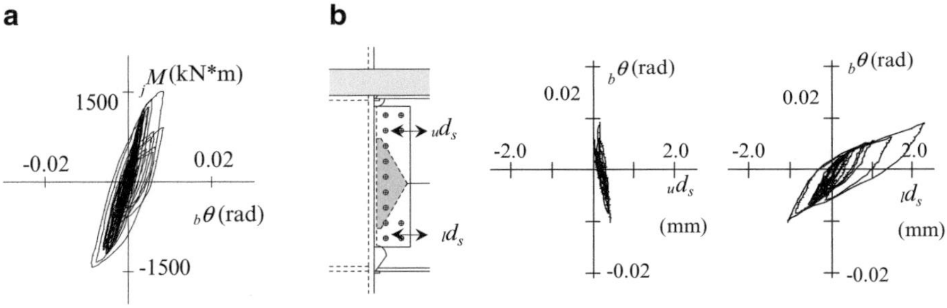


**Fig. 16.8** (**a**) Bending moment versus rotation relations of beam end of field weld WUF-B connection, (**b**) slip of shear plate in WUF-B connection ($_ud_s$: upper level, $_ld_s$: lower level)

In the first test of the SAN wave, fractures of beam ends occurred at three field-weld connections (WUF-B) arranged in the transverse direction. The fracture of WUF-B was located at the weld boundary next to the weld access hole. Several cracks were also observed at the weld boundary next to the weld access hole in other unfractured connections. No obvious damage was observed in the shop weld connections placed in the longitudinal direction, and eventually, two more one-direction tests for the longitudinal direction were conducted for the fracture of the shop weld connection. Figure 16.8a shows bending moment versus rotation relations ($M$-$\theta$ relations) at the beam ends of the field weld WUF-B connection. The bottom flange fractured at about a rotation of 0.01 rad. During positive bending, the resistance decreased due to the fracture, while the resistance by the bolts connecting the shear plate and web remained. The slips of the shear plate are notably larger at the lower level than at the upper level due to the composite effect of the RC floor slabs, as shown in Fig. 16.8b. As the maximum strain value at the bottom flange was confirmed as significantly larger than at the upper flange, the bottom flange fracture was attributed to the amplified strains due to this composite effect as well as the large cumulative inelastic deformations.

## 16.4 Verification of Retrofit for High-Rise Steel Buildings

### *16.4.1 Reinforcement of Beam to Column Connections*

Figure 16.9 shows reinforced WUF-B connections (Chung et al. 2012). As for the supplementary-web weld (SW) connection, welds were applied at the beam web to column face as well as the supplemental fillet welds at the shear tab. The shear tab worked as the backup plate when the welding was performed. As for the wing plate (WP) connection and vertical haunch (VH) connection, because of the presence of RC floor slabs, retrofit was limited to the bottom flange. The width of the wing plate was extended to the column width to enlarge the bottom flange section as much as

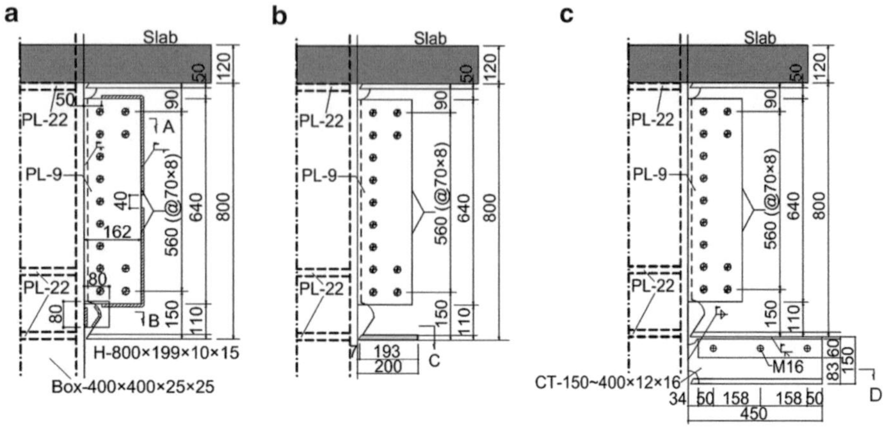


**Fig. 16.9** Reinforced field weld beam-to-column connections: (**a**) supplementary-web weld (SW) connection, (**b**) wing plate (WP) connection, (**c**) vertical haunch (VH) connection (unit: millimeters)

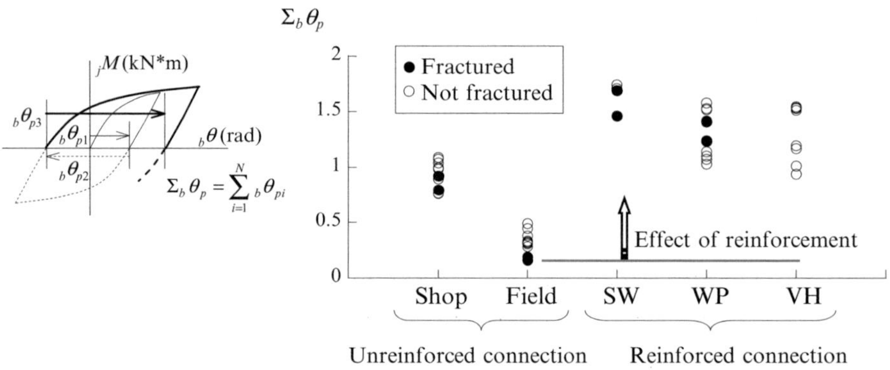


**Fig. 16.10** Cumulative plastic rotation capacities of unreinforced shop-weld and field-weld connections of the 2008 test and reinforced field-weld connections of the 2009 test

possible. The vertical haunch was first bolted to the shear plate that was pre-welded to the bottom flange. Then the welds at the haunch-to-column and the haunch-to-beam bottom flange were applied. The haunch flange was enlarged to the column width to transfer the force to the column. The length of haunch along the beam direction (450 mm) was configured such that the shear force transfer between the beam bottom flange and haunch web would be secured when the welds between the haunch flange and column face reached the maximum tensile force. The haunch size was minimized, and the weight was about 30 N, which could be carried by a person. The deformation capacities of these connections were verified by sequential tests continued until their fractures. Figure 16.10 shows the test results of the reinforced

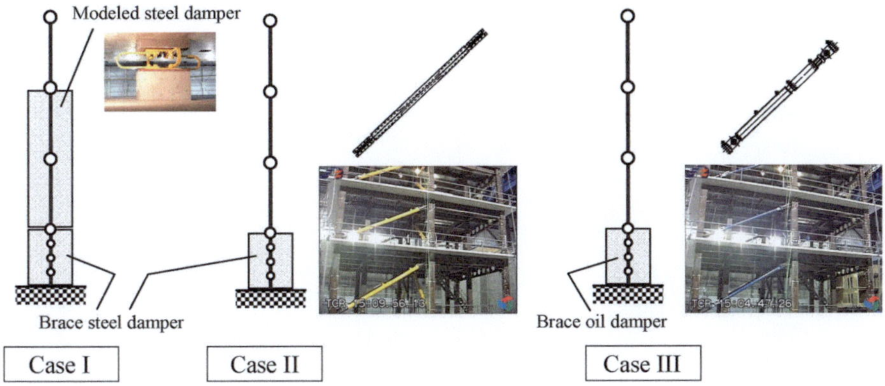


**Fig. 16.11** Test series on retrofit with dampers

connections as well as the unreinforced connections. The deformation capacities of the field weld (originally WUF-B) connections were greatly enhanced by these reinforcement methods.

## *16.4.2 Retrofit with Dampers*

Figure 16.11 shows the concept of the tests on retrofit with dampers. In Case-I and Case-II for retrofit with steel dampers, buckling-restrained brace dampers were incorporated in the lower steel frame, and modeled steel dampers were utilized for the substitute layers. In Case-III for retrofit with oil damper, oil dampers in diagonal braces were incorporated in the lower steel frame. Figure 16.12 shows the longitudinal force versus deformation relations ($F_d$-$\Delta_d$ relations) of the steel and oil dampers. In the SAN wave, the maximum story drift angles of the specimen were reduced to less than 0.01 rad by the hysteretic energy dissipations of the dampers.

Figure 16.13a shows the energy spectrum of input waves. The spectrum was given by the elastic SDOF with a damping ratio of 10 %. The SAN wave exhibits predominant magnitudes at around 3 s, while the EL2 wave had a flat shape. The total input energies of the test specimens characterized by Case-I, Case-II and Case-III as well as with no dampers correspond to the estimations at the individual natural periods in this format. That is, the seismic demand in long period ground motions is reasonably estimated by energy spectra. Each input energy was mostly distributed in the test frame. As shown in Fig. 16.13b, in the test frame, more than 70 % of the energy was absorbed by the steel dampers or the oil dampers. As for the BRB damper, the cumulative inelastic strain capacity was estimated to be about ten times larger than the seismic demand in the SAN wave.

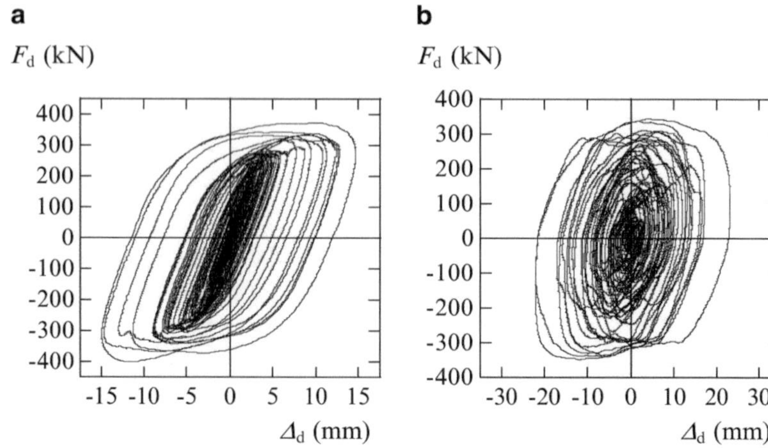


**Fig. 16.12** Hysteretic behavior of damper in SAN wave: (**a**) steel damper (Case II), (**b**) oil damper (Case III)

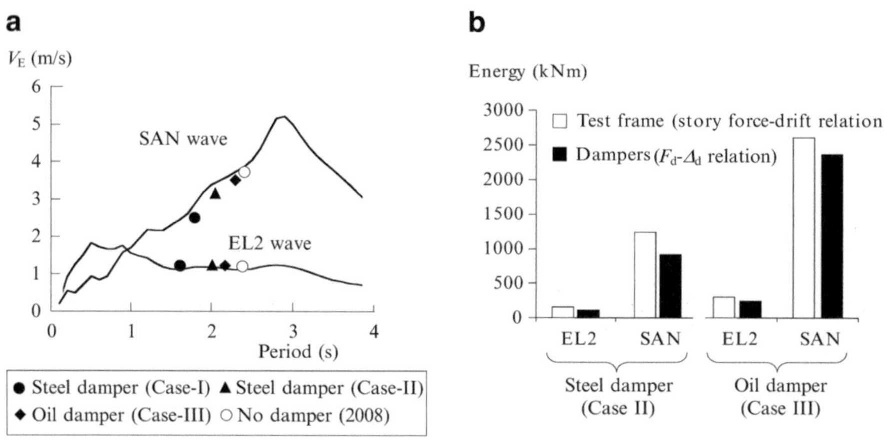


**Fig. 16.13** (**a**) Energy spectra, (**b**) energy dissipation in test frame and dampers

## 16.5 Safety of Rooms in High-Rise Buildings

The behaviour of furniture in high-rise buildings was also studied. The test for furniture was set by focusing on enlightenment concerning disaster prevention. Long-period, long-duration shaking may hit high-rise buildings and produce very large floor responses. Such responses would cause serious damage to non-structural elements, furniture and other building contents, particularly in the upper floors. This test specimen produces a large floor response corresponding to the nineteenth floor at the roof level of substitute layers. Figure 16.14a shows an overview of the shaken

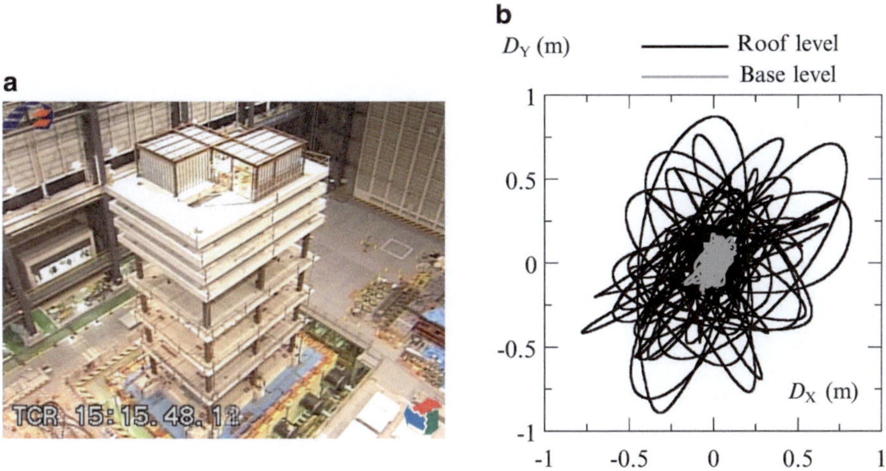


**Fig. 16.14** (**a**) Overview of test specimen, (**b**) floor responses of roof and base levels

**Fig. 16.15** Test result of residential room: (**a**) initial condition of unprepared room at 0 s, (**b**) damage of unprepared room at 130 s, (**c**) damage of prepared room at 130 s

test specimen, and Fig. 16.14b shows the SAN wave input motion of the base level and floor responses of the roof level. Almost identical pairs of test rooms except for the seismic preparation were setup on the roof level of the test specimen. The displacement amplitude of the roof level became five times that of the base level as shown.

Figure 16.15 shows the notable test results of prepared and unprepared residential rooms. The unprepared room suffered significant damage to its contents, while rooms prepared with special tools had very slight damage. In the floor response of high-rise buildings, slender shelves and refrigerators with a high aspect ratio overturned due to the maximum acceleration as well as large velocity, while wooden furniture with small friction coefficient and furniture supported by casters continued sliding extensively due to the large floor displacement. From these test results, the critical need to clamp furniture against overturning and sliding became evident. The strategy adopted here is to disseminate information regarding such unknown factors

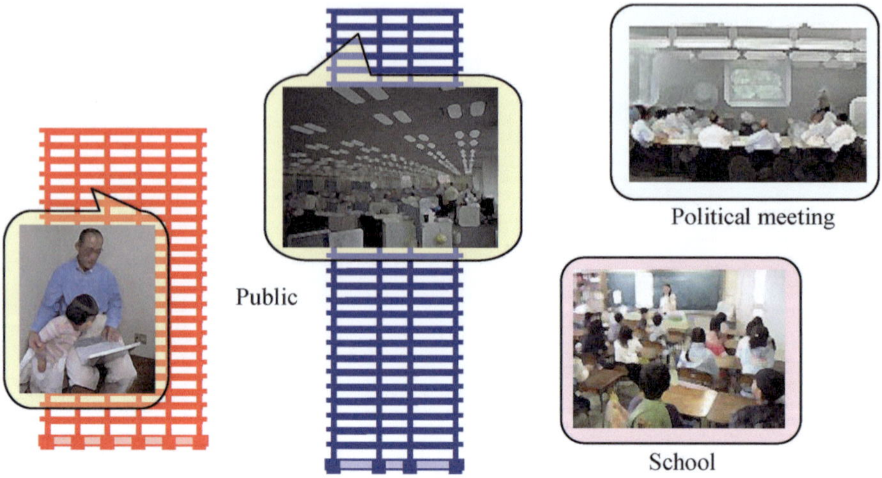


Fig. 16.16 Use of video data for disaster prevention enlightenment and education

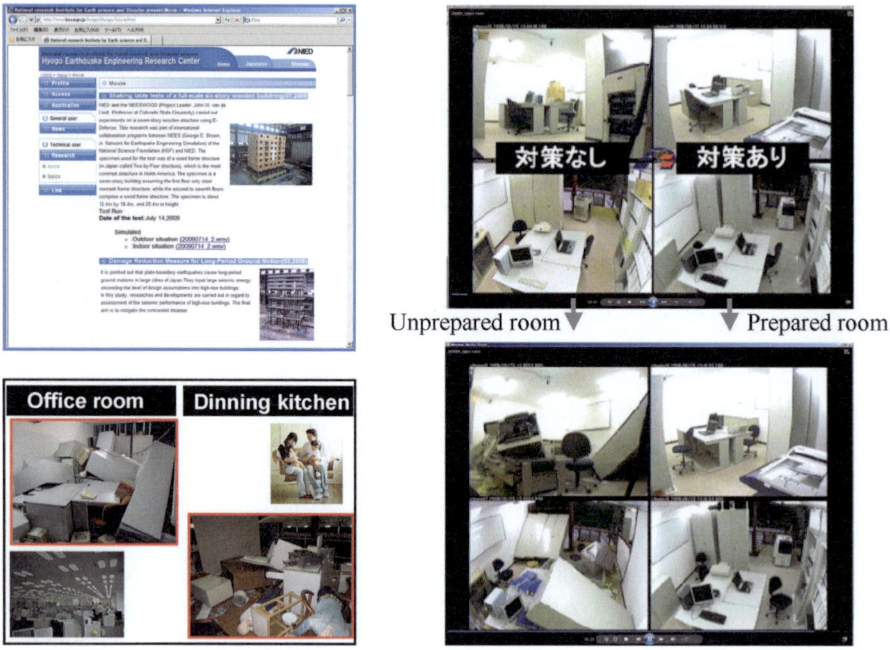


Fig. 16.17 Downloadable video files at http://www.bosai.go.jp/hyogo/movie.html

to the public as shown in Fig. 16.16. The contrast between prepared rooms and unprepared rooms represented in the videos will strongly highlight the need for preparations. Such data files are edited for use in schools, at conventions and so on. The data files are now open at the NIED web site (Fig. 16.17).

## 16.6 Summary

The pattern of previous earthquakes suggests that Japan is most likely to be hit by another large ocean-ridge earthquake by the middle of this century. One serious concern about such events is long-period, long-duration shaking that may hit large cities. Long-period ground motion may induce very large floor response characterized by large velocities and displacements, to hundreds of high-rise buildings. Their structures will sustain a number of cyclic inelastic deformations. In the project based on E-Defense, a substructure test method was employed for the large-scale tests of high-rise buildings.

Focusing on structural performance, a steel moment frame having real connection details was tested with the test system. In the long-period ground motions, a number of cyclic deformations were applied to the test frame. The capacities of the beam to column connections were identified in terms of cumulative inelastic deformation. The cumulative inelastic deformation finally caused fractures at beam-to-column connections. The bottom flange fracture was attributed to amplified strains due to the composite effect of the RC floor slabs as well as the large cumulative inelastic deformations. The introduced reinforcement of such beam-to-column connections drastically enhanced the deformation capacity in terms of cumulative inelastic deformation. Dampers incorporated into the test specimen effectively dissipated input energy and drastically reduced the demand imposed on the test frame.

Another test focusing on the safety of rooms was conducted using the same test specimen. The roof of the test specimen corresponding to the nineteenth floor reproduced a large floor response characterized by large velocities and displacements. In the rooms set up there, the slender shelves overturned immediately, and the heavy copy machine supported on casters moved extensively and experienced repeated collisions. The contrast between the prepared and unprepared rooms was visually verified and the videos were successfully recorded. The edited video files are now open at the NIED web site. Such data begins to contribute in a practical manner to the seismic safety of Japanese society.

A series of tests provided a set of unknown data on the seismic behaviour of high-rise buildings subjected to long period ground motions. According to these tests, checking the seismic performance of existing high-rise buildings is urgently needed. If the capacity to resist long-period ground motions is lacking, such buildings need to be appropriately retrofitted. The preparation of the rooms is not difficult and costs relatively little. This should be applied to all high-rise buildings immediately. The Tokyo metropolitan government prepared strict regulations regarding the preparation of office rooms after seeing the E-Defense test videos.

## References

Akiyama H (2002) Collapse modes of structures under strong motions of earthquake. Ann Geophys 45(6): 791–798

Building Research Institute (2005) The research report about the influence and seismic retrofit technology to the building subjected to long-period ground motion. Tokyo, Japan (in Japanese)

Chung Y, Nagae T, Hitaka T, Nakashima M (2010) Seismic resistance capacity of high-rise buildings subjected to long-period ground motions: E-defense shaking table test. J Struct Eng ASCE 136(6):637–644

Chung Y, Nagae T, Matsumiya T, Nakashima M (2012) Seismic capacity of retrofitted beam-to-column connections in high-rise steel frames when subjected to long-period ground motions. Earthq Eng Struct Dyn 41(4):735–753

Kamae K, Kawabe H, Irikura K (2004) Strong ground motion prediction for huge subduction earthquakes using a characterized source model and several simulation techniques. In: Proceedings of the 13th world conference earthquake engineering, pp 655–666

Ogawa N, Ohtani K, Katayama T, Shibata H (2001) Construction of a three-dimensional, large-scale shaking table and development of core technology. Philos Trans R Soc Lond 359:1725–1751

# Chapter 17
# Pseudo-dynamic Performance Evaluation of Full Scale Seismic Steel Braced Frames Using Buckling-Restrained and In-Plane Buckling Braces

**Keh-Chyuan Tsai, Pao-Chun Lin, Ching-Yi Tsai, and An-Chien Wu**

**Abstract** A series of hybrid and cyclic loading tests were conducted on a 3-story single-bay full-scale frame specimen in the Taiwan National Center for Research on Earthquake Engineering (NCREE) in 2010. There were total three hybrid tests conducted on this frame specimen. Two different lateral force resistant systems including buckling-restrained braced frame (BRBF) and special concentrically braced frame (SCBF) were tested separately. The newly developed thin and welded end-slot buckling-restrained braces (BRBs) were adopted for the first two BRBF hybrid tests. The in-plane (IP) buckling braces were installed in the SCBF for the last hybrid test. The BRBF or the SCBF was designed to sustain a design basis earthquake in Los Angeles and ground motion LA03 was used as the input ground motion for these tests. The inter-story drift reached near 3 % and 4 % radians in the BRBF and SCBF hybrid tests, respectively. The maximum base shear also reached more than 2,000 kN in these tests. Test results indicate that both frame systems performed satisfactorily. This chapter presents the seismic performance of the BRBF and SCBF hybrid tests.

**Keywords** Pseudo-dynamic test • Buckling-restrained brace • Special concentrically braced frame • Seismic performance

K.-C. Tsai (✉) • C.-Y. Tsai
Department of Civil Engineering, National Taiwan University, Taipei, Taiwan
e-mail: kctsai@ntu.edu.tw; d98521004@ntu.edu.tw

P.-C. Lin • A.-C. Wu
NARLabs, National Center for Research on Earthquake Engineering, Taipei, Taiwan
e-mail: pclin@ncree.narl.org.tw; acwu@ncree.narl.org.tw

M. Fischinger (ed.), *Performance-Based Seismic Engineering: Vision for an Earthquake Resilient Society*, Geotechnical, Geological and Earthquake Engineering 32,
DOI 10.1007/978-94-017-8875-5_17, © Springer Science+Business Media Dordrecht 2014

## 17.1 Introduction

The BRBF or SCBF has been evolved into a very effective system for severe seismic applications. They are widely adopted in a number of new buildings and retrofit construction projects all over the world. Recently, novel BRB and IP buckling brace connection designs were investigated in Taiwan. To demonstrate the seismic performances of these brace frame designs (Tsai et al. 2010; Lumpkin 2009; Structural Engineering Institute 2010; Lin et al. 2012), a series of hybrid and cyclic loading tests were conducted on a 3-story single-bay full-scale frame specimen in NCREE in 2010.

### *17.1.1 Taiwan-US Collaborative Experimental Program on Steel Braced Frames*

The properties of SCBF systems have the potential to meet multiple performance objectives. However, previous researches have shown that the current design procedures can lead to soft stories, inadequate gusset plate connections, unexpected failure modes, brittle welds and premature brace failure. To improve the performance and to meet the engineering needs of future seismic load resisting systems, an international research team has been working to develop Tomorrow's Concentric Braced Frame (TCBF) systems. The "NEES-SR SG International Hybrid Simulation of Tomorrow's Braced Frames" is an international collaborative research among NCREE, University of Washington (UW), University of California, Berkeley (UCB) and University of Minnesota (UM). In this project, two series of large-scale braced frames have been constructed and tested at NCREE.

The first test series, entitled TCBF1, a total of five tests were conducted on a reusable, two story test frame which had an identical brace configuration of a multi-story X-brace and composite concrete slabs on each floor. The first three test specimens were designed with out-of-plane (OOP) buckling hollow structural sections (HSS) and wide flange brace shapes (Tsai et al. 2010). The fourth and fifth test specimens were designed with BRBs and IP buckling wide flange braces, respectively. All of these tests were conducted for one degree-of-freedom cyclic loading tests applied on the top floor.

The second test series was entitled TCBF2, composed of seven tests on a reusable full scale, three story, multi-level X-brace in the lower two stories and a chevron brace configuration in the third story. The test setup of first three test specimens was very similar to the one that is described for TCBF1. More details can be found in reference (Lumpkin 2009). The next section presents the fourth and fifth tests conducted using novel BRBs and hybrid test procedures. Then the paper introduces the hybrid test conducted on the 3-story SCBF using IP buckling wide flange braces.

**Fig. 17.1** The connection detail of the (**a**) thin BRB, (**b**) the welded end-slot BRB, and (**c**) the local bulging failure of Specimen 1BRB-1S

### *17.1.2 Hybrid Simulation Test Programs*

There were total three hybrid tests conducted on this frame specimen. Two different lateral force resistant systems including BRBF and SCBF were tested separately. In each of the BRBF hybrid tests, six BRBs using thin (Fig. 17.1a) and welded end-slot (Fig. 17.1b) profiles were adopted. Finally, the IP buckling braces were adopted in the re-used frame in the follow-up SCBF hybrid test. The BRBF or the SCBF was designed to sustain a design basis earthquake in Los Angeles. The acceleration record LA03 was used as the input ground motion for these tests. The steel beam and column members sent from the U.S. were used to fabricate the frame specimen in Taiwan. Prior to the hybrid tests introduced in this paper, the frame had been utilized in the cyclic loading tests for three different designs of SCBF (Lumpkin 2009).

## 17.2 Design of BRBF and SCBF Specimens

### *17.2.1 BRBF Specimen*

The 3-story frame specimen is 9.27 m tall and 6 m wide. It consists of two A992 W12 × 106 (327 × 310 × 15 × 25 mm) columns, one W24 × 94 (617 × 230 × 13 × 22 mm) top beam, and the W21 × 68 (537 × 210 × 11 × 17 mm) section for the middle and lower beams (Fig. 17.2a, b). Welded moment connections were adopted for both ends of the top and middle beams and a bolted web connection was made for each end of the lower beam. A 200 mm thick concrete slab was constructed for each floor to effectively transfer the lateral actuator forces on the specimen. The floor lateral displacements were imposed by two 961 kN, MTS-243 static actuators, and monitored by two temposonic transducers installed for each floor. During the tests, the analytical predictions and the key experimental results were broadcasted on the internet (http://exp.ncree.org/cbf).

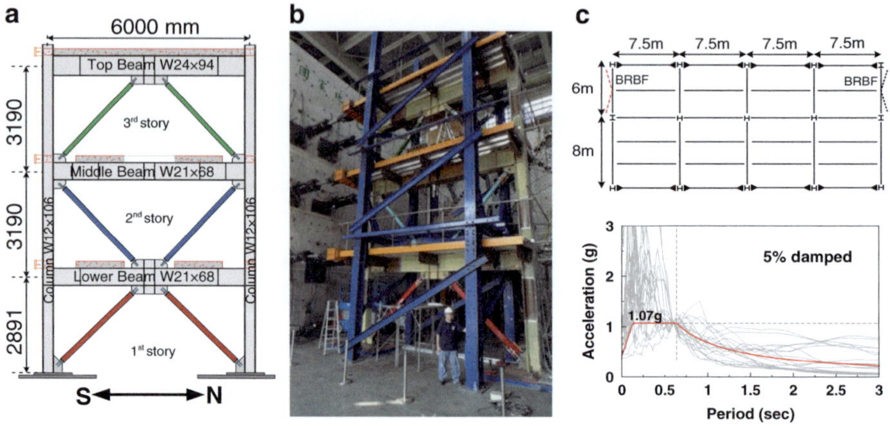


**Fig. 17.2** (**a**) The BRBF elevation, (**b**) the test setup, (**c**) floor framing plan, and (**d**) design acceleration response spectrum and LA01 ~ LA20 acceleration response spectra

**Table 17.1** The design detail of the BRB specimens

| Specimen | Location | Steel casing (mm) |
|---|---|---|
| 3BRB | 3rd story | HSS 125 × 125 × 4.5 |
| 2BRB | 2nd story | Pipe 139 × 4 |
| 1BRB-1 (hybrid test1) | 1st story | HSS 150 × 100 × 6 |
| 1BRB-2 (hybrid test2) | 1st story | HSS 150 × 100 × 9 |

The area of each floor in the 3-story prototype building is 420 m$^2$ (Fig. 17.2c). The design dead load 750 kgf/m$^2$ and the reduced live load 125 kgf/m$^2$ are assumed uniformly distributed over the floor slab. The lateral force resisting system is two steel moment resisting frames and two steel BRBFs in the longitudinal and transverse directions, respectively. The BRBF design is to let the fundamental vibration period in the transverse direction of the prototype building be about 0.6 s commonly found in a typical low rise BRBF. As prescribed in seismic building requirements (Structural Engineering Institute 2010), the 0.6 s elastic spectral acceleration associated with the design base earthquake (DBE) is 1.07 g as shown in the design response spectrum (Fig. 17.2d). The design base shear force for each BRBF is 618 kN. The design of the BRBF follows the procedures prescribed in the model seismic steel building provisions (Structural Engineering Institute 2010). The six BRBs all have the identical core cross sectional area of 1,110 mm$^2$ (15 × 74 mm) with a nominal yield strength of 383 kN. The newly developed welded end-slot BRB (Tsai et al. 2014) and thin profiles were adopted in the test (Lin et al. 2012). The two BRBs in the 1st story resist about 80 % of the design base shear of 618 kN, while each reaches about 90 % of the nominal yield strength in both tension and compression. Table 17.1 shows the details of the BRB designs.

The input earthquake was chosen from the 20 ground motions, adopted in the SAC join research project (Gupta and Krawinkler 1999) and all scaled to the 10/50

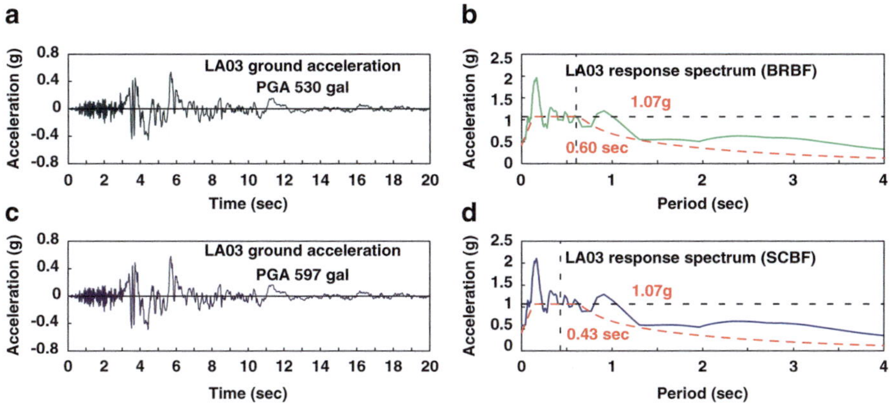


**Fig. 17.3** (**a**) LA03 ground acceleration, and (**b**) the 5 % damped elastic response spectrum for BRBF test; (**c**) LA03 ground acceleration, and (**d**) the 5 % damped elastic response spectrum for SCBF test

hazard level (475 years return period). The ground acceleration record LA03 scaled to a peak ground acceleration (PGA) of 530 gal was chosen as the input earthquake accelerations from a series of nonlinear response history analyses using the PISA3D program (Lin et al. 2009). Figure 17.3 shows the LA03 ground accelerations and the elastic response spectrum. The analysis indicated that the peak inter-story drift of about 0.04 rad could sufficiently reflect a rather severe demand on the frame and BRB specimens.

## *17.2.2 SCBF Specimen*

There was no evident failure found on beams and columns of the frame specimen after the BRBF tests. Thus, the gussets and the BRBs were removed and replaced by six wide flange (H150 × 150 × 7 × 10) braces. This SCBF represents one of the lateral force resistance frames in the 3-story prototype office building with a floor framing plan similar to Fig. 17.2c but the floor area increased from 420 to 765 $m^2$. The fundamental period of the SCBF is 0.43 s found from the numerical model. The elastic spectral acceleration is 1.07 g, and the design base shear force is 952 kN. The ground motion record LA03 scaled to a PGA of 597 gal was chosen as the input earthquake from a series of nonlinear response history analyses using the OpenSees program (McKenna 1997). Figure 17.3 shows the scaled LA03 ground accelerations and the associated elastic acceleration response spectrum for the SCBF test. Table 17.2 summarizes the floor areas, design periods, design base shears, ground motion PGAs, fundamental periods and damping ratios for the BRBF and SCBF. More details about the SCBF specimen can be found in Tsai et al. (2013).

**Table 17.2** Summary of key design parameters for BRBF and SCBF

| Specimen | Floor area ($m^2$) | Design period (s) | Design shear (kN) | PGA (gal) | Period (s) | Damping ratio (%) |
|---|---|---|---|---|---|---|
| BRBF | 420 | 0.60 | 618 | 530 | 0.588 | 1.10 |
| SCBF | 765 | 0.43 | 952 | 597 | 0.460 | 0.30 |

## 17.3 Hybrid Tests of BRBF

To evaluate the actual vibration period and the inherent damping of the BRBF specimen before the hybrid tests, a free vibration hybrid test was conducted by first pulling the initial displacements of 3.45 mm, 6.94 mm, and 9.00 mm for the 1st to the 3rd floor, respectively. These initial displacements were based on the mode shape computed from the modal analyses. The BRBF specimen system's fundamental vibration period was found 0.588 s and the damping was about 1.10 %.

### *17.3.1 BRBF Hybrid Test1*

The LA03 ground accelerations with a PGA 530 gal were applied in the hybrid test1 which was successfully completed within 4 h. Figure 17.4 shows the frame responses. It could be found that the LA03 ground accelerations imposed a pulse-like effect on the BRBF during the shaking time from about 4.0 to 6.0 s. The maximum inter-story drifts and story shears are given in Table 17.3. The maximum base shear reached 2,134 kN and the maximum inter-story drift of 2.93 % radian was found in the 2nd story.

At about 6.0 s earthquake time, evident bulging out of the steel casing (Fig. 17.1c) was observed in the south BRB in the 1st story (1BRB-1S). It then fractured after 2.5 s more earthquake time. The maximum core strains of the BRB specimens were 3.9 %, 3.3 %, and 1.0 % for the 1st to the 3rd story BRBs, respectively. The BRB's cumulative plastic deformation (CPD) experienced in the hybrid test1 are listed in Table 17.4. The maximum CPD of 220 is found in the north BRB in the 1st story (1BRB-1 N). It is evident that the local bulging of the BRB steel casing was the major structural failure in the hybrid test1. The detailed investigation to prevent the steel casing bulging failure can be found in Lin et al. (2012). The CPDs for the BRBs in the 2nd and 3rd story were relatively small. There was no evident failure or crack on the gusset plate or welds. Thus, after replacing the two BRBs in the 1st story, the same ground acceleration record with a reversed direction was applied following the hybrid test1.

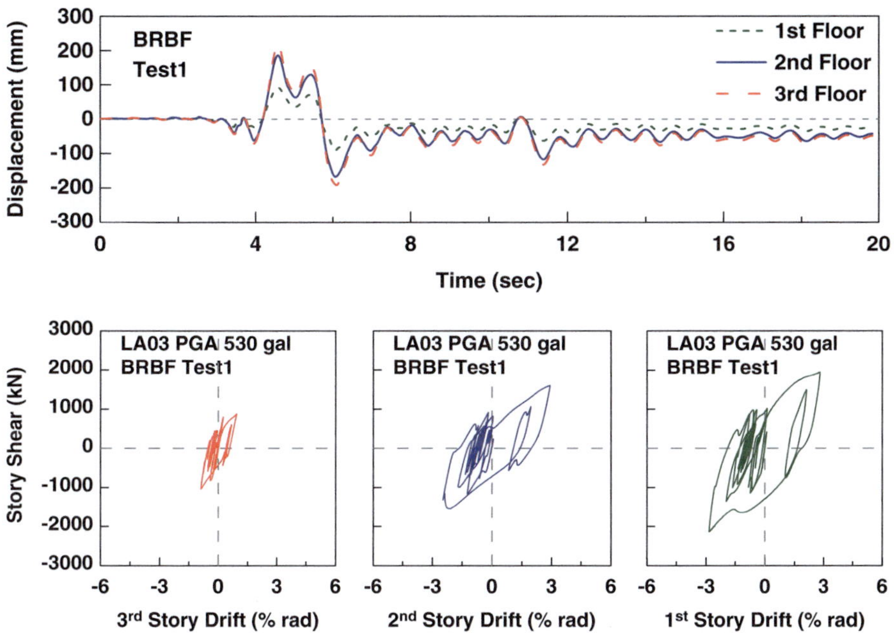


**Fig. 17.4** Floor lateral displacement histories and story shears versus inter-story drifts in BRBF hybrid test1

**Table 17.3** The maximum responses in the BRBF hybrid tests and the SCBF hybrid test

| | BRBF hybrid test1 | | BRBF hybrid test2 | | SCBF hybrid test | |
|---|---|---|---|---|---|---|
| Story | Inter-story drift (% rad) | Story shear (kN) | Inter-story drift (% rad) | Story shear (kN) | Inter-story drift (% rad) | Story shear (kN) |
| 3rd | 0.95 | 1,036 | 0.89 | 1,067 | 0.54 | 1,677 |
| 2nd | 2.93 | 1,605 | 3.27 | 1,701 | 4.00 | 1,992 |
| 1st | 2.83 | 2,134 | 2.80 | 2,207 | 4.94 | 2,539 |

**Table 17.4** BRB specimen's CPD gained in the hybrid test1 and test2

| | Hybrid test1 | Hybrid test2 | Total |
|---|---|---|---|
| 3BRB-N | 40 | 39 | 79 |
| 3BRB-S | 44 | 41 | 85 |
| 2BRB-N | 149 | 132 | 281 |
| 2BRB-S | 172 | 137 | 309 |
| 1BRB-1N | 220 | – | 220 |
| 1BRB-1S | 128 | – | 128 |
| 1BRB-2N | – | 166 | 166 |
| 1BRB-2S | – | 173 | 173 |

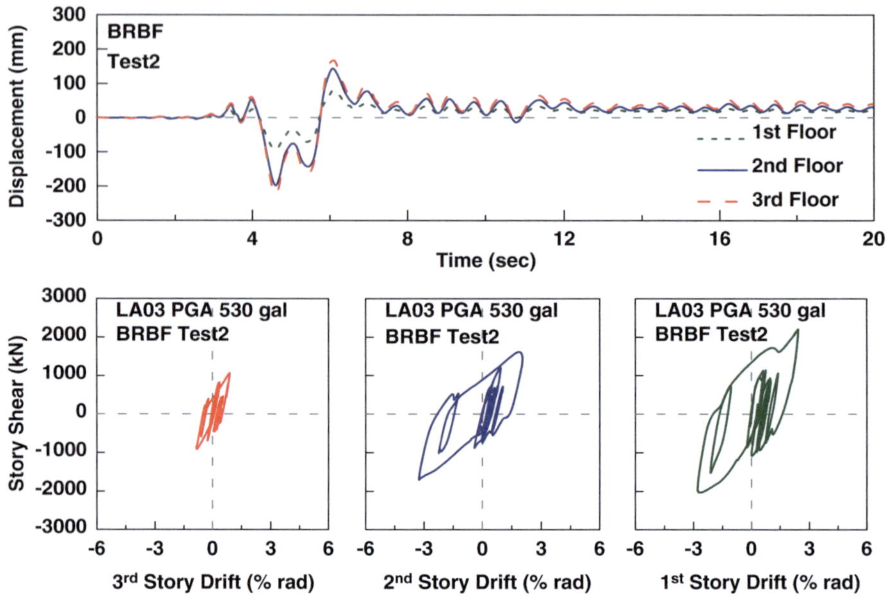


**Fig. 17.5** Floor lateral displacement histories and story shear versus inter-story drift relationships in BRBF hybrid test2

## *17.3.2 BRBF Hybrid Test2*

After the hybrid test1, the BRBs and the gusset plates in the 1st story were removed. The residual displacements were reduced first by properly pushing and/or pulling the BRBF several times. The brand new 1BRB-2 N and 1BRB-2S used the same material and size of core plate as those BRBs in the hybrid test1 but the steel casing wall thickness had been increased from 6 to 9 mm. The two new BRBs were installed in the 1st story after installing the new gusset plates. The same ground motion record LA03 but with reversed direction was applied. Figure 17.5 shows the key BRBF hybrid test2 responses. During the test, no evident damage of the BRBs could be observed.

However, a crack of about 20 mm was found at the weld toe in the 2nd story north upper gusset plate to column flange joint at 5.6 s earthquake time. Another crack of 10 mm was found later at the same joint but in the southern gusset plate. These crack lengths did not propagate further until the end of hybrid test2. The experimental maximum inter-story drifts and story shears are shown in Table 17.3. The magnitudes of the inter-story drifts and story shears are similar to those observed in the hybrid test1. All six BRBs demonstrated stable hysteretic responses and none of them buckled or fractured. The CPDs of the BRB experienced in the hybrid test2 are shown in Table 17.4. The CPDs gained in the 3rd story BRBs during the hybrid test2 are similar to those gained in test1. In the hybrid test2, the maximum CPD of 173 is found in the south BRB in the 1st story (1BRB-2S) while the CPDs for the BRBs in the 2nd and 3rd story were smaller. Since all six BRBs were not failed after hybrid test2, it was decided to conduct the cyclic loading test on the BRBF specimen without repairing the cracks of the welds near the gusset edges. The details of the cyclic loading test results are in Lin et al. (2012).

## 17.4 Hybrid Tests of SCBF

After the free vibration hybrid test, the fundamental vibration period and the damping ratio of the SCBF specimen were found 0.46 s and 0.3 %. Figure 17.6 shows the SCBF experimental responses in the hybrid test. The floor lateral displacement responses show that the LA03 ground accelerations imposed a pulse-like effect on the SCBF during the shaking time from about 4.0 to 6.0 s.

At 3.29 s into the record (Point 1 in Fig. 17.6), the 1st-story south brace buckled, then the 1st-story north brace buckled at 3.58 s (Point 2). The 2nd-story north and south braces buckled at 3.89 s (Point 3) and 4.66 s (Point 4), respectively. At this moment, the roof drift was 2.78 % radian. Except the 3rd-story braces, other braces were all in the inelastic range. The test was paused at 5.67 s (Point 5) into the acceleration record to inspect and photograph the specimen's damage conditions. Figure 17.7 shows that the inner flange of the 2nd-story south column top near the gusset was completely fractured at this point. This column flange fracture could have occurred before or right at the peak roof lateral displacement (Point 4 in Fig. 17.6). It should be noted that the beam and column framing of this specimen had been tested 6 times before this test. In each of these tests, gusset plates were flame cut near the column and beam flanges. The flange surfaces were then ground smooth before a new gusset was re-welded. At this point, the deformation of the knife plate and local buckle found in the 1st-story north brace are shown in Fig. 17.8a, b, respectively. No failures other than the column fracture were found at this point and the test was resumed. A second pause (Point 6) was made at 6.45 s into the test record, and local

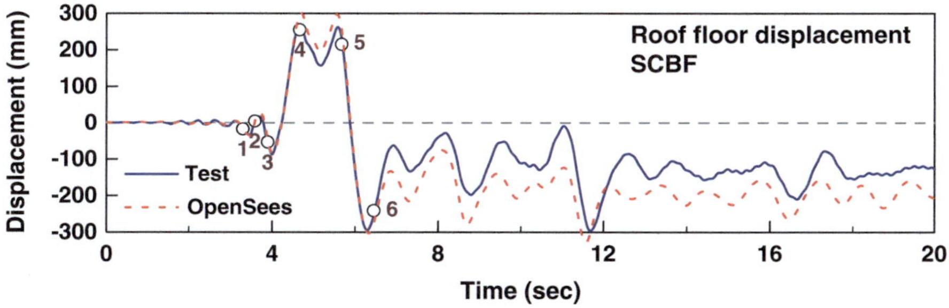


**Fig. 17.6** Test and predicted results of the roof displacement history in the SCBF

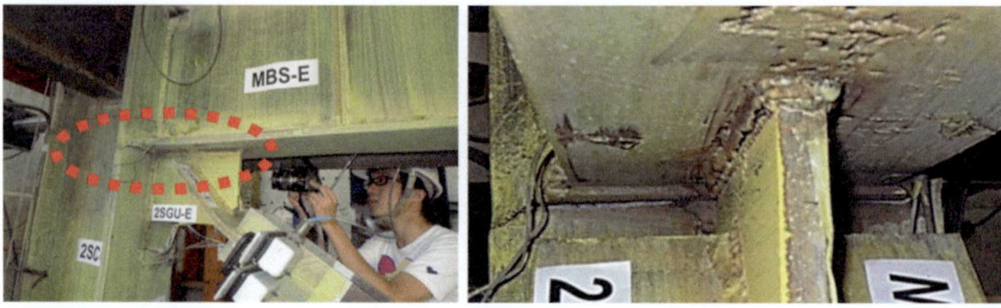


**Fig. 17.7** Fracture of inner flange of the 2nd story south column top near the gusset at the earthquake time 5.67 s

**Fig. 17.8** (**a**) The deformed condition of the knife plate, (**b**) local buckling of the 1st-story south brace, and (**c**) weld crack at the gusset edge

buckling of the 1st-story south brace was observed and gusset edge cracking was observed near the 1st story column flange weld toe (Fig. 17.8c). The hybrid test was resumed and completed in about 4 h.

Figure 17.6 compares experimental and OpenSees analytical roof lateral displacement histories. It is evident that the OpenSees model accurately predicted the experimental peak responses, but not the residual displacement. However, it should be noted that the moment resisting frame specimens had been used six times before this hybrid test, and the OpenSees model was incapable of predicting the consequences of the column flange fracture. Nevertheless, the OpenSees analysis satisfactorily predicted the maximum negative peak roof displacement at 6.29 s into the acceleration record. The displacement history in each floor was shown in Fig. 17.9. The 2nd floor displacement is similar to one in the 3rd floor. It means that the inter-story in the 3rd story is very small. Figure 17.9 also shows the three experimental story shear verse inter-story drift relationships. It is evident that the lateral frame deformations concentrated in the 1st and 2nd stories, while the 3rd story remained essentially elastic. During the LA03 hybrid test, the maximum IP buckling displacement of the 1st story brace reached 508 mm. The fundamental period of the SCBF specimen changed from 0.46 to 1.15 s after the LA03 hybrid test.

Table 17.3 summarizes the peak story shears and inter-story drifts. The peak roof drifts are 3.35 % and 3.07 % radians in the south and north directions, respectively.

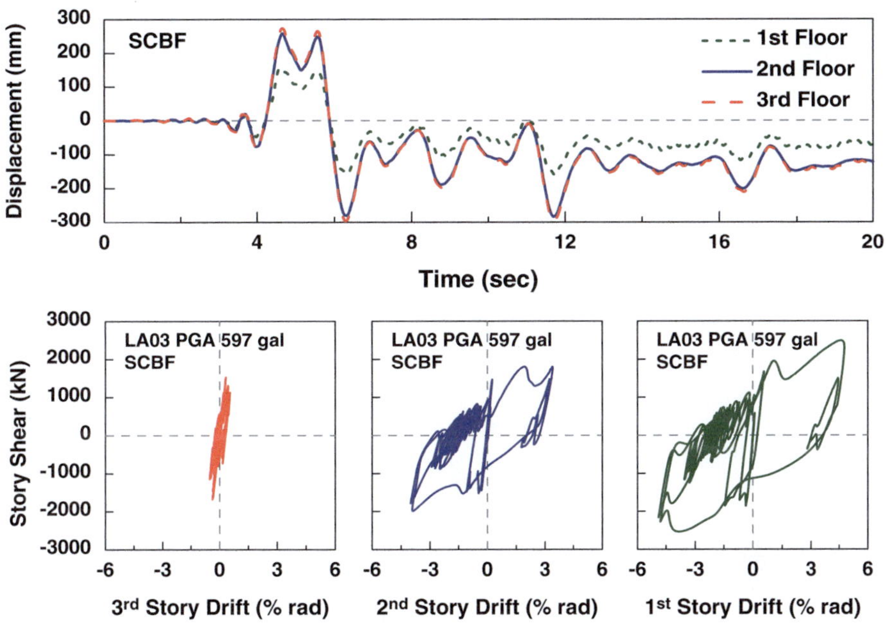


**Fig. 17.9** Floor lateral displacement histories and story shear versus inter-story drift relationships in SCBF hybrid test

The second floor displacements are 155.9 mm (58 % of the roof displacement), and 160.4 mm (54 % of the roof displacement) in the north and south directions, respectively. The maximum inter-story drift for the 2nd story is 4.94 % and 4.76 % radians in the south and north, respectively. The 3rd inter-story drift was only 0.54 % radian. This should be attributed from a relatively high lateral stiffness of the 3rd story. The braces in the 3rd story did not visibly buckle during the test. The maximum base shear was 2,539 kN in south and 2,491 kN in north. More details and discussions could be found in Tsai et al. (2013).

## 17.5 Performance Comparison of the 3-Story BRBF and SCBF

Table 17.3 summarized the maximum responses of the hybrid tests of BRBF and SCBF. During the BRBF hybrid test1, the maximum inter-story drifts were 2.83 %, 2.93 %, and 0.95 % radians for the 1st to the 3rd story, respectively. The maximum inter-story drifts were 2.80 %, 3.27 %, and 0.89 % radians during the BRBF hybrid test2. The maximum inter-story drifts during the SCBF hybrid test were 4.94 %, 4.00 %, and 0.54 % radians. It can be found that the deformation was concentrated in the lower two stories of the SCBF specimen. In other words, the distribution of

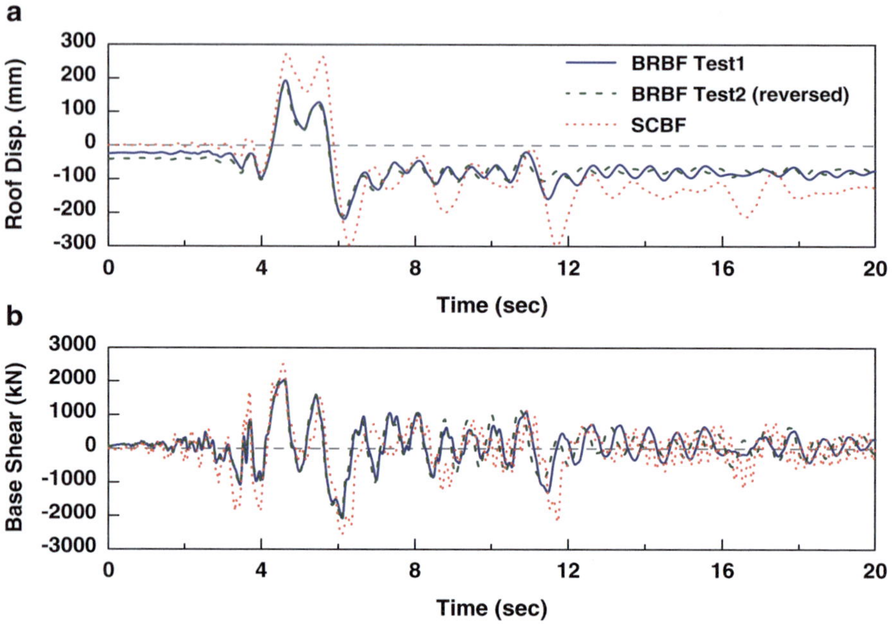


**Fig. 17.10** The comparisons of (**a**) roof displacement, and (**b**) base shear histories in the BRBF and the SCBF tests

the inter-story deformations of the BRBF is more uniform than that in the SCBF. And the SCBF has larger residual deformations than those in the BRBF. The maximum base shears were found 2,134, 2,207, and 2,539 kN during the BRBF hybrid test1, the BRBF hybrid test2, and the SCBF hybrid test, respectively.

The analytical vibration period of the BRBF was 0.60 s, and this value was close to the measured period of 0.588 s as shown in Table 17.2. However, the analytical period and the measured period of the SCBF were 0.43 s and 0.46 s, respectively. The measured damping ratios of the BRBF and the SCBF were 1.1 % and 0.3 %, respectively. Fig. 17.10a compares the roof displacement histories for the BRBFs and SCBF. The comparison of the base shears in the BRBFs and SCBF is shown in Fig. 17.10b.

## 17.6 Summary and Conclusions

The experimental peak inter-story drifts of the 1st and 2nd stories in the SCBF during the DBE were 4.94 % and 4.0 % rad, respectively. The 3rd inter-story drift was only 0.54 % rad. The 2nd story south column top inner flange fractured during the DBE test. The proposed knife-plate to gusset and brace connection performed satisfactorily. All knife plates and IP buckling braces sustained the DBE loads without fracture. Compared to the SCBF, the inter-story drifts were smaller in the BRBF hybrid tests though it reached near 3 %. The performance of the welded end-slot BRB was satisfactory and the local bulging failure could be prevented in the thin BRB design (Lin et al. 2012). This 3-story frame had been re-used and tested for 7 times. The column flange fractured fairly early in the record because of the combined damage, flame cutting, multiple welds and etc. However, the frame still made it through the earthquake loads quite well. It should be noted that the braces in the frame tests tolerate a slightly larger inelastic axial deformations than those observed in the brace component tests (Tsai et al. 2013). This could be due to the "near fault" characteristics of the LA03 record. The tests and the analysis indicate that it had a couple of huge cycles only.

## References

Gupta A, Krawinkler H (1999) Seismic demands for performance evaluation of steel moment resisting frame structures (SAC task 5.4.3), Report No. 132. John A. Blume Earthquake Engineering Center, Department of Civil and Environmental Engineering, Stanford University, Stanford

Lin BZ, Chuang MC, Tsai KC (2009) Object-oriented development and application of a nonlinear structural analysis framework. Adv Eng Softw 40:66–82

Lin PC, Tsai KC, Wang KJ, Yu YJ, Wei CY, Wu AC, Tsai CY, Lin CH, Chen JC, Schellenberg AH, Mahin SA, Roeder CW (2012) Seismic design and hybrid tests of a full-scale three-story buckling-restrained braced frame using welded end connections and thin profile. Earthq Eng Struct Dyn 41(5):1001–1020

Lumpkin JE (2009) Enhanced seismic performance of multi-story special concentrically brace frames using a balanced design procedure. Master thesis, University of Washington

McKenna F (1997) Object oriented finite element programming frameworks for analysis, algorithms and parallel computing. Ph.D. thesis, University of California, Berkeley

Structural Engineering Institute (2010) Minimum design loads for buildings and other structures. American Society of Civil Engineers, Reston

Tsai CY, Tsai KC, Lin CH, Wei CY, Wang KJ, Yu YJ, Wu AC (2010) Cyclic responses of three 2-story seismic concentrically braced frames. Front Archit Civ Eng China 4(3):287–301

Tsai CY, Tsai KC, Lin PC, Ao WH, Roeder CW, Mahin SA, Lin CH, Yu YJ, Wang KJ, Wu AC, Chen JC, Lin TH (2013) Seismic design and hybrid tests of a full-scale three-story concentrically braced frame using in-plane buckling braces. Earthquake Spectra 29(3):1043–1067

Tsai KC, Wu AC, Wei CY, Lin PC, Chuang MC, Yu YJ (2014) Welded end-slot connection and debonding layers for buckling-restrained braces. Earthq Eng Struct Dyn. doi:10.1002/eqe.2423

# Chapter 18
# Theory and Applications of the 3-DOF Modal System for PBSE of Asymmetrical Buildings

**Jui-Liang Lin and Keh-Chyuan Tsai**

**Abstract** Seismic evaluation plays an important role in performance based seismic engineering (PBSE). The modal system is the basis of structural dynamics, which is closely associated with PBSE. This paper shows that the modal system is not necessary to be a single-degree-of-freedom oscillator. Actually, a modal system with three degrees of freedom is even more suitable for representing a single vibration mode of two-way asymmetrical buildings. The proposed three-degree-of-freedom modal system has many advantages of seismic evaluation for inelastic or non-proportionally damped asymmetrical buildings. Furthermore, from the proposed modal system, a novel tuned mass damper has been developed for the modal control of asymmetrical buildings. The study results show that the novel tuned mass damper is effective in reducing two translational and one rotational displacements simultaneously.

**Keywords** Modal system • Asymmetrical buildings • Modal analysis • Tuned mass dampers • Inelastic response spectra • Non-proportional damping

## 18.1 Introduction

The single-degree-of-freedom (SDOF) oscillator, which is usually used to represent a single vibration mode of buildings, is the simplest and the clearest structural model used in structural dynamics. The well-known applications of the SDOF modal

J.-L. Lin (✉)
NARLabs, National Center for Research on Earthquake Engineering, Taipei, Taiwan
e-mail: jllin@ncree.narl.org.tw

K.-C. Tsai
Department of Civil Engineering, National Taiwan University, Taipei, Taiwan
e-mail: kctsai@ntu.edu.tw

M. Fischinger (ed.), *Performance-Based Seismic Engineering: Vision for an Earthquake Resilient Society*, Geotechnical, Geological and Earthquake Engineering 32,
DOI 10.1007/978-94-017-8875-5_18, 

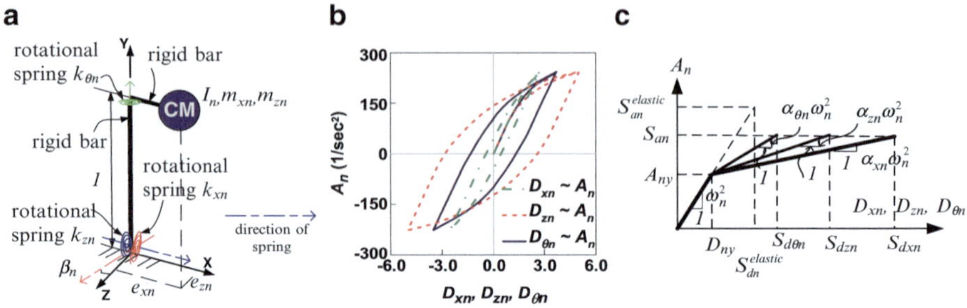


**Fig. 18.1** (**a**) the $n$-th 3DOF modal system; (**b**) The typical one-cycle push-pull curves representing the two roof translation—base shear relationships and one roof rotation—base torque relationship presented in the ADRS format; (**c**) the typical three bilinear pushover curves for a two-way asymmetric-plan building

system are the modal response spectrum analysis and the modal response history analysis. It is noted that even for inelastic structures, the "modal system" is still approximately used, e.g. the inelastic response spectra (Newmark and Hall 1973; Vidic et al. 1992), the capacity-spectrum method (Applied Technology Council 1996) and the modal pushover analysis procedure (Chopra and Goel 2004).

It is a deep-rooted concept that each vibration mode is always represented by a SDOF oscillator, which is corresponding to a SDOF modal equation of motion. Nevertheless, this paper shows how to obtain a modal equation of motion with three degrees of freedom for two-way asymmetrical buildings. Furthermore, a corresponding three-degree-of-freedom (3DOF) modal system can be constructed (Fig. 18.1a). It is noted when we apply the modal inertia force vector to a symmetrical building, there is only one force—displacement curve, which represents the roof displacement versus base shear relationship. However, when we apply the modal inertia force vector to a two-way asymmetrical building, there are three force—displacement curves, which represents the two roof displacement—base shear relationships and one roof rotation—base torque relationship (Fig. 18.1b). The SDOF modal system is sufficient for describing the only one force—displacement relationship of symmetrical buildings, but the SDOF modal system is incapable of describing the three force—displacement relationships simultaneously for asymmetrical buildings. Thus, the 3DOF modal system has the advantage of capturing the non-proportionality between modal translations and modal rotation. In addition, there are other useful applications of the 3DOF modal systems to the seismic evaluation and structural control for asymmetrical buildings.

## 18.2 Theoretical Background

For the sake of completeness and to make it easier to adopt the notations in the rest of this paper, we briefly present the results of the 3DOF modal systems herein. More details can be found in the associated literature (Lin and Tsai 2008a).

### *18.2.1 Three-Degree-of-Freedom Modal Systems*

In order to be consistent with the coordinate system used in the previous study (Lin and Tsai 2008a), the two plan axes are the X- and Z-axis. The Y-axis is upward (opposite to the direction of gravity). The proportionally damped two-way asymmetric-plan buildings with each floor simulated as a rigid diaphragm are considered in this paper. The equation of motion for an $N$-storey two-way asymmetric-plan building subjected to bi-directional ground excitations is

$$\begin{aligned}\mathbf{M}\ddot{\mathbf{u}}+\mathbf{C}\dot{\mathbf{u}}+\mathbf{K} &= -\mathbf{M}\boldsymbol{\iota}_x\ddot{u}_{gx}(t)-\mathbf{M}\boldsymbol{\iota}_z\ddot{u}_{gz}(t)\\ &= -\sum_{n=1}^{3N}\mathbf{s}_{xn}\ddot{u}_{gx}(t)-\sum_{n=1}^{3N}\mathbf{s}_{zn}\ddot{u}_{gz}(t)=-\sum_{n=1}^{3N}\Gamma_{xn}\mathbf{M}\boldsymbol{\varphi}_n\ddot{u}_{gx}(t)\\ &\quad-\sum_{n=1}^{3N}\Gamma_{zn}\mathbf{M}\boldsymbol{\varphi}_n\ddot{u}_{gz}(t)\\ &= -\sum_{n=1}^{3N}\left(\Gamma_{xn}\ddot{u}_{gx}+\Gamma_{zn}\ddot{u}_{gz}\right)\mathbf{M}\boldsymbol{\varphi}_n,\end{aligned} \quad (18.1)$$

in which the displacement vector, $\mathbf{u}$, mode shape, $\boldsymbol{\varphi}_n$, mass matrix, $\mathbf{M}$, stiffness matrix, $\mathbf{K}$, influence vectors, $\boldsymbol{\iota}_x, \boldsymbol{\iota}_z$, modal inertia force distributions, $\mathbf{s}_{xn}$, $\mathbf{s}_{zn}$, and modal participation factors, $\Gamma_{xn}, \Gamma_{zn}$, are

$$\mathbf{M}=\begin{bmatrix}\mathbf{m}_x & \mathbf{0} & \mathbf{0}\\ \mathbf{0} & \mathbf{m}_z & \mathbf{0}\\ \mathbf{0} & \mathbf{0} & \mathbf{I}_0\end{bmatrix}_{3N\times 3N},\quad \mathbf{K}=\begin{bmatrix}\mathbf{k}_{xx} & \mathbf{k}_{xz} & \mathbf{k}_{x\theta}\\ \mathbf{k}_{zx} & \mathbf{k}_{zz} & \mathbf{k}_{z\theta}\\ \mathbf{k}_{\theta x} & \mathbf{k}_{\theta z} & \mathbf{k}_{\theta\theta}\end{bmatrix}_{3N\times 3N}$$

$$\mathbf{u}=\begin{bmatrix}\mathbf{u}_x\\ \mathbf{u}_z\\ \mathbf{u}_\theta\end{bmatrix}_{3N\times 1},\quad \boldsymbol{\varphi}_n=\begin{bmatrix}\boldsymbol{\varphi}_{xn}\\ \boldsymbol{\varphi}_{zn}\\ \boldsymbol{\varphi}_{\theta n}\end{bmatrix}_{3N\times 1},\quad \boldsymbol{\iota}_x=\begin{bmatrix}\mathbf{1}\\ \mathbf{0}\\ \mathbf{0}\end{bmatrix},\quad \boldsymbol{\iota}_z=\begin{bmatrix}\mathbf{0}\\ \mathbf{1}\\ \mathbf{0}\end{bmatrix}$$

$$\mathbf{s}_{xn}=\Gamma_{xn}\mathbf{M}\boldsymbol{\varphi}_n,\quad \mathbf{s}_{zn}=\Gamma_{zn}\mathbf{M}\boldsymbol{\varphi}_n,\quad \Gamma_{xn}=\frac{\boldsymbol{\varphi}_n^T\mathbf{M}\boldsymbol{\iota}_x}{\boldsymbol{\varphi}_n^T\mathbf{M}\boldsymbol{\varphi}_n},\quad \Gamma_{zn}=\frac{\boldsymbol{\varphi}_n^T\mathbf{M}\boldsymbol{\iota}_z}{\boldsymbol{\varphi}_n^T\mathbf{M}\boldsymbol{\varphi}_n} \quad (18.2)$$

When only the force $-\left(\Gamma_{xn}\ddot{u}_{gx}+\Gamma_{zn}\ddot{u}_{gz}\right)\mathbf{M}\boldsymbol{\varphi}_n$ is applied to the building, the equation of motion (Eq. 18.1) becomes

$$\mathbf{M}\ddot{\mathbf{u}}_n+\mathbf{C}\dot{\mathbf{u}}_n+\mathbf{K}\mathbf{u}_n=-\left(\Gamma_{xn}\ddot{u}_{gx}+\Gamma_{zn}\ddot{u}_{gz}\right)\mathbf{M}\boldsymbol{\varphi}_n,\ n=1\sim 3N\,, \quad (18.3)$$

in which $\mathbf{u}_n$ is the $n$-th modal displacement response and $\mathbf{u}=\sum_{n=1}^{3N}\mathbf{u}_n=\sum_{n=1}^{3N}\boldsymbol{\varphi}_n D_n$. $D_n$ is the generalized modal coordinate. By re-defining $\mathbf{u}_n$ equal to

$\begin{bmatrix} \boldsymbol{\varphi}_{xn} & \mathbf{0} & \mathbf{0} \\ \mathbf{0} & \boldsymbol{\varphi}_{zn} & \mathbf{0} \\ \mathbf{0} & \mathbf{0} & \boldsymbol{\varphi}_{\theta n} \end{bmatrix}_{3N\times 3} \begin{bmatrix} D_{xn} \\ D_{zn} \\ D_{\theta n} \end{bmatrix}_{3\times 1}$ and pre-multiplying both sides of Eq. 18.3 by $\begin{bmatrix} \boldsymbol{\varphi}_{xn} & \mathbf{0} & \mathbf{0} \\ \mathbf{0} & \boldsymbol{\varphi}_{zn} & \mathbf{0} \\ \mathbf{0} & \mathbf{0} & \boldsymbol{\varphi}_{\theta n} \end{bmatrix}^T$, the 3DOF modal equation of motion for the $n$-th vibration mode is obtained as

$$\mathbf{M}_n\ddot{\mathbf{D}}_n + \mathbf{C}_n\dot{\mathbf{D}}_n + \mathbf{K}_n\mathbf{D}_n = -\left(\Gamma_{xn}\ddot{u}_{gx} + \Gamma_{zn}\ddot{u}_{gz}\right)\mathbf{M}_n\mathbf{1},\ n = 1 \sim 3N \tag{18.4}$$

where

$$\mathbf{M}_n = \begin{bmatrix} \boldsymbol{\varphi}_{xn}^T\mathbf{m}_x\boldsymbol{\varphi}_{xn} & 0 & 0 \\ 0 & \boldsymbol{\varphi}_{zn}^T\mathbf{m}_z\boldsymbol{\varphi}_{zn} & 0 \\ 0 & 0 & \boldsymbol{\varphi}_{\theta n}^T\mathbf{I}_0\boldsymbol{\varphi}_{\theta n} \end{bmatrix}_{3\times 3},\quad \mathbf{D}_n = \begin{bmatrix} D_{xn} \\ D_{zn} \\ D_{\theta n} \end{bmatrix}_{3\times 1}$$

$$\mathbf{C}_n = \begin{bmatrix} \boldsymbol{\varphi}_{xn}^T\mathbf{c}_{xx}\boldsymbol{\varphi}_{xn} & \boldsymbol{\varphi}_{xn}^T\mathbf{c}_{xz}\boldsymbol{\varphi}_{zn} & \boldsymbol{\varphi}_{xn}^T\mathbf{c}_{x\theta}\boldsymbol{\varphi}_{\theta n} \\ \boldsymbol{\varphi}_{zn}^T\mathbf{c}_{zx}\boldsymbol{\varphi}_{xn} & \boldsymbol{\varphi}_{zn}^T\mathbf{c}_{zz}\boldsymbol{\varphi}_{zn} & \boldsymbol{\varphi}_{zn}^T\mathbf{c}_{z\theta}\boldsymbol{\varphi}_{\theta n} \\ \boldsymbol{\varphi}_{\theta n}^T\mathbf{c}_{\theta x}\boldsymbol{\varphi}_{xn} & \boldsymbol{\varphi}_{\theta n}^T\mathbf{c}_{\theta z}\boldsymbol{\varphi}_{zn} & \boldsymbol{\varphi}_{\theta n}^T\mathbf{c}_{\theta\theta}\boldsymbol{\varphi}_{\theta n} \end{bmatrix}_{3\times 3},\quad \mathbf{1} = \begin{bmatrix} 1 \\ 1 \\ 1 \end{bmatrix}_{3\times 1}$$

$$\mathbf{K}_n = \begin{bmatrix} \boldsymbol{\varphi}_{xn}^T\mathbf{k}_{xx}\boldsymbol{\varphi}_{xn} & \boldsymbol{\varphi}_{xn}^T\mathbf{k}_{xz}\boldsymbol{\varphi}_{zn} & \boldsymbol{\varphi}_{xn}^T\mathbf{k}_{x\theta}\boldsymbol{\varphi}_{\theta n} \\ \boldsymbol{\varphi}_{zn}^T\mathbf{k}_{zx}\boldsymbol{\varphi}_{xn} & \boldsymbol{\varphi}_{zn}^T\mathbf{k}_{zz}\boldsymbol{\varphi}_{zn} & \boldsymbol{\varphi}_{zn}^T\mathbf{k}_{z\theta}\boldsymbol{\varphi}_{\theta n} \\ \boldsymbol{\varphi}_{\theta n}^T\mathbf{k}_{\theta x}\boldsymbol{\varphi}_{xn} & \boldsymbol{\varphi}_{\theta n}^T\mathbf{k}_{\theta z}\boldsymbol{\varphi}_{zn} & \boldsymbol{\varphi}_{\theta n}^T\mathbf{k}_{\theta\theta}\boldsymbol{\varphi}_{\theta n} \end{bmatrix}_{3\times 3} \tag{18.5}$$

$D_{xn}$, $D_{zn}$, and $D_{\theta n}$, are referred to as the X- and Z-directional modal translations and modal rotation respectively. The $n$-th 3DOF modal system corresponding to the 3DOF modal equation of motion is shown in Fig. 18.1a. The elastic properties of the $n$-th 3DOF modal system are as follows:

$$m_{xn} = \boldsymbol{\varphi}_{xn}^T\mathbf{m}_x\boldsymbol{\varphi}_{xn},\quad m_{zn} = \boldsymbol{\varphi}_{zn}^T\mathbf{m}_z\boldsymbol{\varphi}_{zn},\quad I_n = \boldsymbol{\varphi}_{\theta n}^T\mathbf{I}_0\boldsymbol{\varphi}_{\theta n}$$

$$\beta_n = \tan^{-1}\left(\frac{\boldsymbol{\varphi}_{zn}^T\mathbf{k}_{zx}\boldsymbol{\varphi}_{xn}}{\boldsymbol{\varphi}_{xn}^T\mathbf{k}_{xx}\boldsymbol{\varphi}_{xn}}\right)\quad k_{xn} = \frac{\boldsymbol{\varphi}_{xn}^T\mathbf{k}_{xx}\boldsymbol{\varphi}_{xn}}{C^2},\quad k_{zn} = \boldsymbol{\varphi}_{zn}^T\mathbf{k}_{zz}\boldsymbol{\varphi}_{zn} - k_{xn}S^2$$

$$k_{\theta n} = \boldsymbol{\varphi}_{\theta n}^T\mathbf{k}_{\theta\theta}\boldsymbol{\varphi}_{\theta n} - e_{xn}^2 k_{zn} - (e_{xn}S - e_{zn}C)^2 k_{xn}$$

$$\begin{bmatrix} e_{xn} \\ e_{zn} \end{bmatrix} = \begin{bmatrix} k_{xn}SC & -k_{xn}C^2 \\ k_{zn} + k_{xn}S^2 & -k_{xn}SC \end{bmatrix}^{-1} \begin{bmatrix} \boldsymbol{\varphi}_{xn}^T\mathbf{k}_{x\theta}\boldsymbol{\varphi}_{\theta n} \\ \boldsymbol{\varphi}_{zn}^T\mathbf{k}_{z\theta}\boldsymbol{\varphi}_{\theta n} \end{bmatrix} \tag{18.6}$$

where $C = cos\beta_n$ and $S = sin\beta_n$. The inelastic parameters of the $n$-th 3DOF modal system, including the yielding moments $M_{yxn}$, $M_{yzn}$, $M_{y\theta n}$ and the post-yielding stiffness $k'_{xn}$, $k'_{zn}$, $k'_{\theta n}$ of the three rotational springs of the 3DOF modal system (Fig. 18.1a), are

$$M_{yxn} = A_{xny}m_{xn},\ M_{yzn} = A_{zny}m_{zn} \tag{18.7a}$$

$$M_{y\theta n} = A_{\theta ny} I_n + A_{xny} m_{xn} e_{zn} - A_{zny} m_{zn} e_{xn} \tag{18.7b}$$

$$k'_{xn} = \frac{m_{xn}}{\dfrac{\frac{m_{xn}}{k_{xn}} + \frac{(I_n + m_{xn}e_{zn} - m_{zn}e_{xn})e_{zn}}{k_{\theta n}}}{\alpha_{xn}} - \dfrac{(I_n + m_{xn}e_{zn} - m_{zn}e_{xn})\, e_{zn}}{k'_{\theta n}}} \tag{18.7c}$$

$$k'_{zn} = \frac{m_{zn}}{\dfrac{\frac{m_{zn}}{k_{zn}} - \frac{(I_n + m_{xn}e_{zn} - m_{zn}e_{xn})e_{xn}}{k_{\theta n}}}{\alpha_{zn}} + \dfrac{(I_n + m_{xn}e_{zn} - m_{zn}e_{xn})\, e_{xn}}{k'_{\theta n}}} \tag{18.7d}$$

$$k'_{\theta n} = k_{\theta n} \cdot \alpha_{\theta n} \tag{18.7e}$$

where $A_{xny}$, $A_{zny}$ and $A_{\theta ny}$ are the yielding accelerations and $\alpha_{xn}$, $\alpha_{zn}$ and $\alpha_{\theta n}$ are the post-yielding stiffness ratios of the three pushover curves idealized as three bilinear curves in the acceleration-displacement-response-spectra (ADRS) format (Fig. 18.1c). The stated three pushover curves, obtained from the modal pushover analysis of the original building, represent the relationships of the two roof translations versus the two base shears and the roof rotation versus the base torque (Fig. 18.1b). The $n$-th modal response $\mathbf{D}_n$ is obtained by using the step-by-step integration of Eq. 18.4. The total displacement response $\mathbf{u}$ of the original building is obtained as

$$\mathbf{u} = \sum_{n=1}^{3N} \mathbf{u}_n = \sum_{n=1}^{3N} \mathbf{\Phi}_n \mathbf{D}_n = \sum_{n=1}^{3N} \begin{bmatrix} \boldsymbol{\varphi}_{xn} & \mathbf{0} & \mathbf{0} \\ \mathbf{0} & \boldsymbol{\varphi}_{zn} & \mathbf{0} \\ \mathbf{0} & \mathbf{0} & \boldsymbol{\varphi}_{\theta n} \end{bmatrix}_{3N\times 3} \begin{bmatrix} D_{xn} \\ D_{zn} \\ D_{\theta n} \end{bmatrix}_{3\times 1} \tag{18.8}$$

## 18.2.2 The Modal Parameters of Two-Way Asymmetrical Buildings

The independent elastic 3DOF modal parameters are the vibration period $T_n$, the frequency ratios $\Omega_{\theta xn}$ and $\Omega_{\theta zn}$, the modal eccentricity $e_{xn}$, and the normalized modal eccentricity ratio $|e_{xn}|/r_{xn}$ (Lin et al. 2012a, b). Given the values of these five elastic 3DOF modal parameters, all the other elastic 3DOF modal parameters can be determined from following equations:

$$\Omega_{\theta xn} = \frac{\omega_{\theta n}}{\omega_{xn}} = \sqrt{\left(1 + \frac{m_{xn}}{I_n} e_{zn} - \frac{m_{zn}}{I_n} e_{xn}\right)(1 - e_{zn})}$$

$$\Omega_{\theta zn} = \frac{\omega_{\theta n}}{\omega_{zn}} = \sqrt{\left(1 + \frac{m_{xn}}{I_n} e_{zn} - \frac{m_{zn}}{I_n} e_{xn}\right)(1 + e_{xn})}$$

$$T_n = \frac{T_{xn}}{\sqrt{1 - e_{zn}}} = \frac{T_{zn}}{\sqrt{1 + e_{xn}}}, \quad m_{xn} + m_{zn} + I_n = 1 \tag{18.9}$$

The relationship between the strength ratio and the inelastic 3DOF modal parameters is presented as:

$$M_{yxn} = \frac{S_{dn}^{elastic}}{R}(-1 + e_{zn})k_{xn}, \quad M_{yzn} = \frac{S_{dn}^{elastic}}{R}(1 + e_{xn})k_{zn}$$
$$M_{y\theta n} = \frac{S_{dn}^{elastic}}{R}k_{\theta n} \tag{18.10}$$

Given the strength ratio, all the yielding moments of the three rotational springs in a 3DOF modal system (Fig. 18.1a) can be determined by using Eq. 18.10.

## 18.3 Applications of the 3DOF Modal Systems

Equations 18.1, 18.2, 18.3, 18.4, 18.5, 18.6, 18.7a–e, and 18.8 clearly indicate that the 3DOF modal systems can be applied to estimate the seismic response histories of two-way asymmetrical buildings (Lin and Tsai 2008a). Furthermore, because the absorbed energy of the 3DOF modal system can be decomposed into three parts resulting from different types of deformation, the absorbed energy of asymmetrical buildings resulting from translational and rotational deformations can be separately estimated in terms of the 3DOF modal system (Lin and Tsai 2011). Other applications of the 3DOF modal system include the development of the bidirectional coupled tuned mass dampers (BiCTMD) for the modal control of two-way asymmetrical buildings (Lin et al. 2011), the modal response history analyses of non-proportionally damped asymmetrical buildings (Lin and Tsai 2008b) and the construction of the inelastic response spectra for asymmetrical structures, referred to as SAS, under the exertion of bidirectional ground motions (Lin et al. 2012a, b). In the following contents of this paper, we briefly present the last three mentioned applications of the 3DOF modal systems. More details can be found in the associated literatures (Lin et al. 2012a, b, 2011; Lin and Tsai 2008b).

### *18.3.1 The Bidirectional Coupled Tuned Mass Dampers*

When the BiCTMD is designed to control a 3DOF modal system, the properties of the BiCTMD are set proportionally to those of the 3DOF model. That is,

$$\mathbf{M}_{an} = \mu\mathbf{M}_n, \mathbf{C}_{an} = \beta\mathbf{C}_n, \mathbf{K}_{an} = f\mathbf{K}_n \tag{18.11}$$

where $\mathbf{M}_{an}$, $\mathbf{C}_{an}$ and $\mathbf{K}_{an}$ are the mass, damping and stiffness matrices, respectively, of the BiCTMD, expressed as:

$$\mathbf{M}_{an} = \begin{bmatrix} m_{axn} & 0 & 0 \\ 0 & m_{azn} & 0 \\ 0 & 0 & I_{an} \end{bmatrix}, \quad \mathbf{C}_{an} = \begin{bmatrix} c_{axxn} & 0 & c_{ax\theta n} \\ 0 & c_{azzn} & c_{az\theta n} \\ c_{a\theta xn} & c_{a\theta zn} & c_{a\theta\theta n} \end{bmatrix}$$

$$\mathbf{K}_{an} = \begin{bmatrix} k_{axxn} & 0 & k_{ax\theta n} \\ 0 & k_{azzn} & k_{az\theta n} \\ k_{a\theta xn} & k_{a\theta zn} & k_{a\theta\theta n} \end{bmatrix} \tag{18.12}$$

The parameters $\mu$, $\beta$ and $f$ shown in Eq. 18.11 are the mass ratio, the damping ratio, and the frequency ratio, respectively. The value of the mass ratio $\mu$ is selected by the designer and is usually around 0.05. The optimum values of $f$ and $\beta$ for the BiCTMD are:

$$f = \mu f_{0n}^2, \ \beta = \mu f_{0n} \frac{\xi_{an}}{\xi_n} \tag{18.13}$$

where $f_{0n}$ and $\xi_{an}$ are the optimum values of the frequency ratio and the damping ratio of the corresponding conventional TMD controlling a single-degree-of-freedom (SDOF) main system with damping ratio $\xi_n$ and mass ratio $\mu$.

When the BiCTMD for controlling the $n$th vibration mode is placed on the $j$th floor of an actual $N$-story two-way asymmetric-plan building, then the mass, damping, and stiffness matrices of the BiCTMD, respectively denoted as $\mathbf{M}_{an}^s$, $\mathbf{C}_{an}^s$, and $\mathbf{K}_{an}^s$, are:

$$\mathbf{M}_{an}^s = \begin{bmatrix} m_{axn} & 0 & 0 \\ 0 & \left(\frac{\phi_{xn,j}}{\phi_{zn,j}}\right)^2 m_{azn} & 0 \\ 0 & 0 & \left(\frac{\phi_{xn,j}}{\phi_{\theta n,j}}\right)^2 I_{an} \end{bmatrix} \tag{18.14a}$$

$$\mathbf{C}_{an}^s = \begin{bmatrix} c_{axxn} & 0 & \frac{\phi_{xn,j}}{\phi_{\theta n,j}} c_{ax\theta n} \\ 0 & \left(\frac{\phi_{xn,j}}{\phi_{zn,j}}\right)^2 c_{azzn} & \left(\frac{\phi_{xn,j}}{\phi_{zn,j}}\right)\left(\frac{\phi_{xn,j}}{\phi_{\theta n,j}}\right) c_{az\theta n} \\ \frac{\phi_{xn,j}}{\phi_{\theta n,j}} c_{a\theta xn} & \left(\frac{\phi_{xn,j}}{\phi_{zn,j}}\right)\left(\frac{\phi_{xn,j}}{\phi_{\theta n,j}}\right) c_{a\theta zn} & \left(\frac{\phi_{xn,j}}{\phi_{\theta n,j}}\right)^2 c_{a\theta\theta n} \end{bmatrix} \tag{18.14b}$$

$$\mathbf{K}_{an}^s = \begin{bmatrix} k_{axxn} & 0 & \frac{\phi_{xn,j}}{\phi_{\theta n,j}} k_{ax\theta n} \\ 0 & \left(\frac{\phi_{xn,j}}{\phi_{zn,j}}\right)^2 k_{azzn} & \left(\frac{\phi_{xn,j}}{\phi_{zn,j}}\right)\left(\frac{\phi_{xn,j}}{\phi_{\theta n,j}}\right) k_{az\theta n} \\ \frac{\phi_{xn,j}}{\phi_{\theta n,j}} k_{a\theta xn} & \left(\frac{\phi_{xn,j}}{\phi_{zn,j}}\right)\left(\frac{\phi_{xn,j}}{\phi_{\theta n,j}}\right) k_{a\theta zn} & \left(\frac{\phi_{xn,j}}{\phi_{\theta n,j}}\right)^2 k_{a\theta\theta n} \end{bmatrix} \tag{18.14c}$$

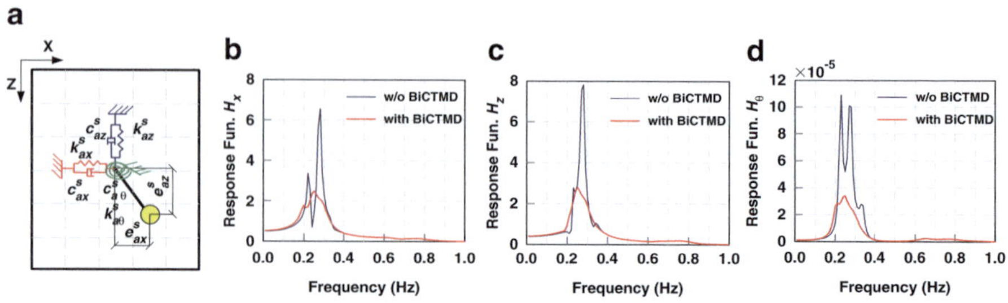


**Fig. 18.2** (**a**) The physical model of the BiCTMD located on a floor plane; the amplitudes of the frequency response functions for the roof displacements in (**b**) the x-translational, (**c**) the z-translational and (**d**) the rotational directions

where $m_{axn}$, $m_{azn}$, $I_{an}$, $c_{axxn}$, $c_{ax\theta n}$, $c_{a\theta xn}$, $c_{azzn}$, $c_{az\theta n}$, $c_{a\theta zn}$, $c_{a\theta\theta n}$, $k_{axxn}$, $k_{ax\theta n}$, $k_{a\theta xn}$, $k_{azzn}$, $k_{az\theta n}$, $k_{a\theta zn}$ and $k_{a\theta\theta n}$ are defined in Eq. 18.12; $\phi_{xn,j}$, $\phi_{zn,j}$ and $\phi_{\theta n,j}$ are the $j$th components of the $n$th mode shape in the translational and rotational directions, respectively. Figure 18.2a shows the physical model of the BiCTMD located on a floor plane. Figure 18.2b–d are the amplitudes of the frequency response functions of the roof displacements in the three directions for a 20-story two-way asymmetrical building (Lin et al. 2011). Figure 18.2b–d indicate that the BiCTMD is effective in reducing the two translational and the rotational displacements simultaneously.

### 18.3.2 The Modal Response History Analyses of Non-proportionally Damped Asymmetrical Buildings

The modal damping matrix, $\mathbf{C}_n$, given in Eq. 18.5 is equal to

$$\mathbf{C}_n = \begin{bmatrix} \boldsymbol{\varphi}_{xn} & \mathbf{0} & \mathbf{0} \\ \mathbf{0} & \boldsymbol{\varphi}_{zn} & \mathbf{0} \\ \mathbf{0} & \mathbf{0} & \boldsymbol{\varphi}_{\theta n} \end{bmatrix}^T_{3N\times 3} \mathbf{C} \begin{bmatrix} \boldsymbol{\varphi}_{xn} & \mathbf{0} & \mathbf{0} \\ \mathbf{0} & \boldsymbol{\varphi}_{zn} & \mathbf{0} \\ \mathbf{0} & \mathbf{0} & \boldsymbol{\varphi}_{\theta n} \end{bmatrix}_{3N\times 3} \tag{18.15}$$

If the original building is a proportionally damped system, i.e.

$$\mathbf{C} = \begin{bmatrix} \mathbf{c}_{xx} & \mathbf{c}_{xz} & \mathbf{c}_{x\theta} \\ \mathbf{c}_{zx} & \mathbf{c}_{zz} & \mathbf{c}_{z\theta} \\ \mathbf{c}_{\theta x} & \mathbf{c}_{\theta z} & \mathbf{c}_{\theta\theta} \end{bmatrix}_{3N\times 3N} = \alpha\mathbf{M} + \beta\mathbf{K}, \tag{18.16}$$

the modal damping matrix would be

$$\mathbf{C}_n = \begin{bmatrix} \boldsymbol{\varphi}_{xn} & \mathbf{0} & \mathbf{0} \\ \mathbf{0} & \boldsymbol{\varphi}_{zn} & \mathbf{0} \\ \mathbf{0} & \mathbf{0} & \boldsymbol{\varphi}_{\theta n} \end{bmatrix}_{3N\times 3N}^{T} (\alpha\mathbf{M}+\beta\mathbf{K}) \begin{bmatrix} \boldsymbol{\varphi}_{xn} & \mathbf{0} & \mathbf{0} \\ \mathbf{0} & \boldsymbol{\varphi}_{zn} & \mathbf{0} \\ \mathbf{0} & \mathbf{0} & \boldsymbol{\varphi}_{\theta n} \end{bmatrix}_{3N\times 3N}$$
$$= \alpha\mathbf{M}_n + \beta\mathbf{K}_n \tag{18.17}$$

Therefore, if the original building is a non-proportionally damped system, i.e.

$$\mathbf{C} \neq \alpha\mathbf{M} + \beta\mathbf{K}, \tag{18.18}$$

the modal damping matrix would also be non-proportional, i.e.

$$\mathbf{C}_n \neq \alpha\mathbf{M}_n + \beta\mathbf{K}_n \tag{18.19}$$

It indicates that a non-proportionally damped two-way asymmetric-plan system will result in $3N$ non-proportionally damped 3DOF modal equations of motion, which are able to take the out-of-phase motions between the modal translations and the modal rotation into account. Thus, the 3DOF modal equations of motion are more appropriate to be used in the modal response history analysis of non-proportionally damped asymmetric-plan structures than the SDOF modal equations of motion. The 3DOF modal equations of motion possess the non-proportionally damped property at the expense of increasing two DOFs in the modal coordinate. The proposed 3DOF modal equations of motion still can be easily computed by commercially available structural analysis programs. On the other hand, the proposed method keeps the clarity and the simplicity of the modal response history analysis in calculating the seismic responses of structures.

The three-story two-way asymmetrical example building with viscous dampers installed in the two directions of each floor was used to verify the effectiveness of the proposed approach (Lin and Tsai 2008b). The results, not shown in this paper, indicate that the displacement histories at all of the three stories are satisfactorily estimated.

### *18.3.3 The Inelastic Response Spectra for Asymmetrical Structures*

The 3DOF modal parameters have been briefly shown in Sect. 18.2.2. In addition, the ranges of the 3DOF modal parameter values have been investigated in the associated literature (Lin et al. 2012a). These completed tasks enable the construction of the inelastic response spectra for asymmetrical structures (SAS). Figure 18.3a shows the constant strength response spectra with the strength ratio equal to 3 and 6. Moreover, the mean values plus one standard deviation are also shown in Fig. 18.3a. It is obvious that the ductility values approach the strength values when the vibration period is very large, which is the same as that observed in the

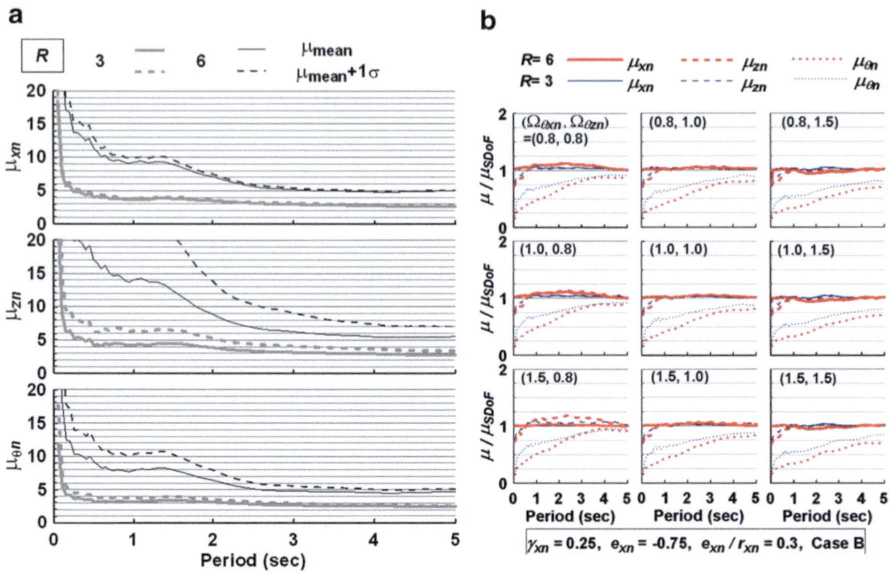


**Fig. 18.3** (**a**) The mean and the mean plus one standard deviation of the constant-strength SAS for all considered 3DOF modal systems under the applied 40 pairs of ground motion records (**b**) $\mu/\mu_{SDoF}$ spectra for 3DOF modal parameter values equal to $\gamma_{xn} = 0.25$, $|e_{xn}|/r_{xn} = 0.3$, $e_{xn} = -0.75$, $\alpha_{xn} = 0.05$, $\alpha_{zn} = 0.08$, and $\alpha_{\theta n} = 0.3$

conventional response spectra. Figure 18.3a shows that the three components of the mean ductility demands for $R = 3$ are close, i.e. $\mu_{xn} \approx \mu_{zn} \approx \mu_{\theta n}$. Nevertheless, $\mu_{xn}$, $\mu_{zn}$ and $\mu_{\theta n}$ are quite different for $R = 6$. This implies that the conventional response spectra gradually become inadequate as the strength ratio increases. That is, it is insufficient to use a SDOF modal system for estimating the three different modal ductility demands of a substantially inelastic asymmetric-plan building.

Figures 18.3b show the ratios of the ductility demands estimated by using 3DOF modal systems to those estimated by using the corresponding SDOF modal systems, denoted as $\mu/\mu_{SDoF}$. The corresponding SDOF modal systems are defined as those having the values of vibration period, strength ratio and post-yielding stiffness ratio same as the corresponding values of $T_n$, $R$ and $\alpha_{xn}$ in the 3DOF modal systems. It is because both the SDOF and 3DOF modal systems possess these three parameters. Figure 18.3b shows that most of the values of $\mu_{xn}/\mu_{SDoF}$ and $\mu_{zn}/\mu_{SDoF}$ are close to one and $\mu_{\theta n}/\mu_{SDoF}$ is less than one. This indicates that the X- and Z-translational ductility demands can be approximately estimated by using the conventional response spectra under the case of the selected parameter values. However, since $\alpha_{\theta n}$ is significantly larger than $\alpha_{xn}$, the rotational ductility demand obtained by using the conventional response spectra is overestimated. The $\mu/\mu_{SDoF}$ spectra illustrated in the associated literature clearly show that the conventional response spectra are very likely to underestimate or overestimate the ductility demands of the asymmetric-plan buildings. The difference between $\mu$ and $\mu_{SDoF}$ increases as the strength ratio increases. Furthermore, it is confirmed that the interaction effects among the three components of modal deformations in the two-way asymmetric-plan structures indeed exist.

## 18.4 Conclusions

For elastic proportionally damped multistory asymmetrical buildings, the 3DOF modal system is the same as the conventional SDOF modal system because of $D_{xn} = D_{zn} = D_{\theta n}$. In addition, the 3DOF modal system itself has only one active vibration mode. The other two vibration modes of the 3DOF modal system are spurious, whose modal participation factors are equal to zero. Nevertheless, for inelastic or non-proportionally damped multistory asymmetrical buildings, the force—displacement relationships of the 3DOF modal system can reflect the non-proportionality between the modal translations and modal rotation. The applications of the 3DOF modal system are the estimation of the seismic responses of inelastic or non-proportionally damped asymmetrical buildings, the development of the BiCTMD for the seismic control of asymmetrical buildings and the construction of inelastic response spectra for asymmetrical structures. These applications based on the 3DOF modal system have the advantages, which cannot be obtained by using the conventional SDOF modal system.

## References

Applied Technology Council (1996) Seismic evaluation and retrofit of concrete buildings, Report ATC-40, Applied Technology Council, Redwood City, CA

Chopra AK, Goel RK (2004) A modal pushover analysis procedure to estimate seismic demands for unsymmetric-plan buildings. Earthq Eng Struct Dyn 33:903–927

Lin JL, Tsai KC (2008a) Seismic analysis of two-way asymmetric building systems under bi-directional seismic ground motions. Earthq Eng Struct Dyn 37:305–328

Lin JL, Tsai KC (2008b) Seismic analysis of non-proportionally damped two-way asymmetric elastic buildings under bi-directional seismic ground motions. J Earthq Eng 12(7):1139–1156

Lin JL, Tsai KC (2011) Estimation of the seismic energy demands of two-way asymmetric-plan building systems. Bull Earthq Eng 9(2):603–621

Lin JL, Tsai KC, Yu YJ (2011) Bi-directional coupled tuned mass dampers for the seismic response control of two-way asymmetric-plan buildings. Earthq Eng Struct Dyn 40(6):675–690

Lin JL, Tsai KC, Yang WC (2012a) Inelastic responses of two-way asymmetric-plan structures under bi-directional ground excitations- PART I: Modal parameters. Earthquake Spectra 28(1):105–139

Lin JL, Yang WC, Tsai KC (2012b) Inelastic responses of two-way asymmetric-plan structures under bi-directional ground excitations-PART II: Response spectra. Earthquake Spectra 28(1):141–157

Newmark NM, Hall WJ (1973) Seismic design criteria for nuclear reactor facilities. In: Building practices for disaster mitigation, Report no. 46. National Bureau of Standards, U.S. Department of Commerce, Washington, DC, pp 209–236

Vidic T, Fajfar P, Fischinger M (1992) A procedure for determining consistent inelastic design spectra. Workshop on Nonlinear Seismic Analysis of RC Structures, Bled, Slovenia, 13–16 July

# Part V
# Vision in Europe

# Chapter 19
# Pushover-Based Analysis in Performance-Based Seismic Engineering – A View from Europe

**Peter Fajfar and Matjaž Dolšek**

**Abstract** In this chapter, it is claimed that pushover-based methods, although subject to several limitations, often represent a rational practice-oriented tool for the estimation of the seismic response of structures. It is shown that the relations between quantities controlling the seismic response can be easily understood if a pushover-based analysis is presented graphically in the acceleration – displacement (AD) format. One of the pushover-based methods, i.e., the N2 method, which is implemented in Eurocode 8, as well as its extensions, is very briefly summarized. Additionally, some recent pushover-based applications are listed. Finally, as an example of the application of pushover analysis, the seismic performance assessment of a multistorey building with consideration of aleatory and epistemic uncertainties is presented.

**Keywords** Seismic analysis • Pushover analysis • Nonlinear analysis • N2 method • Seismic codes • Eurocode 8 • Acceleration-displacement format • IDA • Seismic risk

## 19.1 Introduction

Seismic analysis is an essential part of Performance-Based Seismic Engineering (PBSE). It is needed to obtain estimates of the response of a structure and its contents when subjected to expected ground motions. Since the problem is a dynamic one, and, in majority of cases, inelastic, the theoretically correct analysis method is nonlinear response-history analysis. Moreover, since the ground motion is random,

P. Fajfar (✉) • M. Dolšek
Faculty of Civil and Geodetic Engineering, University of Ljubljana, Jamova 2, SI 1000 Ljubljana, Slovenia
e-mail: pfajfar@ikpir.fgg.uni-lj.si; mdolsek@fgg.uni-lj.si

M. Fischinger (ed.), *Performance-Based Seismic Engineering: Vision for an Earthquake Resilient Society*, Geotechnical, Geological and Earthquake Engineering 32,
DOI 10.1007/978-94-017-8875-5_19, 

and the structural characteristics are uncertain, in principle a probabilistic approach is needed. However, in practice, considering the financial and time constraints, and the level of knowledge of engineers, usually simplified analysis methods are used. Probabilistic approaches have not yet been implemented in structural analysis, with the exception of nuclear power plant structures, where traditionally elastic analyses are performed. In practice, a whole range of deterministic analysis methods is used, depending on the importance of the structure and on the required accuracy. The simplest is the equivalent lateral force static procedure. The standard method (used in Europe) is elastic response spectrum analysis, where the beneficial effects of energy dissipation in the inelastic range and those of overstrength are taken into account by reducing the seismic forces. Over the last decade, methods based on nonlinear static (pushover) analysis have become popular, but they have not yet been widely accepted among the engineers in practice. Linear response-history analysis makes sense only if a structure is required to remain undamaged after a strong earthquake, e.g. in the case of a nuclear power plant. Nonlinear response-history analysis is rarely used in practice, mainly for important structures, and requires very skilful engineers.

In our opinion, there is a dangerous trend that (mostly young) engineers, impressed by the new opportunities provided by the development of computer hardware and software, perform complex analyses of very sophisticated structural models without a deeper understanding of the problem. As Sozen (2002) observed, "Ready access to versatile and powerful software enables the engineer to do more and think less." A traditional engineer is well trained in the deterministic static linear analysis of planar structures. Everything beyond these limitations usually causes problems. In seismic design, there is the additional problem that one should basically think in terms of displacements and not in terms of forces, otherwise the basic concepts of seismic design cannot be understood. In order to overcome these problems, analysis methods which are not used as a black box, but allow the designer to think about the structure and its seismic response, should be used. Moreover, probabilistic considerations should be gradually introduced into practice by means of simplified approaches, which can be easily understood.

Our thesis is that pushover-based analyses, which can be presented graphically in the acceleration – displacement (AD) format, help to better understand the basic relations between seismic demand and capacity, and between the main structural parameters determining the structural response, i.e. stiffness, strength, deformation, and ductility. As in the case of any approximate method, pushover-based methods are based on a number of assumptions. When applying these methods, their limitations should be observed. It cannot be expected that they will accurately predict the seismic demand for any structure and any ground motion. The limitations of pushover-based methods have been discussed e.g. in (Krawinkler and Seneviratna 1998) and (Krawinkler 2006). Nevertheless, in spite of the many simplifications which are involved in pushover-based approaches, and in spite of the many limitations which apply, these methods can provide a lot of important information about the structural response. Most pushover-based methods permit visualization of the response and its progression from small loads to the loads associated with the target displacement. They are a very useful tool for understanding the general structural behaviour. Compared to nonlinear response-history analysis, which usually provides the most reliable information on structural response (if performed correctly), pushover-based methods represent a much simpler and much more transparent tool, which, in most cases, is able to detect the most critical parts of a structure. The input data for a pushover-based analysis are much simpler. An average spectrum is used instead of a number of accelerograms. Detailed data on the hysteretic behaviour of the structural elements are not needed. There are no problems with the modelling of damping. The amount of computation time is only a fraction of that required for a nonlinear response-history analysis, and the use of the analysis results is straightforward.

Of course, the advantages of pushover-based methods listed above have to be paid for through a lower accuracy compared to that obtainable by nonlinear response-history analysis. Their accuracy will most probably not be adequate for final design or for the final assessment of important structures and structures with important higher mode effects. Once again, the limitations of the applicability of pushover-based methods have to be emphasized.

In this chapter, first the relations between quantities controlling seismic response are discussed. It is shown that these relations can be easily understood if a pushover-based analysis is presented graphically in the acceleration – displacement (AD) format. Then, one of the pushover-based methods, i.e., the N2 method which is implemented in Eurocode 8 (EC8) (CEN 2004), as well as its extensions, is very briefly summarized. Additionally, some pushover-based applications are listed. Finally, as an example of the application of pushover analysis, the seismic performance assessment of a multistorey building with consideration of aleatory and epistemic uncertainties is presented. Note that ground motion uncertainty is the only source of uncertainty which is, in this chapter, categorized as aleatory in nature, whereas the epistemic uncertainty represents the uncertainty in seismic response for a given ground motion resulting from selected physical and modelling uncertainty.

## 19.2 Relations Between Quantities Controlling the Seismic Response

Traditionally, elastic analysis has been used in seismic codes. The beneficial effect of inelastic energy dissipation, which takes place in ductile structures, and of overstrength, i.e. the actual strength beyond the design level which is an inherited characteristic of the great majority of structures, is taken into account by reducing the elastic acceleration spectra with the reduction (behaviour ($q$), or response modification ($R$)) factors. The simple chart, provided in the first edition of Clough-Penzien's Dynamics of Structures (Clough and Penzien 1975) (Fig. 19.1) is essential for understanding of the concept of strength reduction factors due to ductility capacity. (Unfortunately, in the second edition of the book, the "Ductility factor method"

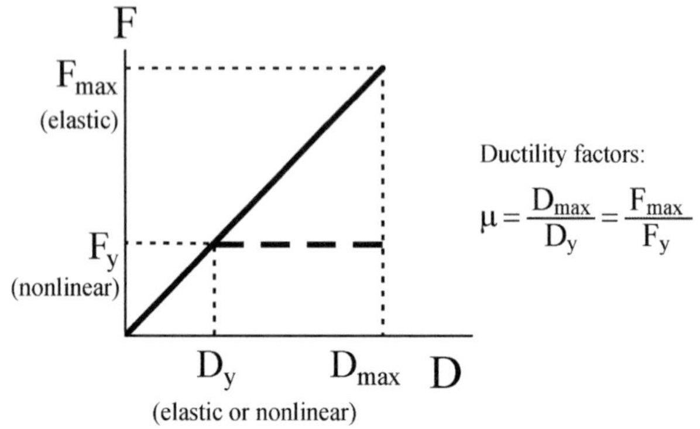


**Fig. 19.1** Explanation of the "Ductility factor method" (Re-plotted from Clough and Penzien 1975)

and Fig. 19.1 disappeared.) It is assumed that the deformations of a structure produced by a given ground motion are essentially the same, whether the structure responds elastically or yields significantly. This assumption represents the "equal displacement rule", originally proposed in (Veletsos and Newmark 1960). A lot of research has been done over the last five decades on the relations between elastic and inelastic demand quantities. Results differ depending on the set of ground motions, and on the structural characteristics used in statistical studies. However, extensive research has not devalued the simple equal displacement rule. On the contrary, at least in the case of SDOF structures on firm sites with a fundamental period in the medium- (velocity controlled) or long-period (displacement controlled) range, with relatively stable and full hysteretic loops, the equal displacement rule has proved to be an adequate assumption. Using the equal displacement rule, Fig. 19.1 suggests that a structure can accommodate an imposed displacement $D_{max}$ not only if its strength is large enough to remain in the elastic range ($F_{max}$), but also if its strength is smaller ($F_y$), provided that it has a sufficient ductility capacity ($\mu$). The ductility dependent reduction factor $R_\mu$, defined as the ratio between the strength of the elastic system ($F_{max}$) and that of an idealized inelastic system with the same stiffness ($F_y$), is equal to the ductility factor $\mu$. The overstrength factor is not presented in Fig. 19.1. It is easy to include it in the figure (Fig. 19.2) and to show that the total strength reduction factor, used in codes, is a product of the ductility-dependent reduction factor $R_\mu$ and the overstrength factor $R_s$ defined as the ratio of the strength at yield ($F_y$) and the design strength ($F_d$).

The educational value of Fig. 19.1 can be much increased by using the acceleration-displacement (AD) format, introduced in (Freeman et al. 1975). Figure 19.1, if plotted in the AD format (force has to be divided by mass), can be combined with demand spectra (Fig. 19.2). The inelastic spectrum in the medium- and long-period ranges in Fig. 19.2 is based on the equal displacement rule. In Fig. 19.2 the quantities relevant for the seismic response of an ideal elasto-plastic SDOF system can be visualized. Seismic demand is expressed in terms

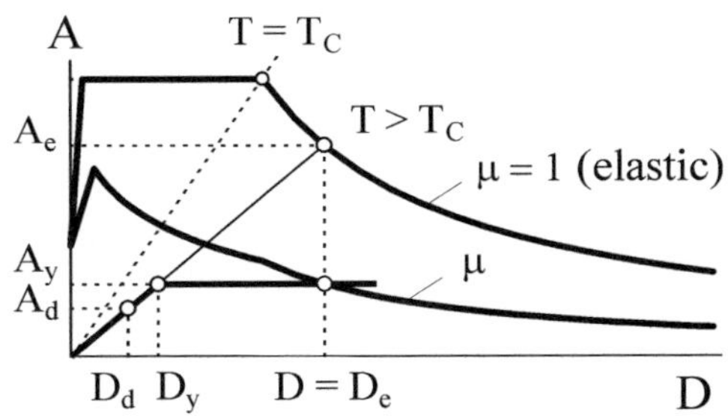


**Fig. 19.2** Elastic and inelastic demand spectra versus capacity curve

of accelerations and displacements, which are the basic quantities controlling the seismic response. Demand is compared with the capacity of the structure expressed by the same quantities. Figure 19.2 helps us to understand the relations between the basic quantities, and to appreciate the effects of changes of parameters. The intersection of the radial line corresponding to the elastic period of the idealised bilinear system $T$ with the elastic demand spectrum $A_e$ defines the acceleration demand (strength) required for elastic behaviour, and the corresponding elastic displacement demand $D_e$. The yield acceleration $A_y$ represents both the acceleration demand and the capacity of the inelastic system. The reduction factor $R_\mu$ can also be expressed as the ratio between the accelerations corresponding to elastic ($A_e$) and inelastic systems ($A_y$). If the elastic period $T$ is larger than or equal to $T_C$, which is the characteristic period of the ground motion, the equal displacement rule applies and the inelastic displacement demand $D$ is equal to the elastic displacement demand $D_e$. Figure 19.2 also demonstrates that the displacements $D_d$ obtained from elastic analysis with reduced seismic forces, and corresponding to the design acceleration $A_d$, have to be multiplied by a reduction factor $R$, which is the product of $R_\mu$ and the overstrength factor, which can also be defined as $A_y$ / $A_d$. The intersection of the capacity curve and the demand spectrum provides an estimate of the inelastic acceleration and displacement demand. This feature allows the extension of visualization to more complex cases, in which different relations between elastic and inelastic quantities and different idealizations of capacity curves are used. However, in such cases the simplicity of relations, which is of paramount importance for practical design, is lost. Note that Fig. 19.2 does not apply to short-period structures.

Figure 19.2 can be used for both traditional force-based design, as well as for the increasingly popular deformation-controlled (or displacement-based) design. In these two approaches, different quantities are chosen at the beginning. Let us assume that the approximate mass is known. The usual force-based design typically starts by assuming the stiffness (which defines the period) and the approximate global ductility capacity. The seismic forces (defining the strength) are then determined, and finally displacement demand is calculated. In direct displacement-based design, the starting points are typically displacement and/or ductility demands. The quantities to be determined are stiffness and strength. The third possibility is a performance evaluation procedure, in which the strength and the stiffness (period) of the structure being analysed are known, whereas the displacement and ductility demands are calculated. Note that, in all cases, the strength corresponds to the actual strength and not to the design base shear according to seismic codes, which is in all practical cases less than the actual strength. Note also that stiffness and strength are usually related quantities. All these approaches can be easily visualised with the help of Fig. 19.2.

The relations apply to SDOF systems. However, they can be used approximately also for a large class of MDOF systems, which can be adequately represented by equivalent SDOF systems.

## 19.3 The N2 Method and Its Extensions

At the University of Ljubljana, a pushover-based method, called the N2 method, has been developed (Fajfar and Fischinger 1987, 1989; Fajfar 1999, 2000). The basic version of this method has been implemented in Eurocode 8. Recently, a number of extensions to the N2 method have been developed, which allow the removal of some limitations of the basic N2 method, but keep the approach conceptually simple. These extensions are as follows.

**Frames with Masonry Infill** The N2 method can be used for the analysis of frames with masonry infill if a multi-linear idealization of the pushover curve is used instead of a bilinear idealization, and if specific inelastic spectra are used (Dolšek and Fajfar 2005).

**Higher Mode Effects Both in Elevation and in Plan** The basic assumption used in pushover-based methods is that the structure vibrates predominantly in a single mode. This assumption is not always fulfilled, especially in the case of high-rise buildings and/or torsionally flexible plan-asymmetric buildings. The extended N2 method, which takes into account higher mode effects both in plan and in elevation, is based on a combination (basically an envelope) of the results of the basic pushover analysis and those of a standard elastic response spectrum analysis. For plan asymmetric buildings see (Fajfar et al. 2005), for higher modes in elevation see (Kreslin and Fajfar 2011), and for a combination of both see (Kreslin and Fajfar 2012).

**Incremental N2 Analysis (IN2)** The IN2 curve, which can be obtained by incremental N2 analysis (in usual cases, only a single N2 analysis is needed) represents an approximation to the IDA curve (Dolšek and Fajfar 2004).

**Probability of Failure** By combining together the SAC-FEMA method, which permits probability assessment in closed form, and the N2 method, which is used for the determination of the "failure" point, an explicit equation for the quick estimation of the annual probability of "failure" (i.e. the probability of exceeding the near collapse limit state) of a structure can be derived, which is appropriate for practical applications, provided that predetermined default values for the dispersion measures are available (Dolšek and Fajfar 2007; Fajfar and Dolšek 2012).

**Web Based Methodology for the Prediction of Approximate IDA Curves** A user-friendly web-based methodology for the prediction of approximate IDA curves, which consists of two independent processes, was recently proposed. The result of the first process is a response database of the single-degree-of-freedom model, whereas the second process involves the prediction of approximate IDA curves from the response database by using n-dimensional linear interpolation. This methodology results in a web application (e.g. http://ice4risk.slo-projekt.info/wida/, Peruš et al. 2013), which can be used within the N2 method in order to obtain a more accurate (compared to the standard procedure) record-specific estimate of the target displacement, as well as an estimate of the dispersion measures.

## 19.4 Examples of Pushover-Based Applications

Pushover-based methods are approximate seismic analysis methods, but they are computationally efficient and in many cases provide results with sufficient accuracy. Even if a more accurate nonlinear dynamic analysis is performed, all the problems associated with the seismic performance assessment of structures cannot be solved, since the nonlinear models, which are used in both types of analyses, are approximate. It therefore makes sense that pushover-based methods are used not only for the approximate assessment of the nonlinear seismic response of structures, but also in combination with more accurate methods of analysis for applications where computational time still represents an important constraint. Several such pushover-based applications have been developed at the University of Ljubljana. Some examples are as follows.

**Sensitivity and Uncertainty Analysis** Sensitivity and uncertainty analyses can be conveniently conducted by using pushover analysis in combination with inelastic spectra or with the nonlinear dynamic analysis of an equivalent single-degree-of-freedom-model (Dolšek 2012; Celarec et al. 2012).

**Structure-Specific Ground Motion Record Selection for Progressive Incremental Dynamic Analysis** Pushover analysis in conjunction with the nonlinear dynamic analysis of an equivalent single-degree-of-freedom model can be used in order to select the most representative ground motion records (Azarbakht and Dolšek 2007, 2011).

**Iterative Pushover-Based Procedures for the Approximate Incorporation of Non-simulated Failure Modes** The approximate simulation of failure modes, which are not directly simulated in the structural model, can be achieved through an iterative procedure which involves pushover analysis, post-processing of the analysis results using limit-state checks of the components, and model adaptation. Such an approach was recently implemented for the probabilistic performance assessment of infilled frames with consideration of the shear failure of columns (Celarec and Dolšek 2012).

**Seismic Risk Assessment** The effect of the structural ageing process on seismic risk can be estimated by using the N2 method for assessing the limit-state intensity over time, taking into account deterioration due to aggressive environmental conditions, such as reinforcement corrosion (Celarec et al. 2011).

## 19.5 Seismic Performance Assessment of a Four-Storey RC Frame with the Consideration of Aleatory and Epistemic Uncertainties

In this sub-chapter, an example of the use of pushover analyses in combination with nonlinear dynamic analyses of simple SDOF systems, aimed at avoiding a large number of computationally demanding nonlinear dynamic analyses of MDOF systems, is presented.

In the case of nonlinear dynamic analysis, the aleatory uncertainty due to the random nature of ground motion is simulated by a set of records, whereas epistemic uncertainty, which is knowledge-based, is usually neglected, although some recent studies have shown that their effects may be important (e.g. Dolšek 2009, 2012). Such simplification is a consequence of the complexity of the nonlinear dynamic analysis, and it can be simply overcome if the seismic response parameters are assessed by pushover-based methods.

A simplified method for seismic risk assessment with consideration of aleatory and epistemic uncertainties was recently proposed (Dolšek 2012). It involves a pushover analysis of a set of structural models, which is defined by utilizing the Latin Hypercube Sampling (LHS) technique, and nonlinear dynamic analysis of the corresponding equivalent SDOF models. The set of structural models captures the epistemic uncertainties, whereas the aleatory uncertainty is, as usual, simulated by a set of ground motion records. Although the method is very simple to implement, it does not include the widely used assumption of independent effects due to aleatory and epistemic uncertainties. Thus, epistemic uncertainty has a potential influence on the median limit-state peak ground acceleration, and not only on dispersion, as has been assumed in some other approximate methods.

The characteristics of the proposed method are summarized through an example of a four-storey reinforced concrete building (Fig. 19.3a, Dolšek 2012), which was designed according to early versions of Eurocodes 2 and 8 for a peak ground acceleration of 0.3 g, soil type B, ductility class high (DCH), and a behaviour factor of q = 5 (Fardis 1996). For this building, the fragility parameters, i.e. the limit-state peak ground acceleration $a_{g,LS}$ and the corresponding dispersion $\beta_{LS}$, were assessed

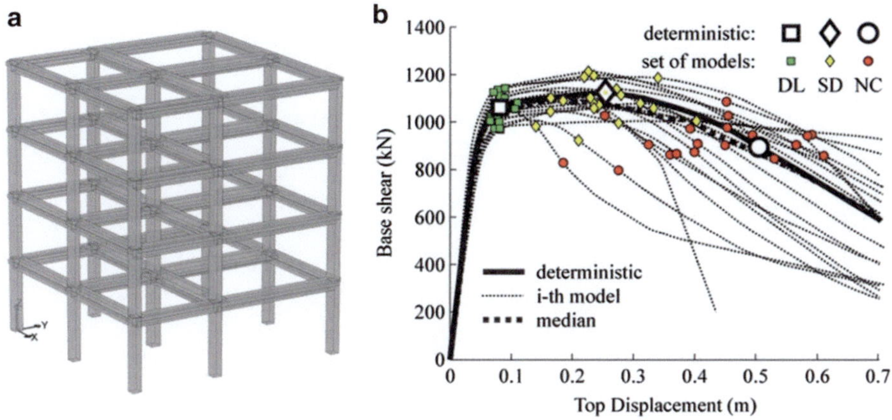


**Fig. 19.3** (**a**) 3D view of the four-storey building and (**b**) the pushover curves corresponding to the set of structural models, the pushover curve corresponding to the deterministic model, and the so-called median pushover curve. The *highlighted points* indicate the DL, SD and NC limit-states

with consideration of the aleatory and epistemic uncertainties, and compared to the case where epistemic uncertainty was neglected. The fragility parameters were assessed for the limit states of damage limitation (DL), significant damage (SD), and near collapse (NC), which were defined according to Eurocode 8. The aleatory uncertainty was modelled by means of a set of 14 ground motion records, which were selected from the European Strong Motion Database (Ambraseys et al. 2000). The selection criteria and a detailed description of the records are presented elsewhere (Dolšek 2012).

A simplified nonlinear structural model was created in the PBEE Toolbox (Dolšek 2010), in conjunction with OpenSees (McKenna and Fenves 2010). The deterministic structural model is based on the median values of the material strength and on other uncertain parameters of the model. In addition to the deterministic model, a set of structural models was created in order to simulate the effects of epistemic uncertainty. The following parameters were considered to be uncertain: mass, strength of the concrete and the reinforcing steel, effective slab width, damping, initial stiffness, and ultimate rotation in the plastic hinges of the beams and columns. All the uncertain parameters were modelled with normal or lognormal random variables. In order to estimate the effects of the epistemic uncertainty a set of twenty structural models was generated based on the LHS technique. The pushover curves for these 20 models are presented in Fig. 19.3b and compared to the corresponding median curve as well as to the pushover curve of the deterministic model. Quite a large scatter of both limit-state displacements and the strength of the building can be observed.

The fragility parameters were determined by performing an incremental dynamic analysis (IDA) of the equivalent SDOF model. The IDA was performed for each equivalent SDOF model and for each ground motion record, respectively, from the

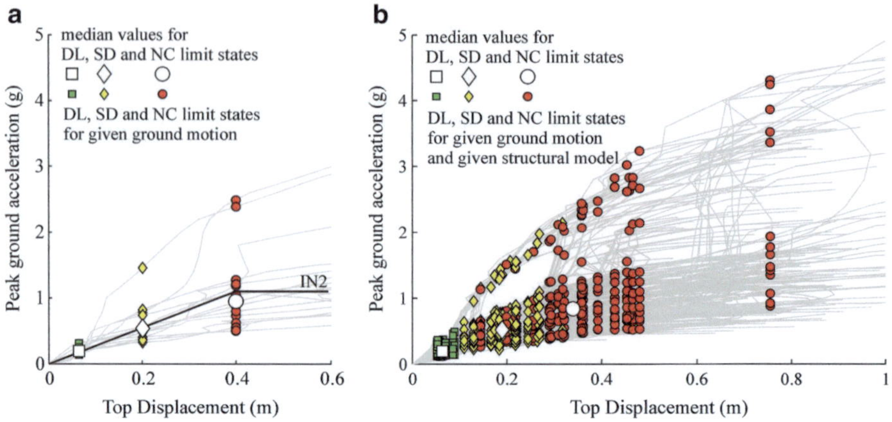


**Fig. 19.4** (**a**) The IDA curves and the IN2 curve of the equivalent SDOF model, which was defined based on the deterministic model, and (**b**) the IDA curves of the equivalent SDOF models determined based on the set of structural models. The *highlighted points* show the DL, SD and NC limit states, as determined according to IDA

**Table 19.1** Fragility parameters for the three limit states estimated with consideration of aleatory uncertainty, and with consideration of both sources of uncertainty

| | Aleatory uncertainty | | | Aleatory and epistemic uncertainty | | | Aleatory and epistemic vs. aleatory uncertainty | | |
|---|---|---|---|---|---|---|---|---|---|
| Parameter | DL | SD | NC | DL | SD | NC | DL | SD | NC |
| $a_{g,LS}$ (g) | 0.19 | 0.54 | 0.95 | 0.19 | 0.52 | 0.84 | 1.00 | 0.96 | 0.88 |
| $\beta_{LS}$ | 0.21 | 0.39 | 0.49 | 0.28 | 0.44 | 0.57 | 1.33 | 1.13 | 1.16 |

set of models and records. For comparison reasons, the IDA and incremental N2 (IN2) were also performed for the equivalent SDOF model, which corresponds to the deterministic structural model. The results of the IDA are IDA curves, which were used for the determination of $a_{g,LS}$. The limit-state peak ground accelerations and the top displacements (limit-state points), as well as the IDA curves, are presented in Fig. 19.4. The IN2 curve is linear up to the NC limit state and constant after this limit state is reached. Since the equal displacement rule applies, only a single point (NC) is needed for the determination of the complete IN2 curve. The values of $a_{g,LS}$ for the DL and SD limit states estimated by the N2 method are practically the same as those determined by IDA, whereas those of $a_{g,LS}$ for the NC limit state are very similar to those determined by IDA.

The fragility parameters (Table 19.1) were calculated based on the limit-state points (determined by IDA) presented in Fig. 19.4. The results in Table 19.1 indicate that consideration of epistemic uncertainty, in addition to aleatory uncertainty, increases the dispersion, and can substantially decrease the limit-state intensities. The latter effect increases with the severity of the limit state.

## 19.6 Disclaimer

In the contribution, the views of the authors are presented, which are based on long experience in research, teaching and consulting work. They reflect the situation in Slovenia. They do not necessarily apply to the whole of Europe. We do not believe that there is a unique vision, in Europe, regarding PBSE in general, and seismic analysis in particular, although there is a common project in earthquake engineering, i.e. Eurocode 8, which is based on the formidable efforts of European researchers and engineers, and which represents a good platform for future developments.

## 19.7 Conclusions

Pushover-based analyses represent a rational practice-oriented tool for the seismic analysis of structures. Compared to the traditional elastic analyses which are employed in seismic codes, they provide a lot of additional important information about expected structural response, e.g., in most cases they are able to detect the most critical parts of a structure. Compared to nonlinear response-history analysis, they represent a much simpler and much more transparent tool. A pushover-based analysis, which can be presented graphically in the acceleration – displacement format, helps us to better understand the basic relations between seismic demand and capacity, and between the main structural parameters determining the structural response, i.e. stiffness, strength, deformation, and ductility. It permits visualization of response and its progression from small loads to loads associated with the target displacement and beyond. It is a very useful tool for understanding the general structural behaviour. Like any approximate method, pushover-based methods are based on a number of assumptions. Their limitations should be observed. They should not be regarded as a replacement for standard elastic analyses, but rather as a complement to them, as a kind of "second opinion", in standard practical applications. Of course, the accuracy and reliability of pushover-based methods cannot be compared with those of nonlinear response-history analysis. They cannot replace nonlinear response-history analysis in cases when enhanced accuracy and reliability are required. Pushover-based methods can also be used in combination with more accurate methods of analysis for applications where computational time still represents an important constraint.

**Acknowledgements** The research presented in this chapter was supported by the Slovenian Research Agency. This support is gratefully acknowledged.

## References

Ambraseys N et al (2000) Dissemination of European Strong-Motion Data. CD-ROM collection. European Council, Environment and Climate Research Programme

Azarbakht A, Dolšek M (2007) Prediction of the median IDA curve by employing a limited number of ground motion records. Earthq Eng Struct Dyn 36(15):2401–2421

Azarbakht A, Dolšek M (2011) Progressive incremental dynamic analysis for first-mode dominated structures. J Struct Eng 137(3):445–455

Celarec D, Dolšek M (2012) Practice-oriented probabilistic seismic performance assessment of infilled frames with consideration of shear failure of columns. Earthq Eng Struct Dyn, Published online in Wiley Online Library (wileyonlinelibrary.com). doi10.1002/eqe.2275

Celarec D, Vamvatsikos D, Dolšek M (2011) Simplified estimation of seismic risk for reinforced concrete buildings with consideration of corrosion over time. Bull Earthq Eng 9(4):1137–1155

Celarec D, Ricci P, Dolšek M (2012) The sensitivity of structural response parameters to uncertain modelling variables of masonry infilled frames under earthquake loading. Eng Struct 35(2):165–177

CEN (2004) Eurocode 8: design of structures for earthquake resistance. Part 1: General rules, seismic action and rules for buildings, EN 1998-1. Euro Commit for Stand, Brussels

Clough RW, Penzien J (1975) Dynamics of structures. McGraw-Hill, New York

Dolšek M (2009) Incremental dynamic analysis with consideration of modelling uncertainties. Earthq Eng Struct Dyn 38:805–825

Dolšek M (2010) Development of computing environment for the seismic performance assessment of reinforced concrete frames by using simplified nonlinear models. Bull Earthq Eng 8:1309–1329. doi:10.1007/s10518-010-9184-8

Dolšek M (2012) Simplified method of seismic risk assessment of buildings with consideration of aleatory and epistemic uncertainty. Struct Infrastruct Eng 8(10):939–953

Dolšek M, Fajfar P (2004) IN2 – A simple alternative for IDA. In: Proceedings of the 13th world conference on earthquake engineering, Vancouver, Canada, paper 3353

Dolšek M, Fajfar P (2005) Simplified non-linear seismic analysis of infilled reinforced concrete frames. Earthq Eng Struct Dyn 34(1):49–66

Dolšek M, Fajfar P (2007) Simplified probabilistic seismic performance assessment of plan-asymmetric buildings. Earthq Eng Struct Dyn 36(13):2021–2041

Fajfar P (1999) Capacity spectrum method based on inelastic demand spectra. Earthq Eng Struct Dyn 28(9):979–993

Fajfar P (2000) A nonlinear analysis method for performance-based seismic design. Earthquake Spectra 16(3):573–592

Fajfar P, Dolšek M (2012) A practice-oriented estimation of the failure probability of building structures. Earthq Eng Struct Dyn 41(14):531–547

Fajfar P, Fischinger M (1987) Non-linear seismic analysis of RC buildings: implications of a case study. Eur Earthq Eng 1:31–43

Fajfar P, Fischinger M (1989) N2 – A method for non-linear seismic analysis of regular buildings. In: Proceedings of the 9th world conference earthquake engineering, vol V. Tokyo, Kyoto, 1988, Maruzen, Tokyo, 1989, pp 111–116

Fajfar P, Marušić D, Peruš I (2005) Torsional effects in the pushover-based seismic analysis of buildings. J Earthq Eng 9(6):831–854

Fardis MN (ed) (1996) Experimental and numerical investigations on the seismic response of RC infilled frames and recommendations for code provisions, ECOEST/PREC 8, Rep. no. 6. LNEC, Lisbon

Freeman SA, Nicoletti JP, Tyrell, JV (1975) Evaluations of existing buildings for seismic risk – a case study of Puget Sound Naval Shipyard, Bremerton, Washington. In: Proceedings of the 1st U.S. National conference on earthquake engineering, EERI, Berkeley, CA, pp 113–122

Krawinkler H (2006) Importance of good nonlinear analysis. Struct Des Tall Spec Build 15(5):515–531

Krawinkler H, Seneviratna GDPK (1998) Pros and cons of a pushover analysis for seismic performance evaluation. Eng Struct 20(4–6):452–464

Kreslin M, Fajfar P (2011) The extended N2 method taking into account higher mode effects in elevation. Earthq Eng Struct Dyn 40(14):1571–1589

Kreslin M, Fajfar P (2012) The extended N2 method considering higher mode effects both in plan and elevation. Bull Earthq Eng 10(2):695–715

McKenna F, Fenves GL (2010) Open system for earthquake engineering simulation. Pacific Earthquake Engineering Research Center, Berkeley, CA, http://opensees.berkeley.edu

Peruš I, Klinc R, Dolenc M, Dolšek M (2013) A web-based methodology for prediction of approximate IDA curves. Earthq Eng Struct Dyn 42(1):43–60

Sozen M (2002) A way of thinking. EERI newsletter, April 2002

Veletsos AS, Newmark NM (1960) Effects of inelastic behavior on the response of simple systems to earthquake motions. In: Proceedings of the 2nd WCEE, vol. 2. Tokyo, pp 895–912

# Chapter 20
# Challenges and Problems in Performance-Based Design of Tall Buildings

**M. Nuray Aydınoğlu**

**Abstract** Tall building design is becoming a major area application of performance-based seismic design, as evidenced by several design guidelines and consensus documents published in the last few years. In general, performance-based earthquake engineering has brought new dimensions to tall building design, leading to a major transformation from the linear strength-based approach to the nonlinear deformation-based design practice. Consequently it becomes possible that the structural restrictions imposed on tall buildings by traditional prescriptive seismic design codes can be removed. However design guidelines have not fully matured yet and there are several issues, on which consensus has not been reached yet. On the other hand, it has to be admitted that the design profession is not prepared yet to fully implement the requirements of the performance-based design. Conceptual transformation from the prescriptive code-based design to a non-prescriptive design based on completely new features including nonlinear modeling, response-history analysis and deformation-based acceptance criteria represents a great challenge. Tall building design engineers are in need of appropriate design tools to help them, at least in the preliminary design stage, for a smooth transition to the performance-based design. The present paper is intended to identify some of the critical problems the design engineers face in the challenging new era of performance-based tall building design.

**Keywords** Tall Buildings Initiative (TBI) • Tall buildings • Prescriptive design • Code-level design • Preliminary design • Minimum base shear • Core wall system • Coupled walls • Coupling beam • Modal Pushover Analysis (MPA) • Incremental Response Spectrum Analysis (IRSA) • Modal capacity diagrams • Tension wall • Compression wall • Diagonal reinforcement • Transfer slab •

---

M.N. Aydınoğlu (✉)
Department of Earthquake Engineering, Kandilli Observatory and Earthquake Research Institute, Boğaziçi University, Çengelköy, 34684 İstanbul, Turkey
e-mail: aydinogn@boun.edu.tr

M. Fischinger (ed.), *Performance-Based Seismic Engineering: Vision for an Earthquake Resilient Society*, Geotechnical, Geological and Earthquake Engineering 32,
DOI 10.1007/978-94-017-8875-5_20, 

Piled foundation • Shear migration • Podium effects • Soil-structure interaction • Kinematic interaction • P-Delta effect • Response Spectrum Analysis (RSA) • Outriggers • Story drift ratio • Shear • Axial force

## 20.1 Introduction

Tall building seismic design has evolved during the last few years to become a major area of application of performance-based earthquake engineering. This development has opened a new door to design engineers who were struggling to overcome the structural restrictions imposed on tall buildings by traditional prescriptive seismic design codes. In a broader sense, performance-based earthquake engineering has brought new dimensions to tall building design, leading to a major transformation from the linear strength-based design to a nonlinear deformation-based design practice.

In this context, special seismic design recommendations/guidelines and consensus documents for tall buildings based on performance-based design principles have been developed and published in the last few years by several institutions. These include Los Angeles Tall Buildings Structural Design Council – LATBSDC (2005, 2008), Structural Engineers Association of Northern California – SEAONC Tall Buildings Task Group (2007), Council on Tall Buildings and Urban Habitat – CTBUH Seismic Working Group (2008). Following this development, a draft version of a tall building design code was prepared in 2008 for the Istanbul Metropolitan Municipality (IMM 2008; Aydınoğlu 2011) where tall building construction is booming. In the meantime Pacific Earthquake Engineering Research Center (PEER) conducted a multi-year collaborative effort, called Tall Buildings Initiative (TBI), to develop more comprehensive performance-based seismic design guidelines for tall buildings (PEER/TBI 2010) along with a supporting document on modeling and acceptance criteria for nonlinear response (PEER/ATC 2010).

In view of the relatively rapid development of the subject, a number of practical problems have emerged in practice, which may be classified into two groups. Firstly, although the above-mentioned recommendations/guidelines have been developed by consensus on most of the challenging design issues, there still remain a number of critical areas where different opinions exist. Secondly, it has to be admitted that the design profession is not prepared yet to fully implement the requirements of the performance-based design. Conceptual transformation from the prescriptive code-based design to a non-prescriptive design based on nonlinear modeling, response history analysis and deformation-based acceptance criteria is not an easy and straightforward process. In this respect it is being observed that tall building designers are in need of appropriate design tools to help them, at least in the preliminary design stage, for a smooth transition to the performance-based design. This paper is intended to examine such problems and identify the critical issues in tall building design practice related to this new, challenging design concept.

## 20.2 Basic Issues

During the rapid development of tall building seismic design guidelines in a relatively short period of time that elapsed between the publication of the first consensus document of LATBSDC (2005) and the latest comprehensive design guidelines of TBI (PEER/TBI 2010), a number of basic problems have emerged on which consensus have not been fully reached. Typical of those problems are discussed in the following.

### *20.2.1 How to Define a Tall Building?*

Defining tall buildings has not been a straightforward issue in the development of design guidelines. Initially a consensus appeared to be reached for a minimum height limit of 160 ft in LATBSDC (2005), SEAONC (2007) and LATBSDC (2005) and 50 m in CTBUH (2008). However a radically different definition was adopted in TBI (PEER/TBI 2010) referring to "*the unique characteristics of tall buildings including (i) a fundamental translational period of vibration significantly in excess of 1 s, (ii) significant mass participation and lateral response in higher modes of vibration, (iii) a seismic force resisting system with a slender aspect ratio such that significant portions of the lateral drift result from axial deformation of walls and/or columns as compared to shearing deformation of the frames and walls.*"

It seems doubtful whether such a more detailed definition of a tall building would be appropriate for practical purposes instead of simply specifying a height limit.

### *20.2.2 The Issue of Minimum Base Shear: Keeping or Leaving Prescriptive Design Requirements?*

In principle, performance-based design approach should not contain prescriptive design requirements. However it may be argued that some of the prescriptive code requirements may be retained during a transition period before the full application of performance-based design approach can be achieved by the profession. Along this reasoning, in the first two attempts of development of tall building design guidelines, namely, in LATBSDC (2005) and SEAONC (2007), the so-called "*code-level design*" stage has been retained including the "*minimum base shear*" requirement. However, a number of modifications on prescriptive code requirements were made such as relaxing height limitations, removal of force amplification (over-strength) and reliability/redundancy factors, etc.

It is interesting to note that prescriptive design requirements are almost completely eliminated in the subsequent development of tall building design guidelines, as reflected in LATBSDC (2008), CTBUH (2008) and TBI (PEER/TBI 2010). The only exception was the retention of the "*minimum base shear*" strength requirement in LATBSDC (2008) with a fixed base shear coefficient of 3 %, which was presented as a "*capacity design*" requirement. The following commentary is excerpted from LATBSDC (2008):

> Admittedly, imposition of a minimum base shear strength requirement is not a performance based design provision . . . . Requiring the same minimum base shear strength corresponding to essentially elastic behavior of the structure, is simply retention of this Los Angeles tall building design tradition.

Although retention of the minimum base shear requirement is attributed to a local design tradition, the real intent is probably expressed in the concluding paragraph of the commentary:

> LATBSDC and its invited advisory group were of the opinion that elimination of prescriptive code evaluation from the current edition of this document justified retaining a minimum base shear strength requirement. As more information is developed on the performance of buildings analyzed and designed according to this document, this limit may be either modified or eliminated.

Although the latest, most comprehensive tall building design guidelines TBI (PEER/TBI 2010) excludes any minimum base shear requirement, it appears that the tall building designers favor such a provision to be specified on the basis of local seismic hazard level.

## 20.3 Preliminary Design Issues

Since a general consensus appears to be reached on leaving traditional prescriptive design approach (LATBSDC 2008, PEER/TBI 2010), preliminary design stage needs to be given a special emphasis for the development of a suitable tall building structural system later to be designed on performance basis through nonlinear seismic analysis.

In this respect, LATBSDC (2008) considers the preliminary design stage as merely equivalent to the application of *Capacity Design Rules* with the additional provision of a minimum base shear strength requirement. On the other hand TBI (PEER/TBI 2010) treats the preliminary design issue in a more detailed fashion, additionally including recommendations on system configuration, wind effects, limiting building deformations, setbacks and offsets, diaphragm demands, outrigger elements, etc.

Capacity design rules are intended to insure that "*structural system for the building has well defined inelastic behavior where nonlinear actions and members are clearly defined and all other members are stronger than the elements designed to experience nonlinear behavior.*" Detailed lists are provided in both LATBSDC (2008) and TBI (PEER/TBI 2010) to identify the "*zones and actions commonly designated for nonlinear behavior*".

### 20.3.1 How to Apply Capacity Design Principles?

When applying capacity design principles, it is stated that in LATBSDC (2008) that "*linear analysis may be used to determine the required strength of the yielding actions*". It is doubtful whether such a simplistic approach would be acceptable for proportioning tall building structural systems. In this regard, a related commentary in TBI (PEER/TBI 2010) cautiously adopts a more rational approach:

> Capacity design concepts are a good starting point when considering desirable system and element actions. While a strict application of capacity design may not be practical or even warranted in the final design, early consideration of these principles will help establish a clear hierarchy to the anticipated building response and will serve to guide the development of the design, which will later be confirmed through nonlinear response history analysis.

A further comment reads:

> Capacity design approaches provide a useful means to configure a structure to produce predictable inelastic behavior. However, the higher-mode response common in tall buildings can lead to inelastic behavior in zones that simplistic approaches to capacity design will be unable to predict...... Traditional engineering practice has focused strictly on the first translational mode when setting strength requirements and lateral force distributions. For tall buildings, the second or even third mode of vibration can be equally, if not more, important to the overall design.

Regarding the proportioning based on capacity design principles, it is finally concluded as:

> Ultimately, engineers must rely on analytical verification of behavior to detect any additional zones of inelastic behavior other than those suggested by initial capacity design proportioning of the structure.

The above quoted paragraphs from TBI (PEER/TBI 2010) clearly signify the problems associated with the application of capacity design principles in the preliminary design stage of tall buildings. The critical question lies in the use of *linear analysis* to determine the required strength of the yielding actions, as recommended in LATBSDC (2008). In this respect, a frequently encountered example is the preliminary design of coupled wall systems and the systems with outriggers. This particular issue is treated in the next subsection.

### 20.3.2 How to Proportion Core Wall Systems with Coupling Beams and/or Outriggers?

Core walls with peripheral columns represent the most common structural system of tall buildings. Frames with down stand beams are rarely used and in many cases, even completely eliminated leading to flat plate systems. Thus, the so-called *dual systems* with moment-resisting frames (back-up systems) are practically discarded. A number of engineers who faithfully provided the back-up systems in all their past prescriptive code applications appear to be hesitant in accepting this new situation. In this respect, it can be argued that coupled walls with sufficiently stiff and strong coupling beams effectively provide a similar back-up action expected from the moment resisting frames of dual systems with cantilever walls.

However engineers often experience difficulty in preliminary sizing of coupled core walls. Reliable practical analysis tools that would help understand the nonlinear seismic behavior of coupling beams and their role in seismic response of coupled wall systems are not available. Both coupled walls and coupling beams generally undergo significant nonlinear response and coupling beams experience excessive plastic deformations throughout the height of the building. The nonlinear behavior of wall pieces is significantly influenced by the stiffness and strength of coupling beams. In the current practice, linear analysis is being employed inevitably in the preliminary design stage to identify the stiffness and strength of coupled wall components and their distribution. Such a procedure would most likely lead to an overdesign of coupling beams with inappropriate and probably heavily congested reinforcement requirements. On the contrary, a preliminary design based on a linear analysis with reduced seismic loads may result in under-designed wall elements especially in terms of their shear strength.

The situation is almost the same in the case of slender core walls systems requiring outriggers for seismic stability. As it is pointed out in TBI (PEER/TBI 2010), "*it is important to consider the impact of the outriggers on the supporting columns and walls under maximum demand levels*". Again, trying to estimate the axial load demands in supporting columns and walls by linear analysis may lead to unreliable design decisions. In order to control those axial load demands, yielding outrigger elements such as buckling restrained braces have been preferably used in recent applications.

The pushover analysis appears to be the only practical analysis procedure that could replace the elastic analysis in the preliminary design stage. However the traditional pushover analysis procedure is impaired by being limited with a single mode, which is not acceptable for tall buildings. On the other hand, the widely used approximate multi-mode pushover procedure Modal Pushover Analysis – MPA (Chopra and Goel 2002) fails to identify the nonlinear deformations correctly, as it is based on individual static nonlinear analyses run for each mode independently, ignoring the joint contribution of modes in the development of nonlinear response, e.g., the formation of plastic hinges. Furthermore modal load patterns of MPA are based on initial elastic mode shapes and they are kept invariant during individual modal pushovers, which become obsolete once yielding develops upward from the bases of walls.

Although still approximate, a reasonably accurate multi-mode pushover method applicable to the coupled wall systems in the preliminary design stage could be the Incremental Response Spectrum Analysis – IRSA Method (Aydınoğlu 2003, 2004). IRSA is an adaptive multi-mode pushover procedure, in which well-known response spectrum technique is utilized in each piecewise-linear step of an incremental analysis. Following the formation of a plastic hinge at a given step, free vibration analysis is repeated for the new system configuration with the new hinge, and the

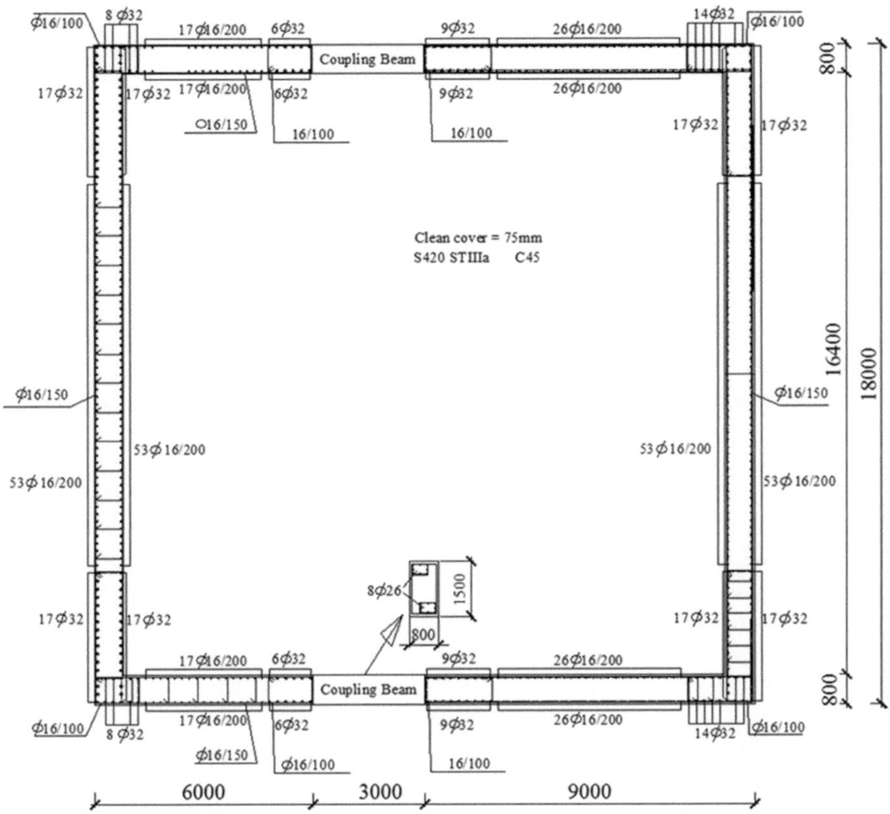


**Fig. 20.1** Non-symmetrical coupled wall core system of a 45 story-180 m tall building with diagonally reinforced coupling beams

distribution of modal displacement increments (or equivalent modal seismic load increments) is updated accordingly. This is particularly important for wall systems where mode shapes are significantly modified following the yielding of wall bases, which could result in drastic changes in the magnitudes and distribution of moments and shears in wall sections (Krawinkler 2006). On the other hand, IRSA method is capable of considering P-Delta effects in a consistent manner.

### *20.3.3 Approximate Nonlinear Response of a Coupled Wall System: A Case Study*

A 45 story – 180 m tall building structural system with a non-symmetrical coupled wall as shown in Fig. 20.1 is analyzed for an X-direction earthquake action with the above-referred Incremental Response Spectrum Analysis – IRSA Method

(Aydınoğlu 2003, 2004). The perimeter columns (not shown) are assumed as gravity columns. Base level dimensions and reinforcement of walls and the coupling beams with diagonal reinforcement are given in Fig. 20.1. A reasonable amount of diagonal reinforcement is provided to the coupling beams on the basis of engineering judgment so as not to create an unduly congestion. Note that this is an extremely important design decision for a realistic preliminary design of coupling beams, which cannot be exercised with a linear analysis. Walls are also reinforced according to minimum requirements of the current Turkish code.

For an approximate nonlinear analysis in the preliminary design stage, both walls as well as coupling beams are modeled as frame elements and their nonlinear behavior is represented by plastic hinges. A P-M hinge is modeled for each wall at every story and M hinges are modeled at both ends of free lengths of coupling beams. No strain-hardening is considered. The effective (cracked) section stiffnesses of walls and coupling beams are taken as 50 % and 10 %, respectively, of the corresponding gross section properties.

The structural system is analyzed under a seismic action of a maximum credible earthquake level in both + X (left-to-right) and −X (right-to-left) directions. The response spectrum is specified with a short-period spectral acceleration of $S_S = 1.55$ g and a 1.0-s spectral acceleration of $S_1 = 0.90$ g. Five modes have been considered in the analysis. Figure 20.2 shows *modal capacity diagrams* of all five modes developed by IRSA under + X direction earthquake in terms of modal displacements and modal pseudo-accelerations, in which P-Delta effects are automatically taken into account. Circles indicate the plastic hinges developed through the system. Superimposed on the same figure is the elastic response spectrum considered in the analysis in ADRS (acceleration-displacement response spectrum) format. It is observed that nonlinearity of the structural system has been mainly confined in the first two modes with gradual tendency to linear response in higher modes.

Figure 20.3 depicts the extent of yielding in walls. Note that only tension walls yielded in both + X and −X earthquake directions, while yielding occurred in all coupling beams. Figures 20.4, 20.5, 20.6, 20.7, 20.8, 20.9, 20.10, 20.11 and 20.12 exhibit valuable information on approximate nonlinear response of the coupled wall system analyzed. Red (bold) and blue lines indicate the response quantities obtained from nonlinear analysis with IRSA and linear analysis with RSA (Response Spectrum Analysis), respectively, calculated under the same earthquake input considering the same section stiffness parameters. Story drift ratio profiles are shown in Fig. 20.4. Plastic rotations of coupling beams (which are almost identical at both beam ends) are given in Fig. 20.5. Cumulative plastic rotations of tension walls (i.e. left- and right-hand walls under + X and −X direction seismic actions, respectively) are presented in Fig. 20.6. Wall bending moment profiles under + X and −X direction earthquakes are given in Figs. 20.7 and 20.8, shear profiles are shown in Figs. 20.9 and 20.10, and finally wall axial forces are depicted in Figs. 20.11 and 20.12, respectively. The dotted lines in the last two figures indicate the wall axial forces due to gravity loads only.

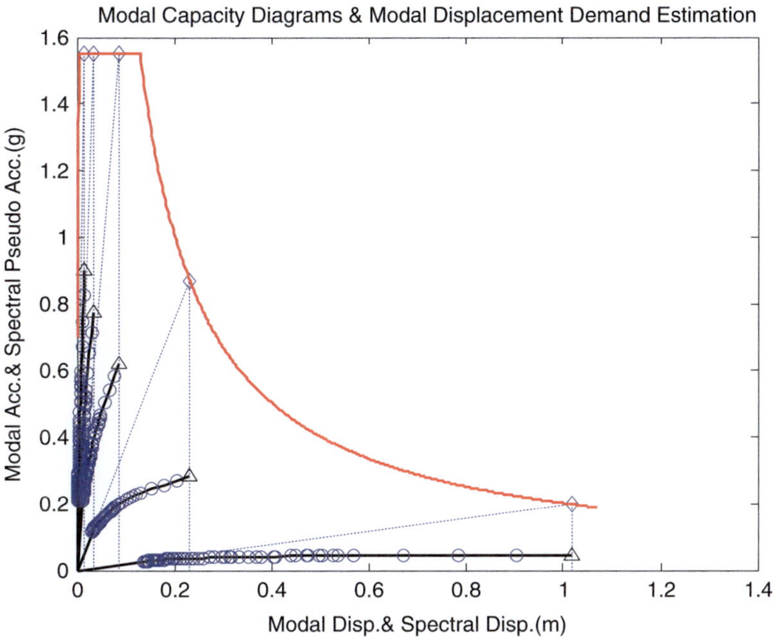


**Fig. 20.2** Modal capacity diagrams of the first five modes and superimposed elastic response spectrum: Modal displacement demand estimation by IRSA through equal displacement rule (+X direction earthquake)

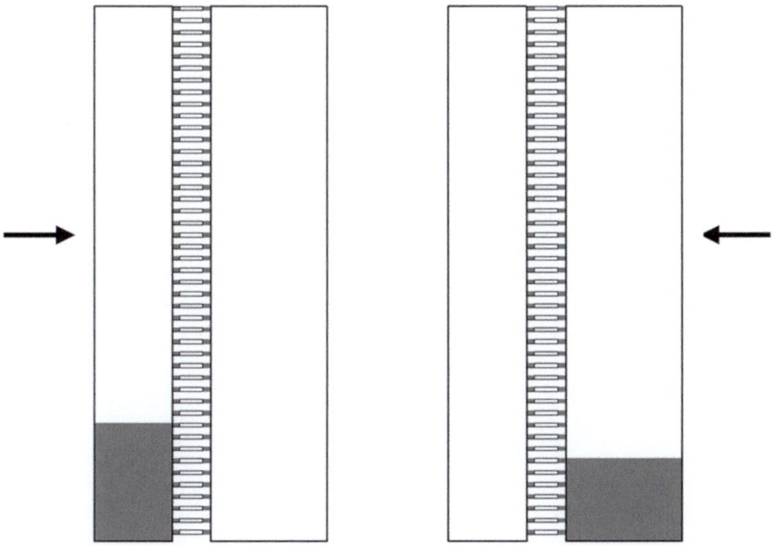

**Fig. 20.3** Plastic hinges of coupling beams and yielding zones of walls (*dark colored*) for + X and − X direction earthquakes

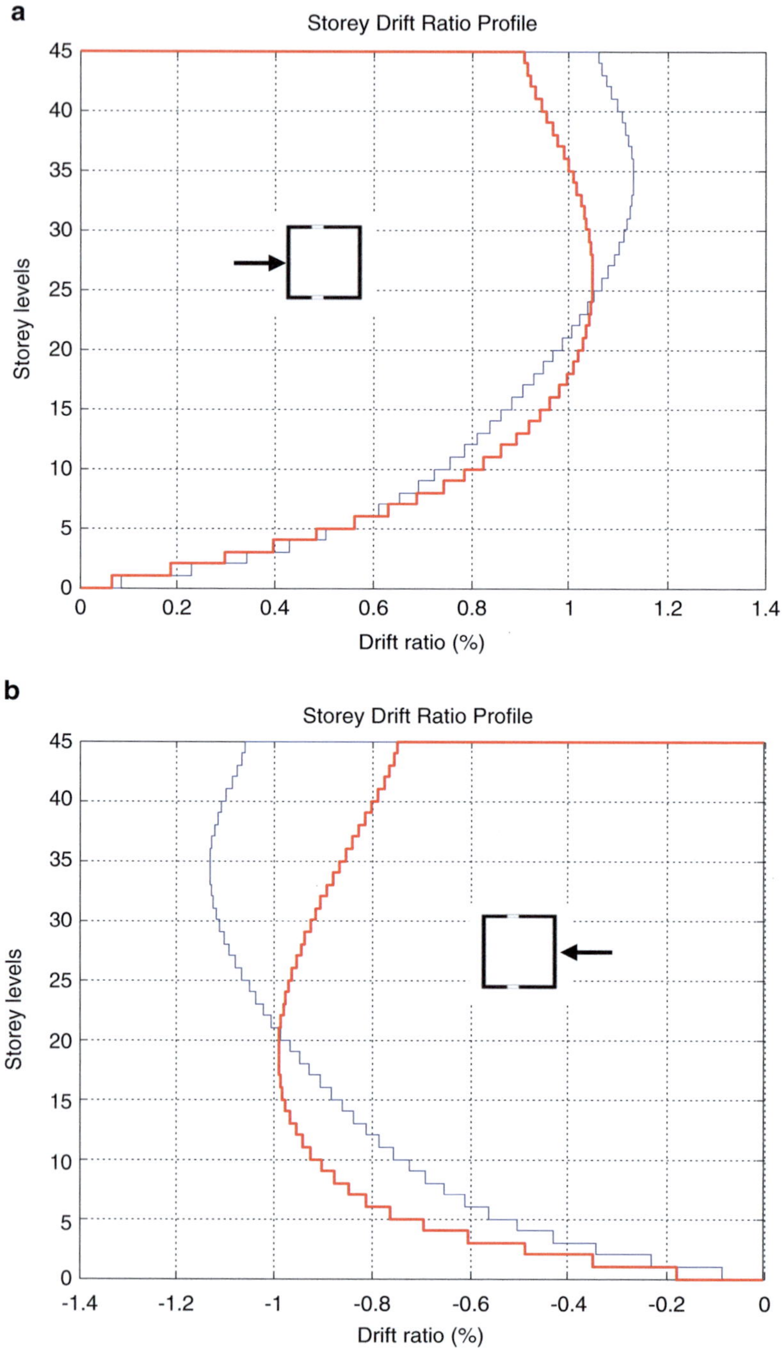


**Fig. 20.4** Story drift ratio profiles for (**a**) + X direction, (**b**) − X direction earthquakes

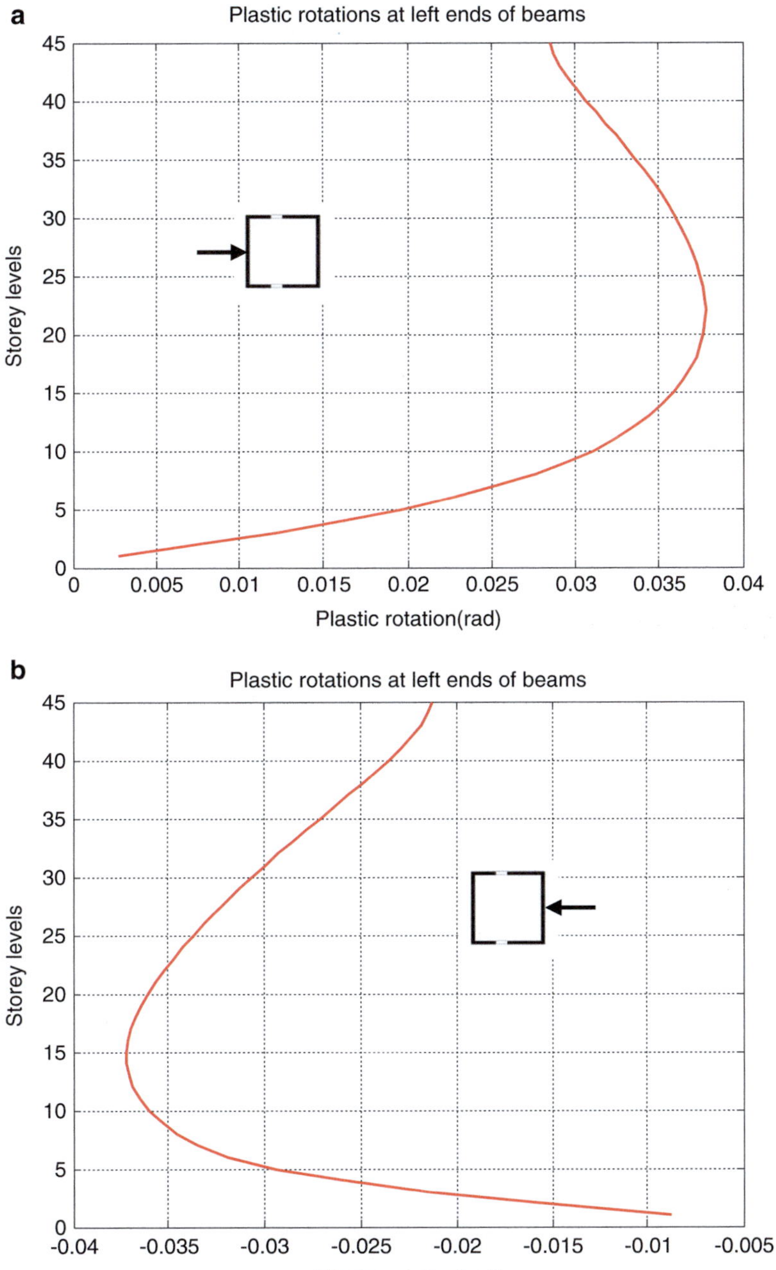


**Fig. 20.5** Plastic rotations of coupling beams for (**a**) + X direction, (**b**) − X direction earthquakes (rotations at *left* and *right ends* are almost identical)

a

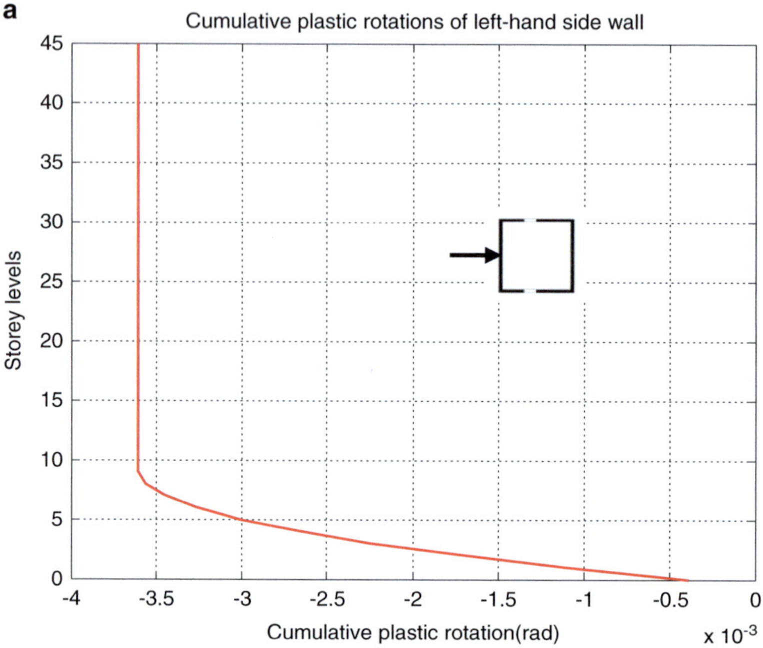


b

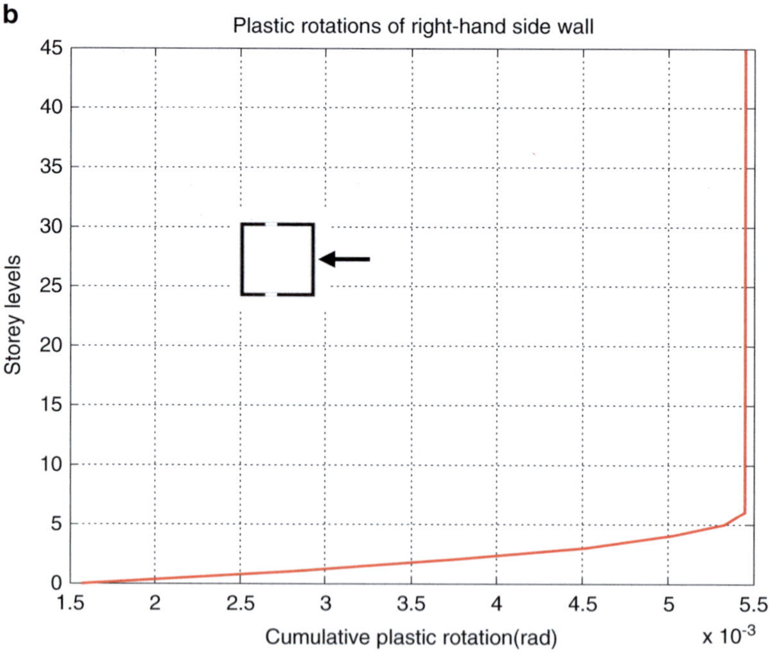


**Fig. 20.6** Cumulative plastic rotations (from *bottom up*) of the tension wall for (**a**) + X direction, (**b**) − X direction earthquakes

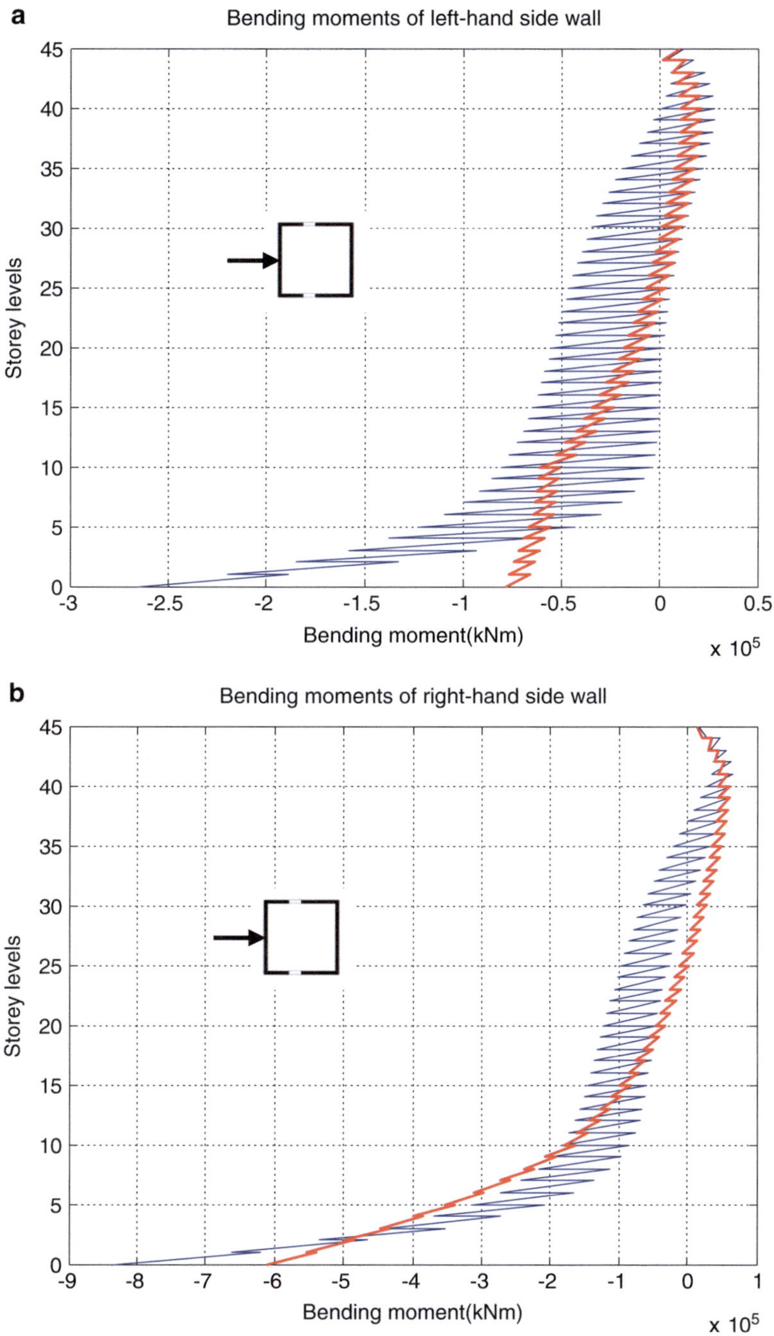


**Fig. 20.7** Bending moments of (**a**) *left-hand side* wall, (**b**) *right-hand side* wall under + X direction earthquake

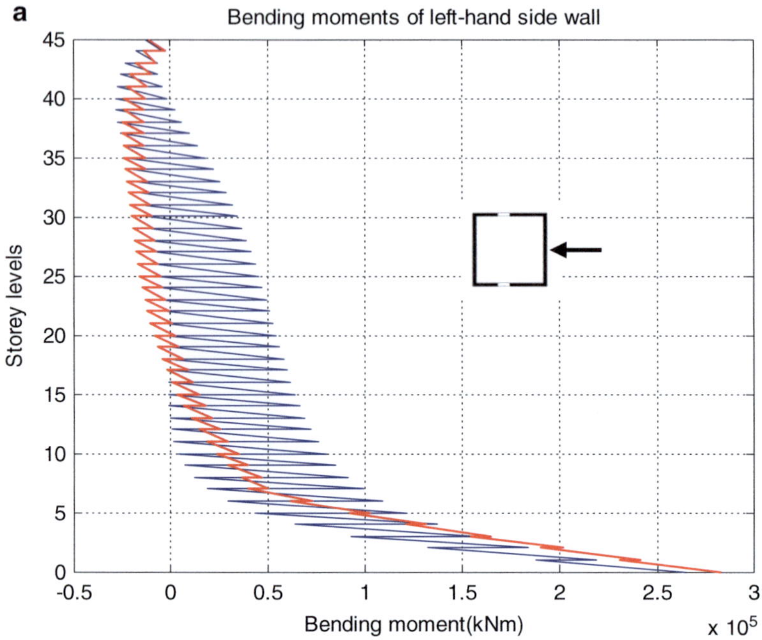


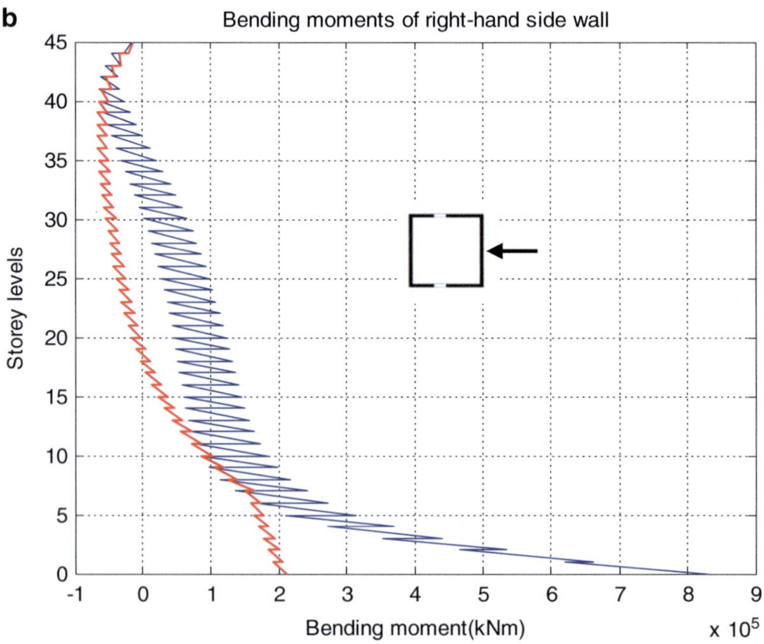


**Fig. 20.8** Bending moments of (**a**) *left-hand side* wall, (**b**) *right-hand side* wall under − X direction earthquake

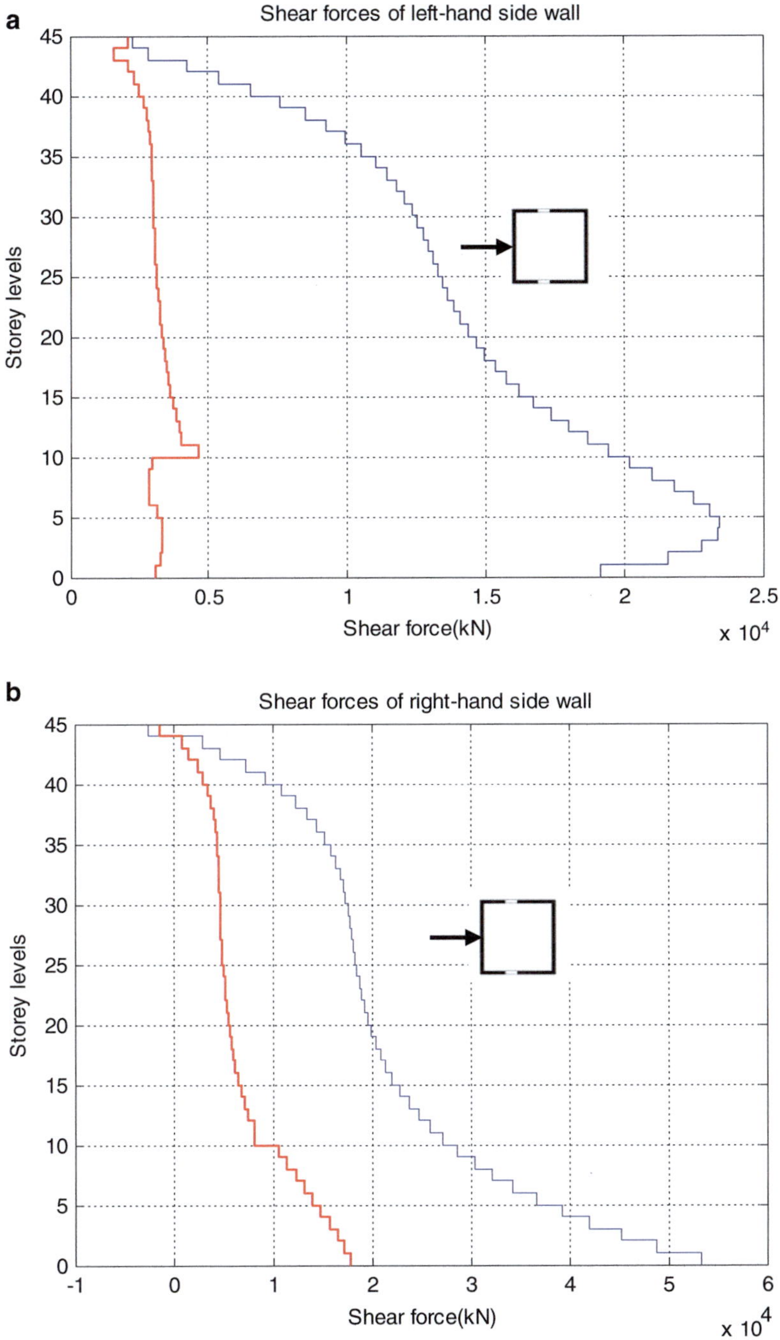


**Fig. 20.9** Shear forces of (**a**) *left-hand side* wall, (**b**) *right-hand side* wall under + X direction earthquake

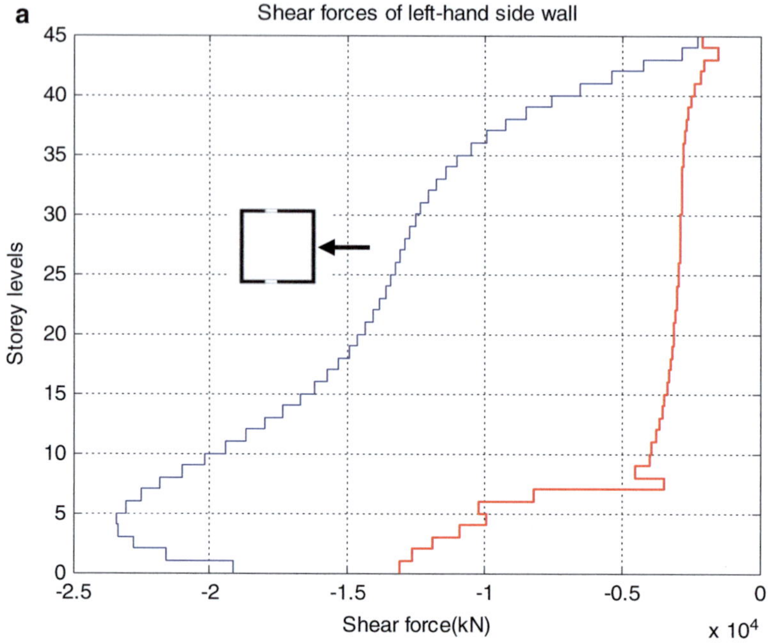


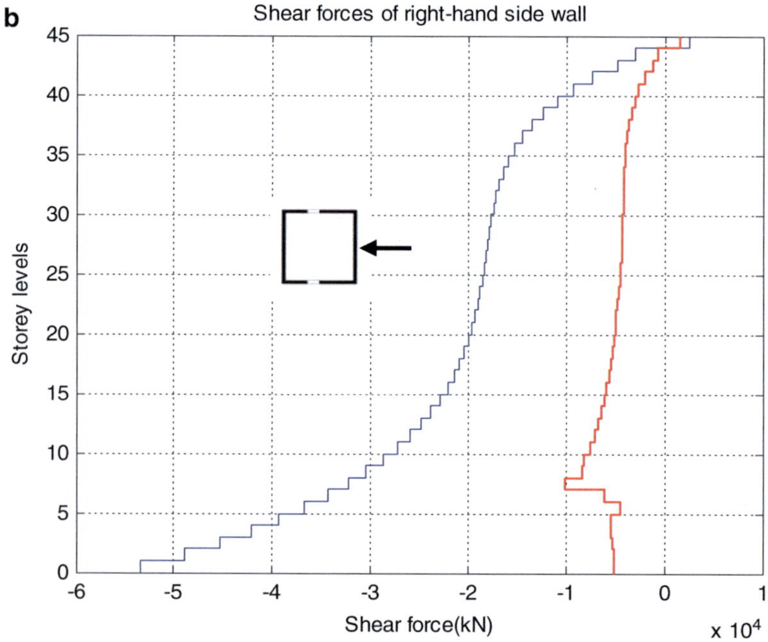


**Fig. 20.10** Shear forces of (**a**) *left-hand side* wall, (**b**) *right-hand side* wall under −X direction earthquake

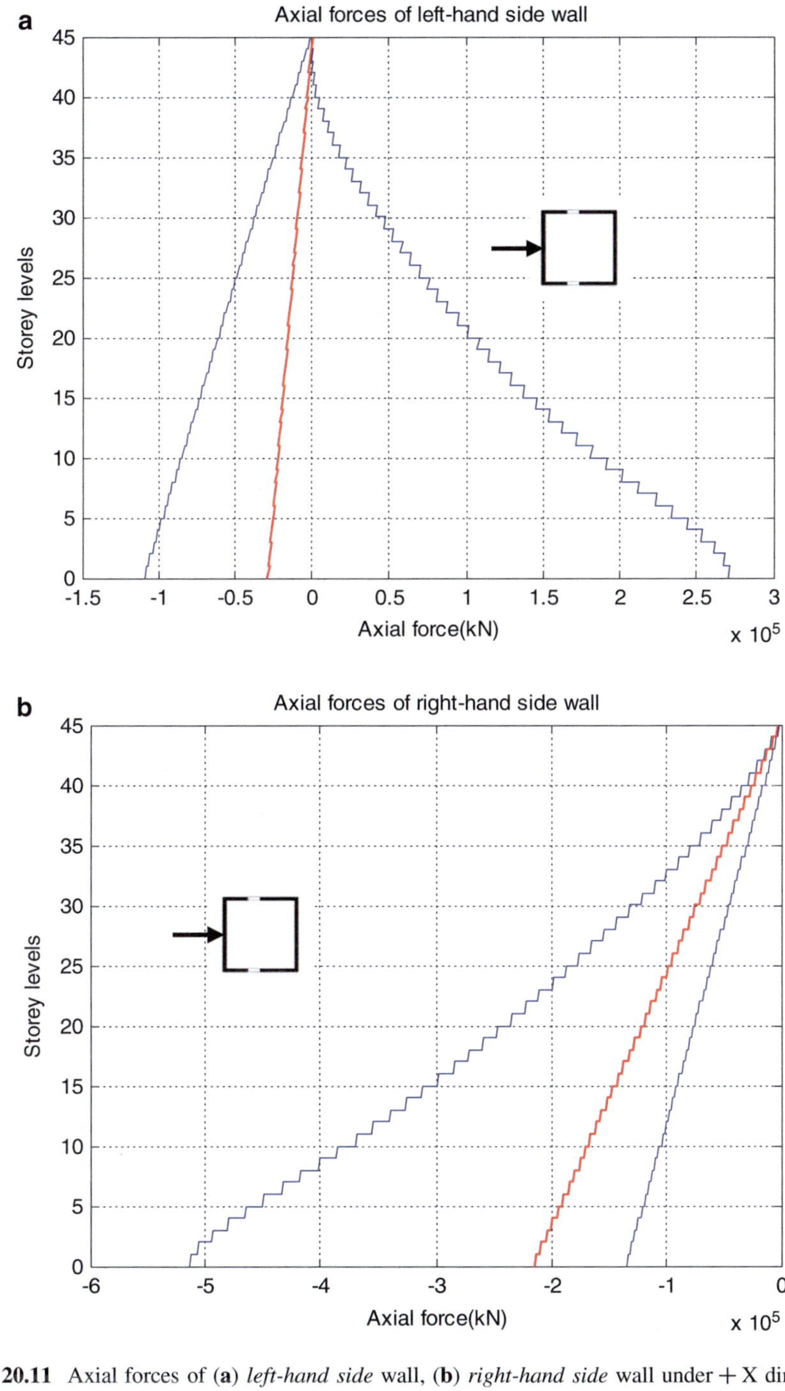


**Fig. 20.11** Axial forces of (**a**) *left-hand side* wall, (**b**) *right-hand side* wall under + X direction earthquake (+ tension, − compression)

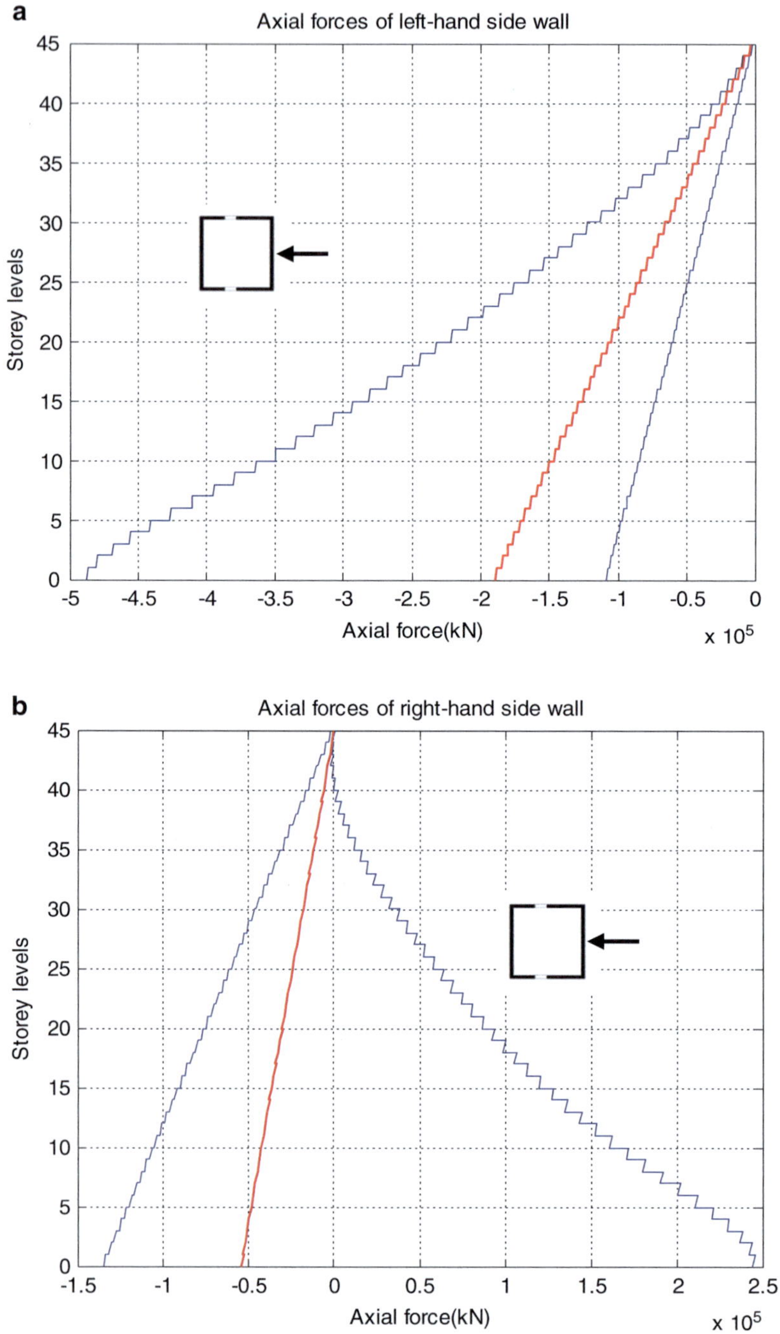


**Fig. 20.12** Axial forces of (**a**) *left-hand side* wall, (**b**) *right-hand side* wall under − X direction earthquake (+ tension, − compression)

It is observed that, as expected, coupling beams experience excessive plastic rotations throughout the height of the building. The distributions of elastic and inelastic wall bending moments and shears confirm the expected trends. Note that reductions in elastic bending moments and shears due to inelastic behaviour do not exhibit a uniform trend (note the scale differences). Further it is observed that inelastic wall axial forces could change sign with respect to elastic forces, as shown in Figs. 20.11 and 20.12. Regarding the wall shears a notable behavior is observed, which would never be detected by a linear analysis in a preliminary design stage: The weaker tension wall, which yields along a significant height from the base, transfer its post-yield shear all along the yielded zone to the stronger compression wall (see Figs. 20.9 and 20.10, note the scale differences), which may be called *shear migration*. In the case of a + X direction earthquake, this may be attributed to the smaller size of the tension wall. Yet, a similar migration is again observed in the case of a −X direction earthquake from the larger size tension wall to the smaller size compression wall. *Shear migration* may play a very important role in the preliminary design of not only non-symmetrical coupled walls, but at the same time symmetrical ones.

The advantage of such a simple nonlinear analysis is that similar nonlinear evaluations can be made for various design options in a very short period of time. Sensitivity analysis can also be performed easily, for example, by changing the strengths (reinforcement ratios) as well as the stiffness reduction parameters of the cracked sections of walls and coupling beams.

## 20.4 Other Structural Design Issues

Several other design problems can be cited as critical questions for the tall building designers. Among them, podium effects and performance-based design of foundations including soil-structure interaction problem are discussed as typical problems in the following sections.

### *20.4.1 How to Model Podium Effects?*

Podiums can be identified as plan-wise enlarged lower portions of tall buildings, which may be constructed above-ground and/or underground. In the latter case lateral stiffness and strength of the podium structure is generally controlled by the peripheral walls, which lead to an abrupt change in the lateral load transfer mechanism from the tall superstructure to the podium. The base-shear of the tall superstructure (story shear of the first story above the podium) is forced to change the load path through the ground floor slab, which is generally called a “transfer slab”. This slab may partially undergo nonlinear behavior. The load transfer mechanism may become even more complex involving the stiffness and strength contribution of one or more slabs below as well as the interaction of the perimeter walls with the surrounding soil. Thus, the soil medium enters to the scene as an additional element for modeling.

It has to be admitted that currently design engineers do lack appropriate tools to model such a complex and partially nonlinear behavior of such interacting structural components. In the near future, significant research effort needs to be directed towards this problem where the current design practice is based rather on past experience and engineering judgment.

### 20.4.2 Performance-Based Design of Foundations

Although there is a growing trend in academic circles to extend the performance-based design concept to the geotechnical/foundation engineering field, it seems doubtful whether the geotechnical engineers are ready to leave so much accustomed concepts such as "allowable bearing pressure".

While structural engineers are striving to estimate the inelastic deformation demands for the tall building superstructure, attempting to do the same in foundation/soil medium attracts a lesser interest. Consequently deformation-based acceptance criteria for foundation design are not included in consensus documents/guidelines on tall building seismic design.

### 20.4.3 Dynamic Soil-Structure Interaction: Need for Practical Procedures for an Overly Complex Problem

Majority of design engineers feel themselves comfortable with the perception that seismic soil-structure interaction can be handled in a straightforward manner by simply specifying an appropriate set of soil springs at the soil-structure and/or soil-pile interface. While such an over-simplifying approach remains as a serious problem to be corrected, the engineers who are aware of the significance and relative complexity of the dynamic soil-structure interaction complain about the lack of practical tools in performance-based earthquake engineering practice.

The performance-based seismic design of tall buildings with piled foundations in weak soil conditions inevitably requires the consideration of soil-pile-structure interaction in the nonlinear range under strong earthquakes. In particular, the analysis of the so-called kinematic interaction effects is vitally important in estimating the seismic demands on piles. In spite of the fact that advanced nonlinear analysis software packages are available separately for tall buildings (e.g. CSI 2006) as well as for soil-pile systems (e.g. Itasca 2011), a single combined analysis software that is capable of handling the nonlinearities of such a combined soil-pile-structure model is not readily available for practical tall building design applications. Therefore the preferred approach is to utilize readily available software independently for the tall building superstructure and the foundation-soil-pile system, respectively, and combine their analysis capabilities within the framework of the well-known Substructure Method of soil-structure interaction (Aydınoğlu 1993). Such a practical application is recently reported for an actual case study performed in İzmir, Turkey, where a number of tall buildings are being constructed on thick alluvial deposits (Aydınoğlu et al. 2012).

## 20.5 Concluding Remarks

Application of performance-based seismic design approach represents a new and challenging development in tall building design. A number of critical issues are highlighted in this contribution, mainly dealing with the difficulties experienced by the design engineers in the transition (or transformation) stage from the prescriptive code design to a more liberal/non-prescriptive performance-based design.

In this regard, the question of keeping or completely leaving at least some of the prescriptive design rules, such as the issue of minimum base shear, is discussed. Preliminary design issues for performance-based design are emphasized with special reference to coupled wall systems. Modeling of podiums is indicated as a critical problem in terms of nonlinear behavior of transfer slabs coupled with the interaction of peripheral walls with the surrounding soil medium.

Performance-based seismic design of foundations including practical methods for dynamic soil-structure interaction remind the necessity of structural and geotechnical engineers speaking the same language of performance-based engineering to jointly achieve the improved analysis and design tall buildings.

## References

Aydınoğlu MN (1993) Consistent formulation of direct and substructure methods in nonlinear soil-structure interaction. Soil Dyn Earthq Eng 12:403–410

Aydınoğlu MN (2003) An incremental response spectrum analysis based on inelastic spectral displacement for multi-mode seismic performance evaluation. Bull Earthq Eng 1:3–36

Aydınoğlu MN (2004) An improved pushover procedure for engineering practice: Incremental Response Spectrum Analysis (IRSA). International workshop on "Performance-based seismic design: concepts and implementation", Bled, Slovenia, PEER Report 2004/05, pp 345–356

Aydınoğlu MN (2011) Draft seismic design code for tall buildings in Istanbul metropolitan area. U.S.-Iran-Turkey seismic workshop on "Seismic risk management in urban areas", 14–16 Dec 2010, Istanbul, Turkey – PEER Report 2011/07, pp 55–63

Aydınoğlu MN, Siyahi B, Akbaş B, Fahjan Y (2012) Soil – pile – structure interaction: a practical approach for performance-based seismic design of tall buildings. In: Second international conference on performance-based design in earthquake geotechnical engineering, Taormina, Italy, 28–30 May 2012

Chopra AK, Goel RK (2002) A modal pushover analysis for estimating seismic demands for buildings. Earthq Eng Struct Dyn 31:561–582

CSI (2006) Perform – components and elements for perform-3D and perform-collapse, version 4. Computers and Structures Inc., Berkeley, CA

CTBUH (2008) Recommendations for the seismic design of high-rise buildings – a consensus document. Council on Tall Buildings and Urban Habitat, Seismic Working Group, Chicago

IMM (2008) Istanbul seismic design code for tall buildings, Draft version IV. Istanbul Metropolitan Municipality, Istanbul

Itasca (2011) FLAC – Fast Lagrangian Analysis of Continua, $FLAC^{3D}$ version 4.0. Itasca Consulting Group, Minneapolis, MN

Krawinkler H (2006) Importance of good nonlinear analysis. Struct Des Tall Spec Build 15:515–531

LATBSDC (2005) An alternative procedure for seismic analysis and design of tall buildings located in the Los Angeles region – a consensus document. Los Angeles Tall Buildings Structural Design Council, Los Angeles

LATBSDC (2008) An alternative procedure for seismic analysis and design of tall buildings located in the Los Angeles region – a consensus document. Los Angeles Tall Buildings Structural Design Council, Los Angeles

PEER/ATC (2010) Modeling and acceptance criteria for seismic design and analysis of tall buildings. PEER/ATC 72–1, Applied Technology Council, Redwood City – Pacific Earthquake Engineering Center, Berkeley, CA

PEER/TBI (2010) Guidelines for performance-based seismic design of tall buildings, version 1.0. Pacific Earthquake Engineering Research Center, PEER Report 2010/05, Nov 2010. Prepared by the Tall Buildings Initiative (TBI) Guidelines Working Group, Berkeley, CA

SEAONC (2007) Recommended administrative bulletin on the seismic design & review of tall buildings using non-prescriptive procedures. Tall Buildings Task Group, Structural Engineers Association of Northern California, April 2007, San Francisco, CA

# Chapter 21
# Performance Based Earthquake-Resistant Design: Migrating Towards Nonlinear Models and Probabilistic Framework

**Adnan Ibrahimbegovic, Luc Davenne, Damijan Markovic, and Norberto Dominguez**

**Abstract** In this work we present our recent works that follow two modern tendencies in modelling and design of engineering structures for extreme loading such as earthquakes: (i) fine scale models for providing the simplest, fine-scale interpretation of inelastic damage mechanisms at the origin of energy dissipation and damping phenomena, as opposed to coarse scale of stress resultants; (ii) the role of probability in this kind of modelling approach. We consider application of these ideas first to structures, especially irreplaceable structures, such as nuclear power plants, and move onto the complex systems such as water networks.

**Keywords** Irreplaceable structures • Heterogeneities • Numerical models • Nonlinear analysis • Performance-based design • Probability framework

A. Ibrahimbegovic (✉)
Ecole Normale Supérieure, LMT-Cachan, 61 avenu du president Wilson, 94235 Cachan, France
e-mail: adnan.ibrahimbegovic@ens-cachan.fr

L. Davenne
University Paris 10, Paris, France
e-mail: luc.davenne@u-paris10.fr

D. Markovic
EDF, SEPTEN, Lyon, France
e-mail: damijan.markovic@edf.fr

N. Dominguez
Ecole Polytechnique, Civil Engineering Master Program, Mexico City, Mexico
e-mail: ndominguez@ipn.mx

M. Fischinger (ed.), *Performance-Based Seismic Engineering: Vision for an Earthquake Resilient Society*, Geotechnical, Geological and Earthquake Engineering 32,
DOI 10.1007/978-94-017-8875-5_21, 

## 21.1 Introduction

The classical design procedure for structures submitted to seismic risk, as defined in current design codes (e.g. Eurocode 8), possesses several crucial ingredients inherited from linear analysis. The case in point is a relatively simple seismic analysis technique based on the pushover analysis of a multi-degree-of-freedom model and the response spectrum analysis of an equivalent single-degree-of-freedom system, which is the basis of N2 design method; e.g. see (Fajfar and Fischinger 1987, 1988) for earlier and (Fajfar 2007) for more recent works. The main advantage of this current design approach with main steps based upon linear analysis is in its efficiency, and in a clear result interpretation when used in the acceleration–displacement format. However, there are also severe restrictions when using such a design method for 3D case, which lead to loss of efficiency (Fajfar 2007).

It has been long recognized that the modal analysis that serves as the basis of the current design approach might lead to less predictive results, especially for the case of very complex structures and systems where the global failure mode for push-over is not unique. However, it is only since recent times that we have acquired the modelling capabilities; e.g. see (Benkemoun et al. 2010; Dominguez et al. 2005; Ibrahimbegovic and Brancherie 2003; Ibrahimbegovic and Delaplace 2003; Ibrahimbegovic and Markovic 2003; Ibrahimbegovic and Melnyk 2007; Ibrahimbegovic 2009; Ibrahimbegovic et al. 2008, 2010; Jehel et al. 2010; Markovic and Ibrahimbegovic 2004) to provide a more refined interpretation of the local failure mechanisms rather than the global, push-over-type models. We have also very well advanced in the long process, from early works (Ibrahimbegovic and Wilson 1990, 1992) to more recent works (Ibrahimbegovic and Markovic 2003) or (Ibrahimbegovic et al. 2008) in development of the robust computational procedure that is able to isolate the most severely damaged zone and deliver the corresponding scenario on the interplay of different local mechanisms in building the global failure mechanisms for structures and systems of any complexity. We have also managed to place such a computational procedure within the framework of probability in order to account for material heterogeneities, and thus construct the corresponding probabilistic bounds (Hautefeuille et al. 2009) as well as to obtain the interpretation of the dominant failure mechanism with respect to structure size, or so-called size effect (Ibrahimbegovic et al. 2011).

Much of this work was carried out in close collaboration with nuclear industry champions in France, EDF, CEA and IRSN, which was motivated by the safety of irreplaceable structures and components in nuclear power plants, when subjected to extreme conditions such as earthquakes. In this paper, we provide a short description of these developments. Of special interest for our works are reinforced-concrete structures, since this is by far the most dominant construction technology in France. Hence, a very clear illustration can be given to different scales we describe (see Fig. 21.1), starting from complex system scale (structure with its environment),

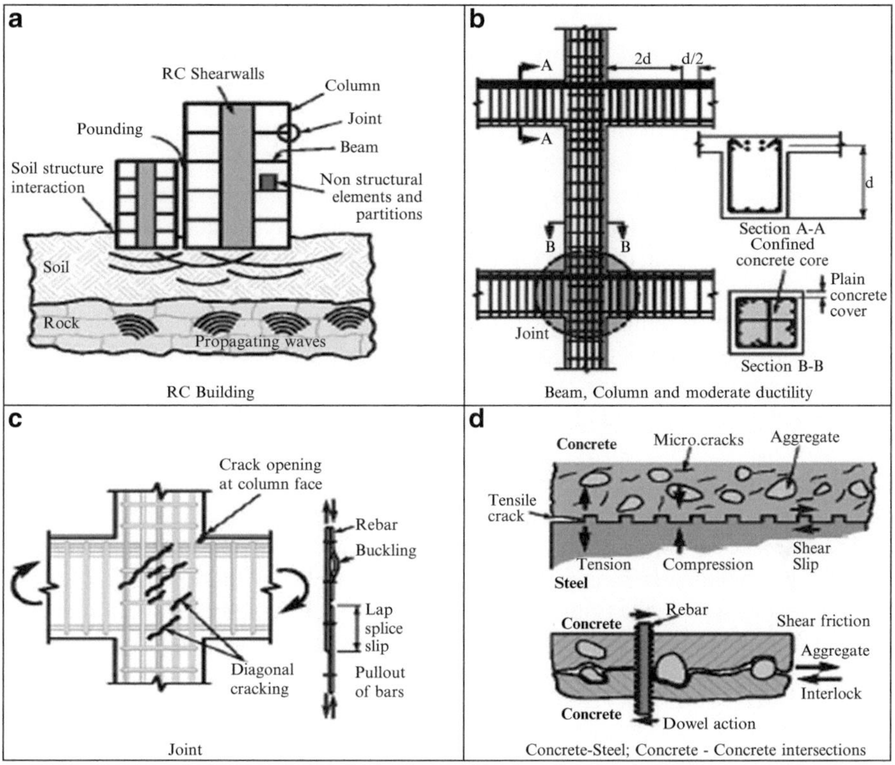


**Fig. 21.1** Different scales in constructing the models for various damage mechanisms for reinforced concrete structures, from complex system scale (structure with its environment), over structural scale (illustrated for reinforced concrete frame structures) to material scale for concrete

over to structure scale (here illustrated for reinforced concrete frames) and finally to material scale (here illustrated as two-phase representation of concrete).

The paper outline is as follows. In the next two sections we first discuss the structural point of view related to providing the most robust computational procedure for seismic analysis as the basis of performance based design approach, along with its extension towards complex systems. Subsequent section considers the material point of view, which provides the most reliable interpretation of inelastic damage mechanisms, which is important for damping effect. The same point of view provides the natural framework to quantify the uncertainties of the modelling and computations due to material heterogeneities and probabilistic interpretation of size effect. The next section carries on with uncertainties in loading conditions. The results of several numerical simulations are integrated in each section to further illustrate the developments to follow. The concluding remarks are given at the end of the paper.

## 21.2 Structure-Scale Interpretation of Damage Induced Damping and Efficient Computational Procedure

Early approach to performance based design considered the design strategy where the isolation devices protected the structure leading only to localized nonlinearities. One example taken from this work considers the piping system in a nuclear power plant (see Fig. 21.2), with the damage significantly reduced when damping devices are placed on the structure. The dedicated solution methods have been developed for this class of problems, to allow very efficient computations by using model reduction (Ibrahimbegovic and Wilson 1990, 1992), with negligible loss of result accuracy, which is the main requirement for performance based design (see Table 21.1 for typical results of this kind).

In terms of ensuring the computational efficiency, the main remaining difficulty concerns the problems where the extreme event would damage the system (especially its irreplaceable component) beyond the possibility of reparation, and would not lead only to limited nonlinearities. This should happen for extreme event, such as a very severe earthquake that can be followed by other extreme loading conditions (e.g. fires, tsunamis). In France where nuclear industry is by far dominant source of electric power supply, the tragic event of Fukushima has reminded the authorities to what extent the nuclear plants can be sensitive to extreme loading conditions of this

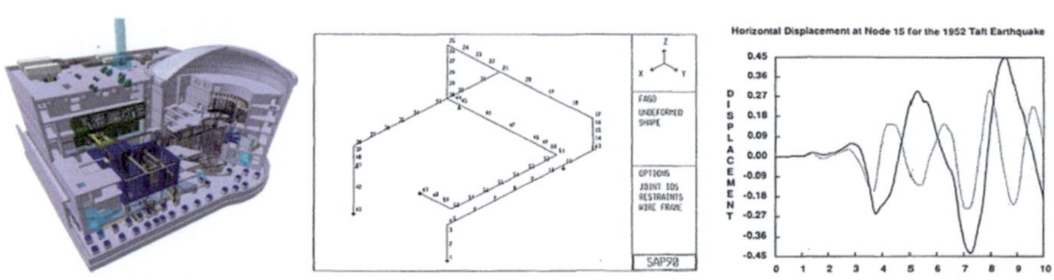


**Fig. 21.2** Nuclear power plant, FE model of its piping system with energy dissipation devices and corresponding response reduction due to energy dissipation (- - - without energy absorbing device, .... with energy absorbing device)

**Table 21.1** Computational effort for computing dynamic response of piping system

| Method | CPU time [s] |
|---|---|
| Global coord: | |
| Standard nonlinear | 1767.47 |
| Proposed direct | 281.83 |
| Standard linear | 174.03 |
| Modal coord: | |
| Proposed dyn. subs. 5 vec. | 27.07 |
| Proposed dyn. subs. 15 vec. | 40.13 |
| Proposed mode sup. 15 vec. | 7.63 |
| Proposed mode sup. 25 vec. | 11.48 |
| Standard mode sup. 15 vec. | 5.18 |

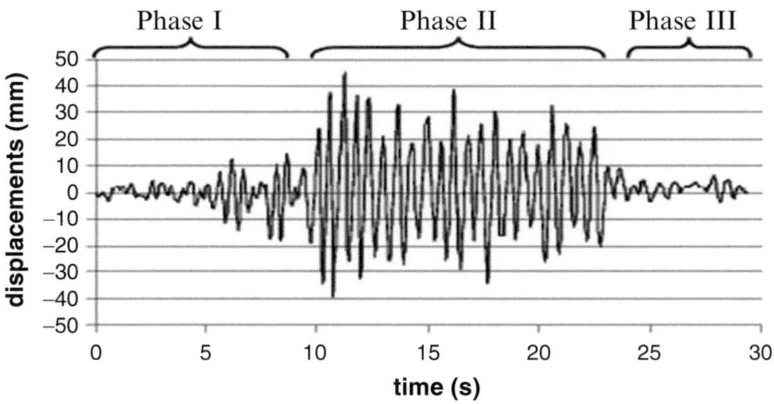


**Fig. 21.3** Three phases of dynamic response due to earthquake:(i) small vibrations of undamaged structure, (ii) strong shaking causing severe damage in forced vibrations, (iii) free vibrations of already damaged structure

kind. This has further spurred renewed interest in safety check of existing power plants and performance based design of new ones.

The main goals in modelling structure performance are as follows: We would need to avoid, if possible, or at least to reduce the exclusive use of Rayleigh damping, either proportional (Clough and Penzien 1996) or non-proportional (Ibrahimbegovic and Wilson 1989), and replace it by more reliable interpretation of different damage induced dissipation mechanisms. The most comprehensive model of this kind should be capable of interpreting three different phases of earthquake response (see Fig. 21.3), with a single set of parameters: (i) phase of forced vibrations for very low level of damage, (ii) forced vibrations for strong-shaking response resulting in considerable damage, (iii) free-vibrations of damaged structure.

In a number of recent works, we have strived to develop the models and methods for earthquake resistant performance based design. With a first glance on this kind of method, one can conclude that the procedure quickly becomes complicated, with difficult issues pertaining to: non-linear dynamics methods, multi-scale models and stochastic analysis. Moreover, these methods impose very high computational cost, which makes them perhaps prohibitively expensive for conventional structures. However, these difficulties and computational cost faded away when it came to very important structures from nuclear industry in France, which justified from the start the use of more refined models and methods in order to reduce the risk.

Much of the progress has been made recently in computational mechanics community in dealing with nonlinear behavior of concrete and reinforced concrete, which can be brought to bear upon the current engineering design practise. More precisely, concerning the models of RC structures, we have managed to provide more sound interpretation and rigorous thermodynamical basis as opposed to macro models, such as rotating crack stress-resultant model, (Ibrahimbegovic and Frey

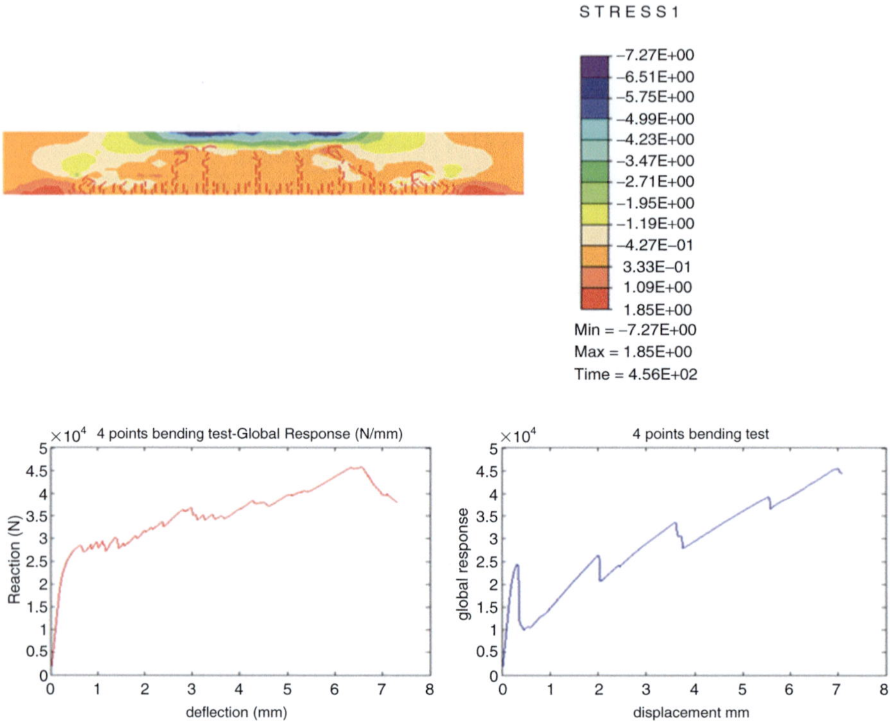


**Fig. 21.4** Reinforced concrete model of beam in four-point bending test capable of predicting crack spacing and opening, computed force-displacement diagrams for strong reinforcement and for weak reinforcement

1993), and we can now better combine plasticity, damage, and different hardening or softening laws (Ibrahimbegovic 2009; Ibrahimbegovic and Brancherie 2003; Ibrahimbegovic et al. 2010). The main advantage of this kind of model is starting at 'material' scale to build the corresponding interpretation of governing failure mechanics. This can be accomplished in a multi-scale fashion scale (Ibrahimbegovic and Markovic 2003, 2004) or yet (Ibrahimbegovic and Delaplace 2003), choosing the most appropriate scale to provide the most reliable interpretation of failure mechanisms. Using the multi-scale approach implies that the model parameters have clear physical interpretations and are easier to define. Of particular importance is the model ability to quantify the main failure mode of localization deformation caused by softening with respect to the amount of energy, so-called fracture energy, (Ibrahimbegovic 2009) that has to be supplied until complete failure, which provides the possibility for unifying displacement based with force based failure criteria. Moreover, the model of this kind is capable of solving some outstanding problems, such as shear failure in reinforced-concrete or crack spacing and opening (see Fig. 21.4).

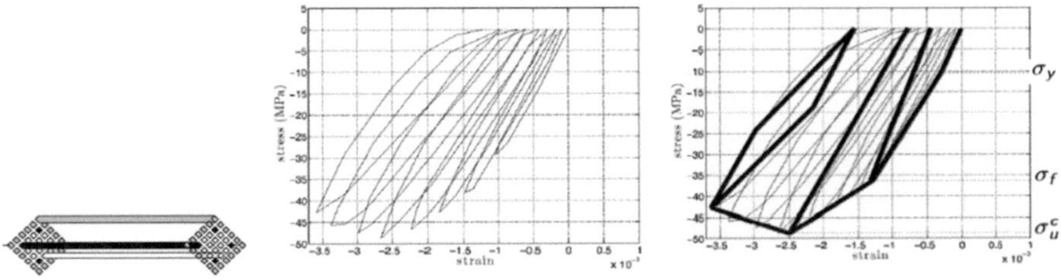


**Fig. 21.5** Hysteresis loops for concrete structure fiber in compression: experimental result on the *left* and computed results on the *right*

All these concepts, initially developed for quasi-statics applications, have been recently brought to bear upon dynamics providing very robust computational methods, both implicit and explicit (Ibrahimbegovic et al. 2008; Delaplace and Ibrahimbegovic 2006). The list of mechanisms that must be represented when trying to model dynamic response for three different phases of earthquake is rather long: (i) plasticity for residual deformation, (ii) damage for modification of elastic response, (iii) viscosity for rate sensitivity, (iv) different hardening phenomena and (v) softening phenomena for final failure mode. A typical example, for dynamic response of concrete in compression where any of these mechanisms will play an important role is given in Fig. 21.5. In the same figure we provide the hysteresis loops for the same loading program as produced by the proposed model (Jehel et al. 2010). The key idea of the proposed model, which employs a set of eight internal variables, is to only use the linear evolution laws for each internal variable for the corresponding evolution of local damage. Moreover, we employ direct stress interpolations combined with multi-fiber beam model, which allows keeping additive split of each inelastic mechanism contributing to total dynamic response. We thus do not need any local iterative procedure for computing the evolution of internal variable (typically, such a computation is not controlled by the code users), which contributes a great deal to the code computational robustness.

## 21.3 Complex System-Scale Interpretation of Damage Mechanisms

In this section we would like to illustrate that the modelling and computations of complex-system dynamic response is still not at the desired level of robustness. The case in point considers a RC beam-column connection subjected to a cyclic loading. The model is constructed by using three different model ingredients: plane elements with the Mazars model for damage in concrete (Mazars 1986), the classical plasticity model (Ibrahimbegovic 2009) for the steel bar, and the bond-slip model proposed in (Dominguez et al. 2005). The reference results are provided in experimental work carried out by (Alamedinne and Ehsani 1991).

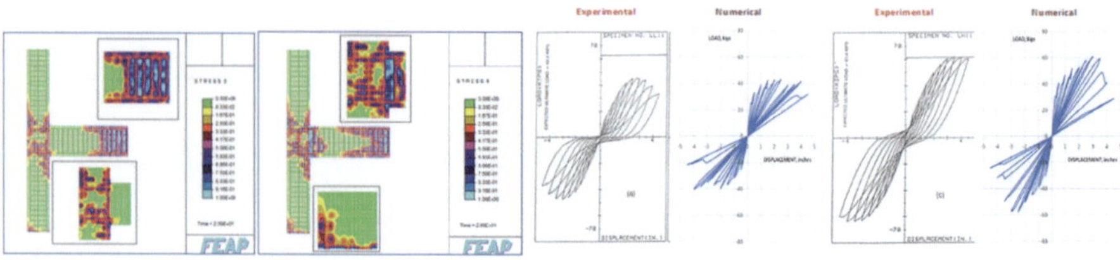


**Fig. 21.6** Reinforced concrete structure beam-column connection: damage variable contours for high and for low confinement, comparison between experimental and numerical hysteresis loops

The comparison between the experimental and the numerical response of two specimens with different reinforcement providing low and high confinement is presented in Fig. 21.6. The curves on the left side of the figure correspond to the low confinement reinforced specimen, in which the maximal load capacity was reached between 40 and 45 for a displacement close to 2. For the high confinement reinforced specimen, the curves on the right side show a maximal load capacity close to 60, and close to displacement equal to 3. It can be observed that some values are very similar (maximal load capacity associated to the lateral displacement), but it is not possible to reproduce numerically any dissipative loop or permanent deformation. The reason is in the use of damage model of (Mazars 1986), which includes only the slope variation of the elastic unloading, without any accumulated permanent deformation as observed in the experiments. It can also be noted that the high confinement reduces the damage to the connection, by shifting the most damaged domain toward the beam. The latter can be of interest in trying to enforce the strong-column weak-beam design concept for this kind of structures.

Another example of complex system found in infrastructure are water networks, which play a very important role in ensuring resilience since their destruction could delay the rescue action (e.g. extinguishing the fires) and increase the number of post-earthquake victims (e.g. spreading disease due to lack of clean water). In order to provide computational efficiency, we need reduced models (Cremer et al. 2002; Davenne et al. 2003) that can capture all pertinent nonlinear phenomena with the smallest possible computational expense. The recent computational results (Halfaya et al. 2008) have shown that the damage is concentrated either in the branching zones or in the pipe junctions with buildings or pumping stations.

## 21.4 Material-Scale Interpretation of Damage Mechanisms and Related Probability Aspects

In this section we briefly present the most reliable interpretation of damage mechanisms that are at the origin of energy dissipation and damping, which can be provided at material or micro-scale, the finer scale that is not as much affected by material heterogeneities. In particular, we study a two-phase representation

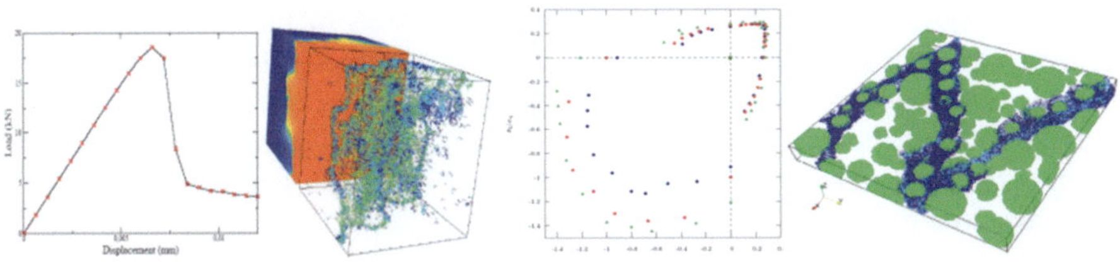

**Fig. 21.7** Failure modes representation of concrete by multi-scale model: force-displacement diagram and crack-pattern in uniaxial tension; yield surface as a function of aggregate volume fraction in biaxial compression

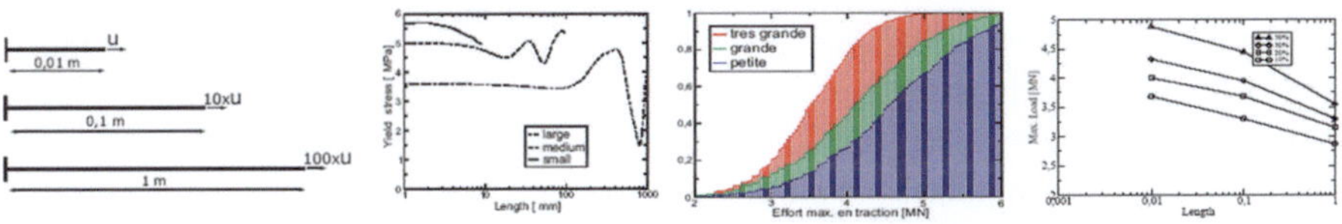


**Fig. 21.8** Different specimen sizes (small, medium and large) with respect to correlation length, typical realizations of yield stress, cumulative distribution functions and corresponding size effect showing that for small and homogeneous specimen dominant failure mechanism is fracture process zone while for large and heterogeneous specimen dominant failure mechanism is macro-crack

of concrete separating the aggregate and cement paste. The former is chosen of spherical form (other forms are currently investigated) according to the concrete curves specifying the chosen mix. The models of this kind (Ibrahimbegovic and Delaplace 2003; Benkemoun et al. 2010) can provide a realistic description of different failure mechanisms for different stress states; some illustrative results for simple tension or biaxial compression are provided in Fig. 21.7. We also show in Fig. 21.7 that the computed results are comparable to the classical choice of Rankin criterion for failure of concrete in tension along with the Drucker-Prager failure criterion in compression. However, the Drucker-Prager like criterion in compression can capture the increase of failure resistance corresponding to increase in volume fraction of aggregate.

Another important feature of the proposed model is that it can provide the reliable estimate of uncertainty description related to material heterogeneities. The latter is very important for structural failure, since for vast majority of cases the failure process starts from the weakest part of a heterogeneous structure. In particular, we use the computational procedure (Hautefeuille et al. 2009) where the mechanical properties of each phase remain deterministic, but the geometry of the specimen allows for random position of aggregates distributed by Gibbs process, while keeping the aggregate sizes between the largest and smallest, according to the chosen concrete mix. The results of those computations are then used to construct probability distribution of material parameters variations, which allows us to replace the usual deterministic values by random fields represented by Karhunen–Loève expansion (see Fig. 21.8).

The proposed multi-scale approach where the results of fine scale are used to construct the corresponding probabilistic bounds on material properties have already provided a very recent breakthrough of interpretation of size effect (Ibrahimbegovic et al. 2011), with the interplay between dominant failure mechanisms, either fracture process zone or macro-crack, is changing with the specimen size. The extension of this procedure is still to be done for dynamics.

## 21.5 Probability Framework for Safety Assessment

Presented probability framework for safety assessment under uncertainties introduced by material heterogeneities ought to be completed by considering other sources of uncertainty, related to earthquake induced extreme loading. Natural hazards such as earthquakes are particularly difficult to tackle because of the limited knowledge about probability of occurrence that stems from contrast between the geological and human time scales. This task becomes extremely challenging for very rare but devastating events.

We distinguish two types of safety assessment approaches: deterministic and probabilistic. The former approach requires confirming level of structural safety for any accident scenario, whereas the latter quantifies level of safety by a probability description of different scenarios that accepts occurrence of a severe accident. The current French legislation for nuclear safety imposes the deterministic approach, where probabilistic analysis is mainly used for completion of a deterministic safety demonstration. We firmly believe that such a practise is likely to evolve more towards probability, in order to capture rare events.

There exists a probabilistic approach for seismic risk assessment, already introduced in USA by Electric Power Research Institute (EPRI 2003), considering Seismic Probabilistic Safety Assessment that can be applied to any industrial installations. The kind of probabilistic approach uses a similar scenario based concept as in the deterministic approach. However, its analytic ingredients are not systematically conservative, but are based upon best-estimate methods enhanced with corresponding probabilistic models of related uncertainties. The probabilistic approach can thus lead to more efficient design, because it allows to identify the most important risk contributors, and thus to optimize risk mitigation efforts. At the end of the process we can calculate the probability of a severe accident (core damage) and/or of significant radioactive releases, expressed usually in terms of probability per reactor per year. However, one of the major difficulties of the Seismic Probabilistic Safety Assessment is in trying to correctly evaluate the uncertainties of the applied models (thus the importance of the break-through pertinent to material-scale model uncertainty, presented in the previous section). They can be a real drawback, since in most cases the uncertainties estimations can have an important impact on the final result.

The standard probabilistic approach is carried out in four stages: seismic hazard assessment, complex-system analysis, fragility assessment and risk quantification.

The goal of the Seismic Hazard Analysis is to calculate the annual probability of exceeding in terms of a given loading parameter value. The most commonly used choice is the Peak Ground Acceleration measured in units of $g$, the earth gravity acceleration. Due to earthquake complexities and seismological measurements scattering, the seismic hazard assessment is usually the most critical stage of the whole probabilistic risk assessment, which defines the seismic loading applied to a nuclear power plant. The large uncertainty typical of this phase is especially characteristic of a region (like France) with normally small seismic activity, or any other region with respect to rare events. Moreover, the choice of the loading parameter is also very important. In particular, even though the Peak Ground Acceleration (PGA) is the most commonly used parameter it is also well established that it is not equally relevant for all types of earthquakes. Namely the experience has shown that any so-called near-field low-magnitude earthquake with a relatively large PGA would cause much less damage than a far-field high-magnitude earthquake with the same value of the PGA. Hence, another loading parameter, so-called Cumulated Absolute Velocity, is often considered to be a more judicious choice. However, since most of the seismic risk assessment approaches are based on PGA, the transition to CAV (or even further treating the earthquake as a random process) can still be considered as the next big goal for industrial applications. Even more ambitious extension in that respect would be considering probabilistic representation of the entire ground-motion time history, which can be constructed based on a stochastic model that depends on seismic source parameters. This is not in very near future since not only it would require the correct description of earthquake event uncertainty but also very elaborate computations treating the earthquake as a random field (Jalayer and Beck 2008).

The seismic analysis of a complex-system applied to the nuclear power plant, considers internal events establishing an 'event tree' and a 'failure tree'. The former considers what can trigger an earthquake provoked accident and the latter analyzes the corresponding accident scenarios. This typically results in providing the Seismic Equipment List to establish the safety related equipment to be analyzed in detail. For each piece of equipment in a Nuclear Power Plant related to safety insurance, we have to establish its fragility. This can be achieved either by experiments, (e.g. shaking table tests) or by numerical simulation, with former currently more preferred choice, except for plants' piping systems. The fragility of equipment is presented in terms of the probability of failure for a given level of acceleration. For reliable assessment of equipment fragility, it is also important to define the equipment position within the complex-system assembly, in order to correctly transfer the PGA effects for this particular location.

The final stage of the Seismic Probabilistic Safety Assessment corresponds to the assembling the hazard analysis and the fragility curves to get the probability of failure of a given complex system or its particular component (e.g. a reactor unit, as the most important safety issue in a Nuclear Power Plant).

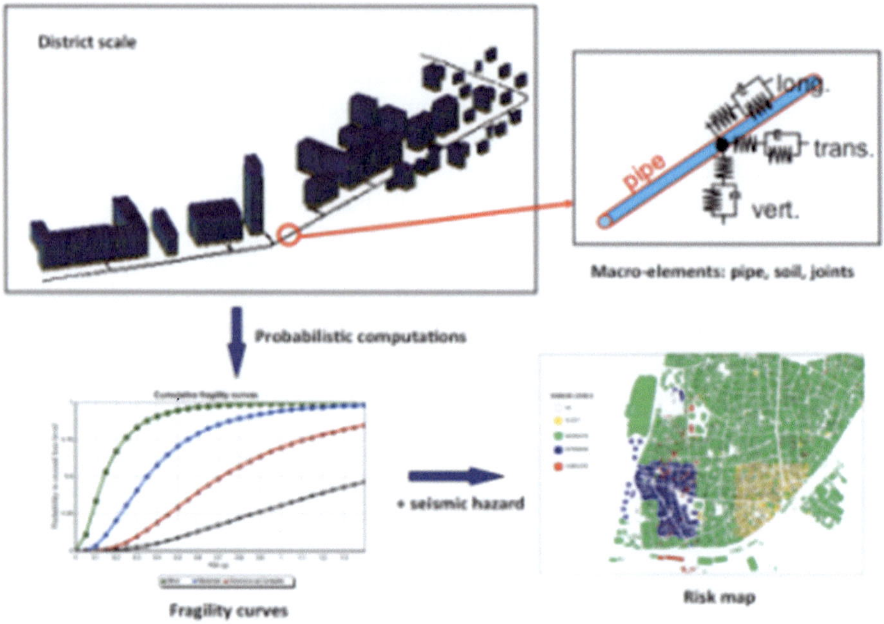


**Fig. 21.9** Development of risk maps, or seismic vulnerability indices maps, made from fragility curves and seismic hazard for complex system of urban area and its component of water network

Another example of fragility curves computations for complex system concerns the seismic vulnerability of a water supply network, which is calculated with a three step procedure (Seyedi et al. 2010) by exploiting the reduced models described in Sect. 21.3. The typical results are illustrated in Fig. 21.9.

## 21.6 Conclusions

Quantifying different sources of uncertainty and their combined effect is needed to provide optimized risk assessment and risk mitigation tools. There exists a very strong demand for this kind of tools, on one side from the society to reduce any industrial risk and on the other side from the industry to ensure its competitiveness and operation under severe economic constraints. In terms of complex system resilience, the performance of a risk mitigation system can be defined by protection of human lives and limitation of the impact on the environment. In this context, an efficient design of structures and systems is of the key importance, but not the only criterion to take into account.

Modifications of the current practice approaches to assess seismic risk by placing them in probability framework, the most direct extension is the corresponding computation of fragility curves for complex-system components. Hence, it becomes very important that the signal transmitted from the ground motion to the particular component is adequate, i.e. below a certain level. It is thus clear that a sufficiently good understanding of the structural response, especially in its inelastic regime, is a very worthy scientific goal. The novel results we presented in this paper, related to both modelling issues and to uncertainty quantification at the material scale, should provide a significant contribution to achieving this goal.

**Acknowledgements** This work was supported by French Ministry of Research (ANR project 'SISBAT'), and industrial collaborations with EDF, CEA and IRSN. This funding is gratefully acknowledged.

## References

Alamedinne F, Ehsani MR (1991) High-strength RC connections subjected to inelastic cycling loading. ASCE J Struct Eng 117:829–850
Benkemoun N, Hautefeuille M, Colliat J-B, Ibrahimbegovic A (2010) Failure of heterogeneous materials: 3D meso-scale FE models with embedded discontinuities. Int J Numer Methods Eng 82:1671–1688
Clough RW, Penzien J (1996) Dynamics of structures. McGraw Hill, New York
Cremer C, Pecker A, Davenne L (2002) Modelling of nonlinear dynamic behaviour of a shallow strip foundation with macro-element. Earthq Eng 6:175–211
Davenne L, Ragueneau F, Mazars J, Ibrahimbegovic A (2003) Efficient approach to earthquake engineering analysis. Comput Struct 81:1223–1239
Delaplace A, Ibrahimbegovic A (2006) Time-integration schemes for dynamic fracture problem using the discrete model. Int J Numer Methods Eng 65:1527–1544
Dominguez N, Brancherie D, Davenne L, Ibrahimbegovic A (2005) Prediction of crack pattern distribution in Reinforced Concrete by coupling a strong discontinuity model of concrete cracking and a bond-slip of reinforcement model. Eng Comput 22:558–582
EPRI (2003) Electric Power Research Institute. Seismic risk assessment implementation guide– final report 1002989
Eurocode 8: Design of structures for earthquake resistance, European Community, 1998
Fajfar P (2007) Seismic assessment of structures by a practice-oriented method. In: Ibrahimbegovic A, Kozar I (eds) Extreme man-made and natural hazards in dynamics of structures. Springer, Dordrecht, pp 257–284
Fajfar P, Fischinger M (1987) Non-linear seismic analysis of RC buildings: implications of a case study. Eur Earthq Eng 1(1):31–43
Fajfar P, Fischinger M (1988) N2 - A method for non-linear seismic analysis of regular buildings. In: Proceedings of the ninth world conference on EE, Tokyo-Kyoto, 5, pp 111–116
Halfaya FZ, Davenne L, Bensaibi M (2008) Seismic vulnerability assessment of water supply network. In: Ibrahimbegovic A, Zlatar M (eds) Proceedings NATO-ARW 'Damage assessment and reconstruction after natural disasters and previous military activities. Springer, Sarajevo, pp 322–328
Hautefeuille M, Melnyk S, Colliat JB, Ibrahimbegovic A (2009) Probabilistic aspects of localized failure of massive heterogeneous structures. Eng Comput 26:166–184
Ibrahimbegovic A (2009) Nonlinear solid mechanics: theoretical formulations and finite element solution methods. Springer, Berlin
Ibrahimbegovic A, Brancherie D (2003) Combined hardening and softening constitutive model for plasticity: precursor to shear slip line failure. Comput Mech 31:88–100

Ibrahimbegovic A, Delaplace A (2003) Microscale and mesoscale discrete models for dynamic fracture of structures built of brittle materials. Comput Struct 81:1255–1265

Ibrahimbegovic A, Frey F (1993) Stress resultant finite element analysis of reinforced concrete plates. Eng Comput 10:15–30

Ibrahimbegovic A, Markovic D (2003) Strong coupling methods in multi-phase and multi-scale modeling of inelastic behavior of heterogeneous structures. Comput Methods Appl Mech Eng 192:3089–3107

Ibrahimbegovic A, Melnyk S (2007) Embedded discontinuity finite element method for modeling of localized failure in heterogeneous materials with structured mesh: an alternative to extended finite element method. Comput Mech 40:149–155

Ibrahimbegovic A, Wilson EL (1989) Simple numerical algorithms for mode superposition analysis of discrete systems with non-proportional damping. Comput Struct 33:523–531

Ibrahimbegovic A, Wilson EL (1990) A methodology for dynamic analysis of linear structure-foundation systems with local nonlinearities. Earthq Eng Struct Dyn 19:1197–1208

Ibrahimbegovic A, Wilson EL (1992) Efficient computational procedures for systems with local nonlinearities. Eng Comput 9:385–398

Ibrahimbegovic A, Jehel P, Davenne L (2008) Coupled damage-plasticity constitutive model and direct stress interpolation. Comput Mech 42:1–11

Ibrahimbegovic A, Boulkertous A, Davenne L, Brancherie D (2010) Modeling of reinforced-concrete structures providing crack-spacing based on XFEM, ED-FEM and novel operator split solution procedure. Int J Numer Methods Eng 83:452–481

Ibrahimbegovic A, Colliat JB, Hautefeuille M, Brancherie D, Melnyk S (2011) Probability based size effect representation for failure of civil engineering structures built of heterogeneous materials. In: Papadrakakis M, Fragiadakis M, Stefanou G (eds) Computational methods in stochastic dynamics. Springer, Berlin, pp 289–311

Jalayer F, Beck JL (2008) Effects of two alternative representations of ground-motion uncertainty on probabilistic seismic demand assessment of structures. Int J Earthq Eng Struct Dyn 37:61–79

Jehel P, Davenne L, Ibrahimbegovic A, Leger P (2010) Towards robust viscoelastic-plastic-damage model with different hardening-softening capable of representing salient phenomena in seismic loading applications. Comput Concr 7(4):365–386

Markovic D, Ibrahimbegovic A (2004) On micro-macro interface conditions for micro-scale based FEM for inelastic behavior of heterogeneous materials. Comput Methods Appl Mech Eng 193:5503–5523

Mazars J (1986) A description of micro- and macroscale damage of concrete structures. J Eng Fract Mech 25:729–737

Seyedi M, Gehl P, Douglas J, Davenne L, Mehzer N, Ghavamian S (2010) Development of seismic fragility surfaces for reinforced concrete buildings by means of nonlinear time-history analysis. Earthq Eng Struct Dyn 39:91–108

# Chapter 22
# Seismic Fragility of RC Buildings Designed to Eurocodes 2 and 8

**Alexandra Papailia, Georgios Tsionis, and Michael N. Fardis**

**Abstract** Fragility curves are constructed for prototype regular RC frame and wall-frame buildings designed and detailed per EC 2 and EC 8. The aim is to evaluate how the Eurocodes achieve their seismic performance goals for RC buildings designed to them. These goals seem to be met in a consistent and uniform way across all types of buildings considered and their geometric or design parameters, except for concrete walls of Ductility Class Medium, which may fail early in shear despite their design against it per EC 8. In fact they do not perform much better than those in braced systems per EC 2 alone.

**Keywords** 2nd order effects • Capacity design • Chord rotation • Damage state • Dual buildings • Eurocode 2 • Eurocode 8 • Fragility curves • Frame buildings • Probability – conditional probability – probability of exceeding a damage state • RC buildings – braced buildings – unbraced buildings • Seismic assessment • Seismic design • Shear walls • Slenderness limit

## 22.1 Introduction

The aim of the paper is to evaluate how the European structural design standards (the Eurocodes, ECs) achieve their stated seismic performance goals for Reinforced Concrete (RC) buildings. To this end, the seismic performance of prototype plan- and height-wise regular RC frame and dual buildings designed to EC 2 and 8 (CEN 2004a, b) is assessed using the analysis and evaluation tools provided by EC 8 itself in its part devoted to (performance- and displacement-based) assessment of existing buildings (CEN 2005). The performance is assessed for two damage states

A. Papailia • G. Tsionis • M.N. Fardis (✉)
Department of Civil Engineering, University of Patras, 26504 Patras, Greece
e-mail: alepap@teemail.gr; tsionis@upatras.gr; fardis@upatras.gr

M. Fischinger (ed.), *Performance-Based Seismic Engineering: Vision for an Earthquake Resilient Society*, Geotechnical, Geological and Earthquake Engineering 32,
DOI 10.1007/978-94-017-8875-5_22, 

of members: (a) yielding and (b) ultimate (taken as a 20 % drop in peak flexural or shear resistance). Instead of carrying out a deterministic performance assessment, fragility curves are derived for the generic members of the individual EC-designed buildings.

Seismic fragility curves depict the probability of a specific damage state been exceeded, conditional on a seismic intensity measure (IM). They first appeared in the form of damage probability matrices, e.g. Braga et al. (1982), Spence et al. (1992), empirically derived from seismic damage data. The main drawback of the empirical approach is the lack of data in certain ranges of seismic intensity and its dependence on the features of the earthquakes and the building stock in the database. It is in principle overcome by analytical methods that use the capacity spectrum, e.g. Kircher et al. (1997), or nonlinear dynamic analyses of single- and multi-degree-of-freedom systems, e.g. Masi (2003), Singhal and Kiremidjian (1996). A hybrid method has also been proposed (Kappos et al. 1998), combining analytical and observational data to compensate for the lack of the latter. Most existing fragility curves have been produced for classes of buildings. In a more recent methodology they are constructed for a specific building as a function of its stiffness, strength and ductility properties (Jeong and Elnashai 2007). The fragility curves in the present paper also refer to individual prototype buildings and not to classes thereof. Unlike other analytical fragility curves constructed without recourse to Monte Carlo simulation, the present ones are not based on a global dispersion parameter $\beta$ with prescribed value, but are built point-by-point from the conditional probability of exceeding the damage state.

## 22.2 Scope of Fragility Analyses

Prototype regular RC-frame or RC wall-frame (dual) buildings are studied. The parameters considered and their values are (Papailia 2011):

- the number of storeys: 5, 8 and for frames 2 as well;
- the level of seismic design:
  - Design for gravity loads only – not even for wind – with EC 2 alone;
  - Seismic design per EC 8 for DC L (Low), M (Medium) or H (High) and various levels of design peak ground acceleration (PGA) and for the EC 8 Type 1 standard spectrum for soil type C – firm soil (see first two columns of Table 22.1 for the considered combinations of DC and design PGA, incorporating the Soil factor *S* of 1.15 for soil type C).
- For dual systems: the fraction of the seismic base shear taken by the walls.

All storeys have the same height, $h_{st} = 3$ m. Slab thickness is 150 mm. The buildings are rectangular in plan, with the same geometric parameters and member sizes in both horizontal directions. Bay length is the same throughout the plan

**Table 22.1** Beam and interior column depth in frames. Base case: $f_c = 25$ MPa, $f_y = 500$ MPa, $l_b = 5$ m

| | | 2-storey frame | | 5-storey frame | | 8-storey frame | |
|---|---|---|---|---|---|---|---|
| Design PGA (g) | DC | $h_b$ (m) | $h_c$ (m) | $h_b$ (m) | $h_c$ (m) | $h_b$ (m) | $h_c$ (m) |
| 0.0 (EC 2) | – | 0.40 | 0.45 | 0.40 | 0.55 | 0.40 | 0.65 |
| 0.10 | L | 0.40 | 0.45 | 0.40 | 0.55 | 0.40 | 0.65 |
| 0.15 | L, M | 0.40 | 0.45 | 0.40 | 0.55 | 0.40 | 0.65 |
| 0.20 | M, H | 0.40 | 0.45 | 0.40 | 0.55 | 0.40 | 0.65 |
| 0.25 | M, H | 0.40 | 0.45 | 0.45 | 0.55 | 0.45 | 0.65 |
| 0.30 | M, H | 0.40 | 0.45 | 0.45 | 0.60 | 0.45 | 0.70 |
| 0.35 | H | 0.40 | 0.45 | 0.50 | 0.65 | 0.50 | 0.75 |

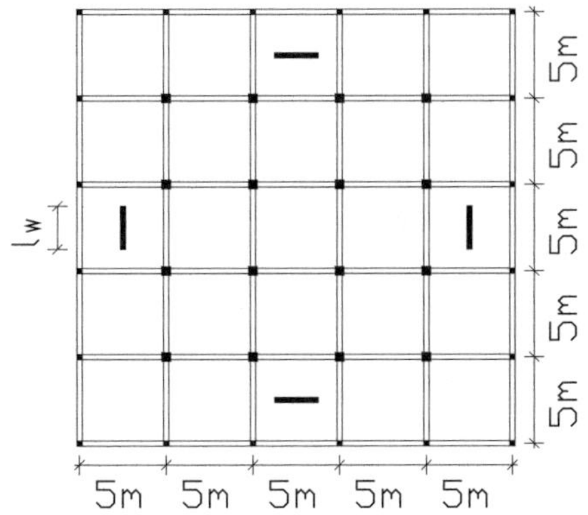


**Fig. 22.1** Plan of prototype wall-frame (dual) building

($l_b = 5$ m in the base case). Wall-frame buildings have columns on a 5 m × 5 m grid and two parallel rectangular walls in each horizontal direction per 5 × 5 bays of the building plan, as shown in Fig. 22.1. For simplicity and generality, no beams are considered to frame into the walls: the walls just share with the frames the same floor displacements, with the floor diaphragms assumed rigid. The column size and the width of beams ($b_w = 0.3$ m) are constant in all storeys. Beam depths are constant in each storey, but in frame buildings they may be different in different storeys. Interior columns are square and all of the same size. Permanent loads, including the dead weight of the structure, finishings, partitions and façades, amount to 7 kN/m$^2$. The nominal value of occupancy loads is 2 kN/m$^2$.

Perimeter columns and beams are taken with about one-half the rigidity, *EI*, of interior ones at the same storey, by reducing their depth by 5 or 10 cm as appropriate. Then it may be assumed that seismic moments and chord rotation demands are the same at all beam ends in a storey and over all its interior columns; exterior columns develop one-half the elastic seismic moments of same-storey interior ones but have the same seismic chord rotation demands. Besides, the axial force variation due to the seismic action may be neglected in interior columns. Another simplification is that vertical elements are fixed at ground level and the columns are taken to have the point of inflection at storey mid-height. Bending moments in columns or walls due to gravity loads are neglected. P-δ effects due to the seismic action are always taken into account. All these simplifying assumptions are made both for the seismic design with conventional cracked rigidities and for the assessment of the seismic response and the estimation of the fragility using the secant-to-yield-point rigidity as elastic rigidity per CEN (2005).

The discussion in this paper is limited to the design and the fragility curves of interior columns and beams only. More details may be found in Papailia (2011) and Fardis et al. (2012).

The depths of beams and interior columns of frames listed in Table 22.1 are chosen iteratively as the minimum reasonable values necessary to meet all requirements of EC 2 and 8 – if the later applies – including EC 8's 0.5 % storey drift limit under the damage limitation seismic action (which is taken as 50 % of the design seismic action). The size of columns in gravity-only-designs per EC 2 is controlled by the slenderness limit below which EC 2 allows neglecting at the two lowest storeys 2nd-order effects under factored gravity loads (the "persistent-and-transient" design situation per Eqs. 6.10a, b in EN 1990). Frame buildings are considered in this respect as unbraced and wall-frame (dual) ones as braced. The minimum column size per EC 2 governs in most EC 8-designed columns as well.

The depths of columns and beams given in Table 22.2 for dual buildings are chosen as the minimum ones meeting all requirements of EC 2 and 8 – if the later applies, but with EC 8's 0.5 % storey drift limit for the damage limitation seismic action met thanks also to the walls (see below). The length of the wall section, $l_w$, in gravity-only-designs is chosen as the minimum necessary to meet EC 2's lateral bracing condition for negligible 2nd-order effects in braced frames. In ductile seismic designs of dual buildings, $l_w$ and the column depths are chosen together so that EC 8's 0.5 % storey drift limit for the damage limitation seismic action is met and at the same time the fraction of the building's total base shear, $V_{tot,base}$, taken by the two walls, $V_{wall,base}$, covers a wide range of values. The reason is that EC 8 uses the ratio of stiffness, $V_{wall,base}/V_{tot,base}$, instead of that of strengths, to categorise buildings as wall systems (those having $V_{wall,base}/V_{tot,base} \geq 0.65$), frame-equivalent dual (those with $0.35 \leq V_{wall,base}/V_{tot,base} \leq 0.5$) or wall-equivalent dual ($0.5 \leq V_{wall,base}/V_{tot,base} \leq 0.65$), which have different behaviour factor values $q$ in EC 8 and follow different design rules. To cover this range of $V_{wall,base}/V_{tot,base}$ in most dual buildings the wall length $l_w$ falls short of the EC 2 bracing requirement for braced frames.

For a design PGA up to 0.20 g in a 5-storey building, or to 0.15 g for a 8-storey one, interior columns and beams have the minimum depth meeting the EC2 slenderness limit for braced systems. At higher design PGAs the width of most DC H walls increases from 0.25 m to 0.50 m, owing to the more stringent shear design rules for DC H walls; besides, larger frame members are needed, to control the drift in the upper storeys where the walls are ineffective.

In the base case the nominal material properties are $f_c = 25$ MPa and $f_y = 500$ MPa (with very ductile steel of Class C per EC 2). In addition to the base

**Table 22.2** Column, beam and wall depth in dual buildings; fraction of base shear taken by walls and fundamental period

| | | 5-storey dual building | | | | | 8-storey dual building | | | | |
|---|---|---|---|---|---|---|---|---|---|---|---|
| Design PGA (g) | DC | $h_c$ (m) | $h_b$ (m) | $l_w$ (m) | $V_{wall,b}$ (%) | $T^b$ (s) | $h_c$ (m) | $h_b$ (m) | $l_w$ (m) | $V_{wall,b}$ (%) | $T^b$ (s) |
| 0.0 (EC 2) | – | 0.40 | 0.40 | 3.2 | 75 | 0.8 | 0.45 | 0.40 | 5.0 | 85 | 1.0 |
| 0.10 | L | 0.40 | 0.40 | 1.5 | 37 | 1.0 | 0.45 | 0.40 | 2.0 | 45 | 1.5 |
| | | | | 2.0 | 53 | 1.0 | | | 2.5 | 57 | 1.4 |
| | | | | 2.5 | 65 | 0.9 | | | 3.5 | 73 | 1.3 |
| 0.15 | L, M | 0.40 | 0.40 | 1.5 | 37 | 1.0 | 0.45 | 0.40 | 2.0 | 45 | 1.5 |
| | | | | 2.0 | 53 | 1.0 | | | 2.5 | 57 | 1.4 |
| | | | | 2.5 | 65 | 0.9 | | | 3.5 | 73 | 1.3 |
| 0.20 | M, H | 0.40 | 0.40 | 1.5 | 37 | 1.0 | 0.45 | 0.45 | 2.0/– | 42/– | 1.3/– |
| | | | | 2.0 | 53 | 1.0 | | | 3.0/3.0[a] | 63/73 | 1.2/1.1 |
| | | | | 2.5 | 65 | 0.9 | | | 4.0/– | 76/– | 1.1/– |
| 0.25 | M, H | 0.45 | 0.45 | 2.0 | 44 | 0.8 | 0.45 | 0.50 | 2.0/– | 40/– | 1.2/– |
| | | | | 2.5 | 57 | 0.8 | | | 3.0/– | 61/– | 1.1/– |
| | | | | 3.5/3.5[a] | 73/81 | 0.7/0.6 | | | 4.0/5.5[a] | 74/90 | 1.0/0.7 |
| 0.30 | M, H | 0.50 | 0.50 | 2.0 | 36 | 0.7 | 0.50 | 0.50 | 2.5/– | 47/– | 1.1/– |
| | | | | 3.0/2.5 | 59/49 | 0.6/0.6 | | | 3.5/– | 64/– | 1.0/– |
| | | | | 4.0/4.5[a] | 73/85 | 0.5/0.4 | | | 4.0/7.0[a] | 70/92 | 0.9/0.6 |
| 0.35 | H | 0.50 | 0.55 | 2.5 | 46 | 0.6 | 0.55 | 0.55 | 9.5[a] | 96 | 0.4 |
| | | | | 5.5[a] | 89 | 0.3 | | | | | |

If DC M and H walls have different dimensions two $l_w$ and $T$ values are given, the first for DC M, the second for DC H; if a blank is given as second value, the first one applies only to DC M

[a]The width of the wall is $l_w = 0.5$ m; in all other cases it is $l_w = 0.25$ m

[b]Computed with 50 % of the rigidity of the uncracked gross section

case (with $l_b = 5.0$ m also), parametric studies of 2- and 8-storey frame designs per EC 8 were carried out, to see how sensitive the results and conclusions are to the values of $f_c$, $f_y$ and $l_b$.

## 22.3 Methodology of Fragility Analysis

For the damage states of member yielding or ultimate condition in flexure (conventionally taken as a 20 % drop in peak resistance), the Damage Measure (DM) is the chord rotation at a member end. For the member ultimate condition in shear, it is the shear force outside the plastic hinge or in it (considered then alongside the value of the rotation ductility factor). Although spectral displacement, $S_d(T_1)$, is an efficient and informative Intensity Measure (IM) for ductile failure modes (in flexure) and spectral acceleration, $S_a(T_1)$, for brittle ones (i.e., in shear), peak ground acceleration at the top of the soil (PGA) is taken as IM, as it tunes better with the use of design PGA as the main seismic design parameter of the building.

The estimation of the damage measures as a function of the excitation PGA and the construction of fragility curves takes place with the analysis methods and assumptions in Part 3 of Eurocode 8 (CEN 2005). These analyses are deterministic, using mean values of material properties. Once plastic hinges start forming in the frame, shear forces in beams and columns are calculated from the plastic mechanism and the yield moments of the sections that have already yielded. Once a plastic hinge forms at a wall base in a dual system, the shears all-along the wall are amplified for inelastic higher mode effects after yielding according to Eibl and Keintzel (1988), adopted in Part 1 of Eurocode 8 for DC H walls.

For given Intensity Measure-IM (PGA of the excitation), the deterministic analysis per CEN (2005) gives the mean values of DM demands (chord rotations at member ends, shear forces in or outside plastic hinges, rotation ductility factor). The mean values of the capacities corresponding to these DMs for the two damage states of member yielding and ultimate are determined again according to Eurocode 8, Part 3 (CEN 2005) and the way it accounts for flexural failure due to steel or concrete, confinement, etc. Note that the usual approach in fragility analysis without Monte Carlo simulation is to: (a) find the IM-value at which the mean (or median) DM-demand equals the mean (or median) DM-capacity, (b) establish from it the median of the lognormal distribution of IM describing the fragility curve and (c) supplement it with a default value for its coefficient of variation (normally $\beta = 0.6$). By contrast, here non-parametric fragility curves are established point-by-point, from the conditional-on-IM probability that the (random variable) DM-demand for given IM exceeds the (random variable) DM-capacity. The mean (or median) values of these two random variables are established according to the first part of this paragraph. Their variances are estimated from the Coefficients of Variation (CoV) itemized in Table 22.3. The CoV-values listed for the chord rotation demands for given spectral value at the fundamental period are based on extensive past comparisons of inelastic chord rotation demands in height-wise regular multistorey

**Table 22.3** Values of coefficients of variation for the fragility curves

| Demand | CoV | Capacity | CoV |
|---|---|---|---|
| Beam chord rotation demand, for given spectral value at fundamental period | 0.25 | Beam or column chord rotation at yielding | 0.33 |
| Column chord rotation demand, for given spectral value at fundamental period | 0.20 | Beam or column ultimate chord rotation | 0.38 |
| Wall chord rotation demand, for given spectral value at fundamental period | 0.25 | Shear resistance in diagonal tension (inside or outside plastic hinge) | 0.15 |
| Beam shear force demand, for given spectral value at fundamental period | 0.10 | Wall chord rotation at yielding of the base | 0.40 |
| Column shear force demand, for given spectral value at fundamental period | 0.15 | Wall ultimate chord rotation at the base | 0.32 |
| Wall shear force demand, for given spectral value at fundamental period | 0.20 | Wall shear resistance in diagonal compression | 0.175 |
| Spectral value, for given PGA and fundamental period | 0.25 | | |

buildings to their elastic estimates (Kosmopoulos and Fardis 2007; Panagiotakos and Fardis 1999). Those listed for the shear forces are based on limited parametric studies. The CoV-values of the capacities reflect the uncertainty in the models used for the estimation of their mean values (drawn from Biskinis and Fardis (2010a, b), Biskinis et al. (2004), CEN (2005)), as well as the dispersion of material and geometric properties (Biskinis and Fardis 2010a, b; Biskinis et al. 2004). For multiplicative or additive functions in the derivation of the DM-demand or the DM-capacity from the basic random variables, lognormal or normal distributions of the individual random variables are assumed, respectively.

Fragility results are obtained and presented separately for each type of member and storey in the building. They account for mechanical interaction of damage states between different elements only in a mean sense: as the analysis is deterministic and based on mean properties, seismic demands are computed assuming that member yielding has been reached, only if it has taken place with a conditional-on-IM probability of at least 50 %. The fragility curve of a given member at the ultimate damage state is taken as the maximum among its possible ultimate conditions: of the plastic hinge in flexure or shear, and outside the hinge in shear. This presumes full correlation between these different failure modes. If full correlation is assumed between members of the same type (i.e., beams or columns) in a storey, the fragility curve given for a single interior member of this type may be taken to apply to the entire ensemble of such members in the storey.

## 22.4 Indicative Results and Conclusions

The conclusions are based on the full results in Papailia (2011), of which only a sample are shown here. Figures 22.2, 22.3, 22.4, 22.5 and 22.6 (*left*) for frames come in sets of four figures each: the first row is for yielding; the second for ultimate. The

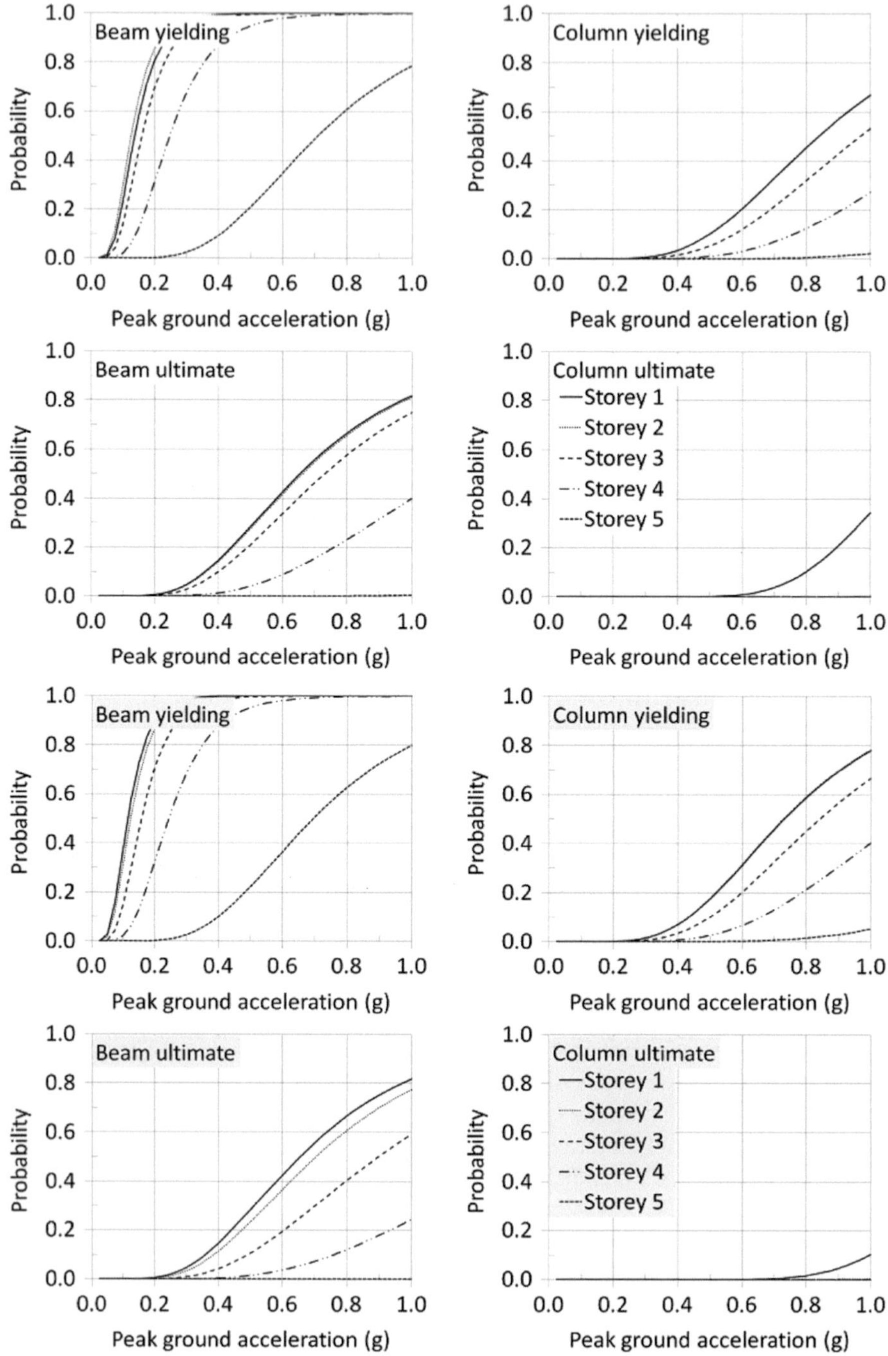


**Fig. 22.2** Fragility curves of 5-storey EC 8 frames designed for PGA = 0.2 g and DC M (*top*) or H (*bottom*)

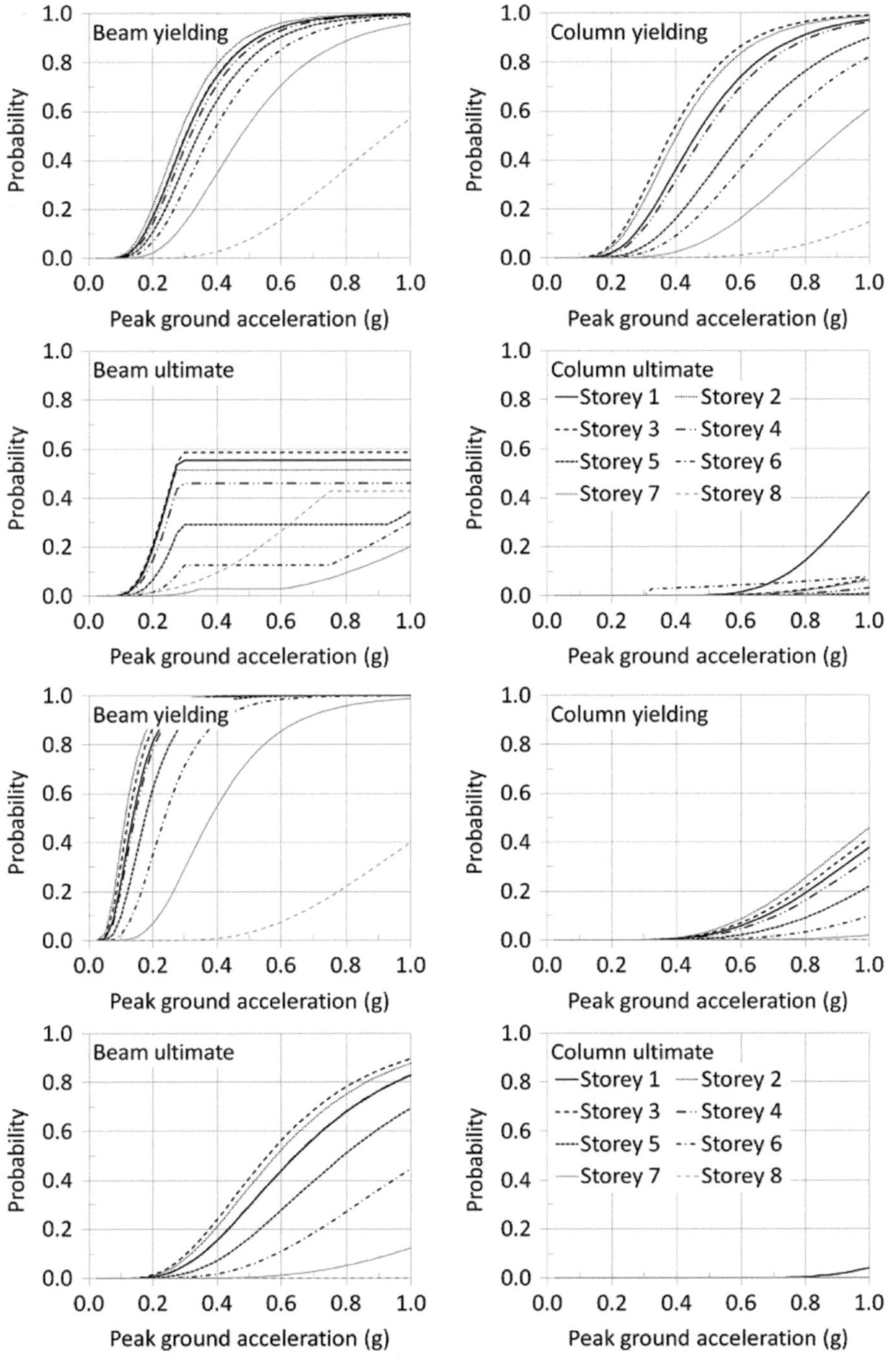


**Fig. 22.3** Fragility curves of 8-storey EC 8 frames designed for PGA = 0.15 g and DC L (*top*) or M (*bottom*)

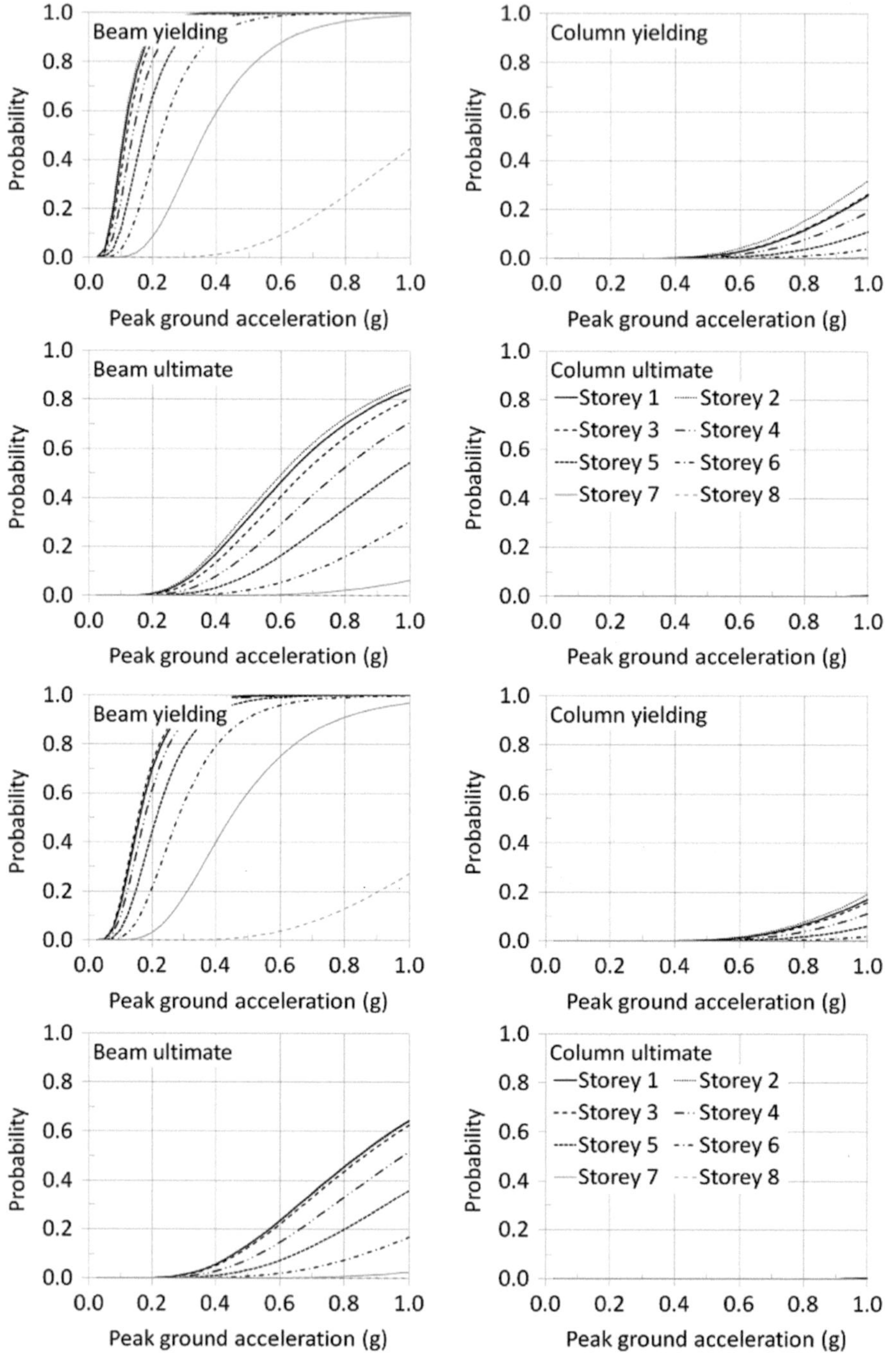


**Fig. 22.4** Fragility curves of 8-storey EC 8 DC H frames designed for PGA = 0.2 g (*top*) or PGA = 0.35 g (*bottom*)

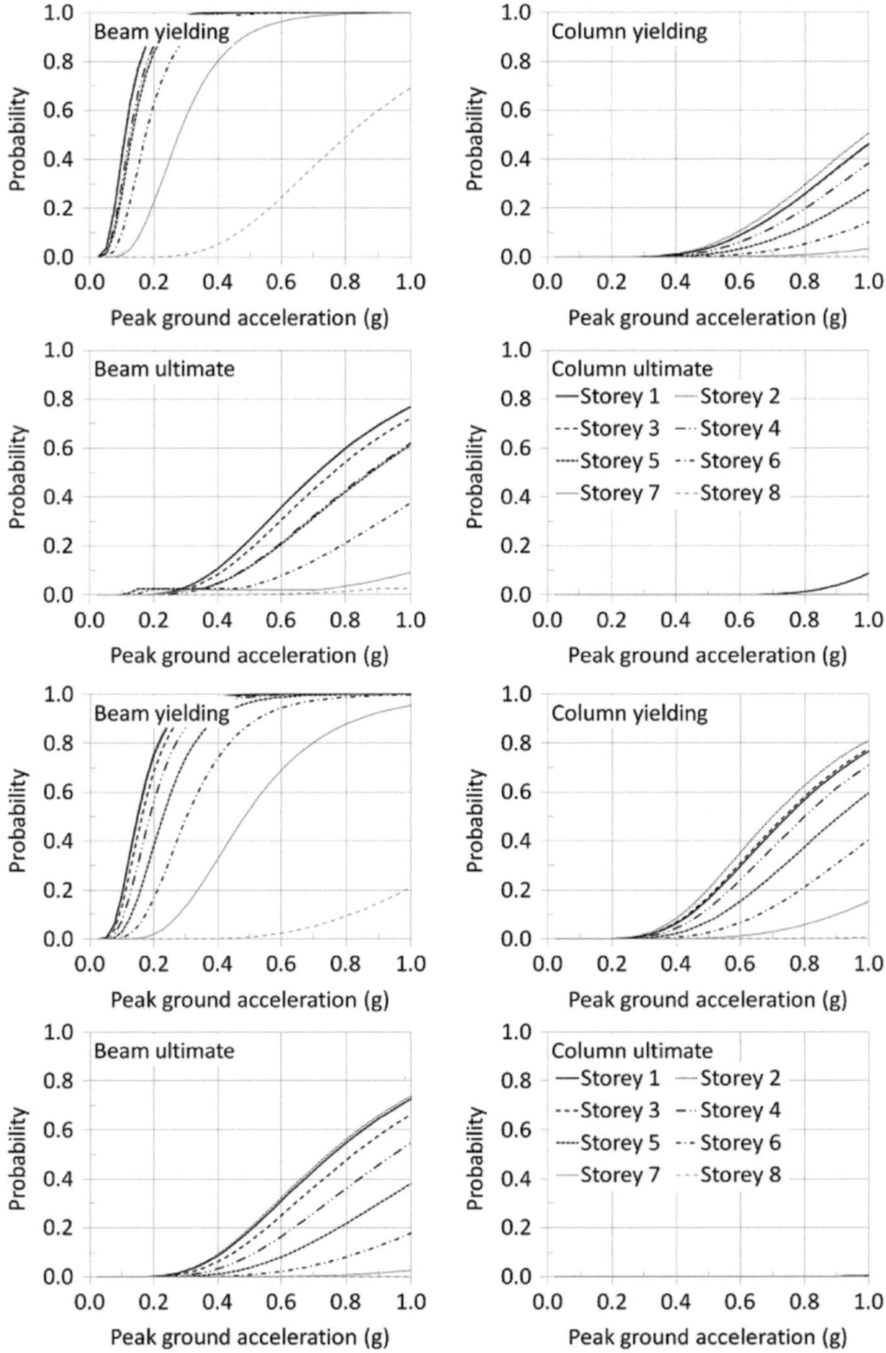


**Fig. 22.5** Fragility curves of 8-storey EC8 frames designed for PGA = 0.25 g and DC H (*top*): $l_b = 6$ m, $f_c = 20$ MPa, $f_y = 400$ MPa; (*bottom*): $l_b = 4$ m, $f_c = 40$ MPa, $f_y = 500$ MPa

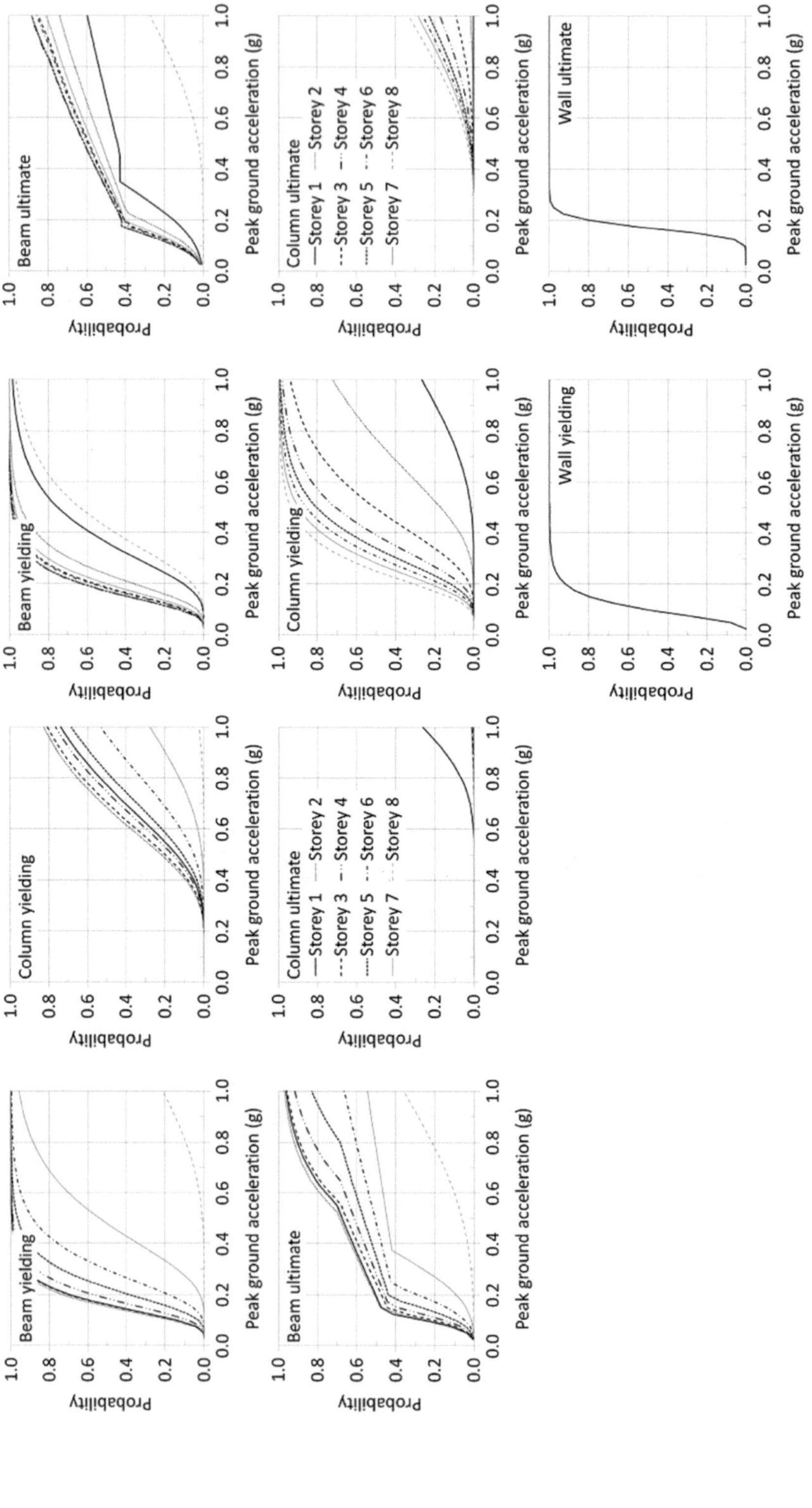


**Fig. 22.6** Fragility curves of 8-storey buildings designed to EC 2 only for gravity: *(left)* frame building designed as unbraced; *(right)* wall building, designed as braced

first column refers to beams; the second to columns. Different curves in each figure concern different storeys. Conclusions for the frames are:

- Nonductile (i.e. DC L) frames designed for PGA above the ceiling of 0.10 g recommended by EC 8 for DC L (Fig. 22.3(*top*)) do not perform well: their (non-capacity-designed) beams possibly fail in shear even before they yield – but are unlikely to do so below the design PGA. The likelihood of beam shear failure stops increasing after plastic hinges form (e.g., the plateau in Fig. 22.2).
- Except for the point above, frames designed to EC 8 give very satisfactory fragility results, even well beyond their design PGAs. Their performance is rather insensitive to their geometric and design parameters, as long as EC 8 is applied.
- The first beams to yield in DC M or H frames are very likely to do so between the damage-limitation and the design PGA; all columns are very likely to stay elastic well beyond that range.
- Beams are much more likely to reach the ultimate condition than columns.
- Design to DC H instead of M does not have a systematic or marked effect.
- Design for higher PGA reduces the fragility (as, e.g., in Fig. 22.4) but the benefit is disproportionately low.
- There is no systematic effect of the number of storeys on fragility (compare Fig. 22.2(*bottom*) with Fig. 22.4(*top*) for different 5 or 8-storeys).
- As shown by the two extreme combinations of nominal concrete or steel grade, $f_c$, $f_y$ and beam span in Fig. 22.5, these parameters do not affect much the fragility and in fact sometimes contrary to expectations, e.g., for column yielding.
- Nonductile frames show higher beam fragilities if designed to EC 2 for gravity only (Fig. 22.6(*left*)) than if designed to EC 8 for DC L and a PGA of 0.10 or 0.15 g (Fig. 22.3 (*top*)). However, their columns have lower fragilities.

Figures 22.6(*right*), 22.7, 22.8 and 22.9 for wall-frame buildings come in sets of six. The first row is for beams, the second for columns, the third for walls. Different curves in the first two rows are for different storeys. The first column per set of six figures is for yielding; the second for ultimate. Conclusions are (see also Papailia (2011) and Fardis et al. (2012)):

- Walls are the most critical elements at both damage states. Once the base of DC M walls yields, the inelastic amplification of their shear forces increases them sufficiently to precipitate shear failure, even below the design PGA! Note that in DC H walls the amplification of shear forces after the base yields is taken into account in design in the detailed way per Eibl and Keintzel (1988), but in DC M walls the seismic shears from the analysis are increased by 50 %. Besides, the design shear resistance of DC H walls is reduced by 60 % for load cycling, whereas that of DC M walls is not. As a result, the median PGA at which DC H walls fail is 1.5–2.0 times higher than the design PGA.
- The walls of nonductile wall buildings designed to EC 2 only for gravity loads are not markedly more fragile at either damage state than in EC 8 wall buildings or wall-equivalent dual systems of DC M, but their columns and beams of all storeys are (cf Figs. 22.6(*right*), 22.7, 22.8 and 22.9(*right*)).

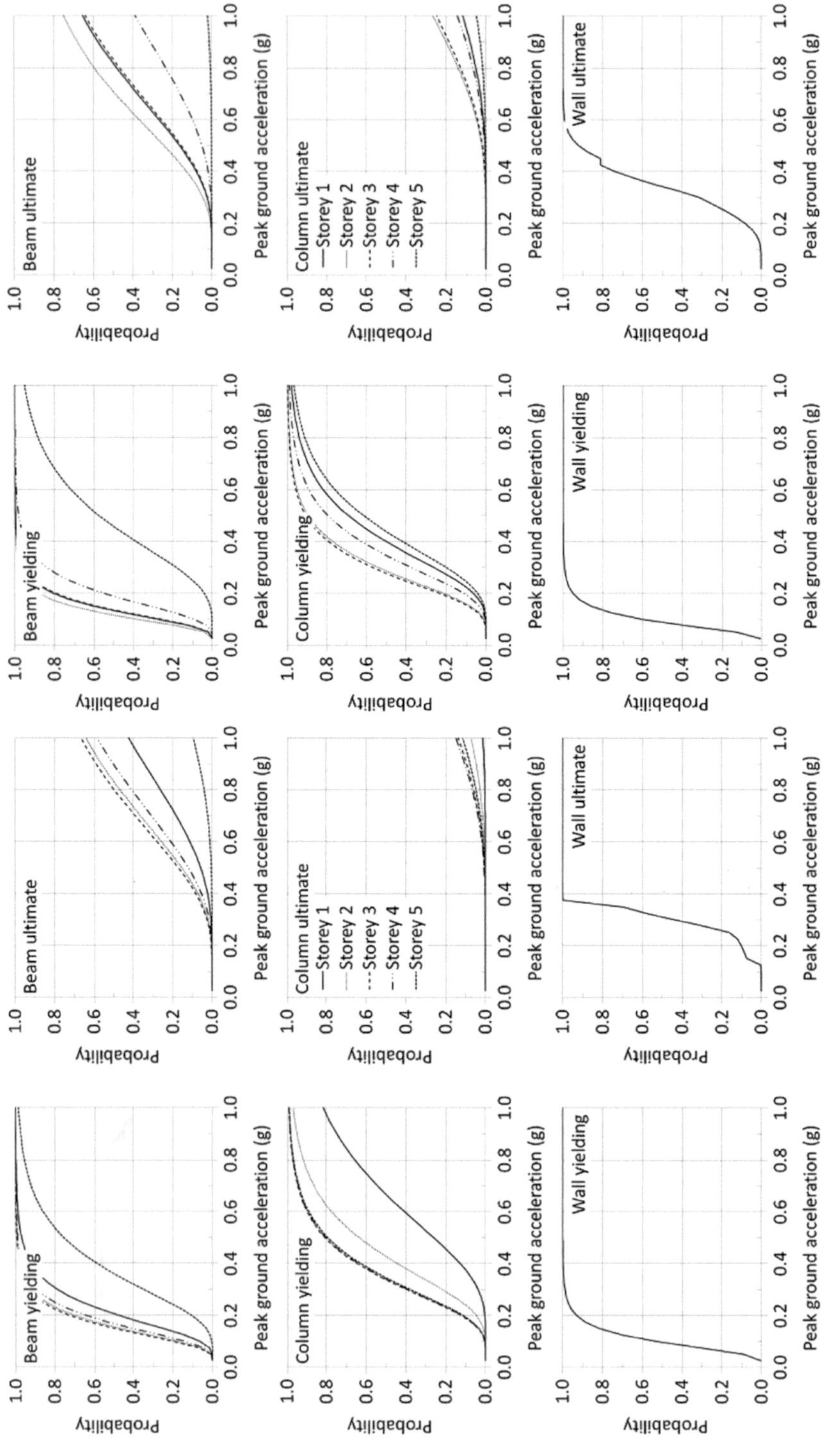


**Fig. 22.7** Fragility curves of 5-storey dual buildings designed to EC 8 for DC H and PGA = 0.2 g (*left*) wall system, (*right*) frame-equivalent dual system

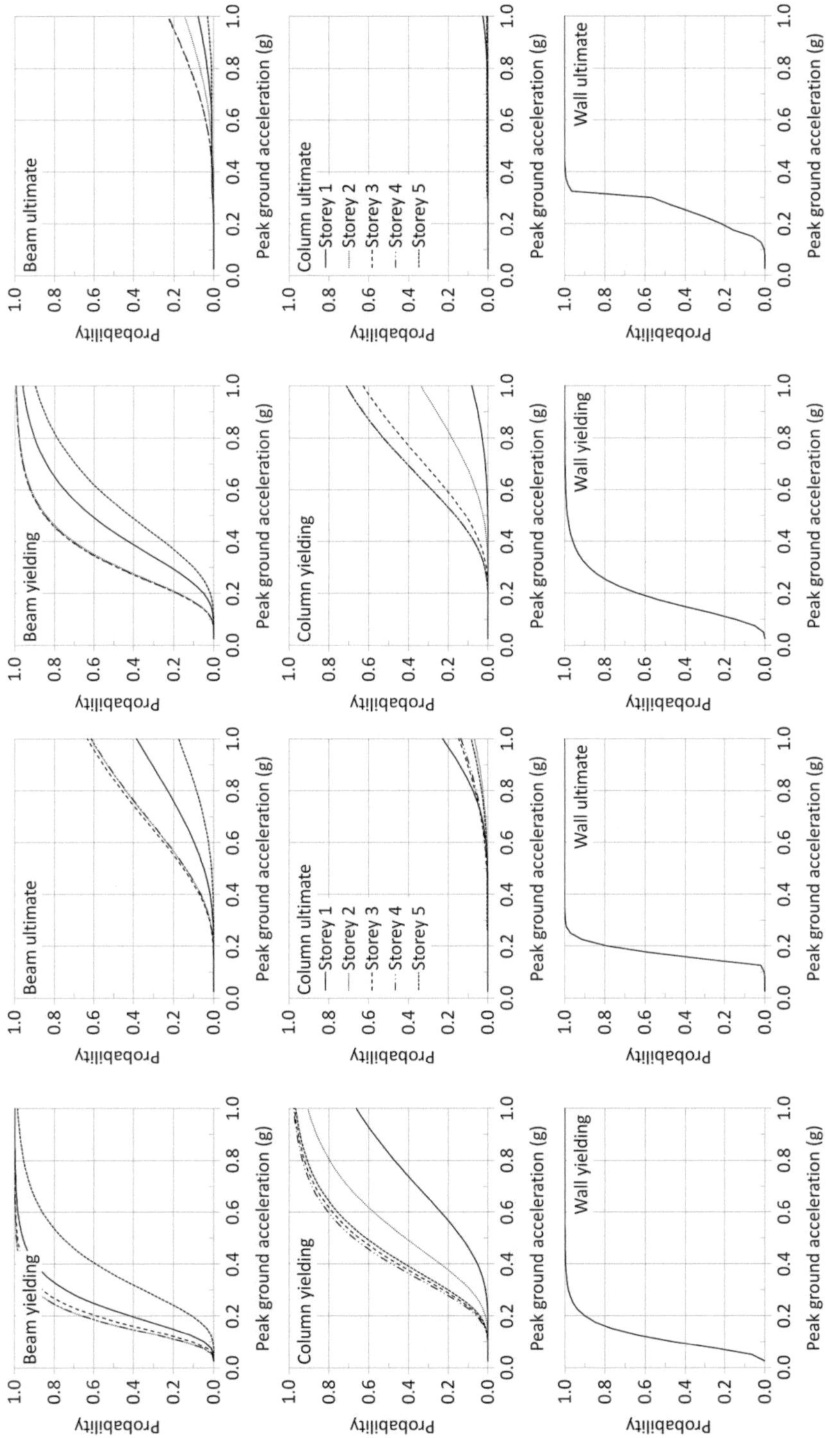


**Fig. 22.8** Fragility curves of 5-storey wall buildings designed to EC 8 for DC M and PGA = 0.2 g (*left*) or PGA = 0.3 g (*right*)

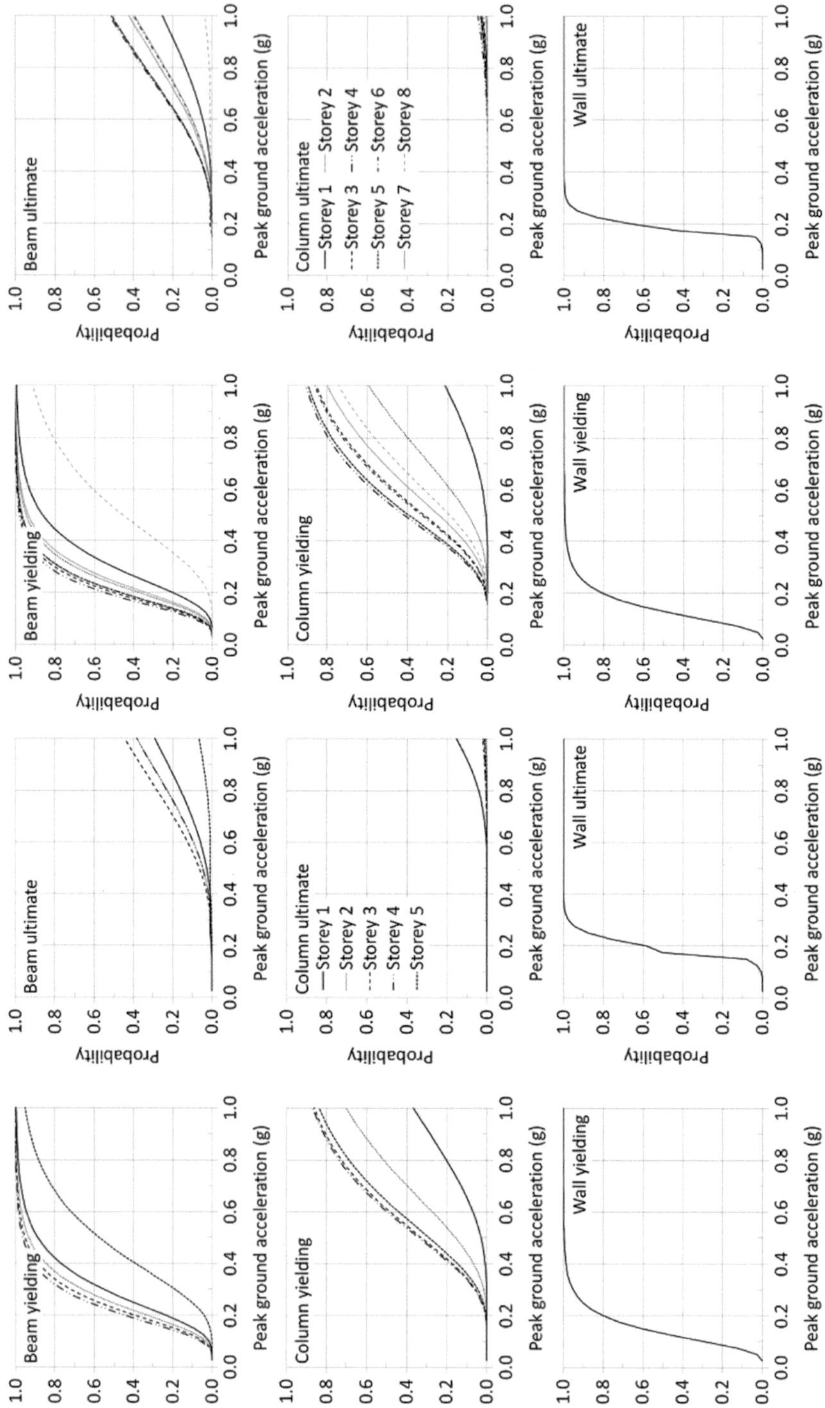


**Fig. 22.9** Fragility curves of (*left*) 5-storey and (*right*) 8-storey wall-equivalent dual buildings designed to EC 8 for DC M and PGA = 0.3 g

- EC 8 wall-equivalent dual and wall buildings have similar fragilities, higher than frame-equivalent for walls, lower for beams or columns (see Fig. 22.7 for a frame-equivalent and a wall building; wall-equivalent duals are in-between, closer to wall systems). Columns and, to a lesser extent, beams of dual systems have higher fragility than in pure frames (cf Fig. 22.7 to Fig. 22.2 (*bottom*)).
- Design to higher PGA (as, e.g., in Fig. 22.8) reduces the fragility of walls and frames.
- Design to higher DC (cf Fig. 22.7(*left*) and 22.8(*left*)) improves the fragility of frame members and very decisively that of walls.
- Taller buildings exhibit only slightly higher fragilities (see, e.g., Fig. 22.9).

**Acknowledgments** The research leading to these results receives funding from the European Community's 7th Framework Programme (FP7/2007-2013) under grant agreement no 244061.

## References

Biskinis DE, Fardis MN (2010a) Flexure-controlled ultimate deformations of members with continuous or lap-spliced bars. Struct Concr 11(2):93–108

Biskinis DE, Fardis MN (2010b) Deformations at flexural yielding of members with continuous or lap-spliced bars. Struct Concr 11(3):127–138

Biskinis DE, Roupakias G, Fardis MN (2004) Degradation of shear strength of RC members with inelastic cyclic displacements. ACI Struct J 101(6):773–783

Braga F, Dolce M, Liberatore D (1982) A statistical study on damaged buildings and an ensuing review of the MSK-76 scale. In: 7th European conference on earthquake engineering, Athens

CEN (2004a) EN 1992–1–1 Eurocode 2: Design of concrete structures – Part 1–1: General rules and rules for buildings. European Committee for Standardization, Brussels

CEN (2004b) EN 1998–1 Eurocode 8: Design of structures for earthquake resistance – Part 1: General rules, seismic actions and rules for buildings. European Committee for Standardization, Brussels

CEN (2005) EN 1998–3 Eurocode 8: Design of structures for earthquake resistance – Part 3: Assessment and retrofitting of buildings. European Committee of Standardization, Brussels

Eibl J, Keintzel E (1988) Seismic shear forces in cantilever shear walls. In: 9th world conference on earthquake engineering, Tokyo/Kyoto

Fardis MN, Papailia A, Tsionis G (2012) Seismic fragility of RC framed and wall-frame buildings designed to the EN-Eurocodes. Bull Earthq Eng 10(6):1767–1793

Jeong SH, Elnashai AS (2007) Probabilistic fragility analysis parameterized by fundamental response quantities. Eng Struct 29(6):1238–1251

Kappos AJ, Stylianidis KC, Pitilakis K (1998) Development of seismic risk scenarios based on a hybrid method of vulnerability assessment. Nat Hazards 17(2):177–192

Kircher CA, Nassar AA, Kustu O, Holmes WT (1997) Development of building damage functions for earthquake loss estimation. Earthquake Spectra 13(4):663–682

Kosmopoulos A, Fardis MN (2007) Estimation of inelastic seismic deformations in asymmetric multistory RC buildings. Earthq Eng Struct Dyn 36(9):1209–1234

Masi A (2003) Seismic vulnerability assessment of gravity load designed R/C frames. Bull Earthq Eng 1(3):371–395

Panagiotakos TB, Fardis MN (1999) Estimation of inelastic deformation demands in multistory RC buildings. Earthq Eng Struct Dyn 28:501–528

Papailia A (2011) Seismic fragility curves for reinforced concrete buildings. Master's Dissertation, University of Patras

Singhal A, Kiremidjian AS (1996) Method for probabilistic evaluation of seismic structural damage. ASCE J Struct Eng 122(12):1459–1467

Spence R, Coburn AW, Pomonis A (1992). Correlation of ground motion with building damage: the definition of a new damage-based seismic intensity scale. In: 10th world conference on earthquake engineering, Madrid

# Chapter 23
# Performance-Based Assessment of Existing Buildings in Europe: Problems and Perspectives

**Paolo Emilio Pinto and Paolo Franchin**

**Abstract** It is by now well recognized that existing structures built before a proper knowledge of seismic hazard was acquired and according in most cases to inadequate seismic design provisions represent by far the major contributor to the total seismic risk. It is equally well known that guidance documents for the assessment of the seismic safety of these structures have lagged behind the development of documents for the seismic design of new structures. In Europe the reference document, Eurocode 8 Part 3 (CEN 2005) is only a few years old. The document is aligned with the recent trends regarding performance requirements and check of compliance in terms of displacements, providing also a degree of flexibility to cover the large variety of situations arising in practice. Nonetheless, in spite of the efforts made to make it rational and to introduce into it results from purposely made original research, the fact remains that EC8-3 could not enjoy at the time of release the support coming from a sufficiently long experience of use. Hence, it comes to no surprise that the widespread use ongoing in a few Countries is already providing suggestions for improvements. The contribution of the paper is two-fold. To provide an overview of the most relevant aspects dealt with in EC8-3, together with remarks coming from use. To indicate how the current state of progress of probabilistic assessment methods can provide today a feasible alternative that overcomes the problems identified in the deterministic codified procedure.

**Keywords** Eurocode 8 Part 3 • Probabilistic approach • Seismic risk analysis • Uncertainty • Logic tree

---

P.E. Pinto (✉) • P. Franchin
Department of Structural and Geotechnical Engineering, University of Rome "La Sapienza", Via Antonio Gramsci 53, 00197 Rome, Italy
e-mail: pinto@uniroma1.it; Paolo.franchin@uniroma1.it

M. Fischinger (ed.), *Performance-Based Seismic Engineering: Vision for an Earthquake Resilient Society*, Geotechnical, Geological and Earthquake Engineering 32,
DOI 10.1007/978-94-017-8875-5_23, 

## 23.1 The Present Situation in Europe: Eurocode 8 Part 3

The European design codes (Eurocodes) consist of a collection of more than 50 documents dealing with practically all types of structures and building materials. Their production started in the late 1980s on the conceptual basis of the pre-normative documents from CEB-FIP denominated Model Codes, from which Eurocodes have inherited the limit-state design approach (known outside Europe as performance-level approach). Eurocode 8 (CEN 2004), the first unified European document on seismic design (first released in 1994, final edition issued in 2004), is no exception and it adopted from the beginning the limit-state design approach. The document does not prescribe an explicit verification of ductility but, rather, it ensures that the structure possesses an adequate ductility by means of rather stringent capacity design procedures. Full-scale experimental testing has typically shown in several occasions how the code produces structures that have a wide margin of safety with respect to design actions.

As it occurred everywhere else in the world the attention shifted to the problem posed by existing structures only recently. As a result Part 3 of Eurocode 8 (EC8-3), dealing with the assessment and retrofit of existing buildings, prepared on the basis of international guidelines such as FEMA 356, was released only in 2005. Hence, in spite of the efforts made to make it rational and to introduce into it results from purposely made original research, EC8-3 could not enjoy at the time of release the support coming from a sufficiently long experience of use. It therefore comes to no surprise that the widespread use ongoing in a few countries is already providing suggestions for improvements.

### *23.1.1 On the Meaning of the "Performance-Based" Attribute*

Before giving an overview of the main aspects of EC8-3, it may be worth spending a few words on an aspect that is seldom explicitly discussed. This is related to the obvious differences in the application of the performance-based procedures for the design of new structures and the assessment of existing ones.

For new structures the required performances are established at the outset as a target and experience shows that design according to current codified procedures, such as e.g. those in the Eurocodes, in general ensures compliance with the requirements, without the need for an explicit proof.

On the contrary, for an existing structure the performance is unknown and its assessment is the goal of the analysis. Achievement of any given level of performance needs to be quantitatively demonstrated in terms of an appropriate metric. Further, considering the amount and nature of the uncertainties characterizing the assessment problem, such an appropriate metric should not be a single-valued quantity but, rather, it should include at least a measure of the dispersion in the assessed performance level.

**Table 23.1** Performance objectives in Eurocode 8

| Hazard level | Performance level |
|---|---|
| $T_R = 2{,}475$ years (2 % in 50 years) | Near collapse (NC): heavily damaged, very low residual strength and stiffness, large permanent drifts but still standing |
| $T_R = 475$ years (10 % in 50 years) | Significant damage (SD): significantly damaged, some residual strength and stiffness, non-structural components damaged, uneconomic to repair |
| $T_R = 225$ years (20 % in 50 years) | Limited damage (LD): only lightly damaged, damage to non-structural components economically repairable |

All Eurocodes claim to be performance-based, and this is certainly true for the documents for the design of new structures. Whether this claim is also acceptable for EC8-3 is subject to debate, as it will be clarified in the following.

### *23.1.2 Performance Requirements*

Similarly to what is done by other international documents (FEMA guidelines among others), EC8-3 is structured with a number of performance-levels, spanning the range of possible damage states from light to collapse, and associated hazard-levels, which are the same as for new structures, specified in terms of average return period with a maximum of about 2,500 years (see Table 23.1).

A remark is in order: a dichotomy exists between the description of performance levels in a loose qualitative form with reference to a global state of damage to the whole structural system, and the way verifications of the compliance with the performance objectives are specified at the member level. The relevance of this dichotomy cannot be understated, since it leaves the burden and responsibility of making a global judgement based on the local results to the analyst. The degree of arbitrariness is an important source of dispersion in the final assessment outcome. This adds up with the other sources of uncertainty, which, as it has already been observed, call for a measure of performance that cannot consist of a single-valued quantity but requires at least an estimate of the range of possible outcomes.

### *23.1.3 Reliability Format*

As far as the reliability approach is concerned, EC8-3 adopts and extends the partial factors format. Beside the usual partial factors on loads and materials, EC8-3, following the approach in the FEMA guidelines, introduces a further factor, called Confidence Factor (CF) whose value is linked to the amount of information available at the time of assessment. This amount is discretized into three, so-called Knowledge Levels (KL), and the corresponding values of CF are used mainly to depress the material strengths.

It can be observed how this format could be valid to cover only a part of the total uncertainty in the problem i.e. that related to the material properties but it is conceptually inadequate to deal with uncertainties related to unavoidable gaps in knowledge on the structural system and on the modelling of its response/capacity. Actually, it can be shown how the uncertainty on the material properties plays in many occasions a secondary role (see e.g. with reference to RC frame structures Franchin et al. 2010). As a consequence the emphasis that the code poses on the extension of material tests to increase the KL seems not justified. Indeed, since only the mean material properties are used in the analysis, much information on the variability is lost. Extensive investigations on the materials can give a false sense of confidence in the assessment outcome.

In fact, it is well known how the construction details are much more influential than material properties on the capacity of the members. Further, it is not uncommon in older buildings (the majority in Europe) the case where no construction drawings are available and hence the very structural layout is affected by uncertainty (not to speak of the cross-section dimensions). The code implicitly assumes that the structural organism is known at the time of assessment. Fact is that quite often this level of knowledge is not reachable, and the code does not give an indication on what lower level of information is still acceptable, or on how to treat the associated uncertainty.

### 23.1.4 Analysis Methods and Modelling

Concerning the analysis methods, the same classical options offered for new structures are proposed for existing ones, without providing a clear hierarchy of accuracy. This choice is recognizedly not appropriate for existing structures, for which one cannot a priori assume a stable dissipative behaviour and the absence of defective response mechanisms. Also in consideration of the considerable relevance, both from an economic and a safety point of view, of an inaccurate verdict, the code should favour the use of more refined analytical tools ("adequately sophisticated" as stated by Priestley et al. 1996) than for the design of new structures. This statement nowadays translates into the requirement of a generalized use of nonlinear methods of analysis, as implicitly recognised for instance with reference to bridges in recent documents such as the comprehensive one being prepared within TG11 of EAEE (Kappos et al. 2012).

### 23.1.5 Verifications

A critical issue that had to be faced by all endeavours to set up documents for the seismic assessment of existing structures has been the unavailability of capacity models for old, non-seismically designed/detailed structural elements due to the fundamental lack of knowledge on their mechanical behaviour. This was obviously also the case of EC8-3. In particular, since this document is displacement-based, the need was related to formulas for the ultimate deformation capacity for the verification of ductile failure modes, and to formulas for the shear strength for the verification of the brittle failure mode.

The current situation is the result of a focussed effort to aggregate and homogenise, to the extent of the possible, existing test data to set up a data base of deformation and strength capacity. Owing to the non homogeneity of the data base (different definitions of "ultimate", incomplete documentation of the tests, different test protocols, etc.) the obtained expressions exhibit a considerable scatter. It is intended to improve them through new purposely made experimental campaigns, like those sponsored by the European Commission (e.g. SERIES 2013).

The formulas, given in an "informative annex" (the form used in the Eurocodes for material that is non mandatory), are due to (Panagiotakos and Fardis 2001) for the ultimate deformation $\theta_u$ and (Biskinis et al. 2003) for the shear strength $V_u$:

$$\theta_u = 0.01(0.3)^{\nu}\left[\frac{\max(0.01;\omega')}{\max(0.01;\omega)}f_c\right]^{0.225}\left(\frac{L_s}{h}\right)^{0.35}25^{\alpha\rho_{sx}\frac{f_{yw}}{f_c}}1.25^{100\rho_d}$$

where $\nu$, $L_s$, $\omega$, $\omega'$, $h$, $\rho_s$ and $\rho_d$ are the normalized axial force, the shear span, the tension and compression mechanical longitudinal reinforcement ratio, the section clear height, the transverse and diagonal geometrical reinforcement ratios, respectively; and:

$$V_u = 0.85\left[\frac{h-x}{2L}(N;0.55A_c f_c) + \left(1-0.55\min\left(5;\mu_{\Delta}^{pl}\right)\right)\right.$$
$$\left.\times\, 0.16\max(0.5;100\rho_{tot})\left(1-0.16\min\left(5;\frac{L_s}{h}\right)\right)\sqrt{f_c}+V_w\right]$$

where $x$, $\mu_{\Delta\mathrm{pl}}$, $\rho_{tot}$, $V_w$ are the neutral axis depth, the plastic part of the displacement ductility demand on the member, the total geometric longitudinal reinforcement ratio and the classical (truss) transverse steel contribution to shear strength. Both formulas are unbiased (i.e. they predict the median) and have coefficients of variation of 40 % and 15 %, respectively.

The above empirical formulas refer essentially to test conducted with in-plane state of deformation. On the contrary the standard situation for most members is that of bi-directional deformation. On this important aspect the code does not give indications. Based on non-exhaustive empirical information, however, Fardis (2006) has proposed the elliptical interaction domain shown in Fig. 23.1.

### 23.1.6 The Latitude of Results

Not unexpectedly, given its avowed partly experimental character, one finds in EC8-3 a number of alternative choices allowed to the analyst, which should

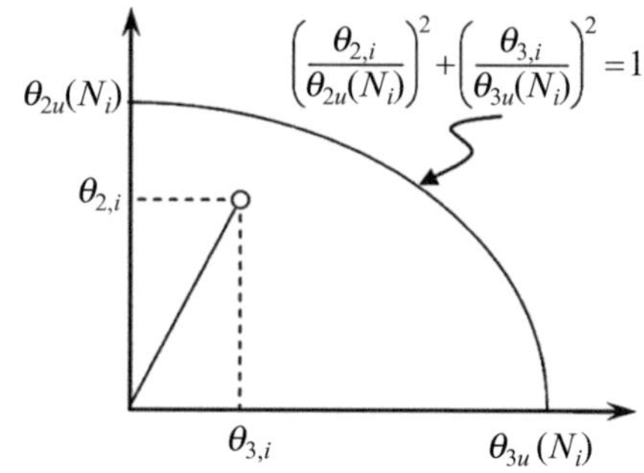


**Fig. 23.1** Elliptical interaction domain for the ultimate deformation capacity under normal force and biaxial bending

be in principle considered as equivalent, but in fact are not and may lead in general to a considerable scatter in the range of outcomes that different analysts, all working within the boundary of the code, can obtain. This is shown in this section through a simple example.

The seismic assessment of a six-storey three-bays plane RC frame (Franchin et al. 2010) is performed fictitiously by a number of distinct analysts. Each analyst is assumed to make independent choices on a number of aspects. For the sake of this illustration not all the admissible choices are considered within this example. They refer only to response analysis, to the input data and to the shear strength capacity model. In particular, five choices are considered:

- Response: both non-linear static (NLS) and dynamic (NLD) analyses are considered (larger variability in the response might have been observed in case linear would also be included). Dynamic analyses have been carried out with a suite of seven spectrum-compatible records (Franchin et al. 2010) that match the response spectrum used for the static analyses (dynamic results are the average over the seven records);
- Response: use of a standard fibre model with stable hysteretic behaviour, called basic modelling (B), versus use of a plastic hinge with section stress resultant-deformation degrading laws in both flexure and shear (the hinges drop load when flexural deformation reaches $\theta_u$ or shear deformation exceeds $\gamma_u$, which is where residual post-peak strength is attained), denominated advanced modelling (A) The latter modelling option allows to follow the sequence of local failures and their consequences on the global behaviour;
- Response: inclusion (T) or exclusion (NT) from the model of non-structural infill panels strength/stiffness (non-linear modelling with equivalent bilinear compression-only struts with degrading behaviour);
- Input data: two values ($\rho_{min}$ and $\rho_{max}$) for the geometric percentage of longitudinal reinforcement in the columns (values that are supposed to represent outcomes from two quantitatively equivalent but differently planned test/inspections campaigns);

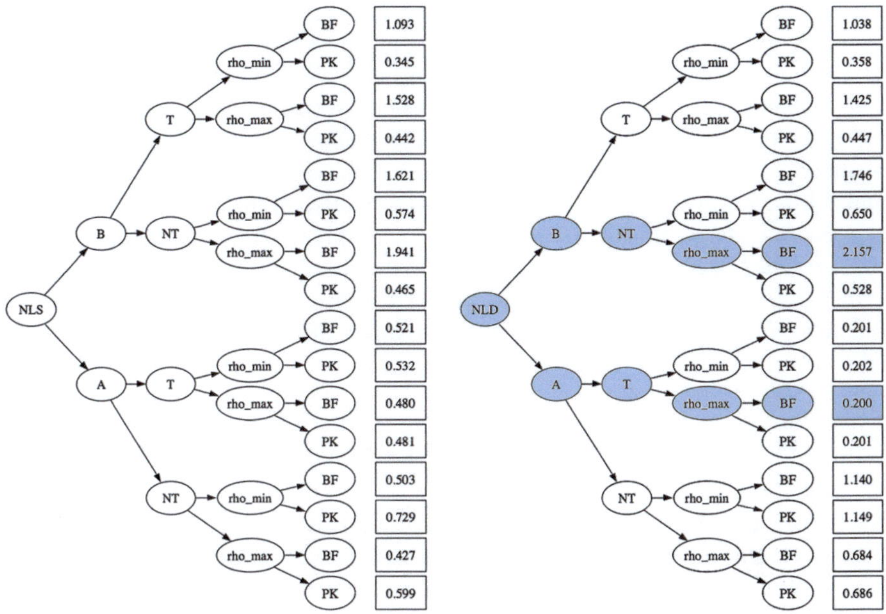


**Fig. 23.2** Tree of analyses: for convenience of representation the full tree is separated into a non linear static portion (NLS, *left*) and a dynamic portion (NLD, *right*). *Ovals* represent different choices, while the *rectangles* report the final outcome of the assessment (global D/C ratio)

- Shear strength capacity model: use of two different models, the one by Biskinis et al. (2003) (BF), given in the informative (non mandatory) Annex of EC8-3, and the other the well-known model by Kowalsky and Priestley (2000) (PK).

It is apparent how a large importance has been attributed to uncertainty stemming from response-determination, as three out of the five considered choices are related to it. The motivation for this weight comes from practical applications that have shown how often, at nominal parity of information on the structure and modelling options, changing the analysis method, or within the same method, changing the modelling options, leads to non negligible differences.

It can be observed that several more sources of uncertainty could have been included, such as, e.g., geometrical dimensions of members, joint reinforcement patterns and joint response and capacity models, floor slab mass, damping model and amount, etc. Finally, it should be noted that the results have been obtained without changing the analysis software, which in all cases is OpenSEES (McKenna and Fenves 2007). More realistically, to reflect personal preferences of the analysts, the different modelling options and analysis methods should have been associated also to different analysis packages. This is recognized to be a major source of uncertainty.

The $2^5 = 32$ different assessment outcomes are represented in graphical form as a tree in Fig. 23.2. Each branch corresponds to a different analyst and the corresponding set of choices. The outcome is expressed as the maximum over the structure of the member demand to capacity ratio over all failure modes (deformation and shear for modelling B, and deformation only for modelling A). This amounts to considering the structural system as a series system of its individual members for the purpose of determining the global failure as a function of the members' failures. This is the default choice that most likely all analysts will make. It is, however, a quite conservative interpretation of the code definition of the significant damage limit state, as a state of widespread structural damage leaving some residual lateral strength and stiffness to the structure.

It is immediate to observe a large variation in the assessment outcomes, which fall in the interval [0.200, 2.157]: the extreme values differ by an order of magnitude.

## 23.2 Perspective: Beyond Current Codified Procedures

As it has been shown, when using the deterministic procedure in the code, the many different uncertainties characterizing the seismic assessment problem may lead to a considerable dispersion in the results of assessments of the same structure. A first step beyond the current state of EC8-3 could be that of asking the analyst to perform a number of assessments under different sets of options, and to evaluate at least a mean and a dispersion of the outcomes. Something along the lines of the previous example. This section goes one step further and presents a probabilistically consistent approach to the treatment of all uncertainties in the assessment problem, before a "practical" proposal is finally put forward in §23.3.

### *23.2.1 Uncertainties*

The uncertainties entering into the problem can be usefully classified in two groups, one amenable of modelling with continuous random variables, the other with discrete ones, whose "states" are associated with alternative choices of the analyst, with the probability masses being subjective measures of the analyst degree of belief.

Nowadays, simple, well-established methods, like the *conditional* simulation methods such as e.g. cloud analysis, multiple-stripe analysis, incremental dynamic analysis (IDA), etc. (see for instance Jalayer and Cornell 2009), requiring a minimum of specialized knowledge, are available to achieve a probabilistic measure of seismic performance of structures when the uncertainties belong of the first group. Uncertainties of the second can be consistently treated by arranging combinations of the discrete variables states into a logic tree. The overall procedure amounts to repeating the evaluation of the probabilistic performance measure for every branch in the tree. The final outcome is a discrete distribution of the performance measure (probability of probability).

Objections to such a straightforward approach could be related to the increased computational effort. Based on experience, the largest proportion of the burden of a nonlinear analysis lies in setting up and gaining confidence in the model. The effort associated with the repetition of the analysis for multiple ground motion time-series and different sets of the parameters can by now be considerably down-played.

## 23.2.2 Performance Measure

The problem of checking the attainment/exceedance of a globally defined limit-state is not overcome by the adoption of a probabilistic assessment method. As already stated in §23.1.6, the default choice of treating the structure as a series system is a conservative one. Consistently with the use of a probabilistic approach one should avoid conservatism searching for the most accurate approximation of reality. In theory this would be possible if the response-determination capabilities had reached the necessary level of maturity, with analyses where members failed in flexure or shear were progressively removed with the following load redistribution taking place. Analyses of this level refinement can be retrieved in the research literature, but are limited to simple structures and require specialized skills that are currently out of reach of the profession.

A possible if approximate way of coping with the limitations of reliable analysis tools within reach of the average analyst, would be, for instance, that outlined in (Jalayer et al. 2007), where at least the series system approach is replaced with a so-called cut-set approach. In such an approach global failure occurs as a series of parallel sub-systems failures: e.g. more than one column in a floor must fail before the floor fails and causes global failure.

Once a satisfactory quantitative definition of the limit state exceedance is chosen, the result of a probabilistic analysis carried out according to one of the mentioned conditional simulation methods is the mean annual frequency of exceedance of the limit state $\lambda_{LS}$. The latter is obtained by convolving the conditional probability of exceedance $\lambda_{LS|IM}$, determined via non linear dynamic analysis repeated for carefully selected ground motion time series, with the corresponding hazard curve $\lambda_{IM}$, i.e. the mean annual frequency of exceedance of the intensity measure IM.

Clearly a value of $\lambda_{LS}$ is obtained for every combination of states of the discrete variables of the second group. The final outcome of the analysis is then obtained by statistical post-processing of the $\lambda_{LS}$ values. The procedure is illustrated in the next section through an example.

## 23.2.3 An Example

Probabilistic assessment of the 15-storeys plane RC frame in Fig. 23.3 is carried out along the lines indicated in the previous section. The figure shows overall frame

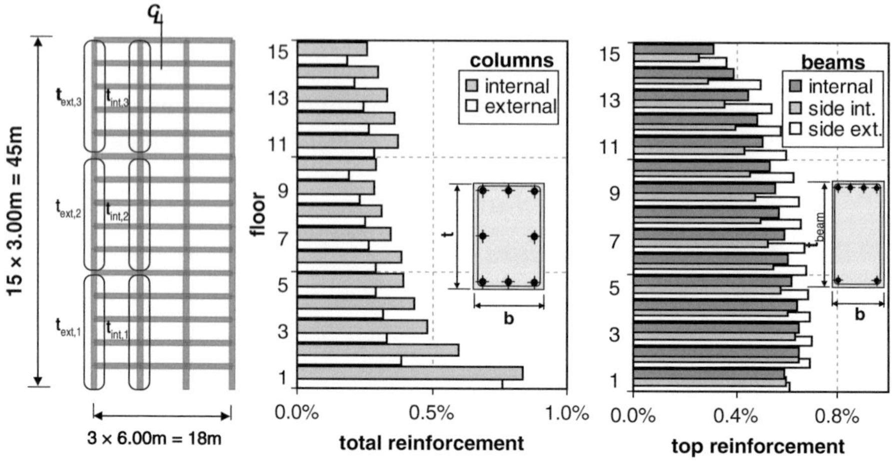


**Fig. 23.3** Fifteen-storeys example frame from Franchin and Pinto (2012). Columns taper every five floors

dimensions, the reinforcement layout of beams and columns, and the floor-wise reinforcement ratios. Details of the (probabilistic) design procedure, as well as cross-section dimensions can be found in Franchin and Pinto (2012).

Uncertainties of the first group modelled in the example include: concrete compressive strength and ultimate deformation, steel yield strength, and the model error term of the ultimate deformation capacity formula. Figure 23.4 shows the theoretical distributions adopted (lognormal or uniform) together with the parameters, and the histogram of relative frequency of the 20 values sampled for the analyses (each set of values has been univocally associated with 1 of the 20 records employed for incremental dynamic analysis).

Uncertainties of the second group include: the choice of the records set to be used for inelastic dynamic analysis, the ultimate deformation of RC columns and the floor mass. In particular, two alternative options (states) are considered for each of the three variables (resulting in $2^3 = 8$ combinations): two independently selected sets of 20 records each, two deformation capacity formulas (here represented in a simplified way through two different medians equal to 2 and 2.5 %, with the same logarithmic standard deviation of 40 %) and two floor mass values of 46 and 63 t.

For each combination of states of the discrete variables of the second group, an IDA is carried out for 20 record-structure pairs, where the values of structural properties are sampled from their respective distributions as shown in Fig. 23.4, yielding a distribution of intensity values that lead to the attainment of the limit state. Post-processing of these values gives one of the eight $\lambda_{LS|IM}$ curves, and, after convolution with the corresponding hazard, one of the eight $\lambda_{LS}$ values.

The whole procedure is illustrated in Fig. 23.5. Branches in the logic tree are attributed probabilities that represent the analyst confidence in the corresponding

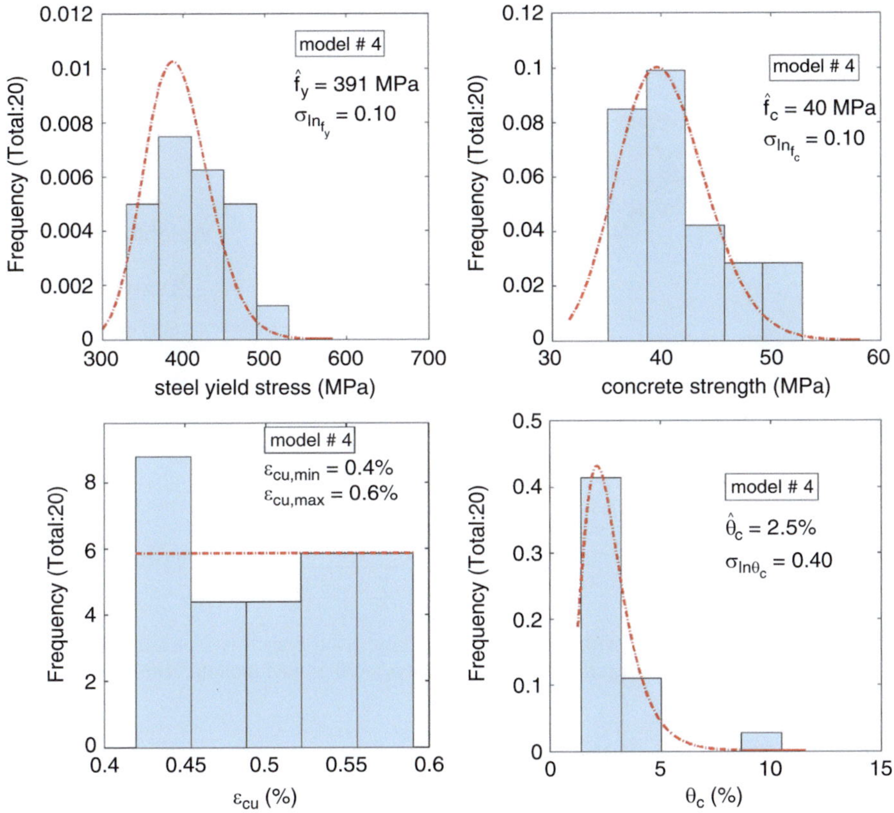


**Fig. 23.4** Uncertainty of the first group: theoretical distribution, parameters, and histogram of sampled values of the model parameters that are treated as continuous random variables

choice. The product of the probabilities in each branch is the probability $p_i$ associated with the $\lambda_{LS}$ value at the end of the branch. Even with the relatively mild variations in the mass, the median deformation capacity and the records (the two independent records selections employed the same criteria), one observes a non-negligible variation of $\lambda_{LS}$, with the largest value being about 2.5 times larger then the smallest one. The last step consists in reporting the expected value and the standard deviation of $\lambda_{LS}$. This value accounts for intrinsic as well as epistemic uncertainty.

One may conclude with a comment on the epistemic variables to be included in the analysis. As shown by the weights attributed to the branches, which coincide with 0.5 or are very close to it, the only variables that make sense to introduce are those for which the analyst has an approximately equal degree of belief (including variables with states associated with 0.9–0.1 weights would be a useless waste of time).

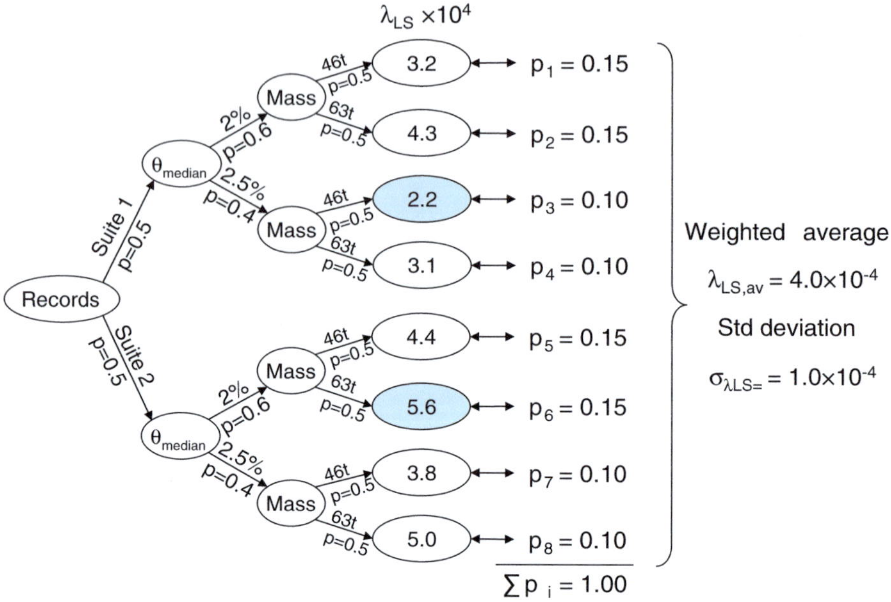


**Fig. 23.5** Logic tree to deal with epistemic uncertainty. Each branch is a probabilistic seismic assessment performed with incremental dynamic analysis (20 ground motion records)

## 23.3 A "Practical" Proposal?

The previous section presented a procedure that enables performing seismic assessment of an existing structure accounting in a consistent, sound and yet conceptually simple manner the different uncertainties entering into the problem.

This said, it is clear that such an analysis still requires several hundreds of inelastic dynamic analyses, which is hardly something that can be done in practice when assessing a real building outside a research context. Furthermore, the computational burden is only one of the two main difficulties arising in connection with the probabilistic demand evaluation based on inelastic dynamic analysis. The other is the issue of selecting proper ground motion records, a task that requires specialized knowledge not necessarily part of the structural engineer's background. One possibility to retain the overall conceptual framework, while making the procedure affordable for practical application, is to replace the IDA with a static pushover analysis. Practical tools to perform the conversion of a static pushover curve into the median IDA curve are already available (Vamvatsikos and Cornell 2004); (Fajfar and Dolšek 2010). Of course there is a price to be paid for such a simplification, due to the approximate nature of the conversion, and to the fact that dispersion in the demand must be assumed based on experience, rather than be calculated through analysis. Once the median IDA and the dispersion are known, the rest of the procedure remains unchanged.

As far as the authors are concerned the only difficulty for a generalized adoption of such a higher level approach to seismic safety assessment of existing structures stays in the time required for disseminating it to a wider audience.

## References


Biskinis D, Roupakias G, Fardis MN (2003) Cyclic deformation capacity of shear–critical RC elements. Fib 2003 symposium: concrete structures in seismic regions, Athens

CEN (2004) Eurocode 8: design of structures for earthquake resistance. Part 1: general rules, seismic action and rules for buildings. EN 1998–1, Euro Commit for Stand, Brussels

CEN (2005) Eurocode 8: design of structures for earthquake resistance. Part 3: assessment and retrofitting of buildings. EN 1998–3, Euro Commit for Stand, Brussels

Fajfar P, Dolšek M (2010) A practice-oriented approach for probabilistic seismic assessment of building structures. In: Advances in Performance-based earthquake engineering, Geotechnical, geological, and earthquake engineering, vol 13, Part 2. Springer, Dordrecht/New York, pp 225–233. ISBN 978-90-481-8745-4

Fardis M (2006) Acceptable deformations of RC members at different performance levels under bidirectional loading. LessLoss Deliverable Report 64. http://www.lessloss.org

Franchin P, Pinto PE (2012) A method for probabilistic displacement-based design of RC structures. ASCE J Struct Eng 138(5):585–591

Franchin P, Pinto PE, Rajeev P (2010) Confidence factor? J Earthq Eng 14:989–1007

Jalayer F, Cornell CA (2009) Alternative non-linear demand estimation methods for probability-based seismic assessments. Earthq Eng Struct Dyn 38:951–972

Jalayer F, Franchin P, Pinto PE (2007) A scalar damage measure for seismic reliability analysis of RC frames. Earthq Eng Struct Dyn 36:2059–2079

Kappos AJ, Saiidi M, Aydinoglu N, Isakovic T (eds) (2012) Seismic design and assessment of bridges: inelastic methods of analysis and case studies. Springer, Dordrecht/New York, 233 pp. ISBN 978-94-007-3943-7

Kowalsky M, Priestley MJN (2000) Improved analytical model for shear strength of circular reinforced concrete columns in seismic regions. ACI Struct J 97(3):388–396

McKenna F, Fenves GL (2007) Open system for earthquake engineering simulation. Pacific Earthquake Engineering Research Center, Berkeley. http://opensees.berkeley.edu

Panagiotakos TB, Fardis MN (2001) Deformations of reinforced concrete members at yielding and ultimate. ACI Struct J 98(2):135–148

Priestley MJN, Seible F, Calvi GM (1996) Seismic design and retrofit of bridges. Wiley, New York

SERIES: Seismic Engineering Research Infrastructure for European Synergies, http://www.series.upatras.gr/

Vamvatsikos D, Cornell CA (2004) Direct estimation of seismic demand and capacity of multidegree-of-freedom systems through incremental dynamic analysis of single degree of freedom approximation. ASCE J Struct Eng 131:589–599

# Chapter 24
# Inelastic Shear Response of RC Walls: A Challenge in Performance Based Design and Assessment

**Matej Fischinger, Klemen Rejec, and Tatjana Isaković**

**Abstract** The large inelastic shear modification factors proposed in Eurocode for ductile RC walls have been verified and modified. Due to this large amplification, which has, in the past, been ignored, and still is, by many designers, RC walls with insufficient shear resistance have been designed and built. In order to study the seismic vulnerability of such walls, a model was proposed, which takes into account both inelastic shear behaviour and inelastic shear-flexural interaction. It is based on the multiple-vertical-line-element macro model. An additional shear spring, which accounts for aggregate interlock, dowel action and horizontal reinforcement resistance, is incorporated into each of the vertical springs. The model successfully simulated the response of a five-storey coupled wall that was tested on the shaking table under bi-axial excitation. The shear resisting mechanisms within the cracks were adequately modelled up until the tension shear failure of both piers.

**Keywords** RC walls • Coupled wall • Inelastic shear • Inelastic shear/flexural interaction • Shear magnification factors • Eurocode • Multiple-vertical-line-element model

## 24.1 Introduction

For decades, existing numerical models have served the engineering community well. However, the vision of earthquake resilient structures and society itself call for more elaborate and complex tools, which should be able to represent more realistically all possible near-collapse mechanisms. One of the many problems to be solved involves the need for better models and methods for the estimation of the

M. Fischinger (✉) • K. Rejec • T. Isaković
Faculty of Civil and Geodetic Engineering, University of Ljubljana,
Jamova 2, SI 1000 Ljubljana, Slovenia
e-mail: matej.fischinger@fgg.uni-lj.si; klemen.rejec@fgg.uni-lj.si; tisak@ikpir.fgg.uni-lj.si

M. Fischinger (ed.), *Performance-Based Seismic Engineering: Vision for an Earthquake Resilient Society*, Geotechnical, Geological and Earthquake Engineering 32,
DOI 10.1007/978-94-017-8875-5_24, 

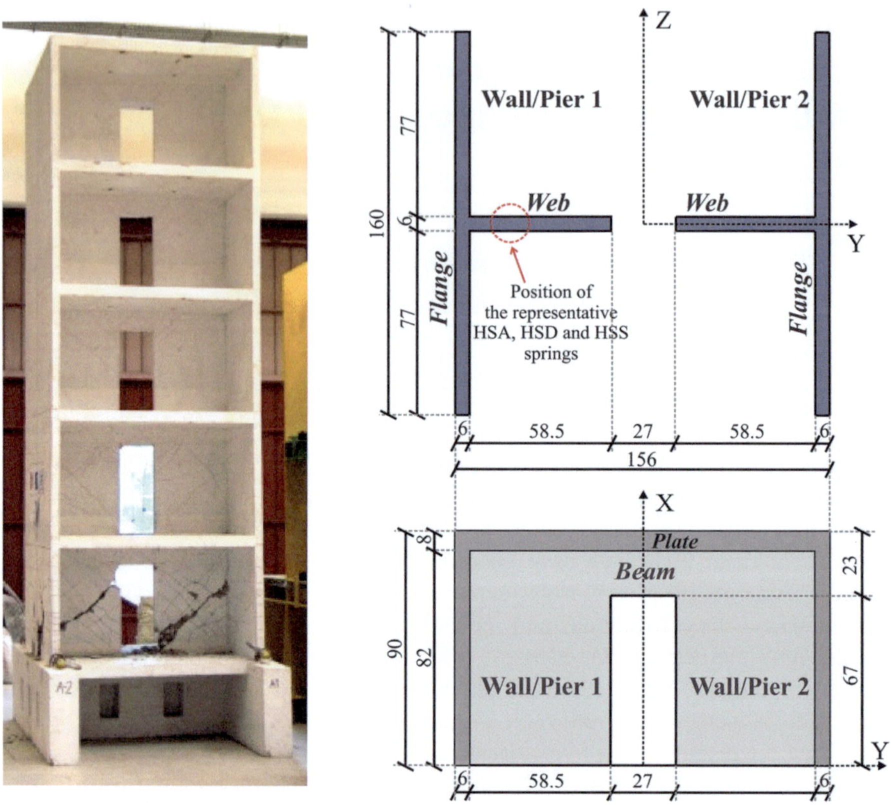


**Fig. 24.1** The wall designed according to Slovenian practice and tested on the shaking table at LNEC in Lisbon (Fischinger et al. 2006)

inelastic shear demand and capacity of reinforced concrete (RC) structural elements. This problem is particularly complex in the case of RC structural walls. The related analyses and observations which are presented in the paper are illustrated by the results of a large-scale shaking table experiment that was performed on a low-rise RC coupled wall (Fig. 24.1). This wall was designed according to the past Slovenian engineering practice (which is similar to that used in Chile).

The wall was modelled and designed to represent part of a typical multi-storey building with structural walls (Fig. 24.2).

Such buildings have been extensively built all over the world, for example in Europe and Chile. This building system is characterized by thin walls and a large wall-to-floor ratio (the structural walls also serve as partition walls). Although, during the recent 2010 Chile earthquake (Boroschek and Bonelli 2014), many compression and shear-compression failures of such walls were observed, the authors believe that this was predominantly due to the misuse of the system beyond its acceptable engineering limits (in particular due to the increasing of the height of the building while keeping the small thickness of the walls unchanged, and the

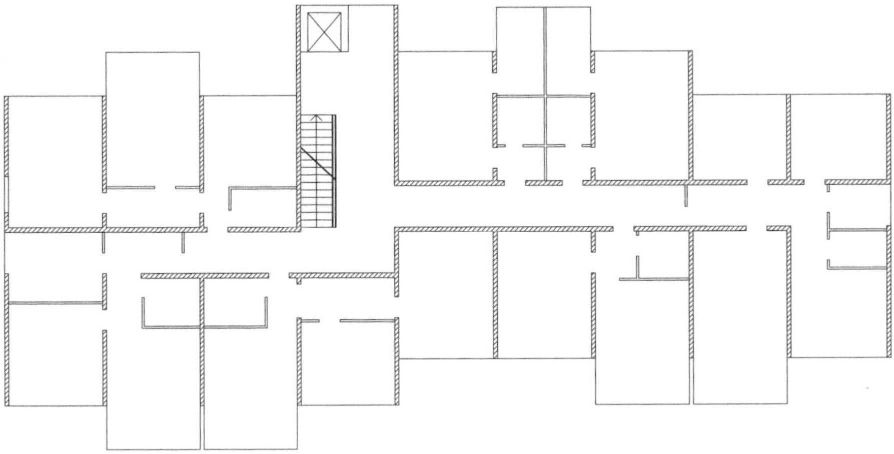

**Fig. 24.2** Typical multi-storey apartment building whose load-bearing structure consists of structural walls

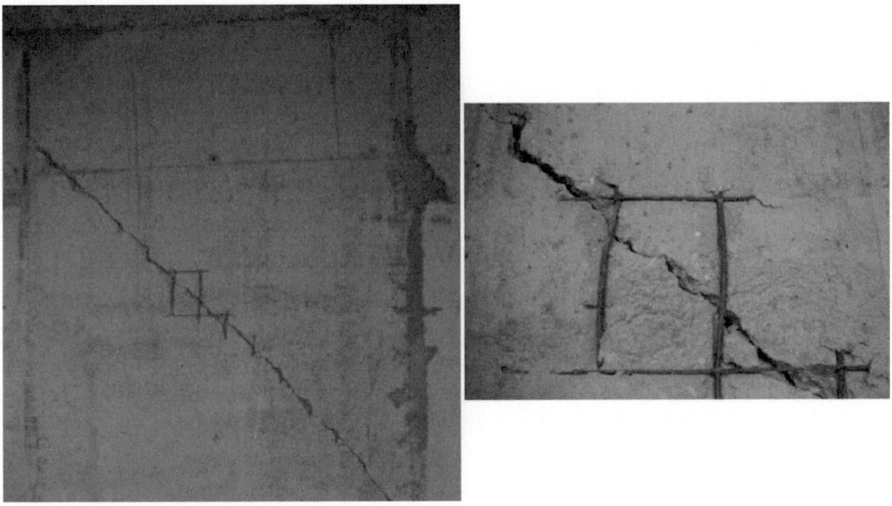

**Fig. 24.3** A pier in a coupled wall damaged during the 1979 Montenegro earthquake

edges lightly confined). If such a system is used for buildings that are not much higher than ten stories, and/or built in moderate seismic regions, the behaviour of the walls should be good, as was observed during several earthquakes in the past (including that which took place in Chile, in 1985) (see Wallace and Moehle 1993). The predominant type of rather rare failures has, in the past, been shear-tension failure (Wood 1991), as is demonstrated in Fig. 24.3, a photograph which was taken after the 1979 Montenegro earthquake. Please note that the common construction practice during the 1970s was to use very weak horizontal reinforcement.

## 24.2 Inelastic Shear Strength Demand in the Design of Cantilever (Wall) Structures

The problem of insufficient shear resistance is not limited just to walls in older buildings. Although a large shear magnification during inelastic response was, long ago, pointed out by Blakeley et al. (1975), even today many designers are not fully aware of this phenomenon and only a few codes, like those used in New Zealand or Eurocode 8 (CEN 2004), consider this magnification explicitly. Eurocode 8 requires that the shear forces obtained by an equivalent elastic analysis $V_{Ed}'$ are multiplied (over the entire height of the wall) by a shear magnification factor ε, in order to obtain the design shear forces $V_{Ed}$:

$$V_{Ed} = \varepsilon \cdot V_{Ed}' \tag{24.1}$$

In the case of ductility class high (DCH) walls, the shear magnification factor is determined from the expression (24.2), which was originally proposed by Keintzel (1990):

$$\varepsilon = q \cdot \sqrt{\left(\frac{\gamma_{Rd}}{q} \cdot \frac{M_{Rd}}{M_{Ed}}\right)^2 + 0.1 \cdot \left(\frac{S_e(T_C)}{S_e(T_1)}\right)^2} \begin{cases} \leq q \\ \geq 1.5 \end{cases} \tag{24.2}$$

where:

$q$ is the behaviour (seismic force reduction) factor used in the design;
$M_{Ed}$ is the design bending moment at the base of the wall;
$M_{Rd}$ is the design flexural resistance at the base of the wall;
$\gamma_{Rd}$ is a factor which is used to increase the design value of resistance, accounting for various sources of overstrength;
$T_1$ is the fundamental period of vibration of the building in the direction of action of the shear forces;
$T_C$ is the upper limit period of the constant spectral acceleration region of the spectrum;
$S_e(T)$ is the ordinate of the elastic response spectrum.

In the derivation of this formula, Keintzel assumed that modal combination can also be applied in the inelastic range, and that only the contribution of the first two modes is important:

$$V_{Ed} = \sqrt{(V_{Ed,1})^2 + (V_{Ed,2})^2} \tag{24.3}$$

Keintzel further assumed that energy dissipation could be associated only with the first mode response (within the hinge location at the base, the flexural moment

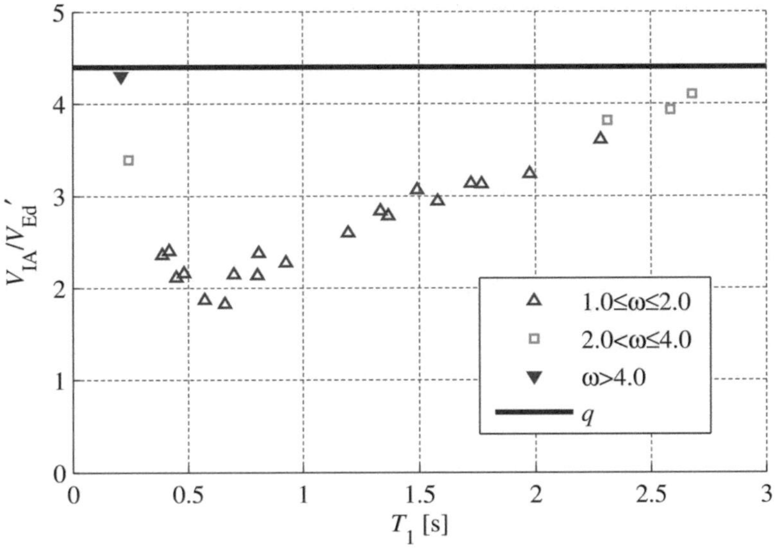


**Fig. 24.4** Ratio of the inelastic shear $V_{IA}$ and the design shear $V_{Ed}'$ (ω is the flexural overstrength)

due to higher modes is practically negligible). For this reason only the contribution of the first mode should be reduced by the factor $q$, whereas the contribution of the second mode should be elastic/unreduced ($q \cdot V_{Ed,2}'$):

$$V_{Ed} = \sqrt{\left(V_{Ed,1}'\right)^2 + \left(q \cdot V_{Ed,2}'\right)^2} \tag{24.4}$$

Considering only the contribution of the first mode to the flexural overstrength (see the previous paragraph), and the ratio of the base shear contributed by the second and the first mode of $\sqrt{0.1 \cdot S_e(T_2)/S_e(T_1)}$ (in the elastic range), the expression (24.2) was derived.

It should be stressed and clearly understood from the presented derivation that the shear magnification factor ε, which was proposed by Keintzel (and included in Eurocode 8), should be applied considering only the base shear due the first mode. However, following the ambiguous notation in Eurocode 8, it is most likely that many designers erroneously apply ε to the total base shear (usually given by commercial computer codes used in design offices).

Recently, a systematic parametric study of the inelastic response of cantilever walls was performed (Rejec et al. 2012) with the aim of studying the adequacy of this shear magnification factor, which had been opposed by many designers as over-conservative. However, the very large increase in shear forces (up to the value of the seismic force reduction factor $q$) was reconfirmed by this study (Fig. 24.4).

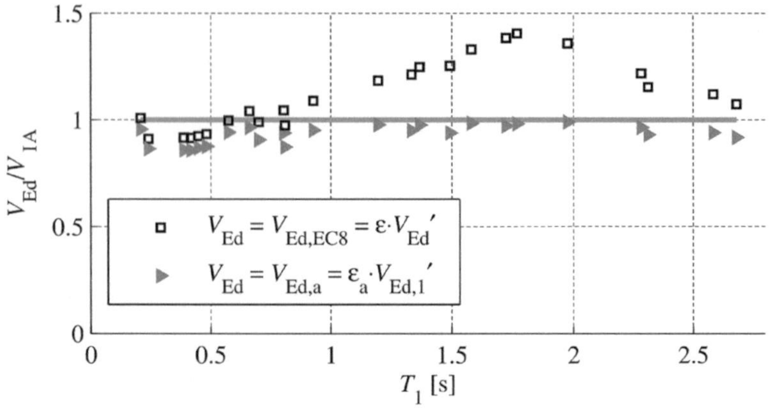


**Fig. 24.5** Design shear forces $V_{Ed}$ compared to the inelastic shear forces $V_{IA}$. The points marked by squares represent the results obtained by the formula given in Eurocode 8, whereas the points marked by triangles represent the results obtained by the proposed formula (24.6)

In (Rejec 2011) it was also demonstrated that:

(a) In general Keinzel's formula (used in Eurocode 8) works fine if it is applied correctly (the shear magnification factor is applied to the base shear contributed by the first mode $V_{Ed,1}'$ only).
(b) However, the upper bound of the shear force should be related to the total base shear force ($V_{Ed,max} = q \cdot V_{Ed}'$) and not only to that defined by the first mode contribution ($V_{Ed,max} = q \cdot V_{Ed,1}'$), as was assumed by Keintzel. This yields an upper bound of the shear magnification factor $\varepsilon_{upper}$, which is even higher than the seismic force reduction factor ($\varepsilon_{upper} > q$):

$$\varepsilon_{upper} = \sqrt{q^2 + 0.1 \cdot \left(q \cdot \frac{S_e(T_C)}{S_e(T_1)}\right)^2} \tag{24.5}$$

and finally:

$$\varepsilon_a = q \cdot \sqrt{\left(\min\left[\frac{\gamma_{Rd}}{q} \cdot \frac{M_{Rd}}{M_{Ed}}; 1\right]\right)^2 + 0.1 \cdot \left(\frac{S_e(T_C)}{S_e(T_1)}\right)^2} \geq 1.5 \tag{24.6}$$

Figure 24.5 graphically illustrates these two observations. If the expression for ε proposed by Keintzel is applied to the total base shear (as it is understood from the ambiguous notation in Eurocode 8), the results are in general over-conservative (see the points marked by squares in Fig. 24.5). However, in the long period region the upper bound for the magnification factor applies, yielding a good match with the results of the inelastic analysis. On the other hand, the properly applied and

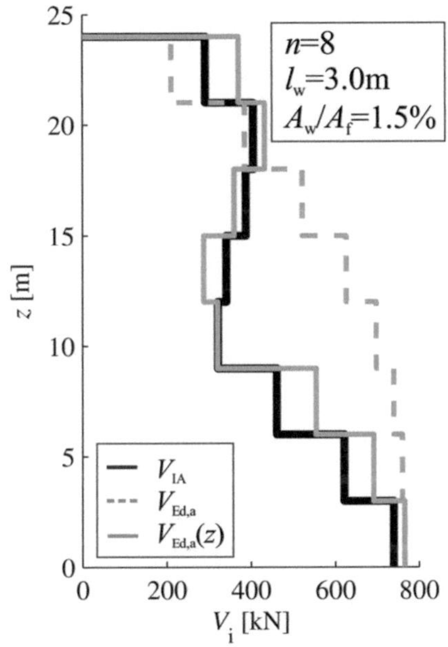


**Fig. 24.6** Distribution of shear forces over the height of the eight-storey wall as given by Eurocode 8 (the *dashed line*), the proposed formula (the *solid grey line*), and inelastic response analysis (the black line) (More data about the inelastic response analysis can be found in Rejec et al. 2012)

modified formula ($\varepsilon_a$) yields a good correlation with the results of the inelastic analysis over the entire span of the periods involved (see the points marked by triangles in Fig. 24.5).

Eurocode 8 suggests that the same shear magnification factor should be used along the entire height of the wall. As expected, this could result in a substantial overestimation of the shear forces at mid-height, and an underestimation of the shear forces at the top, where the contribution of the higher modes is more pronounced than at the base. This observation is illustrated by means of a dashed line in Fig. 24.6 for one of the eight-storey walls analysed in the parametric study (the length of the wall $l_w$ was 3 m, and the assumed wall-to-floor area ratio was 1.5 %).

To account for this variation along the height of the wall, it was proposed (Rejec et al. 2012) that the constant ratio between the contribution of the higher modes and the contribution of the first mode ($\sqrt{0.1}$), which is approximately valid at the base of the wall, should be replaced by a variable ratio along the height – *m(z)* (Eq. 24.7).

$$\varepsilon_a(z) = q \cdot \sqrt{\left(\min\left[\frac{\gamma_{Rd}}{q} \cdot \frac{M_{Rd}}{M_{Ed}}; 1\right]\right)^2 + m(z)^2 \cdot \left(\frac{S_e(T_C)}{S_e(T_1)}\right)^2} \geq 1.5 \qquad (24.7)$$

It was assumed that the distribution of this ratio $m(z)$ was the same as in the case of the elastic flexural cantilever beam (fully realizing that this is only an approximation in the inelastic range, and that it is applicable only to regular walls with no plastic hinges in the upper storeys).

For the chosen eight-storey wall (for more complete results, see Rejec 2011) the results $V_{Ed,a}(z)$ obtained by using $\varepsilon_a(z)$ in combination with $V_{Ed,1}{}'$ are presented in Fig. 24.6. The results are compared with the shear envelopes obtained by using inelastic response history analyses $V_{IA}$ and the design shears obtained by multiplying $V_{Ed,1}{}'$ with the constant $\varepsilon_a(z=0)$ along the entire height (as suggested in Eurocode 8).

## 24.3 Numerical Modelling of the Inelastic Shear Response and Shear-Flexural Interaction in RC Structural Walls

### *24.3.1 Background*

A reliable model for inelastic seismic shear response is still to be defined. For this reason many researchers ignore or try to avoid this problem. They frequently assume that shear failure in newly built walls is automatically prevented by capacity design. However, as has been shown in the previous section, the shear magnification factors have not yet been clearly defined, and many designers/codes even do not use them at all. In the case of the walls of older buildings, researchers try to avoid the problem by assuming elastic shear behaviour, and then making post-analyses checks. However, ignoring inelastic shear-flexural interaction makes the results of such analyses questionable. This is particularly true in the case of seismic risk analyses, where structures are analysed up to the near collapse stage. Improved models for inelastic shear response are therefore needed.

Some other models for inelastic shear-flexural interaction have already been proposed and experimentally verified, e.g. those proposed by Kabeyasawa (1997), Chen and Kabeyasawa (2000), Orakcal et al. (2006), and Kim et al. (2011). However, refinements in the description of the cyclic behaviour are still needed. Another concern is the complexity of some of the proposed models, which makes them difficult to apply to realistic structures.

In general, the research group at the University of Ljubljana has trust in macro models, even in the case of complex behaviour. Macro models are defined here as models which monitor force-displacement rather than stress–strain relationships. In the particular case of structural walls, the authors have used the multiple-vertical-line-element model – MVLEM (Fig. 24.7).

The model has been consistently proved to be efficient in the cases of a predominantly flexural response. For example, it was used in the case of the benchmark prediction for the "San Diego" wall that was awarded the "best prediction" recognition (EERI 2006). However, the research group has still not been able to completely understand and define the inelastic behaviour of shear springs and, first of all, the inelastic interaction of the shear and flexural springs in the model. This lack of knowledge was demonstrated during the "ECOLEADER" test of a coupled wall (Figs. 24.1 and 24.8) (Kante 2005; Fischinger et al. 2006). Due to

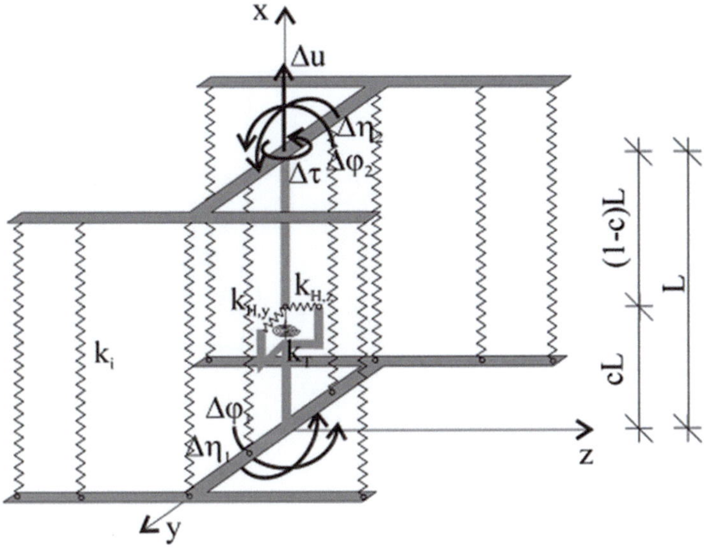


**Fig. 24.7** 3D multiple-vertical-line-element

**Fig. 24.8** Shear failure of the piers in the ECOLEADER wall (see Fig. 24.1)

the overstrength of the coupling beams, large axial forces were induced in the piers, which subsequently failed due to shear-tension interaction. After the test, the use of the compression field theory (Vecchio and Collins 1986) substantially improved the analytical results (Kante 2005). However, the theory was found to be incomplete in the case of cyclic response.

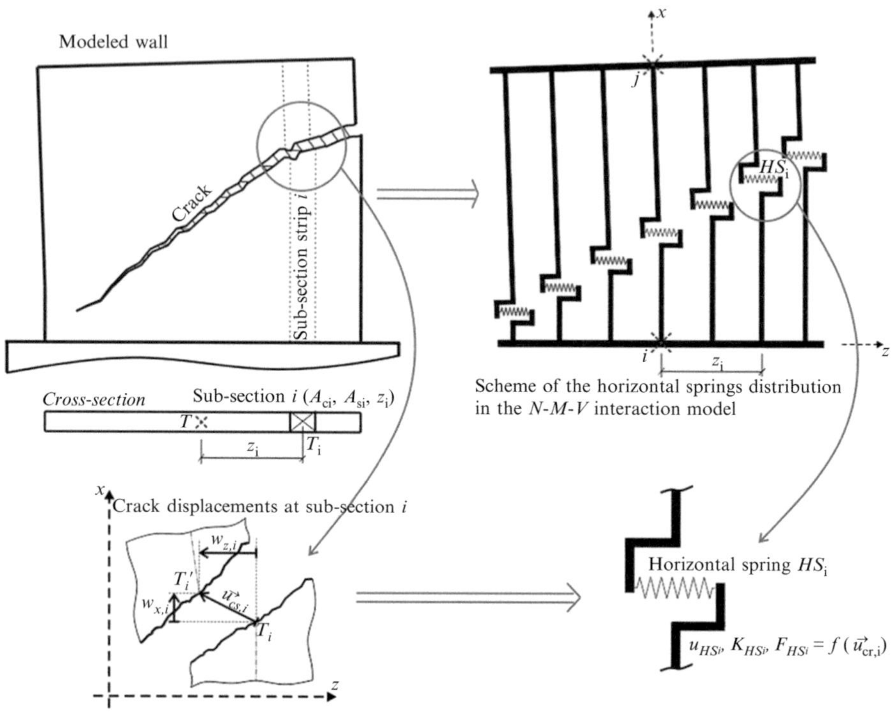


**Fig. 24.9** Model accounting for inelastic shear and shear-flexural interaction in structural walls (the vertical springs are not shown)

## 24.3.2 *Proposed Numerical Model*

In order to account for the inelastic shear behaviour and the axial force – bending moment – shear force (*N*-*M*-*V*) interaction better, the MVLEM was modified (Rejec 2011), and incorporated into the OpenSees program (McKenna and Fenves 2007). The modified element is illustrated in Fig. 24.9 (only the 2D element is shown in order to make the illustration clearer). In principle, one additional shear spring has been introduced into each of the vertical strips (springs), as proposed by Wallace (Orakcal et al. 2006).

The following key assumptions were considered in the development of the model:

- Cracks are straight and equally spaced. The (constant) spacing between cracks should be evaluated according to empirical procedures.
- The shear displacements of the element caused by the compressive deformation of the diagonal struts are neglected. It is assumed that the tensile and shear deformations in the cracked strips are localized in the cracks.

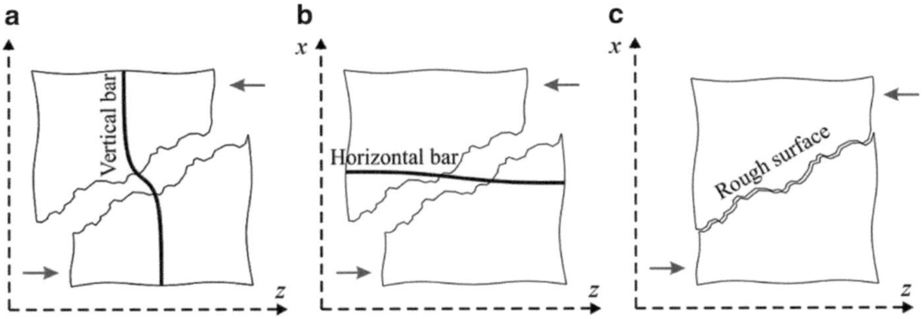


**Fig. 24.10** Mechanisms of shear force transfer across the cracks: (**a**) the dowel effect of the vertical reinforcement; (**b**) the axial resistance of the horizontal/shear reinforcement, and (**c**) aggregate interlock in the crack

**Fig. 24.11** Each horizontal spring consists of three components to account for aggregate interlock (HSA), the dowel effect (HSD), and the shear/horizontal reinforcement (HSS) mechanisms

- Along the height of the wall segment the inclination of the cracks and the displacement within different cracks is assumed to be constant.
- The current crack inclination is evaluated according to the average current strain state in the element, and is updated at every load step (the rotating-crack model).

The above assumptions have been empirically verified, and they are valid for walls with low to moderate compressive axial forces (typical for the European practice, see the Introduction). In other cases the compression strut is additionally checked.

The shear behaviour and resistance modelled by the horizontal springs depend on the mechanisms that transfer the shear force across the cracks (Fig. 24.10). The mechanisms consist of (a) the dowel effect of the vertical bars, (b) the axial resistance of the horizontal/shear bars, and (c) aggregate interlock, i.e. the interlocking of aggregate particles in the crack. The capacity of the latter is highly dependent on the width of the cracks.

Thus, each spring has three components (Fig. 24.11): HSA to account for aggregate interlocking, HSD to account for the dowel action, and HSS to account for the axial resistance of the shear reinforcement. The current characteristics of each component depend on the deformations/displacements at the crack within the individual strip. The displacements are linked to the current displacements of the nodes of the element.

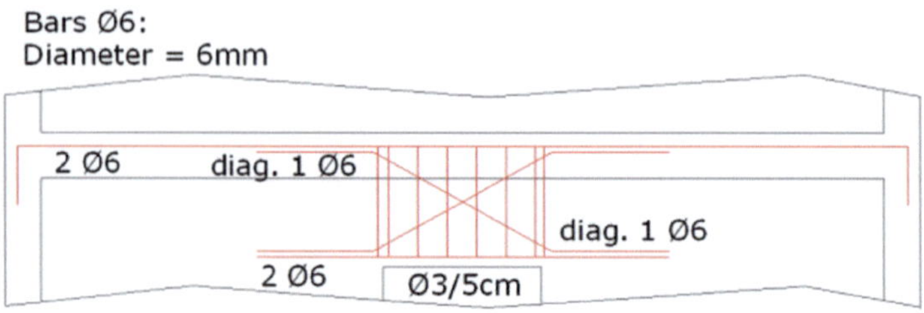


**Fig. 24.12** Reinforcement in the coupling beams

The constitutive relations for the individual springs are based on the semi-empirical relations found in the literature (a detailed description is given in Rejec 2011). Aggregate interlock is modelled by the Lai-Vecchio model (Vecchio and Lai 2004), dowel action by the expressions proposed by Dulacska (1972), and by Vintzeleou and Tassios (1987). The force-displacement relation for the HSS springs is based on the bar-slip model proposed by Elwood and Moehle (2003).

## *24.3.3 Experimental Verification of the Proposed Model*

The inelastic response of the wall presented in the introduction (Fig. 24.1) was analysed in order to verify the suitability of the proposed model. The 1:3 model of a five-storey wall (Fig. 24.1) consisted of two coupled T-shaped piers (Fischinger et al. 2006, 2008, 2010). The piers were reinforced by very light (minimum) reinforcement, according to the Slovenian building practice. The distributed mesh reinforcement amounted to 0.25 % of the cross-section in both directions. Note that the small diameter bars (3 mm) used for the reinforcement mesh in the model were very brittle (their ultimate strain was only 1.5 %). The coupling beams, too, were lightly reinforced (Fig. 24.12). A heavy additional mass was added due to the reduced scale, and to account for the mass in adjacent fields in realistic structures. This required a relatively thick slab, i.e. one with a thickness of 8 cm, which would be equal to 24 cm in the prototype structure.

The shaking table test was performed at LNEC in Lisbon, Portugal within the scope of the ECOLEADER project, which was coordinated by University of Ljubljana team. The Tolmezzo accelerogram, recorded during the 1976 Friuli earthquake, was used in two directions in a series of tests with increasing intensity. In the last of the series of the tests (the 6th run) the table acceleration in the direction of the web wall with openings was $a_{g,max,X} = 1.02$ g, and the acceleration in the direction of the flange walls was $a_{g,max,Y} = 0.52$ g. Failure occurred in the direction of the web (see Fig. 24.8 in Sect. 24.3.1). Typical shear failure of the wall piers was observed. The flange walls were only lightly damaged. Some damage was observed at the unconfined edges, and due to punching caused by the web wall. To the surprise of observers, the supposedly weak coupling beams were practically undamaged.

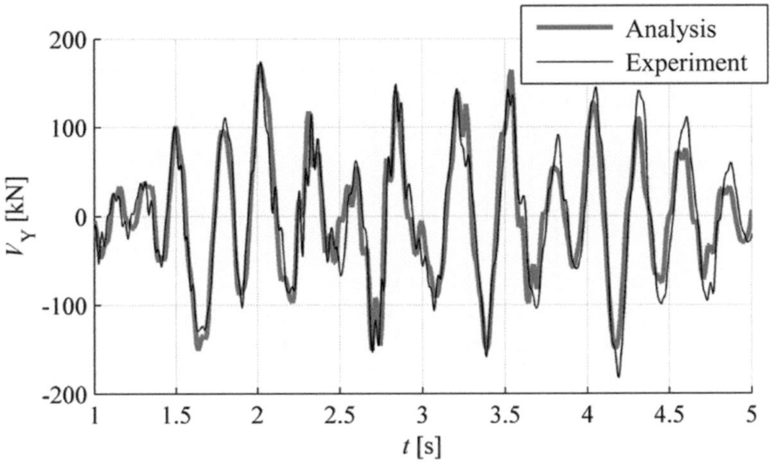


**Fig. 24.13** Base shear response in the direction of the web on the 5th run. Comparison of the numerically and experimentally obtained results

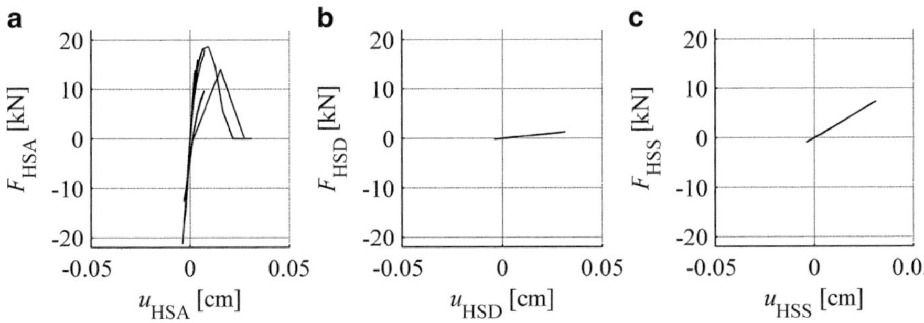


**Fig. 24.14** Response of the shear springs on the 5th run. (**a**) HSA indicates the deterioration of aggregate interlock in one direction. (**b**) the dowel spring HSD and (**c**) the shear reinforcement spring HSS were subsequently activated

No significant inelastic behaviour was detected in the first four runs, neither during the experiment, nor in the numerical model. A moderate inelastic response of the specimen was observed on the 5th run ($a_{g,max,X} = 0.42$ g; $a_{g,max,Y} = 0.73$ g), which was the one before the last. Considerable lifting of the piers due to strong coupling was observed. The vertical bars in the flanges yielded, the cracks in the flanges widened, and shear cracks formed in the webs of both piers in the first storey. The numerical model was able to reproduce the response very well. The nearly perfect match that was obtained in the case of the base shear response history (in the direction of the web) is shown in Fig. 24.13. The behaviour of one of the typical shear springs (the location of the spring is indicated in Fig. 24.1) is analysed in Fig. 24.14. When the web of the pier cracked, the aggregate interlocking mechanism

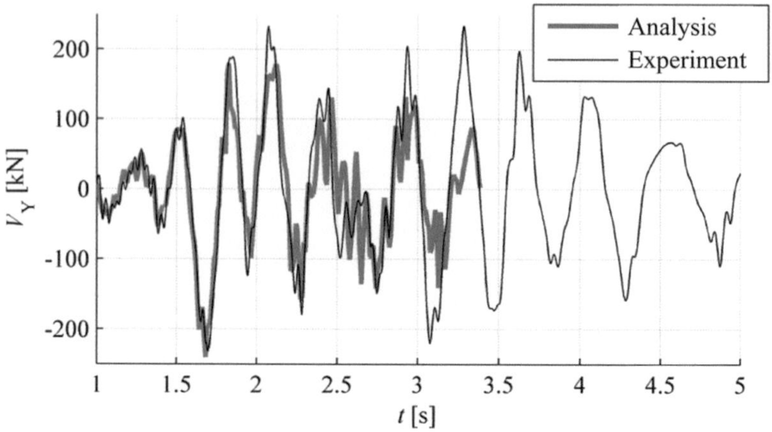


**Fig. 24.15** Base shear response in the direction of the web on the 6th run. Comparison of the numerically and experimentally obtained results

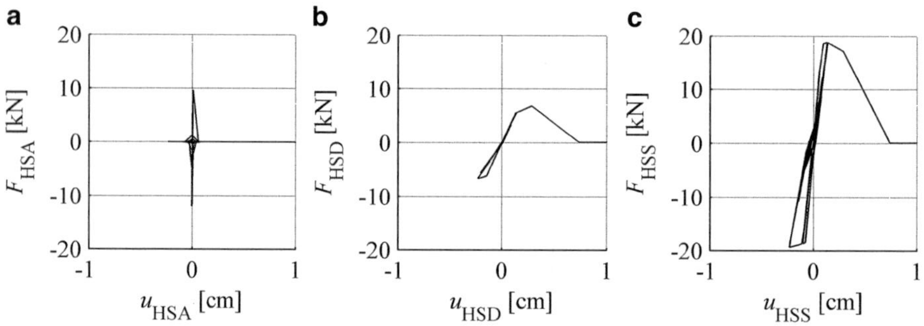


**Fig. 24.16** Response of the shear springs on the 6th run. (**a**) HSA indicates the complete loss of aggregate interlock. (**b**) The dowel mechanism HSD was fully activated and was then completely destroyed. (**c**) The shear reinforcement spring HSS yielded and then soon completely lost resistance (indicating rupture of the very brittle horizontal reinforcement)

was activated (Fig. 24.14a). After this interlocking mechanism had deteriorated, the horizontal reinforcement was activated (Fig. 24.14c). However, it remained elastic. The dowel mechanism, too, was activated, but its contribution was almost negligible (Fig. 24.14b), indicating that the gap within the crack had remained small.

In the last – 6th run both piers failed in shear, and large shear cracks opened up in the flanges of the first floor. This failure was successfully identified and modelled (Figs. 24.15 and 24.16). The aggregate interlocking mechanism, which had considerably deteriorated in the previous run, was completely destroyed (Fig. 24.16a), and the HSS spring indicated the rupture of the very brittle horizontal reinforcement (note the very short yield plateau in Fig. 24.16c), which was actually used in the test specimen. The dowel mechanism was first fully activated, and then failed completely (Fig. 24.16b).

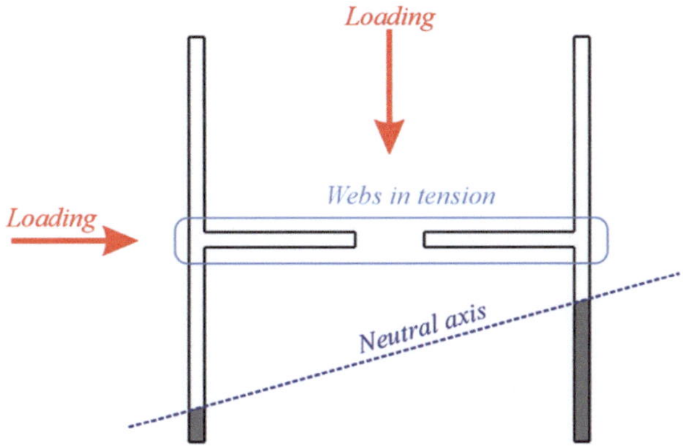


**Fig. 24.17** The identified neutral axis position during the 6th run

The residual resistance of the wall observed in the test (and not numerically verified) can be attributed to the frame action of the flanges and slabs (which was not included in the model). The simultaneous and similar failure of the webs in both piers was attributed to the bi-axial loading. The results of the analysis showed that, at the time of the failure, the webs in both piers were in net tension (Fig. 24.17), which explains the same inclination of the crack in both piers.

## 24.4 Conclusions

During the inelastic response, shear forces in RC structural walls can be much larger than those predicted by equivalent elastic design procedures. The magnification, which is due to overstrength and the effect of higher modes, can be frequently close in size to the seismic force reduction factor.

The shear modification factor $\varepsilon$ proposed in Eurocode 8 for ductile walls (ductility class high) was found to be adequate at the base of walls, providing that it was properly applied (to the base shear contributed by the first mode only) and the upper bound of the modification factor required by the code was increased.

Eurocode 8 assumes constant amplification along the height of the wall, which is conservative at the mid-height and rather unconservative at the top. A variable amplification factor along the height was proposed.

Although the phenomenon of the increasing inelastic shear has been known for a long time, many walls with insufficient shear resistance have been designed in the past and even today. A model is therefore needed to account for inelastic shear behaviour and inelastic shear-flexural interaction.

Such model has been proposed. It is based on the multiple-vertical-line-element macro model. An additional shear spring, which accounts for aggregate interlock, the dowel action, and horizontal reinforcement resistance, is incorporated into each of the vertical springs.

The current characteristics of each component depend on the deformations at the crack (in particular the width of the crack) within the individual strip. The constitutive relations for the individual springs are based on the semi-empirical relations found in the literature. Aggregate interlock is modelled by the Lai-Vecchio model (Vecchio and Lai 2004), whereas the dowel action is modelled by expressions proposed by Dulacska (1972), and Vintzeleou and Tassios (1987). The force-displacement relation for HSS springs is based on the bar-slip model proposed by Elwood and Moehle (2003).

The model successfully simulated the response of a five-storey coupled wall tested on the shaking table under bi-axial excitation. The shear resisting mechanisms within the cracks were adequately modelled up to the tension shear failure of both piers.

## References

Blakeley RWG, Cooney RC, Megget LM (1975) Seismic shear loading at flexural capacity in cantilever wall structures. Bull N Z Natl Soc Earthq Eng 8(4):278–290

Boroschek R, Bonelli P (2014) Lessons from the Chile 2010 earthquake for performance based design and code development. In: Fischinger M, Stojadinović B (ed) Performance-based seismic engineering – vision for an earthquake resilient society. Springer

CEN (2004) Eurocode 8 – design of structures for earthquake resistance. Part 1: general rules, seismic actions and rules for buildings. European standard EN 1998–1, Dec 2004, European Committee for Standardization, Brussels

Chen S, Kabeyasawa T (2000) Modeling of reinforced concrete shear wall for nonlinear analyses. In: Proceedings of the 12th WCEE, New Zealand Society of Earthquake Engineering, Auckland

Dulacska H (1972) Dowel action of reinforcement crossing cracks in concrete. ACI Struct J 69(12):754–757

EERI (2006) News of the membership – blind prediction contest winners. EERI Newslett 40(9):4

Elwood KJ, Moehle JP (2003) Shake table tests and analytical studies on the gravity load collapse of reinforced concrete frames, PEER report 2003/01, University of California, Berkeley, EERI Newsl (2006), 40(9):4

Fischinger M, Isaković T, Kante P (2006) Shaking table response of a thin H-shaped coupled wall. In: Managing risk in earthquake country – 100th anniversary earthquake conference: centennial meeting, San Francisco. Earthquake Engineering Research Institute, Berkeley, CD ROM

Fischinger M, Kramar M, Isaković T, (2008) Using macro elements to predict near-collapse performance of two typical RC building structural systems with lightly reinforced walls and slender precast columns. In: Proceedings of the 14th world conference on earthquake engineering, Beijing, CD ROM

Fischinger M, Rejec K, Isaković T (2010) Seismic behavior of RC structural walls and Eurocode 8 provisions. In: Proceedings of the 9th US National and 10th Canadian conference on earthquake engineering, Oakland, Earthquake Engineering Research Institute, Canadian Association for Earthquake Engineering, Ottawa

Kabeyasawa T (1997) Design of RC shear walls in hybrid wall system. In: Proceedings, Fourth Joint Technical Coordinating Committee, U.S.-Japan cooperative seismic research on composite and hybrid structures, Monterey

Kante P (2005) Seismic vulnerability of RC structural walls (in Slovenian). Ph.D. Dissertation, University of Ljubljana

Keintzel E (1990) Seismic design shear forces in RC cantilever shear wall structures. Eur Earthq Eng 3:7–16
Kim Y, Kabeyasawa T, Matsumori T, Kabeyasawa T (2011) Numerical study of a full-scale six-storey reinforced concrete wall-frame structure tested at E-Defense. Earthq Eng Struct Dyn 41(8):1217–1239. doi:10.1002/eqe.1179
McKenna F, Fenves GL (2007) Open system for earthquake engineering simulation, Pacific Earthquake Engineering Research Center, Berkeley http://opensees.berkeley.edu
Orakcal K, Massone LM, Wallace JW (2006) Analytical modeling of reinforced concrete walls for predicting flexural and coupled shear-flexural responses, PEER report 2006/07. University of California, Berkeley
Rejec K (2011) Inelastic shear behaviour of RC structural walls under seismic conditions (in Slovenian). PhD dissertation, University of Ljubljana
Rejec K, Isaković T, Fischinger M (2012) Seismic shear force magnification in RC cantilever structural walls, designed according to Eurocode 8. Bull Earthq Eng 10(2):567–586. doi:10.1007/s10518-011-9294-y
Vecchio FJ, Collins MP (1986) The modified compressional-field theory for reinforced concrete elements subjected to shear. ACI J 83(22):219–231
Vecchio FJ, Lai D (2004) Crack shear-slip in reinforced concrete elements. J Adv Concr Technol 2(3):289–300
Vintzeleou EN, Tassios TP (1987) Behavior of dowels under cyclic deformations. ACI J 84(1): 18–30
Wallace JW, Moehle JP (1993) Evaluation of ductility and detailing requirements of bearing wall buildings using data from the March 3, 1985, Chile earthquake. Earthq Spectra 9(1):137–156
Wood SL (1991) Performance of reinforced concrete buildings during the 1985 Chile earthquake: implications for the design of structural walls. Earthq Spectra 7(4):607–638

# Chapter 25
# Masonry Buildings, Seismic Performance, and Eurocodes

**Miha Tomaževič**

**Abstract** The paper summarizes the results of recent experimental studies carried out at Slovenian National Building and Civil Engineering Institute and is aimed at providing information for the evaluation of values of design parameters introduced by Eurocodes. On the basis of the results of shaking table tests and taking into consideration damage limitation and displacement capacity of typical masonry buildings, the range of possible values of structural behavior factor has been assessed. As regards the existing buildings, it has been shown that the simultaneous use of confidence and partial material safety factors in seismic resistance verification procedure is too conservative. Different types of units and a series of masonry walls have been tested to propose a measure for sufficient robustness of hollow clay masonry units.

**Keywords** Masonry • Masonry units • Robustness • Masonry walls • Cyclic shear tests • Shear resistance • Shaking table tests • Structural behaviour factor • Existing buildings • Confidence factor

## 25.1 Introduction

Eurocodes (CEN 2004, 2005a, b), European standards for structural design, provide principles and application rules for earthquake resistant design of new masonry buildings as well as specifications to be considered in the case of structural assessment and redesign of existing ones, including masonry buildings of historic importance. Being included into the family of Eurocodes, the design of masonry

M. Tomaževič (✉)
Slovenian National Building and Civil Engineering Institute, Department of Structures, Dimičeva 12, 1000 Ljubljana, Slovenia
e-mail: miha.tomazevic@zag.si

M. Fischinger (ed.), *Performance-Based Seismic Engineering: Vision for an Earthquake Resilient Society*, Geotechnical, Geological and Earthquake Engineering 32,
DOI 10.1007/978-94-017-8875-5_25, 

structures is following the same contemporary design philosophy as the design of any other type of structures. Taking into consideration specific properties of masonry materials, where the probability of not achieving the required mechanical properties is higher than in the case of other structural materials, it is to understand that partial safety factors are higher and load reduction factors are lower than in the case of other structural types.

To make masonry construction competitive, materials of improved strength and thermal insulation properties have been developed and technologies to simplify and speed-up the construction process have been proposed. As preliminary experimental studies indicated, being developed mainly for the intended use in the non-seismic countries, some of these improvements adversely affect the resistance and displacement capacity of masonry structures in seismic situation. Therefore, relevant specifications to limit the use of such materials and construction technologies have been introduced in the Eurocode 8-1, which covers earthquake resistant design (CEN 2004). However, because of the lack of experimental data, the requirements are mainly qualitative, or ranges of values to be used in the design are recommended. It is expected that National Annexes issued by European Union's member states will provide quantitative limitations and narrow the ranges.

To make contribution and provide part of the missing information needed to adequately amend and complement the requirements of the code, experimental research has been conducted also at Slovenian National Building and Civil Engineering Institute (ZAG) in Ljubljana, Slovenia. Some results of this research will be discussed in the following.

## 25.2 Seismic Load Reduction: Behavior Factor q

Most masonry structures belong to the category of structures with regular structural configuration, in the case of which the seismic resistance can be verified by simple equivalent elastic static analysis. Design seismic loads are evaluated on the basis of the response spectra, considering the structure as an equivalent single-degree-of-freedom system, and reducing the ordinates of the elastic spectrum by a factor called "force (strength) reduction factor", which takes into account the displacement and energy dissipation capacity of the structure under consideration, as well as its overstrength. According to Eurocode 8-1, force (strength) reduction factor is called "behavior factor $q$", and is defined as "an approximation of the ratio of the seismic forces that the structure would experience if its response was completely elastic with 5 % viscous damping to the minimum seismic forces that may be used in the design - with a conventional elastic analysis model – still ensuring a satisfactory response of the structure".

Because of many parameters which influence the reduction (Miranda and Bertero 1994), the evaluation of force reduction factors for earthquake resistant design of structures is a relatively complex process. In the case of masonry structures, limited experimental and analytical research has been so far carried out (e.g. Moroni et al.

1992; Da Porto et al. 2009). Consequently, verification of the conventional, code recommended values of factor $q$ for masonry structures, is still needed.

At ZAG, a number of models of different types of masonry buildings have been tested in the last decades. Using advantage of these experiments, the results have been used to verify the Eurocode 8-1 proposed values of behavior factor $q$. It should be noted that the estimation of values has been simplified: mean values of experimental results have been considered and no additional parametric studies have been carried out.

As follows from the definition given in Eurocode 8-1, the behavior factor is the ratio between the seismic force which would develop in an ideal elastic structure ($S_e$) and the design seismic load (design base shear, $S_d$):

$$q = S_e / S_d, \tag{25.1}$$

In the case of seismic resistance verification, each structural element and the structure as a whole should be verified for:

$$E_d \leq R_d, \tag{25.2}$$

where $E_d$ = design value of action effects, i.e. design load in seismic situation, acting on the element and distributed on the element according to the theory of elasticity, and $R_d$ = design resistance of structural element under consideration. Since the elastic analysis methods do not take into consideration the redistribution of seismic loads after yielding of individual structural elements, and the characteristic material strength values are reduced by partial safety factors, $\gamma_M$, the design resistance of structure, $R_d$, is only an approximation, usually much smaller than the actual maximum resistance, $R_{max}$ (or $R_{max,id}$, obtained by idealizing the actual resistance curve with bi-linear relationship, Fig. 25.1). The ratio between the actual maximum resistance, $R_{max}$ (idealized value $R_{max,id}$), and the design resistance of the structure, $R_d$, is called reserve strength (overstrength), ρ:

$$\rho = R_{max} / R_d \, (\text{or } \rho = R_{max,id} / R_{,d}) \, . \tag{25.3}$$

Assuming that design resistance $R_d$ is equal to design seismic load, $S_d$, and substituting design seismic load $S_d$ in Eq. 25.1 by expression for design resistance resulting from Eq. 25.3, behavior factor $q$ can be expressed in terms of actual maximum resistance $R_{max}$ ($R_{max,\,id}$) and overstrength factor ρ as follows:

$$q = \rho S_e / R_{max} \, (\text{or } q = \rho S_e / R_{max,id}) \, . \tag{25.1a}$$

In other words, the behavior factor can be expressed as a product of two parameters, namely factor $S_e/R_{max}$ (or $S_e/R_{max,\,id}$), which is ductility dependent factor (Fajfar 1995), and overstrength factor ρ. According to definition given in Eurocode 8-1, the overstrength is implicitly taken into account in the values of structural behavior factor q, required for seismic resistance verification of various

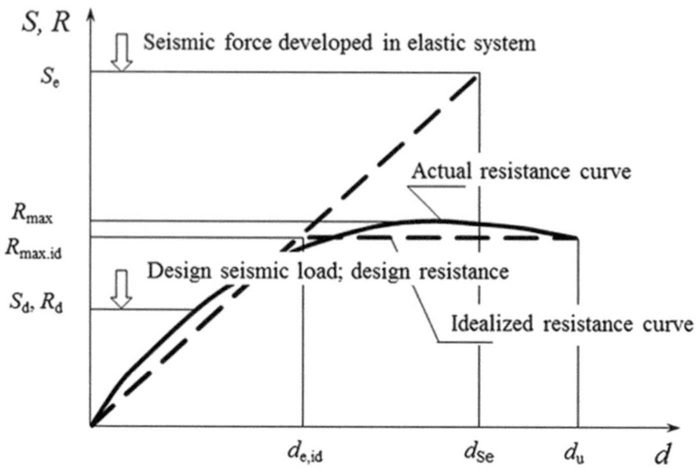


**Fig. 25.1** Comparison of ideal elastic and actual (idealized actual) non-linear behavior of a structure

structural systems. However, no indication whatsoever is given in the code as regards the amount of the expected overstrength, considered when assessing the values of behavior factor proposed for different masonry construction systems. It has been already shown that, if the resistance of a masonry structure is calculated by means of traditional methods of elastic static analysis, significant overstrength can be expected, depending on the structural type and configuration, as well as the method of calculation (Magenes 2006).

The expression for behavior factor, given in Eq. 25.1, is based on the assumption that the maximum displacement response amplitudes of an ideal elastic and equivalent non-elastic rigid structures, subjected to the same ground motion, are equal. Using the same basic definition given in Eurocode, but the assumption of equality of energies (equality of areas below the elastic triangle and actual resistance envelope, Fig. 25.1), the value of behavior factor can be estimated also on the basis of the actual available ductility. For the assessment, the actual resistance curve is idealized as bilinear ideal elastic-ideal plastic relationship. If the assumption of equality of energies is taken into account, structural behavior factor q can be expressed in terms of the global ductility factor of the structure as follows:

$$q_{\mu} = (2\,\mu_u - 1)^{1/2}, \tag{25.4}$$

where $\mu_u = d_u/d_{e,id}$, $d_{e,id}$ = the displacement of the structure at the idealized elastic limit and $d_u$ = the displacement at ultimate limit. In other words, Eq. 25.4 determines the minimum global ductility capacity (ductility demand), which should be ensured if a chosen value of behavior factor $q_{\mu}$ is used for seismic resistance verification. It has been shown (Takada et al. 1988) that the expression is conservative in the ductility range between 1.0 and 10.0, which is the case of all masonry

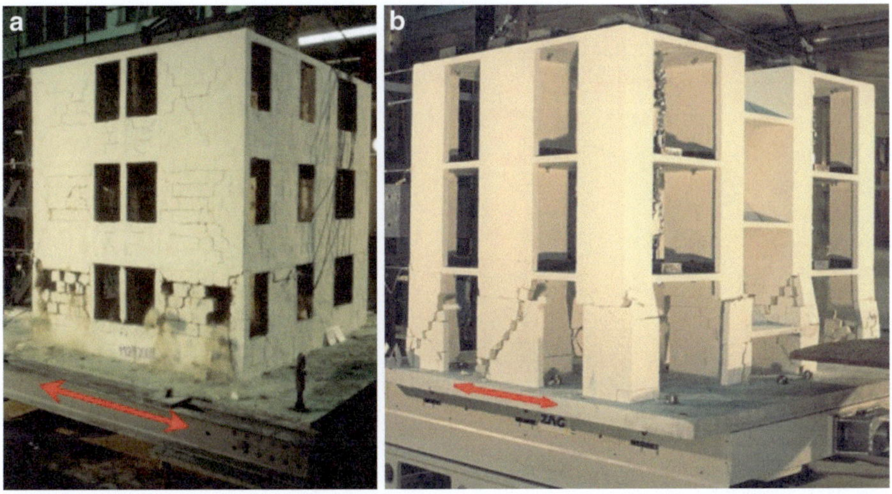


**Fig. 25.2** Confined (**a**) (Tomaževič and Klemenc 1997) and (**b**) plain masonry building models at ultimate state before collapse (Tomaževič and Weiss 2010)

construction systems. The expression does not depend on the vibration period of the structure, however, it has been proposed that a median adjustment factor 1.2 (varying between 1.05 and 1.34) be used for shear buildings (Takada et al. 1988). No adjustment has been considered in this study.

To estimate the possible ranges of values of behavior factor q, the results of a number of shaking table tests of models of confined and unreinforced masonry buildings of different configuration (Fig. 25.2), carried out in the past at ZAG (e.g. Tomaževič and Klemenc 1997; Tomaževič and Weiss 2010; Tomaževič and Gams 2011), have been evaluated.

The measured base shear-first story drift relationships (resistance curves) of each model, idealized as bilinear ideal elastic-ideal plastic relationships, and Eq. 25.4 have been used to evaluate the values (Fig. 25.3). To generalize the data, base shear, *BS*, was expressed nondimensionally in terms of the base shear coefficient, $BSC = BS/W$, where $W$ = the weight of the structure above the base, and interstory drift in terms of rotation angle, $\Phi = d/h$. In the idealization of the experimentally obtained resistance envelope, story drift at the point where the resistance of the structure degrades to 80 % of the maximum, is usually defined as the ultimate [2]. It is assumed that a ductile structure, although severely damaged, will resist such a displacement without risking collapse. However, one of the previous studies (Tomaževič 2007) indicated that the acceptable level of damage to walls (damage Grade 3 according to EMS-98 seismic intensity scale (EMS 1998) occurs at interstory drift equal to approximately $\Phi = 3\Phi_{cr}$, where $\Phi_{cr}$ = interstory drift angle at the damage limit state. Consequently, besides the usual no collapse requirement, expressed by $\Phi_{d,u} = \Phi_{0.8BSC\max}$, damage limitation requirement, expressed by

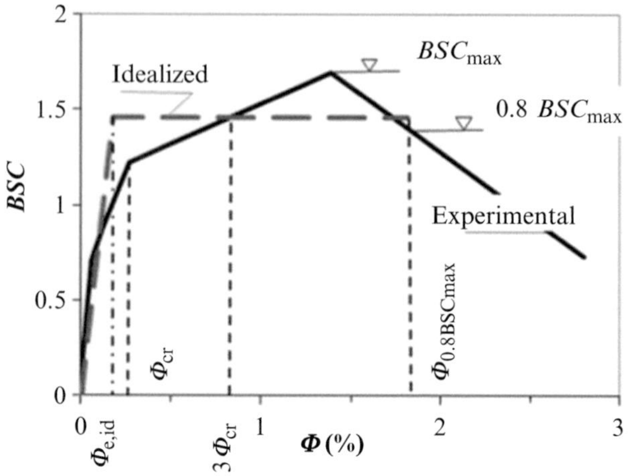


**Fig. 25.3** Evaluation of structural behavior factor on the basis of experimentally obtained resistance curve, after (Tomaževič and Weiss 2010)

**Table 25.1** Values of structural behavior factor $q$, evaluated from the results of model shaking table tests on the basis of ductility and damage limitation requirements (Adopted from Tomaževič and Klemenc 1997; Tomaževič and Weiss 2010; Tomaževič and Gams 2011)

| System | Floors | Materials | $\Phi_{e,id}$ (in %) | $\Phi_{cr}$ (in %) | $\Phi_{0.8BSCmax}$ (in %) | $3\Phi_{cr}$ (in %) | $\mu_u = \frac{3\Phi_{cr}}{\Phi_{e,id}}$ $\mu_u$ | $q$ |
|---|---|---|---|---|---|---|---|---|
| Confined | 3 | Clay block | 0.17 | 0.28 | 2.60 | 0.84 | 4.94 | 2.98 |
| | 3 | | 0.17 | 0.27 | 1.81 | 0.81 | 4.76 | 2.92 |
| | 3 | AAC block | 0.23 | 0.28 | 2.46 | 0.84 | 3.65 | 2.51 |
| | 3 | | 0.36 | 0.48 | 2.27 | 1.44 | 4.00 | 2.65 |
| | 4 | | 0.30 | 0.44 | 2.33 | 1.32 | 4.40 | 2.79 |
| | 3 | Clay block | 0.14 | 0.42 | 1.36 | 1.26 | 9.00 | 4.12 |
| | 3 | | 0.23 | 0.55 | 3.16 | 1.65 | 7.17 | 3.65 |
| Plain | 3 | Calcium silicate | 0.07 | 0.20 | 0.42 | 0.16 | 8.57 | 4.02 |
| | 3 | Clay block | 0.16 | 0.33 | 1.65 | 0.99 | 6.18 | 3.37 |

$\Phi_{d,u} = 3\Phi_{cr}$, is also considered in the evaluation of q factor. The lesser value of the two is taken into account in the evaluation.

Typical results of such evaluation are summarized in Table 25.1. The experimental studies indicated that, in addition to construction system (confined, plain masonry), as assumed by the code, seismic resistance and displacement capacity of masonry buildings depend also on the type of masonry materials and structural configuration. Taking this into account, it can be seen that the values of behavior factor $q$ cannot be assessed by means of only ductility tests of individual structural walls and subsequent numerical analysis of seismic response of the whole structure. Numerical simulation based on the input data obtained by testing of individual walls is usually not enough.

It can be seen, however, that in the case where the values of behavior factor are evaluated by taking into consideration damage limitation requirement ($\Phi_{d,u} = 3\Phi_{cr}$), the differences in the values of factor q, evaluated for different structural types, are not significant. It can be also seen that the values, evaluated in such a way, are well within the range of values, proposed by the code for seismic resistance verification of different masonry construction systems (CEN 2004):

- For unreinforced masonry: $q = 1.5 - 2.5$,
- For confined masonry: $q = 2.0 - 3.0$,
- For reinforced masonry: $q = 2.5 - 3.0$.

As shown by the analysis of experimental results, the values at the upper limit of the Eurocode 8-1 proposed range of values of structural behavior factor $q$ for unreinforced and confined masonry construction systems, i.e. $q = 2.5$ for unreinforced, and $q = 3.0$ for confined masonry structures, are adequate in the case where little or no overstrength, i.e. $\rho \cong 1.0$, is expected. On the basis of the analysis of the expected level of overstrength in the case of typical masonry construction, a proposal has been already made to modify code requirements (Magenes 2006). However, taking into consideration fact the values of factor $q$, presented in this contribution, are mean values, obtained by experiments, additional research and parametric studies are needed to confirm the conclusion and support the proposal.

Pushover methods for calculation of seismic resistance of masonry buildings have been proposed before long (Tomaževič 1978; Magenes et al. 2000). As the correlation between experimentally obtained and calculated results indicates, lateral resistance-displacement characteristics of masonry structures are realistically predicted by these methods (overstrength factor in Eq. 25.1a is $\rho \cong 1.0$). If these methods are used for seismic resistance verification, the values of behavior factor at the upper limit of the Eurocode 8-1 proposed range can be used to determine the design seismic load (resistance demand) needed in seismic resistance verification. However, displacement capacity of the structure should be also verified and compared with displacement (ductility) demand, at the same time.

## 25.3 Material Limitations

### *25.3.1 Masonry Materials and Redesign of Old Buildings*

As recommended by Eurocode 8-3 (CEN 2005b), mean, and not characteristic values of mechanical properties of materials are considered in the redesign of existing buildings, determined either by in-situ testing or by testing specimens, taken from the existing structure, in the laboratory. To obtain the design values, the mean values are reduced with the so called confidence factor, *CF*, the value of which depends on the thoroughness of inspection of the building and reliability of data needed for structural evaluation. In addition, however, the standard requires that partial safety factors for material, $\gamma_M$, be also taken into account to calculate the design values of material strength:

$$f_d = \frac{f}{CF\gamma_M}, \tag{25.5}$$

where $f_d$ = the design value of material strength, $f$ = mean value of material strength, determined by testing, $CF$ = confidence factor, and $\gamma_M$ = partial material safety factor for masonry.

Confidence factor is a function of knowledge level (KL), which, according to Eurocode 8-3 depends on the thoroughness of inspection of the buildings under consideration and the number of tests which have been carried out to assess the state of the structure and material properties. Three knowledge levels are defined in the code. No reduction, i.e. $CF = 1.00$, is needed in the case of the complete structural knowledge (80 % of structural elements inspected, three material specimens tested in each story), $CF = 1.20$ is recommended for the case of the intermediate, and $CF = 1.35$ for the case of the limited knowledge (20 % of elements inspected, one material specimen tested in each story).

According to Eurocode 6-1 (CEN 2005a), the values of partial safety factor for masonry, $\gamma_M$, depend on the factory production control and inspection of works on the site. In normal situation, the values within the range between 1.5 (optimum production control and severe inspection on the site) and 3.0 (no proof regarding the production control and inspection) are considered. In seismic situation, the chosen value can be reduced by 1/3, however in no case $\gamma_M$ should be smaller than 1.5.

Whereas the introduction of confidence factor makes sense and stimulates the amount of inspection and testing of structural materials for the assessment of seismic resistance of old masonry buildings, the reduction by partial material safety factor, as defined in Eurocode 6-1 for the new construction, cannot be accepted. Speaking of structural safety, it is not possible to assess the uncertainties regarding the mechanical properties of old, historic masonry, as is the case of the new construction. At the time of their construction, the modern quality control mechanisms have not yet been established. Consequently, the most unfavorable value should be considered in redesign of such buildings, namely $\gamma_M = 3.0$, which in seismic situation can be reduced to $\gamma_M = 2.0$. This means that only one half of the mean value of masonry strength, obtained by testing the actual materials, can be considered in the redesign – at the best.

As an example of consequence of reduction of experimentally determined values of masonry strength by partial safety factor, $\gamma_M$, on the decision regarding the necessary strengthening measures, the seismic resistance of a series of stone masonry buildings in the region of Posočje, Slovenia, has been analyzed (Table 25.2). The buildings, damaged by the earthquake in 1998 (estimated intensity VIII by European Macroseismic Scale, EMS) (EMS 1998), were strengthened by tying the walls at floor levels and injecting the walls with cementitious grout. Strengthened after the

**Table 25.2** Seismic resistance of typical two-story stone masonry buildings in Posočje, Slovenia, calculated without and with taking into consideration partial safety factor for masonry, $\gamma_M = 2.0$. $CF = 1.0$ in both cases (Adopted from Tomaževič et al. 2000)

| | Wall/floor area (%) | | $f_{t,d} = f_{t,k}$ | | | $f_{t,d} = f_{t,k}/\gamma_M$ | | |
|---|---|---|---|---|---|---|---|---|
| Building type*[a] | x-dir. | y-dir. | $f_{t,d}$ (MPa) | $RC_{dx}$ | $RC_{dy}$ | $f_{t,d}$ (MPa) | $RC_{dx}$ | $RC_{dy}$ |
| a | 10.9 | 6.4 | 0.16 | 0.30 | 0.24 | 0.08 | 0.20 | 0.15 |
| a | 12.0 | 9.1 | 0.16 | 0.28 | 0.28 | 0.08 | 0.21 | 0.19 |
| b | 6.9 | 8.6 | 0.11 | 0.25 | 0.33 | 0.06 | 0.22 | 0.25 |
| b | 12.1 | 11.1 | 0.11 | 0.42 | 0.38 | 0.06 | 0.33 | 0.31 |
| b | 4.7 | 14.6 | 0.11 | 0.19 | 0.47 | 0.06 | 0.17 | 0.33 |
| b | 7.2 | 14.3 | 0.11 | 0.21 | 0.47 | 0.06 | 0.16 | 0.31 |
| b | 15.1 | 13.7 | 0.11 | 0.40 | 0.33 | 0.06 | 0.29 | 0.25 |
| b | 10.5 | 9.5 | 0.11 | 0.39 | 0.29 | 0.06 | 0.31 | 0.25 |
| b | 10.5 | 9.9 | 0.11 | 0.31 | 0.34 | 0.06 | 0.23 | 0.26 |
| b | 10.3 | 10.2 | 0.11 | 0.28 | 0.35 | 0.06 | 0.22 | 0.26 |
| b | 11.9 | 10.3 | 0.11 | 0.29 | 0.34 | 0.06 | 0.28 | 0.29 |
| b | 9.8 | 10.9 | 0.11 | 0.32 | 0.34 | 0.06 | 0.23 | 0.26 |
| b | 8.8 | 8.33 | 0.11 | 0.31 | 0.33 | 0.06 | 0.23 | 0.27 |
| b | 10.6 | 12.0 | 0.11 | 0.35 | 0.36 | 0.06 | 0.28 | 0.28 |
| b | 9.7 | 12.0 | 0.11 | 0.34 | 0.47 | 0.06 | 0.27 | 0.34 |
| b | 7.9 | 4.2 | 0.11 | 0.35 | 0.21 | 0.06 | 0.26 | 0.19 |

[a]Building type a: school, building type b: residential. $BSC_d = 0.375$

earthquake of 1998, they were subjected to another earthquake of the same intensity in 2004. Most buildings remained undamaged in 2004, some of them suffered only minor damage.

In the analysis, shear mechanism model and experimentally obtained values of tensile strength of the strengthened, cement grouted masonry, $f_t$, have been taken into account (Tomaževič et al. 2000). Instead of mean, characteristic values, $f_{t,k}$, of the tensile strength have been taken into account. For easier comparison with the code requested value of the design base shear coefficient, $BSC_d$, the resistance of analyzed buildings is expressed in a nondimensional form of the design seismic resistance coefficient, $RC_d = R_d/W$, where $W$ = the weight of the building.

According to seismic hazard map of Slovenia, design acceleration $a_g = 0.225$ g should be considered in the case of the seismic resistance verification of building structures in the area, built on the firm soil (soil factor $S = 1.0$). The equivalent value of the design base shear, expressed in the nondimensional form of the design base shear coefficient, is $BSC_d = Sa_g\ \beta/q = 1.0{*}0.225{*}2.5/1.5 = 0.375$, where $\beta = 2.5$ = spectral amplification factor for the flat part of elastic response spectra, where typical vibration periods of masonry buildings are located. In the particular case studied, actual ground acceleration records, obtained in 2004 on river deposit, indicated the possibility of even higher seismic loads (Fig. 25.4). However, nonlinear dynamic response analysis of several buildings to recorded ground motion, where the same mechanical properties of stone masonry have been

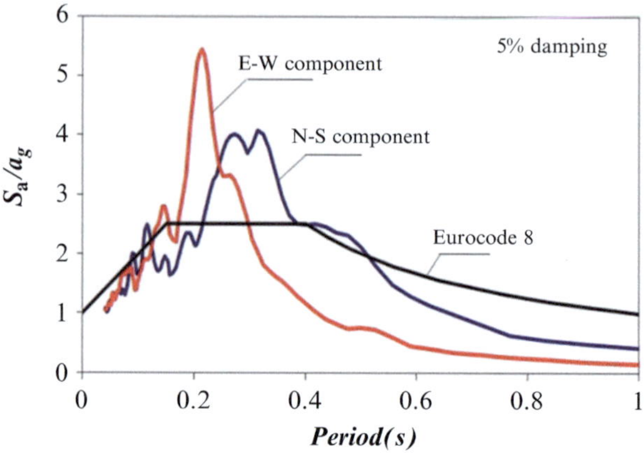


**Fig. 25.4** Eurocode 8 response spectrum in comparison with the response spectra of 2004 Posočje earthquake for 5 % damping (earthquake data obtained from IKPIR)

taken into account as in the case of this analysis, confirmed the realistic assumption of the code as regards the design earthquake (Tomaževič et al. 2000).

As the damage observation analysis after the earthquake indicated, the actual resistance of the analyzed buildings was sufficient to prevent the damage, although the resistance values, calculated by taking into account the characteristic tensile strength of masonry, do not fully comply with the requirements of the code. In the case, where the characteristic values of masonry strength were reduced by partial safety factor, $\gamma_M$, however, the predicted resistance of the analyzed buildings is much lower than required. The results of such analysis would indicate that serious strengthening measures should have been be applied to the analyzed buildings in order to ensure adequate seismic behavior and prevent damage. However, as the actual situation proved, the additional measures were not necessary.

On the basis of this and similar past experiences, it can be seen that, in the case of redesign of old masonry buildings, there is no reason that besides the confidence factors, *CF*, the partial safety factors for masonry, $\gamma_M$, be also considered in determining the design values. As regards the code recommended values of confidence factors, *CF*, belonging to each knowledge level, past experiences show that the code recommended values are too optimistic. The value of $CF = 1.0$ is acceptable, if mechanical properties of masonry are determined either by in-situ tests or in the laboratory by testing specimens, taken from the building under consideration. In such a case, at least one specimen of the specific masonry type should be tested in the building and the composition of the masonry should be verified by removing plaster at least in one location in each story. $CF = 1.35$ should be used if the mechanical properties are obtained by testing at least one specimen in the cluster of buildings of the same typology. Identification of a given type of stone-masonry is carried out by removing plaster and opening the walls at least in one location in each story of the building under consideration. In the case that no tests but identification inspection only is carried out, the recommended value is

$CF = 1.7$. In this case, however, the values of mechanical properties for the specific masonry type are taken from the literature, corresponding to the masonry type under consideration.

## *25.3.2 Robustness of Masonry Units*

Nowadays, solid bricks are replaced by hollow units, the shape and materials of which have been designed to meet the demanding energy saving criteria for buildings with minimum additional thermal insulation layers. Clay units are usually made of specially developed porous clayey materials. They are shaped to have a large percentage of uniformly distributed holes, which requires thin shells and webs. Whereas the load bearing capacity of masonry walls made of such units is adequate for gravity loads, experimental investigations indicated, that the units exhibit local brittle failure if the walls are subjected to a combination of high level of compressive stresses and in-plane horizontal seismic loads at the same time (Tomaževič et al. 2006). As the result of the lack of robustness of masonry units, the behavior of such walls is even less ductile than the behavior of walls built with traditional bricks and mortar (Fig. 25.5a). If masonry is reinforced with steel reinforcement, placed between the units in different ways, and the units are brittle, they are not able to carry the additional compression and shear needed to develop the tension capacity of reinforcement (Fig. 25.5b). Consequently, the design equations, developed on the assumption of solid behavior of units and adequate bond between units, mortar, and reinforcement, do not reflect the actual situation (Tomaževič et al. 2006). As a rule, they are misleading, because they are too optimistic as regards both, lateral load bearing and displacement capacity of reinforced masonry walls. Lack of robustness of the units is the main reason for unreliability.

To avoid local brittle failure of hollow units, requirements are given in most national seismic codes which limit the amount of holes and minimum thickness of shells and webs of the units used for the construction of masonry buildings in seismic zones. In this regard, requirements have been also specified in the draft version of Eurocode 8 (CEN 1994). The void ratio was limited to 50 % of the volume of the units, and the minimum allowable thickness of shells and webs of the units to 15 mm. Since such units no longer exist on the market, the present standard (CEN 2004) requires only that "masonry units should have sufficient robustness in order to prevent local brittle failure." The decision on how to meet the requirement is left to the National Annexes, which "may select the type of masonry units from EN 1996–1, Table 3.1 that satisfy this requirement." However, the decision is not simple, because according to this table, the units to be selected from, are the units, where the volume of holes varies from 25 to 55 % of the gross volume of the unit, and the thickness of shells and webs is not less than 8 mm and 5 mm, respectively.

As an attempt to propose such criteria, the influence of shape of the units on the parameters of seismic resistance of the wall has been investigated at ZAG. For the study, six different types of hollow clay blocks, currently available on the market

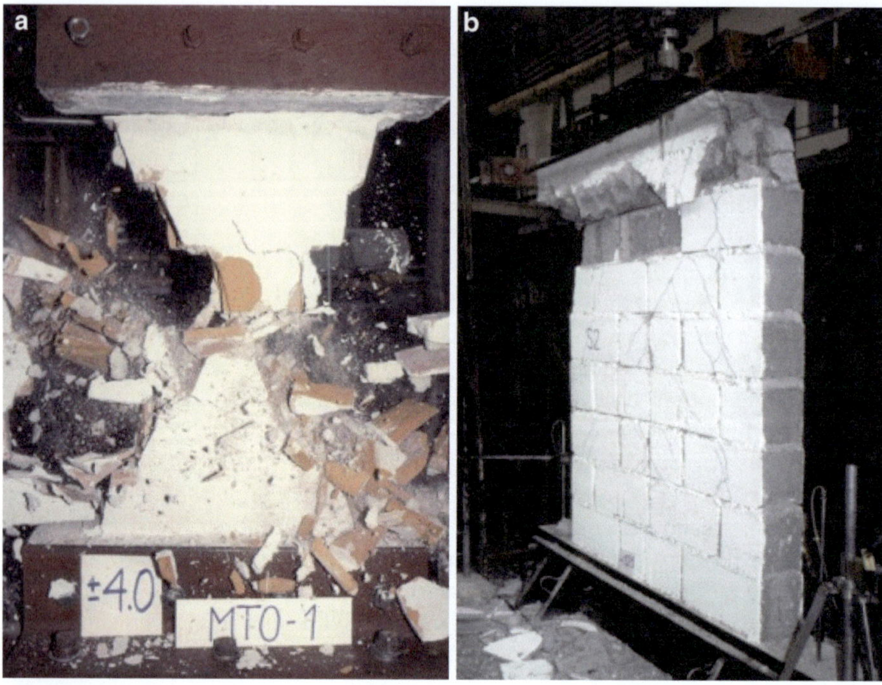


**Fig. 25.5** Brittle shear failure of a highly stressed unreinforced masonry wall (**a**) and (**b**) brittle failure of a complete course of concrete hollow blocks which prevented the activation of reinforcement

**Fig. 25.6** Typical hollow clay blocks tested in the experimental campaign (After Tomaževič and Weiss 2012)

(Fig. 25.6), have been selected. Since all brick producers aim at the same goals, the materials, shape and dimensions of units do not vary substantially. The experimental program consisted of two phases. In the first phase of testing, the mechanical and geometrical characteristics of all types of units have been determined by standardized testing procedures. Then, a series of specific tests has been carried out, by means of which the stress state and failure mechanism of a single unit have been simulated for the case in which the units are part of a shear wall, subjected to

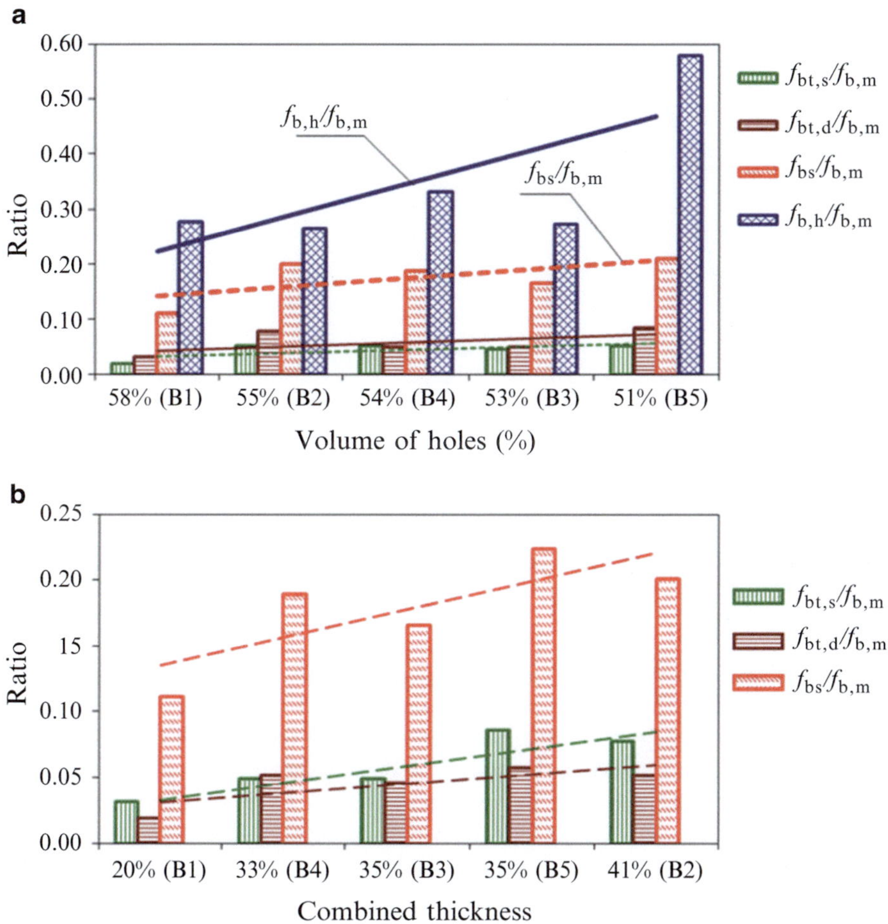


**Fig. 25.7** Correlation between strength parameters and the volume of holes, (**a**) and (**b**) correlation between strength parameters and combined thickness of shells and webs (After Tomaževič and Weiss 2012)

a combination of vertical load and shear in the case of an earthquake. In the second phase, a series of 28 walls, made with all six types of masonry units, have been tested by subjecting them to a combination of constant vertical load and cyclic shear. Two levels of precompression have been chosen to simulate the possible ranges of working stress levels in masonry walls due to vertical loads in actual structures.

Since the quality of masonry blocks is declared by their compressive strength, the correlation between the mean compressive strength of units, normal to the bed joints, $f_{b,m}$, on the one hand, and compressive strength, parallel to the bed joints ($f_{b,h}$), as well as the diagonal tesnile ($f_{bt,d}$), splitting tensile ($f_{bt,s}$) and shear strength ($f_{bs}$), on the other, has been analyzed. The results are shown in Figs. 25.7a, b. In Fig. 25.7a, the ratio between the various strength parameters and compressive

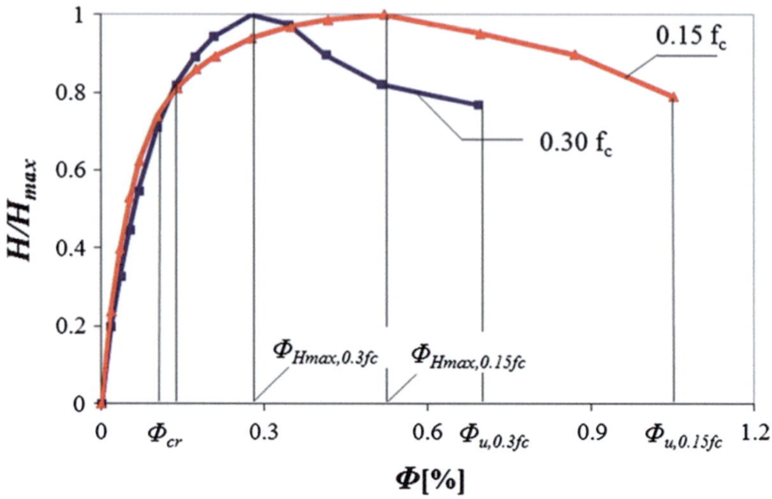


**Fig. 25.8** Comparison of average lateral load – rotation angle envelope curves, obtained by testing the walls at high and low precompression ratio. Lateral resistance is normalized with regard to maximum (After Tomaževič and Weiss 2012)

strength of tested hollow blocks is plotted against the volume of holes. As can be seen, only the ratio between the compressive strength, parallel to the bed joints and compressive strength of units, indicated a trend of increase with the decreased volume of holes. In Fig. 25.7b, however, the dependence of the analyzed strength parameters, normalized by the compressive strength of units, on the combined thickness of shells and webs is shown, which confirms the expectations that the resistance of units to tension and shear depends not only on the quality of materials, but also on the units' shape and amount of holes.

However, the cyclic shear tests of the walls did not confirm this conclusion. All walls failed in shear, as expected, but no difference in resistance and displacement capacity, which could have been attributed to different shapes of the units, has been observed. The resistance envelopes of all walls, tested at the same precompression ratio, were similar in terms of both, resistance and displacements. The coefficient of variation of resistance values at individual displacement amplitudes was 6.5 % for high and 8.4 % for low level of precompression. As can be seen in Fig. 25.8, where the resistance envelopes, averaged for each precompression level, are presented in a nondimensional form, the precompression ratio determined the displacement capacity of the walls. In the figure, the resistance of the walls has been normalized with regard to the maximum and displacements were expressed in terms of drift angle, $\Phi = d/h$ (in %).

Against expectations, the shape and mechanical properties of individual units did not affect the seismic behavior of walls. When subjected to a combination of vertical and cyclic horizontal loads, the working compressive stress/compressive strength of masonry ratio, turned out to be the governing parameter. The behavior of units,

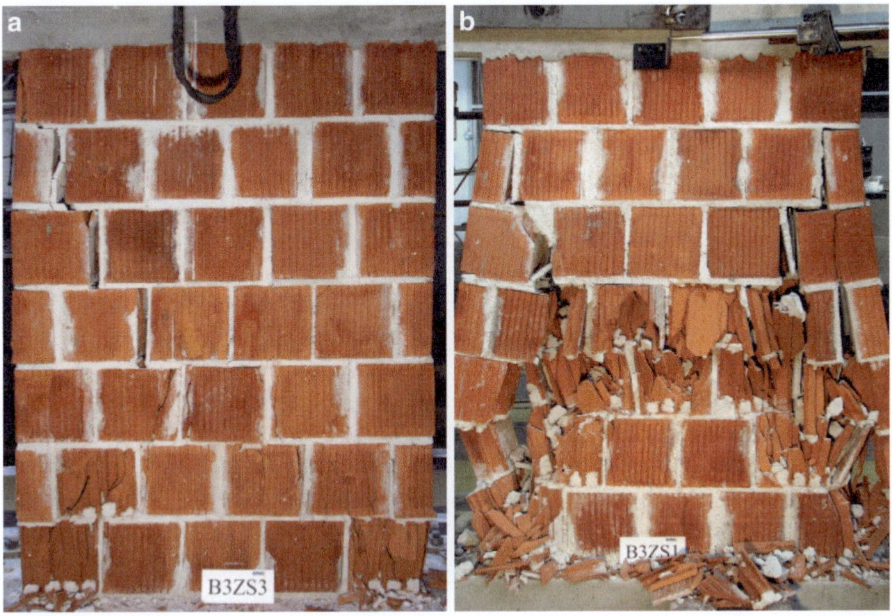


**Fig. 25.9** Damage to the wall at ultimate state at low (**a**) and (**b**) high precompression ratio (After Tomaževič and Weiss 2012)

which exhibited monolithic behavior at low level of precompression, became brittle when subjected to a combination of higher level of precompression and cyclic in-plane shear loads (Fig. 25.9).

In other words, the study indicated that the working stress level in structural walls is the governing parameter. Same units will behave adequately at low, but will exhibit brittle behavior at high level of precompression. Taking into account the same damage limitation criterion as in the case of assessment of behavior factor $q$, it can be proposed that hollow clay units with sectional properties (volume of holes, thickness of shells and webs) at the upper limit of range of values, specified for Group 2 units in Eurocode 6 (CEN 2005a), can be used in seismic zones, if the precompression ratio does not exceed the range of 0.15–0.25. In the case of quasi monolithic units of Group 1, however, the recommended highest precompression ratio can be increased to 0.20–0.25.

## 25.4 Conclusions

On the basis of recent experimental research in seismic behavior of masonry walls and models of buildings, an attempt has been made to quantify the design parameters, for which, due to the lack of experimental data, qualitative requirements are given in Eurocode 8 or relatively conservative values of design parameters are recommended.

Taking into consideration damage limitation criteria and actual displacement capacity of masonry walls and structures as a basis for evaluation, it has been found that the values at the upper limit of the range of values of structural behavior factor $q$, recommended in Eurocode 8-1 for unreinforced and confined masonry buildings, can be used if equivalent static analysis is used for seismic resistance verification. However, taking into consideration fact that the recommendation is based on mean values, obtained by experiments, additional research and parametric studies are needed to confirm the proposal;

In the case of assessment of seismic resistance of existing, historic masonry buildings, there is no need that the experimentally determined properties of masonry materials be reduced by partial material safety factor for masonry, as defined in Eurocode 6-1 for the new construction. The reduction by confidence factor with suggested modification of values for different knowledge levels will provide reliable information;

In order to avoid local brittle failure of contemporary hollow clay masonry units and ensure adequate displacement capacity of structural walls, the design compressive stresses in the walls should be limited to 15–20 % of the compressive strength of masonry.

**Acknowledgment** The paper discusses the results of recent research, which the author and his colleagues carried out at Slovenian National Building and Civil Engineering Institute within the framework of research projects, financed by Slovenian Research Agency. Further details can be found in the referenced, already published works.

## References

CEN (1994) Eurocode 8: Design provisions for earthquake resistance of structures. Part 1–1: general rules for buildings – seismic actions and general requirements for structures. ENV 1998-1-1:1994. CEN, Brussels

CEN (2004) Eurocode 8: design of structures for earthquake resistance, Part 1: general rules, seismic actions and rules for buildings. EN 1998–1:2004. CEN, Brussels

CEN (2005a) Eurocode 6: design of masonry structures – Part 1–1: common rules for reinforced and unreinforced masonry structures. EN 1996-1-1:2005. CEN, Brussels

CEN (2005b) Eurocode 8: design of structures for earthquake resistance, Part 3: assessment and retrofitting of buildings. EN 1998–3:2005. CEN, Brussels

Da Porto F, Grendene M, Modena C (2009) Estimation of load reduction factors for clay masonry walls. Earthq Eng Struct Dyn 38(10):1155–1174

EMS European Macroseismic Scale 1998 (1998) Grünthal G (ed) Proceedings of European Centre for Geodynamics and Seismology, Vol. 15, European Council, Luxemburg

Fajfar P (1995) Design spectra for the new generation of codes: Eurocode 8 achieves the half-way mark. In: Proceedings 10th European conference on earthquake engineering. Balkema, Rotterdam, pp 2969–2974

Magenes G (2006) Masonry building design in seismic areas: recent experiences and prospects from a European point of view. The First European conference on earthquake engineering and seismology. CD-ROM, Geneva, Keynote Address K9: paper 4009

Magenes G, Bolognini D, Braggio C (2000) Metodi simplificati per l'analisi sismica non lineare de edifici in muratura. CNR-Gruppo Nazionale per la Difesa dai Terremoti, Rome

Miranda E, Bertero V (1994) Evaluation of strength reduction factors for earthquake-resistant design. Earthq Spectra 10(2):357–379

Moroni MO, Astroza M, Gomez J (1992) Seismic force reduction factors for masonry buildings. In: Proceedings 10th world conference on earthquake engineering, vol 8. Balkema, Rotterdam, pp 4521—4524

Takada T, Hwang HHM, Shinozuka M (1988) Response modification factor for nonlinear response spectra. In: Proceedings, 9th world conference on earthquake engineering, vol V. Japan Association for Earthquake Disaster Prevention, Maruzen, Tokyo, pp 129–134

Tomaževič M (1978) Improvement of computer program POR. Report ZRMK-IK, Ljubljana (in Slovene)

Tomaževič M (2007) Damage as a measure for earthquake-resistant design of masonry structures: Slovenian experience. Can J Civil Eng 34(11):1403–1412

Tomaževič M, Gams M (2011) Shaking table study and modelling of seismic behaviour of confined AAC masonry buildings. Bull Earthq Eng 10(3):863–893

Tomaževič M, Klemenc I (1997) Verification of seismic resistance of confined masonry buildings. Earthq Eng Struct Dyn 26(10):1073–1088

Tomaževič M, Weiss P (2010) Displacement capacity of masonry buildings as a basis for the assessment of behavior factor: and experimental study. Bull Earthq Eng 8(6):1267–1294

Tomaževič M, Weiss P (2012) Robustness as a criterion for use of hollow clay masonry units in seismic zones: an attempt to propose the measure. Mater Struct 45(4):541–559

Tomaževič M, Klemenc I, Lutman M (2000) Strengthening of existing stone-masonry houses: lessons from the earthquake of Bovec of April 12, 1998. Eur Earthq Eng 14(1):13–22

Tomaževič M, Lutman M, Bosiljkov V (2006) Robustness of hollow clay masonry units and seismic behaviour of masonry walls. Constr Build Mater 20:1028–1039

# Part VI
# Vision in Americas

# Chapter 26
# Performance-Based Earthquake Engineering in the U.S.: A Case Study for Tall Buildings

**Jack Moehle**

**Abstract** Two influential developments in performance-based earthquake engineering in the U.S. are (1) development of the Tall Buildings Initiative *Guidelines for Performance-based Seismic Design of Tall Buildings* and (2) development of the ATC 58 *Guidelines for Seismic Performance Assessment of Buildings*. The content and methods of the two guidelines are summarized. A case-study project uses the Tall Buildings Guidelines to develop tall building conceptual designs for a site in Los Angeles, California, and then uses the ATC 58 Guidelines to explore the performance implications in terms of initial cost and future repair costs considering anticipated future earthquakes. The conceptual designs are done both using a building code prescriptive method and the performance-based method. Earthquake ground motions considered representative of different hazard levels for the site are imposed on an analytical model accounting for nonlinear response characteristics, leading to statistics on engineering demand parameters and associated repair costs. The study identifies apparent shortcomings in the code prescriptive methods as well as benefits associated with the performance-based methods.

**Keywords** RC buildings • Numerical models • Nonlinear analysis • Performance-based earthquake engineering • Repair cost • Building codes

## 26.1 Introduction

Performance-based seismic design methods in the U.S. originated as a practical and effective means to mitigate the seismic risks posed by existing buildings and were later extended to permit development of new buildings designed outside

---

J. Moehle (✉)
Department of Civil and Environmental Engineering, University of California, Berkeley, 775 Davis Hall, Berkeley, CA 94549-1710, USA
e-mail: Moehle@berkeley.edu

M. Fischinger (ed.), *Performance-Based Seismic Engineering: Vision for an Earthquake Resilient Society*, Geotechnical, Geological and Earthquake Engineering 32,
DOI 10.1007/978-94-017-8875-5_26, 

the prescriptive limits of the building code (Moehle et al. 2011a). This practice has become particularly prevalent in the design of very tall buildings in the Western U.S. Initially, engineers adopted ad hoc procedures for performance-based seismic design of tall buildings. Later, documents including SEAONC (2007), LATBSDC (2008), and TBI (2010) formalized these procedures. The earliest of these guidelines (SEAONC 2007) essentially adopted the building code procedures, including strength and drift checks for the Design Earthquake (ASCE 7 2005), but permitted some building code exceptions if adequate performance was demonstrated by nonlinear dynamic analysis. Experience and research (e.g., ATC-72 2010) led to evolution of the procedures with time. In the most recent of these guidelines (TBI 2010), the requirement to check strength and drift for the Design Earthquake is eliminated. Instead, building acceptability is judged based on demonstrated performance for Serviceability Level and Maximum Considered Level seismic demands.

In the aforementioned design guidelines, performance is measured by engineering demand parameters (EDPs) such as building drift, building stability, and local component demands. For a tall building stakeholder, however, performance may be better represented in terms of initial cost and the cost to repair damage from postulated future earthquakes. Advances in defining useful performance metrics, data, models, and analytical tools (e.g., Taghavi and Miranda 2003; Moehle and Deierlein 2004; Yang et al. 2009; ATC-58 2012; Porter et al. 2010;) have enabled practical assessment of expected future repair costs, now embodied in the ATC 58 *Guidelines for Seismic Performance Assessment of Buildings* (ATC-58 2012). Application of such methods to tall buildings designed by alternative methods can provide insight into the performance potential of tall buildings in general, as well as the effectiveness of the alternative design approaches.

The present study examines the design and expected performance of three tall building configurations designed by alternative methods. The study is part of a larger study (Moehle et al. 2011b) that considered alternative design strategies. This paper presents an overview of the design and assessment approaches, results of the case study designs, results of nonlinear dynamic analyses, and financial implications of the designs. The study illustrates a broadly applicable approach for comparing alternative designs in terms of engineering performance and financial measures.

## 26.2 The TBI Guidelines

The TBI (Tall Buildings Initiative) Guidelines present an overview of the recommended design and review process for tall buildings in regions of high seismicity, including detailed procedures to design for serviceability (Serviceability Level) and safety (Maximum Considered Earthquake Level). Serviceability Level seismic demands are obtained from modal response spectrum analysis of a three-dimensional model using a 2.5-percent-damped uniform hazard response spectrum having 43-year return period, with inter-story drift ratios limited to 0.005 and maximum component forces limited to 1.5 times conventional design strengths. In effect, the Serviceability Level check establishes the minimum required building strength, which replaces the strength requirement of the prescriptive building code. The Maximum Considered Earthquake Level check uses nonlinear dynamic analysis of a three-dimensional analytical model subjected to two horizontal components of seven earthquake ground motions scaled to the uniform hazard spectrum representing the Maximum Considered Earthquake Level (ASCE 7 2005). For each pair of horizontal ground motions, the maximum inter-story drift is obtained in each story. For each story, the mean and maximum of the seven drift values are limited to 0.03 and 0.045 (transient), and 0.01 and 0.015 (residual). The Maximum Considered Earthquake Level check intends to demonstrate structural stability during a rare event. Therefore, yielding members are required to respond within limits that can be modeled reliably, and overall strength degradation of the structural system is limited. Provisions also limit the force demands in components with limited ductility (in effect, capacity-protected components). Details of the criteria are found in (TBI 2010).

## 26.3 The ATC 58 Guidelines

The ATC 58 Guidelines for Seismic Performance Assessment of Buildings (ATC-58 2012) describes a general methodology and recommended procedures to assess the probable earthquake performance of individual buildings based on their unique site, structural, nonstructural and occupancy characteristics. Performance measures include potential casualties, repair and replacement cost and schedule, and potential loss of use due to unsafe conditions. The methodology and procedures are applicable to performance-based design of new buildings, and performance assessment and seismic upgrade of existing buildings. The methodology involves many steps, including assembly of a building performance model that defines the building including occupancy; definition of earthquake hazards; analysis of building response; development of a collapse fragility; and various performance calculations. The buildings included in the present study are deemed rugged against collapse for reasonable ground motions. Therefore, in this study we skip the collapse fragility procedure of the methodology. Furthermore, we consider only losses associated with repair cost.

ATC-58 uses a Monte Carlo procedure to explore variability in building performance outcomes given earthquake shaking intensity (Fig. 26.1). First, the building is defined in terms of geometry, occupancy, and performance groups, that is, groupings of similar elements in each story whose performance is likely to affect the overall building performance outcome. An analytical structural model of the building is subjected to a single ground motion to identify maximum values of engineering demand parameters such as inter-story displacement or absolute acceleration. This process is repeated several times to establish expected values and variability of the engineering demand parameters as a function of ground shaking intensity. A

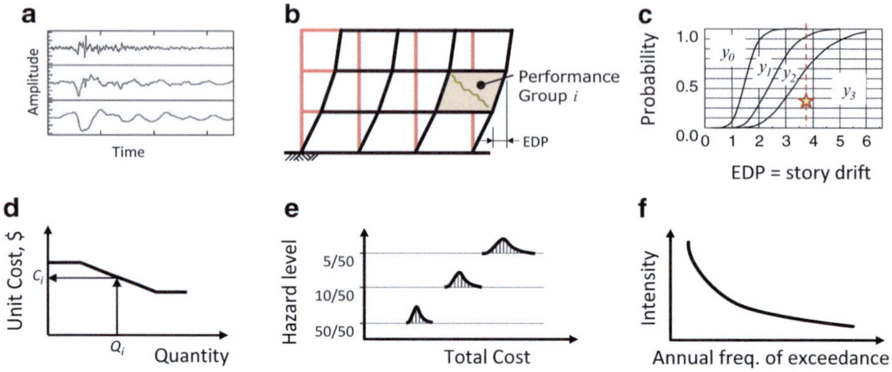


**Fig. 26.1** Capital loss calculation. (**a**) Subject building to ground motions. (**b**) Record engineering demand parameters on damageable performance groups. (**c**) Use random number generator to enter fragility relations and determine damage state. (**d**) Identify repair quantities and costs. (**e**) Repeat many times at each of several hazard levels. (**f**) Integrate with the seismic hazard curve to generate loss measures of interest

statistical technique then is used to generate large numbers of "realizations," each realization representing a plausible response outcome, with the statistics of all the realizations matching that of the smaller set of earthquake response analyses. For each realization, the damage states and repair actions of performance groups are selected based on pre-defined fragility relations. Total building repair cost is then determined based on the total of the building repair quantities and repair actions. By repeating the process a large number of times, the statistics of repair costs are established. The repair costs for each shaking intensity then can be integrated with the seismic hazard curve to establish annual frequencies of exceeding specific repair costs. See ATC-58 for additional details.

## 26.4 Site Seismic Hazard and Representative Ground Motions

An aim of the study was to identify performance characteristics of prototypical tall buildings exposed to the seismic hazard typical of coastal California. Design and subsequent performance analyses required identification of a hypothetical building site. The selected site is in downtown Los Angeles, California (Longitude = −118.25 and Latitude = 34.05). The NEHRP soil site class is C (VS30 = 360 m/s). The site is close to several known faults, including the Puente Hills and San Andreas faults, respectively at closest distances of 1.5 and 56 km from the building site. Figure 26.2 shows a deaggregation of the seismic hazard at 2 and 70 % in 50 years hazard levels (2,475 and 43 years return period, respectively) for vibration periods of 3 and 5 s. The deaggregation results show that for long-period

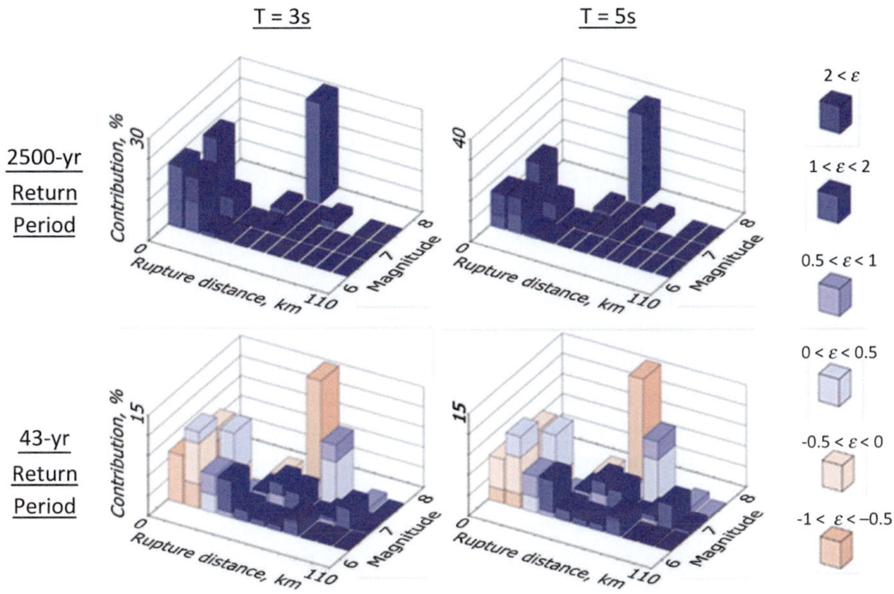


**Fig. 26.2** PSHA deaggregation for the building site for dominant periods of 3 and 5 s; and return periods of 2475 and 43 years (Source: OpenSHA)

structures, T > 3 s, low probability seismic hazard is dominated by two types of earthquakes: (1) Relatively-large-magnitude earthquakes at short distance, and (2) extremely-large-magnitude earthquakes at long distance. Higher probability seismic hazard is dominated by a mixture of seismic events. In the figure, $\varepsilon$ is the number of standard deviations by which an observed logarithmic spectral acceleration differs from the mean logarithmic spectral acceleration of a ground-motion attenuation equation.

These hazard characteristics are used as a basis for design and for selection and modification of ground motions for the study. Code-based designs consider the Design Basis Earthquake (DBE), defined as two-thirds of the Maximum Considered Earthquake (MCE). Performance-based designs consider seismic hazard at the 43-year return period and at the Maximum Considered Earthquake level according to ASCE 7–05. After the buildings were designed, performance was to be assessed at five hazard levels designated SLE25, SLE43, DBE, MCE, and OVE, corresponding to earthquake shaking hazard with return periods of 25, 43, 475, 2,475, and 4,975 years, respectively. The 4975-year level is an extremely rare event usually not considered in design and analysis of buildings in California.

Various options for selecting and scaling earthquake ground motions for response history analysis exist (e.g., ATC-82 2011). Performance-based designs required ground motions scaled to the Maximum Considered Earthquake Level. For this purpose, recorded earthquake ground motions were spectrum-matched by modifying them in the time domain such that the resulting 5 %-damped linear response spectrum closely matched the Maximum Considered Earthquake response spectrum defined by ASCE 7–05 over a period range 0.2–1.5 $T$, where $T$ is the calculated first-mode vibration period. Spectrum matching is permitted by the TBI Guidelines, and was selected because it was judged the most efficient approach for obtaining expected values for design.

For performance assessment of the designed buildings, recorded earthquake ground motions were amplitude-scaled to provide a best fit with site-specific uniform hazard spectra for the various hazard levels. Amplitude scaling was selected to retain more of the record-to-record variability than would be obtained by spectrum matching. Uniform hazard spectra were selected as the targets rather than conditional mean spectra (Baker and Cornell 2006) because of three considerations: (1) Different engineering demand parameters (e.g., drift, acceleration, shear) are affected differently by multiple response modes, such that multiple conditional mean spectra would have been required; (2) The study involved multiple buildings designed by multiple methods, such that there was no single vibration period to which the conditional mean spectrum could be assigned; and (3) The site was strongly affected by multiple types of seismic events. Results obtained using the uniform hazard spectra are likely to exceed those obtained using conditional mean spectra, and likely are conservative. The appropriate use of conditional mean spectra for tall buildings in complex seismic environments continues as an important subject for research at the time of this writing.

For building performance assessments, 15 pairs of horizontal ground motion records were selected from a sub-set of the PEER NGA database (PEER 2005) for each hazard level. The subset database included ground motion records that excluded aftershocks, had a maximum site-to-source distance of 100 km, and were recorded from soil profile with average shear-wave velocity in top 30-m of soil between 180 and 1,200 m/s. The selected records had a long-period filter cutoff significantly longer than the expected fundamental period of the structure. The selected ground motion records were then amplitude-scaled to the target spectrum at each hazard level. The records were limited to have a maximum scaling factor of 5. Although tighter selection and scaling criteria were desirable, this proved not attainable given the available records and the period range for the buildings under consideration.

Available recorded ground motions were insufficient to populate the data set required for the extremely rare OVE hazard level. Thus, eight pairs of amplitude-scaled recorded ground motions were supplemented by seven pairs of synthetic ground motions to complete the OVE set. The latter were obtained from a database of 648 sites from a single simulation of the Puente Hills fault for the Los Angeles region (Graves and Somerville 2006). These ground motions are hybrid broadband signals ($f = 0.0-10.0$Hz); in the low-frequency range ($f < 1.0$Hz) a 3-dimensional finite difference model that simulates fault rupture, wave propagation to a site, and site response is used, whereas for the high-frequency range ($f \geq 1.0$Hz) a stochastic method for ground motion simulation is used. The seven records selected are the best match to the OVE uniform hazard spectrum.

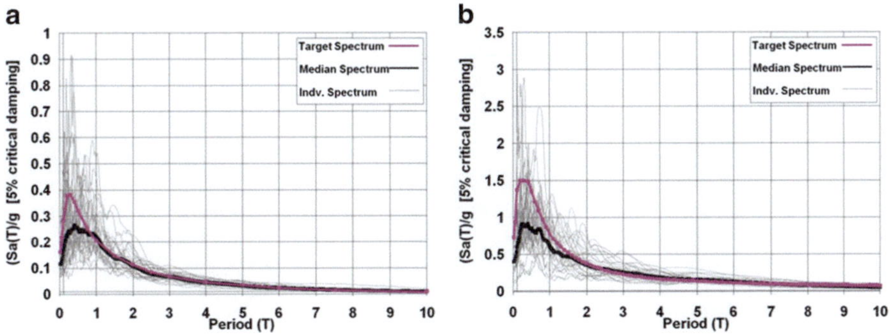


**Fig. 26.3** Target and scaled spectra at (**a**) 43-year return period and (**b**) MCE

Figure 26.3 shows the median of the scaled spectra and target spectra for the SLE43 and MCE hazard levels. The median matches the target spectrum reasonably well in the medium to long period range of interest for all hazard levels.

## 26.5 Case Study Building Designs

Three different high-rise building configurations were selected for study (Fig. 26.4). The 42-story concrete core-wall residential building had four additional stories below grade and used a centrally located concrete core wall with coupling beams for seismic resistance, with unbonded post-tensioned slabs supported by concrete columns and the core wall as the gravity and diaphragm system. The building had typical floor height of 9.67 ft and floor area of approximately 9,000 square feet for each floor above grade. The 42-story concrete dual core wall/frame system had nominally identical configuration except seismic-force-resisting frames replaced gravity slab-column frames along two bays in each principal direction. The 40-story tall steel buckling-restrained braced frame office building had four basement levels and used buckling-restrained braced frames for seismic resistance, with steel gravity framing elsewhere. The building had typical floor height of 13.5 ft and floor area of approximately 18,000 square feet for each floor above grade.

Two alternative designs were carried out. These are designated as Design A (code-based) and Design B (performance-based). Design A uses the prescriptive provisions of the International Building Code (IBC 2006), except the height limit was disregarded. Design B uses the performance-based design approach of the TBI Guidelines (TBI 2010) for seismic design of the structural system. For both Design A and Design B, gravity design, wind design, and nonstructural component seismic design comply with the provisions of IBC (2006). Wind loads are according to ASCE 7 (2005).

The designs were done by structural engineering firms experienced in the seismic design of high-rise buildings. For both Design A and Design B, the structural

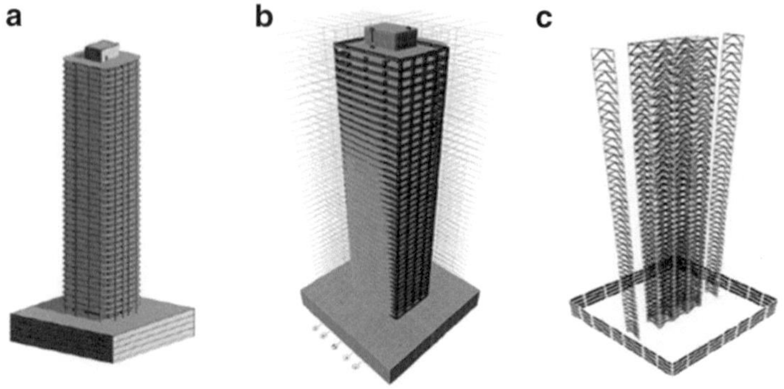


**Fig. 26.4** Case study buildings

engineering firm relied on experience to develop conceptual and preliminary designs for the gravity and lateral force-resisting systems. For Design A, the firms' practices, based on extensive experience, resulted in conservative designs for the seismic force-resisting systems compared with minimum requirements of the building code. For Design B, the preliminary designs required iterations to arrive at acceptable configurations and proportions. The analysis for the Maximum Considered Earthquake Level required development of site-specific response spectra and selection of earthquake ground motions by an engineering seismologist; according to design criteria developed by the project team, these records were spectrum matched to the site-specific spectrum over periods ranging from 0.2 to 1.5 T, where T is the estimated building period (5 s). Nonlinear models of the structural system were implemented in PERFORM-3D (CSI 2009). These models included the structural walls and a representation of the gravity framing in accordance with the modeling recommendations of ATC-72 (2010). The models were subjected to each of the seven pairs of scaled earthquake ground motions to verify acceptable response, without need for additional design iterations to meet the performance criteria. A typical performance-based design of a tall building requires additional design and review effort compared with the prescriptive building code design. It is not unusual to estimate 500 h of additional engineering effort and three to four months of additional design/review time relative to a prescriptive code-based design (Fry and Hooper 2011).

Among many differences in the building designs, the salient features were as follows: For the core-wall building, base shear for Design A (4,600 kips) was lower than that for Design B (8,200 kips) because of the serviceability requirements in the latter case. Because design was governed by wall shear, required wall thickness was greater for Design B (32 in.) than Design A (24 in.). Wall vertical reinforcement also was greater for Design B., Conversely, the allowance in the TBI Guidelines for demand-capacity ratio of 1.5 for ductile actions resulted in weaker coupling beams in Design B than Design A. The overall effect was a stiffer building with stronger wall piers and weaker coupling beams for Design B than Design A. The

**Table 26.1** Initial construction costs (in million $US)

| | Structural cost | | Total cost (structural, nonstructural, contents) | |
|---|---|---|---|---|
| Building | Design A | Design B | Design A | Design B |
| Core wall | 126 | 128 | 140 | 143 |
| Dual core wall/frame | 134 | 159 | 149 | 174 |
| Buckling-restrained braced frame | 276 | 268 | 341 | 333 |

dual core wall/frame system required similar design modifications. In addition, a capacity design approach used in Design B resulted in much larger columns than in Design A. For the buckling-restrained braced frame system, an entirely different structural system involving steel braced frame outriggers was required for Design B that was not required for Design A. However, Design A required very large column sizes because of code prescriptive requirements regarding summing brace forces over the building height to determine column axial forces. For complete details, see the appendices to Moehle et al. (2011b).

The initial construction costs of the designs were estimated by an experienced professional cost estimator in California (Moehle et al. 2011b). The initial construction costs (including design and management fees) are listed in Table 26.1. The values given are for above-grade construction only. In loss studies, reported later in this paper, only the above-grade portions were deemed susceptible to damage.

## 26.6 Analytical Models for Design and Performance Assessment

Analytical models were developed using computer software PERFORM-3D (CSI 2009). The seismic force-resisting systems extended down to the foundation level, which was modeled as rigid. Axial and bending interaction of the core walls was modeled using inelastic fiber elements accounting for longitudinal reinforcement and confined concrete (cover concrete ignored). In-plane shear behavior of the wall was modeled using elastic shear stiffness and an inelastic shear spring with strength equal to $1.5V_n$ (where factor 1.5 is intended to achieve expected strength when applied to nominal strength $V_n$ calculated using ACI 318 (2011)). Coupling beams were modeled using two elastic beam-column elements connected at midspan by a nonlinear shear hinge. Seismic frames were modelled with the usual assumptions considering flexure (nonlinear), shear, and axial flexibilities. Buckling restrained braced frames were modelled considering axial, bending, and shearing flexibilities of beams and columns, and axial flexibilities (nonlinear) of braces. Basement perimeter walls were modeled using elastic shear wall elements with a stiffness reduction factor of 0.8 to account for concrete cracking. A rigid diaphragm was achieved for all levels above ground by "slaving" the horizontal translation degrees of freedom. Slabs in basement levels were modeled using elastic shell elements with a stiffness reduction factor of 0.25 to account for concrete cracking. P-delta effects were taken into account in the model by applying tributary loads to the lateral force-resisting system with remaining loads applied to a dummy column with no lateral stiffness. Floor mass was assigned as lumped mass at each floor. Gravity framing was found to have negligible influence on stiffness and strength, and was excluded from the analytical models.

## 26.7 Dynamic Response of the Case Study Buildings

Analytical models of each building were subjected to the 15 pairs of earthquake ground motions at each of five hazard levels described previously. Figure 26.5 presents a glimpse of the results, in this case showing nominal compressive strain in one corner location of the core wall in the core-only building. In this case, the code-designed building sustains greater compressive strain than does the

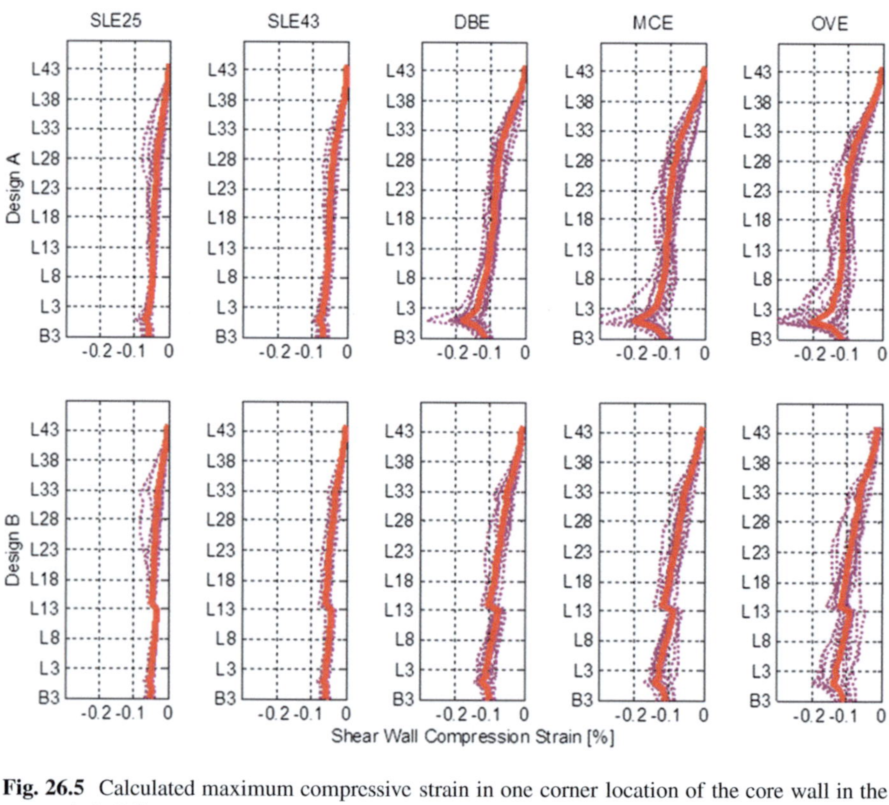


**Fig. 26.5** Calculated maximum compressive strain in one corner location of the core wall in the core-only buildings

performance-based design. The engineering demand parameters were not always "worse" for the code-based design than for the performance-based design, making it difficult to judge the overall performance based solely on engineering demand parameters. The repair cost assessment described in the next section integrates the effects of engineering demand parameters on damage and repair costs, facilitating a more comprehensive performance assessment.

## 26.8 Financial Implications of the Different Design Methods

Using the results from the nonlinear dynamic analyses, the statistical distributions of the structural response were sampled at each of the five hazard levels. The ATC 58 calculation tools were then implemented to generate large numbers of engineering demand parameters (EDPs). The values of the EDPs were then used to assess the damage states of the components. Once the damage states for all components were identified, the repair actions and repair cost for each component were obtained from a look up table (ATC-58 2012). The total repair cost for the entire building was then summed over all the components in the building. Buildings of this size require large amounts of data to define the repair costs. For example, the core wall-only buildings had 1,765 performance groups representing the structural, nonstructural, and contents items deemed susceptible to shaking damage, and each group had to be sampled for each of several thousand realizations to establish repair cost statistics. Fortunately, available software automates the procedures.

It is not possible to show all the repair cost data for all the buildings. Instead, here we focus only on the core wall-only building. Figure 26.6 shows the deaggregation of median total repair costs for each Performance Group type and each hazard level. Performance Groups for core-wall webs and core-wall boundary elements have been combined, and elevators are not shown because the cost is relatively low. For SLE25 shaking intensity level, the repair costs are concentrated in the contents

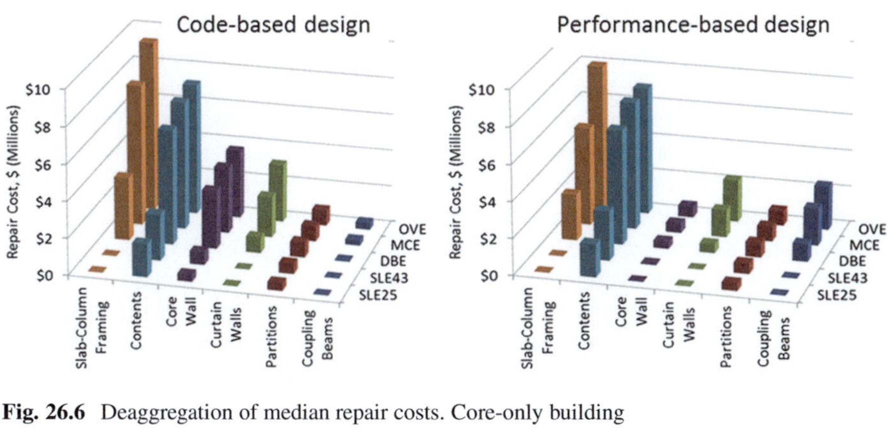


**Fig. 26.6** Deaggregation of median repair costs. Core-only building

**Table 26.2** Median repair costs for the buildings, normalized to initial construction cost

| Building | 43-year return period | | 2475-year return period | |
|---|---|---|---|---|
| | Code design (%) | Performance design (%) | Code design (%) | Performance design (%) |
| Core wall | 2.9 | 2.4 | 15.9 | 12.4 |
| Dual core wall/frame | 2.9 | 2.0 | 12.4 | 10.4 |
| Buckling-restrained braced frame | 0.7 | 0.5 | 3.1 | 2.8 |

and interior partitions, although Design A also sustains some core wall repair costs. As the shaking intensity increases to SLE43, the distribution of repair costs remains similar to SLE25. As shaking intensity reaches DBE, repair costs for slab-column framing and the core wall accelerate for Design A, while repair costs for coupling beams accelerate for Design B. The higher damage for the thinner walls of Design A and higher damage for the weaker coupling beams of Design B is consistent with expectations based on the design results. As the shaking intensity reaches MCE, repair costs for slab-column framing continue to increase for both designs. Curtain wall repair costs also begin to pick up. Contents repair costs have nearly saturated and do not show significant increase with increasing intensity. Trends continue for the OVE.

The relatively high repair costs for slab-column framing is an unexpected result from the analysis. It is possible that fragility relations or repair costs used in the analysis require adjustment. However, the results were reviewed with members of the ATC-58 (2012) project team and confirmed to be reasonable. Assuming the results to be indicative of actual conditions, such results can provide useful information to the structural engineer on a project by indicating where to focus design attention for the purpose of reducing potential damage and repair cost. The results also might suggest revisions in design codes to reduce losses associated with especially vulnerable components, or reduce requirements for more rugged components.

The median repair costs for the buildings at different earthquake hazard levels were calculated for each of the building designs. Results for the 43-year and 2475-year return periods, normalized to initial construction cost, are listed in Table 26.2. Normalized losses for the performance-based designs were, in general, smaller than for the code-based designs, although the results are not drastically different. The buckling-restrained braced frame had the smallest costs. It is noted, however, that residual drift, which might have significant impact to the total repair cost of the buckling-restrained braced frame building, was not included in the calculations.

Table 26.3 summarizes the mean annualized repair cost for the buildings. This number represents the average repair cost per year for all buildings, considering all hazard levels. The results show that the core only building had highest mean annualized repair cost followed by the dual system and then the buckling-restrained braced frame. In general, the mean annualized repair cost decreased as the design shifted from the code-based design to the performance-based design.

**Table 26.3** Mean annualized repair costs for the buildings in $US

| Building | Code design | Performance design |
|---|---|---|
| Core wall | 326,000 | 282,000 |
| Dual core wall/frame | 323,000 | 269,000 |
| Buckling-restrained braced frame | 206,000 | 141,000 |

**Table 26.4** Ratio of total costs

| Building | Code design | Performance design |
|---|---|---|
| Core wall | 1.0 | 1.0 |
| Dual core wall/frame | 1.0 | 0.96 |
| Buckling-restrained braced frame | 1.0 | 0.97 |

If we assume the mean annualized repair costs are equivalent to required insurance premiums to be paid annually, we can calculate the net present value of the insurance premiums given a payment period and assumed time value of money. Here we assumed a period of 50 years and interest rate of 0.03. Total Cost can be defined as the initial construction cost plus the net present value of insurance premiums. Table 26.4 compares Total Costs of the Performance-Based Design relative to the Total Costs of the Code Based Design for each building. The results indicate that Total Costs are equal or less for the performance-based designs. Note that these results are insensitive to the assumed time value of money. Note also that the performance-based designs were not oriented toward optimization of Total Cost, but instead were oriented toward more reliable performance by more explicit representation of the building properties in the design process. Furthermore, there was no intent of the performance-based designs to achieve superior performance but, rather, to more reliably achieve Occupancy Category II performance objectives.

## 26.9 Summary and Conclusions

This paper served to introduce two guidelines recently introduced in U.S. practice, one for performance-based designs of tall buildings and another for the seismic performance evaluation of building designs. The methods are demonstrated through the design and performance assessment of three tall building configurations, each designed according to code-based procedures and according to performance-based procedures. Principal conclusions are:

- The code-based and performance-based designs had notable differences in structural component sizes. The performance-based designs appeared allocate materials more appropriately.

- The code-based and performance based designs had notable differences in seismic performance, but it was difficult to know which performance was superior based solely on the engineering demand parameters and apparent damage. By aggregating the repair costs, it became apparent that the performance-based designs had equal or lesser repair costs. Combining initial costs and annualized repair costs, it was clear that the performance-based designs were slightly more efficient than the code-based designs.

Overall, it is concluded that the performance-based approach is usable and produces acceptable designs. The ATC 58 seismic performance assessment methodology is useful for measuring relative performance of individual buildings and thereby may be useful in deciding among design options. It also can serve as a useful tool to optimize building code requirements.

## References

ACI 318 (2011) Building code requirements for structural concrete (ACI 318–0811) and Commentary. American Concrete Institute, Farmington Hills, 503 pp

ASCE 7 (2005) ASCE 7–05 minimum design loads for buildings and other structures including supplement 1, Reston

ATC-58 (2012) Guidelines for seismic performance assessment of buildings. Applied Technology Council, Redwood City

ATC-72 (2010) Modeling and acceptance criteria for seismic design and analysis of tall buildings. Report No. ATC-72, Applied Technology Council, Redwood City

ATC-82 (2011) Improved procedures for selecting and scaling earthquake ground motions for performing time-history analyses. Applied Technology Council, Redwood City

Baker JW, Cornell CA (2006) Spectral shape, epsilon and record selection. Earthq Eng Struct Dyn 35:1077–1095

CSI (2009) PERFORM-3D. Computers & Structures, Berkeley

Fry JA, Hooper JD (2011) Personal communication from engineers at Magnusson. Klemencic Associates, Seattle

Graves RW, Somerville PG (2006) Broadband ground motion simulations for the Puente Hills thrust system. In: Proceedings, 8th national conference of earthquake engineering, EERI. San Francisco

IBC (2006) International building code. International Code Council, Falls Church

LATBSDC (2008) An alternative procedure for seismic analysis and design of tall buildings located in the Los Angeles region. Los Angeles Tall Buildings Structural Design Council, Los Angeles

Moehle JP, Deierlein GG (2004) A framework methodology for performance-based earthquake engineering. In: Proceedings, 13th world conference on earthquake engineering, Vancouver, Aug 2004, paper no. 679

Moehle JP, Bozorgnia Y, Hamburger RO (2011a) Guidelines for performance-based seismic design of tall buildings. In: Proceedings of the 8th international conference on urban earthquake engineering, Tokyo Institute of Technology, Tokyo, 7–8 Mar 2011

Moehle JP, Bozorgnia Y, Jayaram N, Jones P, Rahnama M, Shome N, Tuna Z, Wallace J, Yang T, Zareian F (2011b) Case studies of the seismic performance of tall buildings designed by alternative means, PEER Report No. 2011/05, Pacific Earthquake Engineering Research Center, University of California, Berkeley

PEER (2005) PEER ground motion database, Pacific Earthquake Engineering Research Center, University of California, Berkeley. http://peer.berkeley.edu/nga/

Porter KA, Johnson G, Sheppard R, Bachman RE (2010) Fragility of mechanical, electrical, and plumbing equipment. Earthq Spectra 26(2):451–472

SEAONC (2007) Administrative bulletin – requirements and guidelines for the seismic design of new tall buildings using non-prescriptive seismic-design provisions AB-083, Structural Engineers Association of Northern California, San Francisco

Taghavi S, Miranda E (2003) Response assessment of nonstructural building components, PEER Report No. 2003/05, Pacific Earthquake Engineering Research Center. University of California, Berkeley

TBI (2010) Guidelines for performance-based seismic design of tall buildings. Report PEER-2010/05, Pacific Earthquake Engineering Research Center, University of California, Berkeley

Yang TY, Moehle JP, Stojadinovic B, Der Kiureghian A (2009) Performance evaluation of structural systems: theory and implementation. J Struct Eng ASCE 135(10):1146–1154

# Chapter 27
# Consideration of Resilience of Communities in Structural Design

**Andrei M. Reinhorn and Gian Paolo Cimellaro**

**Abstract** The paper addresses a design methodology for civil engineering structures, "Resilience-Based Design (RBD)" which evolves from the Performance-Based Design (PBD). Currently engineers approach a structure as if it stands alone, without considering the interaction with the community, which should be regarded as an integrated part of the design process. Indeed, a building structure or a bridge should not be considered anymore alone, but as a group of structures using a "Portfolio Approach" which would allow regional loss analysis. Such approach moves from the concept of "housing units" to the concept of "housing blocks". The goal of RBD is to make communities as "resilient" as possible, developing technologies and actions that allows each structure and/or community to regain its function as promptly as possible. A framework for disaster management, based on open-closed loop control strategy is introduced to integrate the resilience from structures to community and to decision system applied in the design. The fundamental concepts of community resilience are analysed and a common reference framework is established which is based on the acronym "PEOPLES", explained further in the paper. Emphasis is given to hazard intensity measures, engineering demand parameters while a performance matrix defining the performance limit thresholds of RBD is proposed. Some applications are shown incorporating both performance and resilience objectives in order to illustrate the feasibility of the proposed strategy.

A.M. Reinhorn (✉)
Department of Civil Structural and Environmental Engineering, University at Buffalo – State University of New York, Buffalo, NY 14260, USA
e-mail: reinhorn@buffalo.edu

G.P. Cimellaro
Department of Structural, Geotechnical and Building Engineering (DISEG), Politecnico di Torino, 10129 Torino, Italy
e-mail: gianpaolo.cimellaro@polito.it

M. Fischinger (ed.), *Performance-Based Seismic Engineering: Vision for an Earthquake Resilient Society*, Geotechnical, Geological and Earthquake Engineering 32,
DOI 10.1007/978-94-017-8875-5_27, 

**Keywords** Resilience • Performance-based design • Resilience-based design • Reliability • Community resilience • Disaster resilience • Limit states • PEOPLES framework

## 27.1 Introduction

Past earthquakes such as San Fernardo (1906), Loima Prieta (1989), Northridge (1994), Kobe (1995), are often remembered for particular structural problems, however recent earthquakes draw the attention to the effects on communities (2004 Indian Ocean Tsunami, Wenchuan, Haiti, Christchurch, Tohoku earthquake) rather than structures. Natural disasters in general can cause widespread moderate to severe damage that may strain the ability of a community to function. Widespread damage can have a substantial long-lasting social, economic and cultural impact on the well-being and vitality of a city and nation. Some recent examples are the earthquakes in the four continents:

- 2009 L'Aquila earthquake in Italy (Europe);
- 2010 Conception earthquake in Chile (South America)
- 2011 Christchurch earthquake in New Zealand (Oceania);
- 2011 Tohoku earthquake in Japan (Asia);

The 2011 Tohoku Earthquake caused $319 billion in direct damages and $619 billion costs of recovery and other indirect losses, but mostly it showed that post earthquake preparedness and evaluation needed improvements in order to be more systematic, accurate and quicker, trying to limit the resources and the population involved in the recovery process. It definitely showed the lack of robustness in key functions and a limited organized emergency response and rapid recovery. This may be attributed to the lack of any substitute of military organization in their country. In fact, after World War II they were not allowed to have any organization which is able to face disaster with the same military approach such as in other countries.

The 2011 Christchurch earthquake in New Zealand will be remembered for the effect of very close multiple events and aftershocks. Near-field effects generated vertical ground motions that were higher than expected. More stringent limits on columns and walls axial loads for unexpected vertical accelerations appeared to be necessary, as well as more stringent flooring detail. New unmapped faults were found during these events which should have been taken in account in future risk analysis. The presence of all these unknown uncertainties brings to the thought that the use of a Uniform Hazard Spectrum (UHS) might not be appropriate, because it is built using seismogenic earthquake sources defined in term of location, geometry and characteristics which are time and spatially independent. A 6.3 M earthquake aftershock was registered near Christchurch, New Zealand on Monday, February 21, 2011, arriving less than 6 months after a 7.0 M earthquake struck the South Island of New Zealand. These two events cannot be considered as time and spatially independent, therefore the UHS at the site cannot be considered representative of these two events.

The 2009 L'Aquila earthquake in Italy showed a well-organized emergency response in the short time, but the long term reconstruction phase showed malfunctions due to decision system insufficiently based on economic-political-social issues which are also essential for starting the recovery process. It also pointed to the necessity of rehabilitation of old masonry buildings, using either traditional or modern retrofit techniques as preventive risk mitigation, which performed adequately when properly designed and installed, reducing the building vulnerabilities (Cimellaro et al. 2010a).

After these recent extreme events, the international community became more conscious that resilience, i.e. the ability to rebound from loss of functionality, is the key to describe earthquake engineering performance. Designing structural systems to provide the needs of usability and safety before, during, and after extreme events has been the target of centuries. Recent developments lead to processes that set clearly performance as criteria for adequacy of design. The modern design approach follows an iterative process in which the performance of the system is forecast and then adjusted and readjusted to meet the criteria for the desired levels. The performance criteria often include implicitly the balance between usability and safety, in the aftermath of the extreme event. Often prescribed criteria and procedures govern the design process. More recently using more advanced tools and technologies, the criteria can be met with more accuracy and certainty in spite of uncertain and often unknown or unexpected hazards.

The newer concepts of resilience go beyond the initial performance toward the expended process of recovery from losses which govern the system until it is restored to an acceptable condition of its functionality. However, the process of recovery to the desired functionality level becomes a combined engineering-economic issue coupled with organizational-socio-political issues which strongly influence the outcome of the behavior in the aftermath of such extreme events such as strong earthquakes and associated effects. When functionality is defined in terms of usability and safety, and in terms of socio-economic parameters coupled with their organizational capabilities, the design considerations are altered and the iterative process may require different solutions.

The paper presents the proposed concept and the mathematical formulation of the proposed resilience based design (RBD) using a layered approach which is based on the combination of seven dimensions of community resilience. A performance evaluation matrix for defining the performance limit threshold of RBD is also presented.

## 27.2 From Performance Based Design to Resilience Based Design

### 27.2.1 Performance Base Design (PBD) Approach

The PBD is based on seismic hazard, demand, capacity and loss models, and is quantified through the total probability disaggregating the problem into several interim probabilistic models defined as follows (Cornell and Krawinkler 2000; Krawinkler and Miranda 2004; FEMA 1997, 2000, 2006; Moehle 2004):

$$\underbrace{\lambda\left(dv < DV\right)}_{Seismic\ risk} = \iiint \underbrace{G\left(dv|\,dm\right)}_{Loss\ Analysis}\ \underbrace{dG\left(dm|\,edp\right)}_{Damage\ Analysis}\ \underbrace{dG\left(edp|\,im\right)}_{Response\ Analysis}\ \underbrace{|d\lambda(im)|}_{PSHA} \tag{27.1}$$

where $im$ = intensity measure (e.g. $S_a(T_1)$, *epsilon*, $S_{dinelastic}$, *duration* etc.); $dm$ = damage measure (e.g. physical condition & consequences/ramifications); $edp$ = engineering demand parameters (e.g. drift ratio (peak, residual), acceleration, local indices etc.); $dv$ = decision variable (e.g. loss, functionality, downtime, casualties etc.); $\lambda(dv)$ = mean annual frequency of a decision variable (dv); $G(a|b)$ is the probability of exceedance $a > a_0$ given $b$. Each component of Eq. 27.1 needs to be determined statistically.

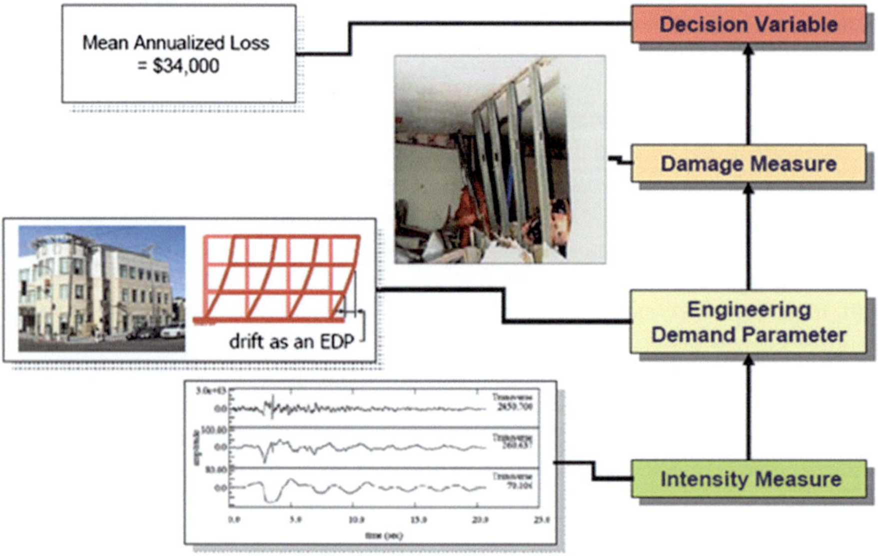


**Fig. 27.1** PEER center methodology for performance-based design (PBD)

The methodology, also known as PEER center methodology, is described clearly in Fig. 27.1. Currently the method has been implemented in ATC-58 (2011) and ATC-63 in the USA, and also in other countries like in China where PBEE has been added to the new version of the code "Seismic Design for Building Structures" (GB50011-2010), to design tall buildings and innovative systems, while majority of buildings could be designed with traditional Response Spectrum Analysis (RSA).

There are parts that the PBD does not cover such as:

- The recovery process, with the resources and decisions needed;
- The portfolio assessment which include functionalities along time;
- Community assessment which addresses the impact of individual asset on the group functionality and vice versa;

The concept of Performance-Based Design/Engineering can be applied to describe the behavior of a single building or structure, but the performance of an individual structure is not governed only by its own performance, but interacts heavily with the performance of other entities within the same community. A clear example of these interdependencies between the building and the community is a hospital, which it will not be able to perform without electricity and water even if the structure has no structural damage. Another example of the limitations of PBD is given by 2009 L'Aquila earthquake, during which the small town of Castelnuovo was completely destroyed, except a single housing unit that was standing after the earthquake and suffered minor damage (Cimellaro et al. 2010a). According to PBD the building would satisfy the expectations, but in RBD the housing unit would be not operational, because it is not able to interact with other entities in the same community.

### *27.2.2 Towards Resilience-Based Design*

Currently design practice approaches a structure as if it stands alone, without considering the interaction with the "community", which should be an integral part of the design process. There is a new fundamental way of looking at all the problem. A building is not considered alone, but as part of a *group of buildings* using a "Portfolio Approach" which allows regional loss analysis. This moves the design process from *housing units* to *housing blocks*. This concept is borrowed from the financial industry, where Modern Portfolio Theory (MPT) was developed in the 1950s through the early 1970s and was considered an important advance in the mathematical modeling of finance (Elton and Gruber 1997). MPT is defined as a theory of investment which attempts to minimize risk for a given level of expected return, by carefully choosing the proportions of various assets. MPT is a mathematical formulation of the concept of *diversification* in investing, with the aim of selecting a collection of investment assets that has collectively lower risk than any individual asset. Analogously the concept of *diversification* can be applied in the field of

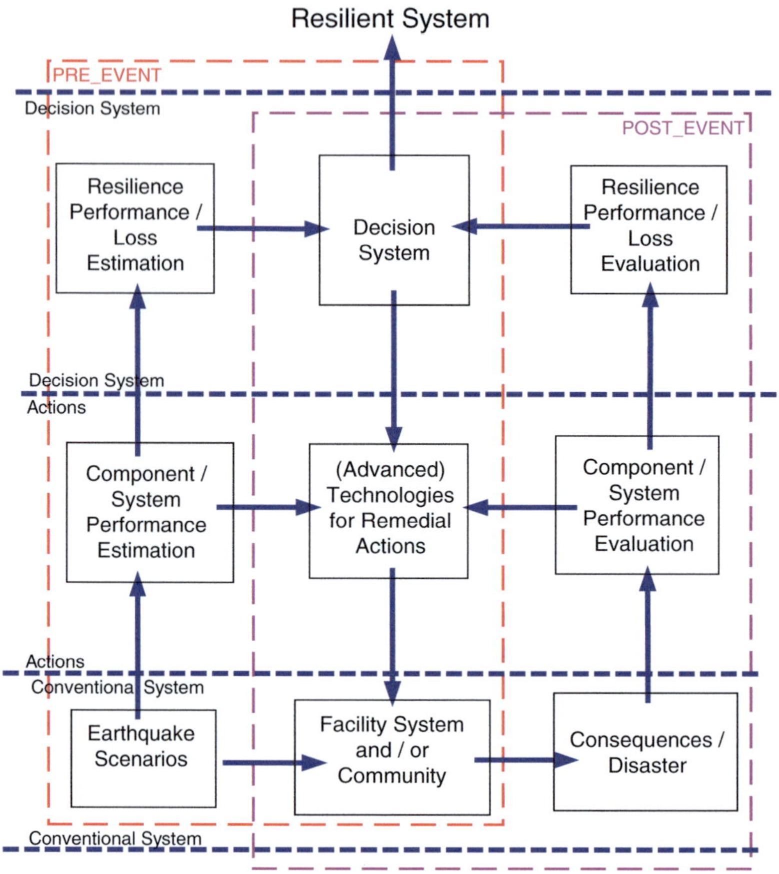


**Fig. 27.2** Design toward resilience

*disaster resilience*, where the diversification in retrofit of different buildings in a given region can increase resilience collectively more than any individual retrofit.

MPT models an asset's return as a normally distributed function, defines risk as the standard deviation of return, and models a portfolio as a weighted combination of assets so that the return of a portfolio is the weighted combination of the assets' returns.

Analogously in RBD, the weight aggregated losses over the building portfolio is equivalent to the portfolio asset's return in MPT, therefore the risk of losses at the community level is the variation of the aggregated losses of the building portfolio. The building portfolio can be viewed as a weight combination of the performance index of each housing unit.

The suggested resilience based design (RBD) follows also an iterative process that covers three layers of a generalized system control diagram (see Fig. 27.2) based on feed-forward (prior to event) and feed-back (post-event) that addresses the resiliency of the system. The first row at the bottom alone does not apply any improvement on the system, while the last two rows of the diagram describe the current concept of performance based design (PBD), which is mostly addressing the integrity of the system alone. The added first row on top together with the other two describes the RBD, in which resilience performance indices are developed and evaluated and then used in a decision support system that selects the appropriate remedial actions or accept the achieved resilient status (Bruneau et al. 2003).

Furthermore, unlike the PBD, the suggested RBD considers the structures and the community as a complex system, where losses as well as the recovery process are coupled dimensionally, which involve several parameters that are not only engineering parameters, such as drift and accelerations, but also other parameters, such as socio-economic environment, gender and age of the population, etc. In this way the recovery process becomes an integral part of the RBD process, that should be planned upfront.

## 27.3 Definition of Resilience and Dimensions

While the resilience is usually described as the ability to rebound from a disaster, the resilience (R) is defined here as a function indicating the capability to sustain a level of functionality, or performance, for a given building, bridge, lifeline networks, or community, over a period defined as the control time $T_{LC}$ that is usually decided by owners, or society, representing the life cycle, or the life span of the system. In detail, Resilience is defined graphically (Fig. 27.3) as the normalized shaded area underneath the functionality function of a system *Q(t)* which is a nonstationary stochastic process and each ensemble is a piecewise continuous function as the one shown in Fig. 27.3. Analytically Resilience is defined as

$$R\left(\vec{r}\right) = \int_{t_{OE}}^{t_{OE}+T_{LC}} Q_{TOT}(t)/T_{LC}\,dt \tag{27.2}$$

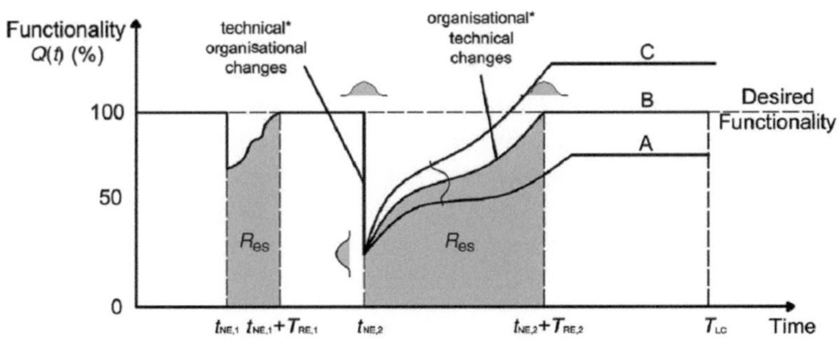


**Fig. 27.3** Resilience (Courtesy Cimellaro et al. 2010b)

where $Q_{TOT}(t)$ is the global *functionality* of the region considered; $T_{LC}$ is the control time of the period of interest; $\vec{r}$ is a vector defining the position within the selected region where the resilience index is evaluated (Cimellaro et al. 2009, 2010b). The global *functionality* is the combination of all *functionalities* related to different facilities, lifelines (Cimellaro et al. 2014), etc. for the case when physical infrastructures resources and services are considered which will be described in the following paragraphs.

### *27.3.1 The Seven Dimensions of Community Resilience*

Disaster resilience is often divided between technological units and social systems (Bruneau et al. 2003). On a small scale, when considering critical infrastructures, the focus is mainly on technological aspects. On a greater scale, when considering an entire community, the focus is broadened to include the interplay of multiple systems – human, environmental, and others – which together add up to ensure the functioning of a society.

In order to emphasize the primary role of the human system in community sustainability, the acronym "PEOPLES" (Renschler et al. 2010, 2011) has been adopted in order to describe the framework that is built on and expands previous research at MCEER linking several previously identified resilience characteristics (technical, organizational, societal, and economic) and resilience attributes (r4: robustness, redundancy, resourcefulness, and rapidity) (Bruneau et al. 2003; Cimellaro et al. 2010b). PEOPLES incorporates MCEER's widely accepted definitions of service functionality, its components (assets, services, demographics) and the parameters influencing their integrity and resilience.

Resilience can be considered as a dynamic quantity that changes over time and across space. It can be applied to engineering, economic, social, and institutional infrastructures, and it can use various geographic scales.

The first step in order to quantify the resilience performance index (R) is to define the *spatial scale* (e.g. building, structure, community, city, region. etc.) of the problem of interest (Fig. 27.4).

It is important to mention also that the entire recovery process is affected by the *spatial scale* of the disaster. Huge disasters will take longer recovery process. The spatial scale will also be used for defining the performance measures that will be considered to define the global functionality of the system. The second step is to define the *temporal scale* (short term emergency response, long term reconstruction phase, midterm reconstruction phase etc.) of the problem of interest. The selection of the control period $T_{LC}$ will affect the resilience performance index, therefore when comparing different scenarios the same control period should be considered. The seven dimensions of the RBD are the following (in random order of their importance):

**P**opulation and Demographics;
**E**nvironment/Ecosystem;

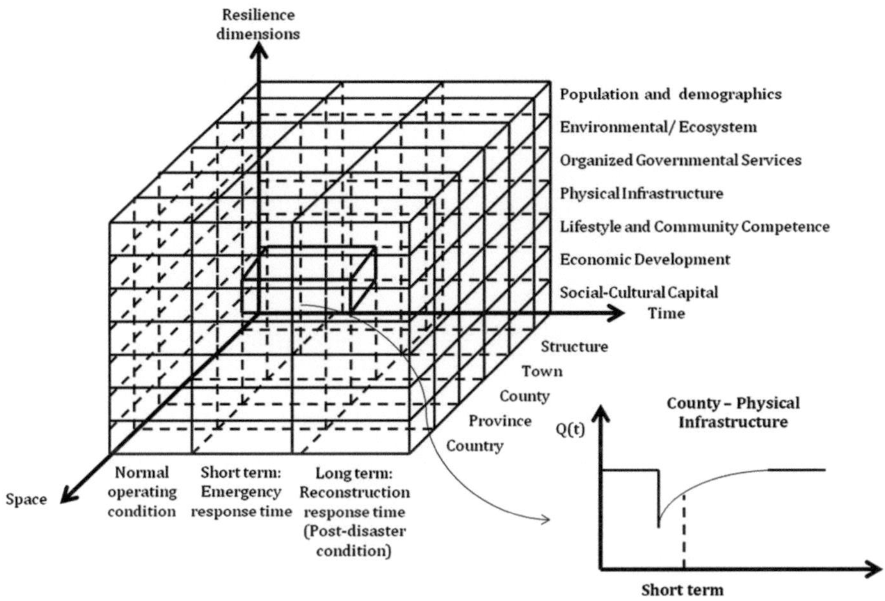


**Fig. 27.4** Spatial and temporal dimension of Resilience-Based Design (RBD) using PEOPLES approach

**O**rganized Government services;
**P**hysical infrastructures;
**L**ifestyle and community competence;
**E**conomic Development;
**S**ocial-Cultural capital

Details about the description of each one of these dimensions can be found in Renschler et al. (2010, 2011). The general framework at the community level is described by the equations below where for each dimension a performance indicator and/or *functionality* is defined which is combined with other *functionality dimensions* as follows

$$Q_{TOT}(t) = Q_{TOT}\left(Q_P, Q_{Env}, Q_O, Q_{Ph}, Q_L, Q_{Eco}, Q_S\right) \tag{27.3}$$

where $Q_{TOT}$ = global functionality; $Q_x$ = functionality of each of the seven dimensions. Within each dimension, functionality is defined as a combination of functionalities of their respective subsystems, for example the functionality of the physical infrastructures is defined as follows

$$Q_{Ph}(t) = Q_{Ph}\left(Q_{Hosp}, Q_{Ele}, Q_{Road}, Q_{Water}, \ldots\right) \tag{27.4}$$

where $Q_{hosp}$ = functionality of the health care facility; $Q_{ele}$ = functionality of the electric network; $Q_{road}$ = functionality of the road network; $Q_{water}$ = functionality

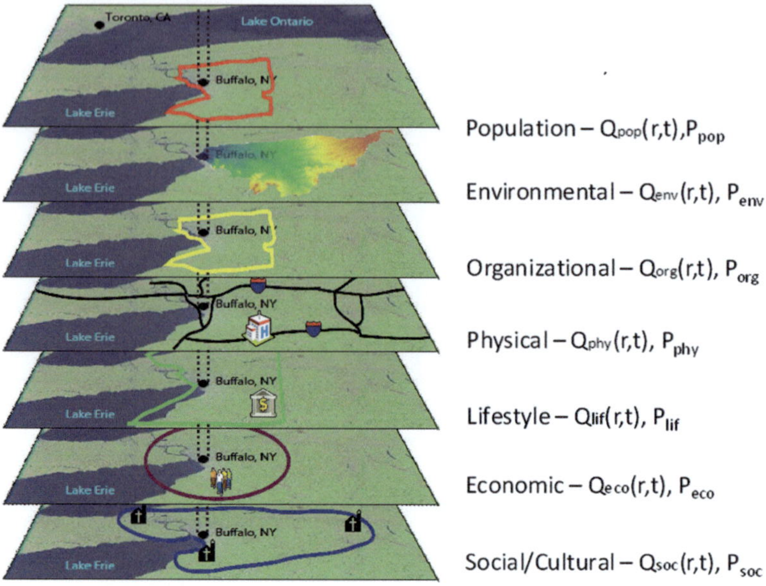


**Fig. 27.5** Layer model of PEOPLES framework

of the water network; etc. Once the geographic scale is defined, it is possible to plot the global functionality $Q_{TOT}$ over the region of interest in a contour plot at a given instant of time $t$, so *time-dependent functionality maps* of the region can be obtained. When also the temporal scale is defined through the control time $T_{LC}$, then the *resilience contour map* of the region of interest can be plotted. The *Resilience maps* will be obtained by integrating over time using Eq. 27.2, therefore they will be *time independent*, but they will vary in space from point to point in the selected region. Finally the community resilience index $R_{com}$ is given by the double integral over time and space as follow

$$R_{com} = \int_{A_C} R\left(\vec{r}\right) / A_C dr = \int_{A_C} \int_{t_{OE}}^{t_{OE}+T_{LC}} Q_{TOT}(t) / \left(A_C T_{LC}\right) dt dr \qquad (27.5)$$

where $\vec{r}$ is a vector defining the position in the selected region where resilience is evaluated; $A_c$ is the total area of the selected region. For each dimension a contour plot can be determined and combined using a layered approach as the one shown in Fig. 27.5. Then a radar graph can be plotted and the area will define the final value of the resilience score (Fig. 27.6) for the region of interest. This will allow identifying gaps as well as priority actions, which will enter in the decision loop.

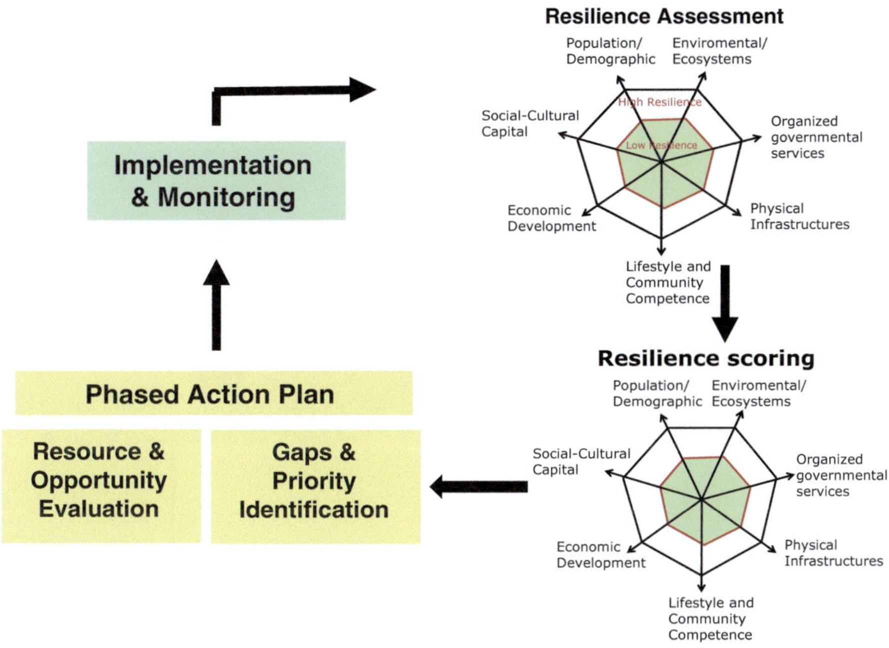


**Fig. 27.6** Resilience scoring using PEOPLES methodology

In summary a schematic step-by-step procedure of the MCEER methodology described in Fig. 27.7 is the following:

1. Define extreme event scenarios (e.g. PSHA, ground motion selection);
2. Define the system model;
3. Evaluate the response of the model;
4. Compute different performance measures (e.g. losses, recovery time, functionality, resilience);
5. Identify remedial mitigation actions (e.g. advanced technologies) and/or resilience actions (e.g. resourcefulness, redundancy, etc.);

This design approach has analogies with the feedback loop taken from control theory. The same framework can be used for a region as well as a single structure (e.g. hospital). In this case the functionality $Q_{hosp}$ reduces to the functionality of a single hospital $Q_{hosp}$ which can be evaluated with the procedure described in the paper of Cimellaro et al. (2011), so the resilience index of a single hospital can be evaluated using Eq. 27.2. However, in Resilience-Based Design (RBD) hospitals or more in general buildings are part of a community, therefore their functionalities cannot be considered independent from the rest of the system (e.g. an housing unit with minor structural damages without water, electric power and roads connecting it to the rest of the world can satisfy the PBD requirements, but it will not satisfy the RBD requirements).

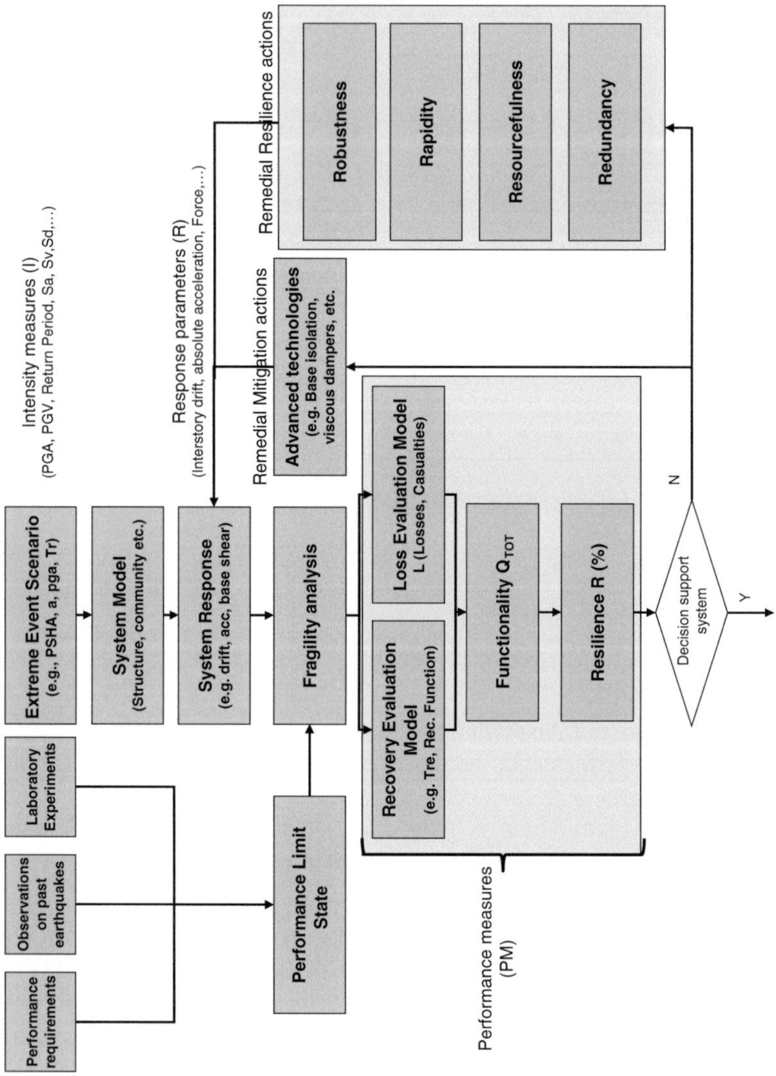


**Fig. 27.7** MCEER methodology for resilience-based design (RBD) based on control (feedback loop) approach

### 27.3.2 Uncertainties in Resilience-Based Design

Either a deterministic, or similarly with the PBD, a probabilistic approach can be used within the RBD methodology with preference to the latter approach when a particular level of confidence of achieving performance objective is of interest. Five random variables are involved in the probabilistic description of the resilience index when uncertainties are included and they are dependent. The joint probability density function of the resilience index is given by the following expression:

$$f_{R,T_{RE},Q,X,I}\left(r,t_{RE},q,x,i\right)=f_{R,T_{RE},Q,X,I}\left(r|\,t_{RE},q,x,i\right)\cdot f_{T_{RE},Q,X,I}\left(t_{RE}|\,q,x,i\right)\cdot f_{Q,X,I}\left(q|\,x,i\right)\cdot f_{X,I}\left(x|\,i\right)\cdot f_{I}(i) \quad (27.6)$$

Therefore the mean of the resilience index which is a random variable is:

$$m_r=E\left\{R\right\}=\int_r\int_{t_{RE}}\int_q\int_x\int_i r\cdot f_{R,T_{RE},Q,X,I}\left(r|\,t_{RE},q,x,i\right)\cdot f_{T_{RE},Q,X,I}\left(t_{RE}|\,q,x,i\right)\cdot f_{Q,X,I}\left(q|\,x,i\right)\cdot f_{X,I}\left(x|\,i\right)\cdot f_{I}(i)\cdot dr\cdot dt_{RE}\cdot dq\cdot dx\cdot di \quad (27.7)$$

or in a more compact form it is given by

$$m_r=E\left\{R\right\}=\int_{-\infty}^{\infty} r\cdot f_{R,T_{RE},Q,X,I}\left(r,t_{RE},q,x,i\right)\cdot dr \quad (27.8)$$

where $I$ = intensity measures; $X$ = response measures; $Q$ = performance measures; $T_{RE}$ = recovery time measures; $R$ = resilience index; $m_r$ = mean resilience index.

The described probabilistic approach is more comprehensive and general, but the information provided to the public (e.g. decision makers, politicians, etc.) should be deterministic (scenario basis or event), because it is more simple and easy to understand.

### 27.3.3 Definitions of Resilience Performance Levels

The objective of Performance Based Seismic Engineering (PBSE) is to design, construct and maintain facilities with better damage control. A comprehensive document has been prepared by the SEAOC Vision 2000 Committee (1995) that includes interim recommendations. The performance design objectives couple expected or desired performance levels with levels of seismic hazard as illustrated by the Performance Design Objective Matrix shown in Fig. 27.8. The performance assessment types according to the ATC 58 (2011) definitions can be divided in: (i) Intensity-based; (ii) Scenario-based; (iii) Time-based. PBD levels focus

| | | Earthquake Performance Level | | | |
|---|---|---|---|---|---|
| | | **Fully Operational** | **Operational** | **Life Safe** | **Near Collapse** |
| Earthquake Design Level | **Frequent** (43 Years) | Basic Objective | Unacceptable | Unacceptable | Unacceptable |
| | **Occasional** (72 Years) | Essential/Hazardous Events | Basic Objective | Unacceptable | Unacceptable |
| | **Rare** (475 Years) | Safety Critical Objective | Essential/Hazardous Events | Basic Objective | Unacceptable |
| | **Very rare** (975 Years) | Not Feasible | Safety Critical Objective | Essential/Hazardous Events | Basic Objective |

**Fig. 27.8** Recommended seismic performance objectives for buildings (SEAOC 1995)

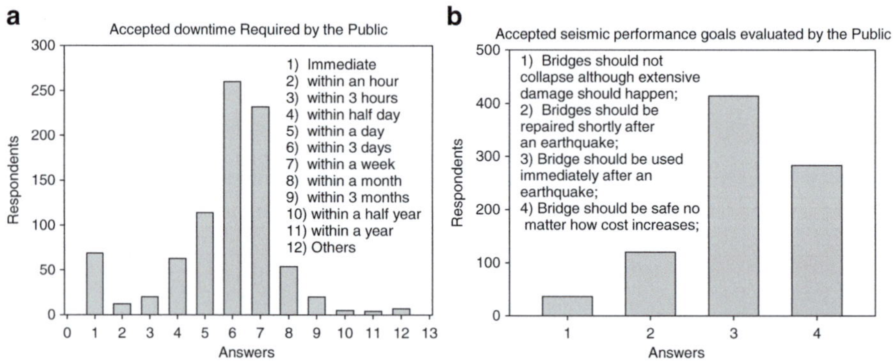


**Fig. 27.9** (**a**) accepted downtime required by the public; (**b**) accepted seismic performance goals evaluated by the public (adapted from presentation of Kawashima et al. at Bled4 workshop, 2011)

on the performances a building can hold during the shaking and are associated to engineering demand parameters such as deformations. More recently SPUR (Maffei 2009) which is the San Francisco planning and Urban Research Association introduced other definitions of performance levels for physical infrastructures based on recovery target states which take in account the safety as well as the recovery time. Five performance measures for buildings have been identified: *(i) Safe and Operational; (ii) Safe and usable during repair; (iii) Safe and usable after repair; (iv) Safe but not repairable; (v) Unsafe;* however a clear quantification of these performance levels has not been provided so far.

In Japan, using public interviews, the accepted downtime and performance goals for bridges have been evaluated (Fig. 27.9) trying to find which are the desired performance goals expected by citizens after an extreme event like an earthquake. This information is going to be very useful to define the boundaries of the different resilience performance levels.

The new proposed Resilience Performance Levels (*RPL*) focus on building performance after the earthquake stops, recognizing the importance of the temporal

| Functionality Performance Levels | | Recovery Time | | |
|---|---|---|---|---|
| | | Short term (Emergency) | Midterm | Long term (Reconstruction) |
| | Fully Operational (Q1) | Basic Objective | Unacceptable | Unacceptable |
| | Operational (Q2) | Essential Objective | Basic Objective | Unacceptable |
| | Partially Operational (Q3) | Critical Objective | Essential Objective | Basic Objective |
| | Near not Operational (Q4) | Not feasible | Critical Objective | Basic Objective |

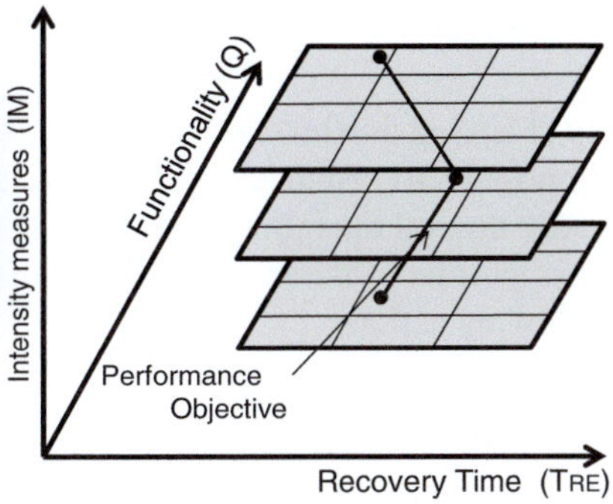


**Fig. 27.10** 3-dimensional resilience performance objectives matrix for structures, communities, systems etc.

dimension (Recovery time $T_{RE}$) in the assessment of the *RPLs* of structures and communities in general (Uma et al. 2006). A 2-D performance domain consisting of Performance Levels *PL(i,j)*, defined by the combination of Functionality (index j) and recovery time (index i) is proposed. By accounting for the effect of the temporal dimension, a 3-dimensional performance matrix (Fig. 27.10) can be visualized as a set of predefined joined performance domains ("masks") for different seismic intensity levels IM and different RPLs.

### *27.3.4 Towards Resilient Communities: The Four "r" Attributes*

Researchers at the Multidisciplinary Center of Earthquake Engineering to Extreme Events (MCEER) (Bruneau et al. 2003; Bruneau and Reinhorn 2007) have identified four attributes along which resilience can be improved. These are *robustness*, *resourcefulness*, *redundancy*, and *rapidity*. Further details about them can be found in Cimellaro et al. (2010b). *Resourcefulness* and *Redundancy* are strongly interrelated. For example, resources, and resourcefulness, can create redundancies that did not exist previously. In fact, one of the major concerns with the increasingly intensive use of technology in emergency management is the tendency to over-rely on these tools, so that if technology fails, or it is destroyed, the response falters. To forestall this possibility, many planners advocate *Redundancy*. Changes in *Resourcefulness* and *Redundancy* will affect the shape and the slope of the recovery curve and the recovery time $T_{RE}$. It also affects *Rapidity* and *Robustness*. It is through *Redundancy* and *Resourcefulness* (as means of resilience) that the *Rapidity* and *Robustness* (the goals of resilience) of an entire system can be improved.

### 27.3.5 *Resilience vs. Sustainability*

By changing the time horizon and the extreme events to normal service loadings, the proposed RBD framework can be used to address the performance of a community under normal operating conditions in order to achieve a more sustainable community. In order to move towards sustainable communities it is necessary to follow the following actions:

- Selection of materials;
- Use of recycled material;
- Consideration of material re-use and disassembly for re-use;
- Energy efficiency;
- Durability and longevity;
- Reparability;
- More efficient and lower impact constructions;
- More efficient design methods, and more efficient structural systems and layouts;
- Integration of structural forms to help achieve the needs of other disciplines;
- Reducing the impact of abnormal events such earthquakes by minimizing the need for repair and disruption of services;

All these actions can be considered as actions to achieve a resilient community therefore they can be described using the proposed RBD framework. All these actions can be measured and quantified using an index that measures the impact on the environmental ecosystem dimensions of the PEOPLES framework, while keeping constant the same structural performance.

## 27.4 Case Study

### 27.4.1 *Hospital System*

One of the requirements of RBD is to achieve redundancy in a given community. In a selected region for example resilience can be improved by increasing the number of hospitals in the region advocating redundancy. However, one of the aspects that should be taken in account in RBD which is not taken in account in the current practice is that if the region is affected by the same earthquake hazard, is it resilient to build the new hospital with the same retrofit technique of the other already existing hospitals? Selecting different retrofit strategies for the new hospital might enhance the probability to survive in case of extreme event of at least one hospital.

Following the same concept an example of design that has followed the current practice of PBD is the C.A.S.E. project realized after 2009 L'Aquila earthquake in

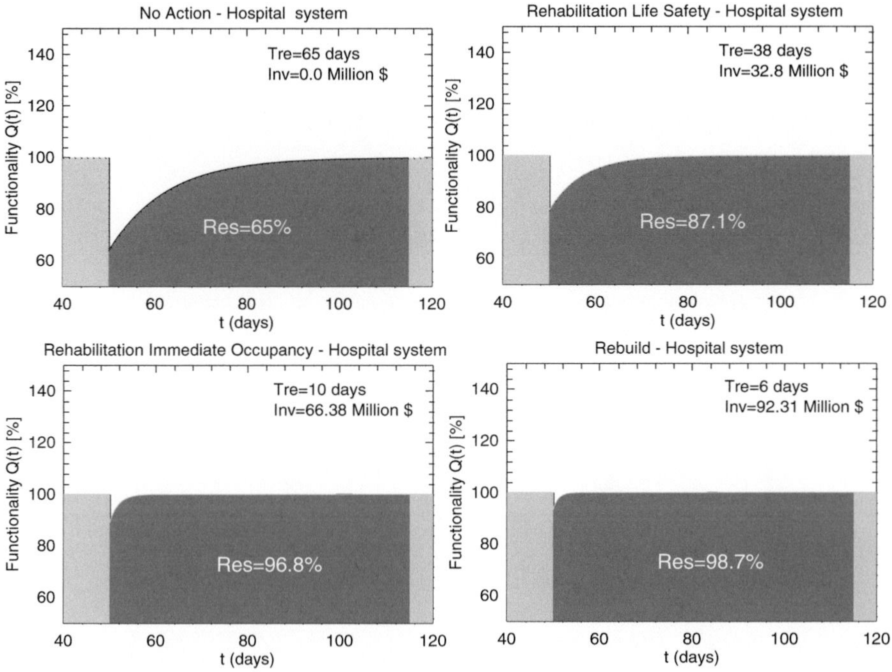


**Fig. 27.11** Resilience-based design of an hospital system according to MCEER methodology (Courtesy Cimellaro et al. 2010b)

Italy. It allowed realizing several housing units with the same type of retrofit that consists in a base isolation system realized with friction pendulum bearings. If, in a very rare scenario, the next earthquake would be characterized by low frequency content, probably all these new housing units would collapse, because they do not have diversity in performance. When looking at all housing units as a system, they are not redundant, because Resilience has not been considered in the design process.

An example based on a series of hospital buildings described by Park et al. (2004) is chosen to illustrate the concept of RBD. They consist of five concrete shear wall systems and one unreinforced masonry bearing system. Four different retrofit actions have been considered from *No Action* to the *Rebuild Option*. If uncertainties in the seismic input are considered by using four different hazard levels, then resilience index can be evaluated using Eq. 27.2 for different rehabilitation strategies are compared as shown in Fig. 27.11. For this case study it is shown that the Rebuild option has the largest disaster resilience of 98.7 %, when compared with the other three strategies, but it is also the most expensive solution ($ 92.3 millions). However, if No Action is taken the disaster resilience is still reasonably high (65.0 %). Further details about this case study can be found in Cimellaro et al. (2009).

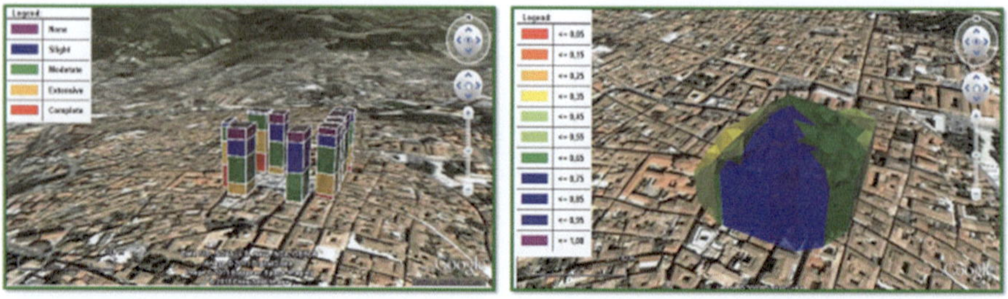

Fig. 27.12 Graphical output of the software implementing the MCEER methodology

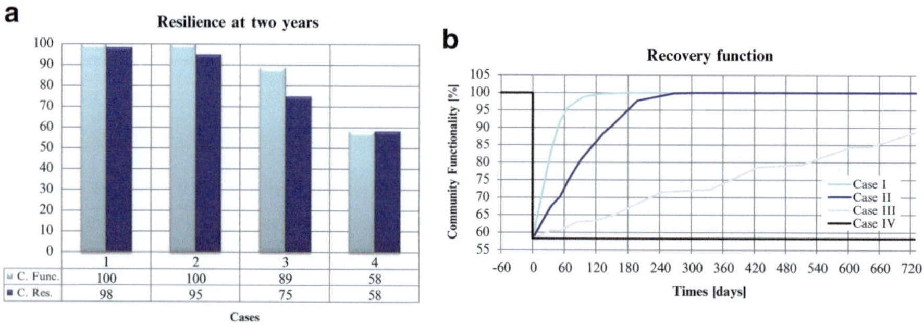


**Fig. 27.13** Graphical output of the software implementing the MCEER methodology (**a**) functionality/resilience of case studies, and (**b**) functionality recovery history

## *27.4.2 Housing Units in L'Aquila Downtown*

The entire framework described in the paper has been implemented in software which uses a Google Earth graphical interface (Arcidiacono et al. 2011). Ten buildings located in the old medieval town of L'Aquila which was affected by 2009 earthquake in Italy, have been selected as case study to show the applicability of the framework within the program. Four different scenarios were considered assuming different recovery actions, based on the number of construction sites working in a day, with the main assumption of no limit to the economic budget.

Functionality as well as Resilience indices were evaluated for two different control periods: at two years and at the end of the construction site work. From the 3D histogram plotted on Google Earth (Fig. 27.12a) is possible to see how the damage states in all buildings are distributed in the selected region of interest. Figure 27.12b shows the contour plot of functionality immediately after the disaster. Figure 27.13a shows the resilience index when the control period is two years, while in Fig. 27.13b are shown the functionality curves for the different scenarios events considered, where it clearly appears that the best scenario is the first one.

## 27.5 Concluding Remarks

After all these recent extreme events worldwide, the international community became conscious that Resilience is the key to describe earthquake engineering performance. A new more general design methodology is presented in this paper, the "Resilience-Based Design" (RBD), developed as an extension of the "Performance Based Design" (PBD).

The goal of RBD is to make individual structures and communities as "resilient" as possible in time, developing technologies and actions that allow each structure and/or community to regain its functionality as promptly as possible. However, even if the performance evaluation of an individual structure is the engineers' goal, the level of performance required should address the effects on and from the entire community.

A Resilience Performance Matrix combining *functionality* loss and recovery time as performance levels and performance objectives at increasing level of seismic intensity has been also presented. The resilience-based approach is illustrated in detail in some applications, showing its efficiency and better description with respect to PBD.

## 27.6 Future Research Trends

In the near future more attention should be given to interdependencies among different dimensions of the framework which if not considered should cause cascade effects which might be dramatic (e.g. physical network interdependencies etc.). Furthermore, in order to coordinate future recovery processes and international collaborations, common formats and tools should be applied for rapid resilience assessments, for example the reports from the recent earthquakes should follow the same common format. New open source community software frameworks should be developed through education and training, in order to develop a common research vision to raise awareness of stakeholders worldwide and local.

**Acknowledgments** The researches leading to these results has received funding from the Multidisciplinary Center for Earthquake Engineering Research (MCEER), from National Institute of Standards and Technology (NIST) and from the European Community's Seventh Framework Programme – Marie Curie International Reintegration Actions – FP7/2007-2013 under the Grant Agreement n° PIRG06-GA-2009-256316 of the project ICRED – Integrated European Disaster Community Resilience.

## References

Arcidiacono V, Cimellaro GP, Reinhorn AM (2011) A software for measuring disaster community resilience according to the PEOPLES methodology. In: Papadrakakis M, Fragiadakis M, Plevris V (eds) COMPDYN 2011 – III ECCOMAS thematic conference on computational methods in structural dynamics and earthquake engineering, Corfù, Greece, 26–28 May 2011

ATC-58-1_draft_75% (2011) Draft guidelines for seismic performance assessment of buildings volume 1 – methodology. Applied Technology Council (ATC-58 Project), 201 Redwood Shores Parkway, Suite 240, Redwood City, California, 94065
Bruneau M, Reinhorn AM (2007) Exploring the concept of seismic resilience for acute care facilities. Earthq Spectra 23(1):41–62
Bruneau M, Chang S, Eguchi R, Lee G, O'Rourke T, Reinhorn AM, Shinozuka M, Tierney K, Wallace W, Winterfelt DV (2003) A framework to quantitatively assess and enhance the seismic resilience of communities. Earthq Spectra 19(4):733–752
Cimellaro GP, Fumo C, Reinhorn AM, Bruneau M (2009) Quantification of seismic resilience of health care facilities MCEER technical report-MCEER-09-0009. Multidisciplinary Center for Earthquake Engineering Research, Buffalo, NY
Cimellaro GP, Christovasilis IP, Reinhorn AM, De_Stefano A, Kirova T (2010a) L'Aquila Earthquake of April 6, 2009 in Italy: rebuilding a resilient city to withstand multiple hazards. MCEER technical report –MCEER-10-0010, MCEER, State University of New York at Buffalo (SUNY), Buffalo, New York
Cimellaro GP, Reinhorn AM, Bruneau M (2010b) Framework for analytical quantification of disaster resilience. Eng Struct 32(11):3639–3649
Cimellaro GP, Reinhorn AM, Bruneau M (2011) Performance-based metamodel for health care facilities. Earthq Eng Struct Dyn 40:1197–1217. doi:10.1002/eqe.1084
Cimellaro GP, Villa O, Bruneau M (2014) Resilience-based design of natural gas distribution networks. J Infrastruct Syst, ASCE. doi:10.1061/(ASCE)IS.1943-555X.0000204
Cornell A, Krawinkler H (2000) Progress and challenges in seismic performance assessment. Peer News April 2000. 3(2)
Elton EJ, Gruber MJ (1997) Modern portfolio theory, 1950 to date. J Bank Finance 21:1743–1759
FEMA (1997) FEMA 273 NEHRP guidelines for seismic rehabilitation of buildings. Federal Emergency Management Agency, Washington, DC
FEMA (2000) FEMA 356 prestandard and commentary for the seismic rehabilitation of buildings. FEMA 356 Federal Emergency management Agency, Washington, DC
FEMA (2006) FEMA 445: next generation performance-based seismic design guidelines. Federal Emergency Management Agency, Washington, DC
Kawashima K, Zafra R, Sasaki T, Kajiwara K, Nakayama M (2011) Seismic performance of existing and new reinforced concrete bridge columns based on full-scale shake table experiment using E-defense (Presentation). Bled4 workshop, performance based seismic engineering – vision for an Earthquake Resilient Society, Bled, Slovenia, 24–27 June 2011
Krawinkler H, Miranda E (2004) Performance-based earthquake engineering. Earthquake engineering: from engineering seismology to performance-based engineering chapter 9(9–1 to 9–59)
Maffei J (2009) Building it right the first time: improving the seismic performance of new buildings. San Francisco Planning Urban Research Association (SPUR), San Francisco, California
Moehle J (2004) A framework for performance-based earthquake engineering. In: Whittaker A (ed) Proceedings of the Tenth U.S.-Japan workshop on improvement of building seismic design and construction practices, report ATC-15-9, Applied Technology Council, Redwood City, CA
Park J, Goodno B, Bostrom A, Craig J (2004) Probabilistic decision analysis for seismic rehabilitation of a regional building system. In: Proceedings of the 13th world conference on earthquake engineering, Vancouver, BC Paper #2254, 1–6 Aug
Renschler C, Frazier A, Arendt L, Cimellaro GP, Reinhorn AM, Bruneau M (2010) Framework for defining and measuring resilience at the community scale: the PEOPLES resilience framework. MCEER technical report –MCEER-10-006, p 91, University at Buffalo (SUNY), The State University of New York, Buffalo, New York
Renschler C, Reinhorn AM, Arendt L, Cimellaro GP (2011) The PEOPLES Resilience Framework: A conceptual approach to quantify community resilience. In: Proceedings of COMPDYN 2011 – 3rd international conference in computational methods in structural dynamics and earthquake engineering, Corfù, Greece, 26–28 May 2011

SEAOC_Vision_2000 (1995) A framework for performance based design, Vol I. Structural Engineers Association of California, Vision 2000 Committee, Sacramento, California
Uma SR, Pampanin S, Christopoulos C (2006) A probabilistic framework for performance-based seismic assessment of structures considering residual deformations First European Conference on Earthquake Engineering and Seismology (1st ECEES), Geneva, Switzerland, 3–8 Sept 2006, paper number: 731

# Chapter 28
# Ground Motion Selection for Performance-Based Engineering: Effect of Target Spectrum and Conditioning Period

**Jack W. Baker, Ting Lin, and Curt B. Haselton**

**Abstract** This chapter presents a study of the impact of conditioning period on structural analysis results obtained from ground motions selected using the Conditional Spectrum concept. The Conditional Spectrum provides a quantitative means to model the distribution of response spectra associated with ground motions having a target spectral acceleration at a single conditioning period. One previously unresolved issue with this approach is how to condition this target spectrum for cases where the structure of interest is sensitive to excitation at multiple periods due to nonlinearity and multi-mode effects. To investigate the impact of conditioning period, we perform seismic hazard analysis, ground motion selection, and nonlinear dynamic structural analysis to develop a "risk-based" assessment of a 20-story concrete frame building. We perform this assessment using varying conditioning periods and find that the resulting structural reliabilities are comparable regardless of the conditioning period used for seismic hazard analysis and ground motion selection. This is true as long as a Conditional Spectrum (which carefully captures trends in means and variability of spectra) is used as the ground motion target, and as long as the analysis goal is a risk-based assessment that provides the annual rate of exceeding some structural limit state (as opposed to computing response

---

J.W. Baker (✉)
Department of Civil and Environmental Engineering, John A. Blume Earthquake Engineering Center, Stanford University, 473 Via Ortega MC 4020, Stanford, CA, USA
e-mail: bakerjw@stanford.edu

T. Lin
Department of Civil, Construction and Environmental Engineering, Marquette University, Engineering Hall 209, 1637 W. Wisconsin Ave, Milwaukee, WI 53233, USA
e-mail: ting.lin@marquette.edu

C.B. Haselton
Department of Civil Engineering, California State University, 209F Langdon Hall, Chico, CA 95929-0930, USA
e-mail: chaselton@csuchico.edu

M. Fischinger (ed.), *Performance-Based Seismic Engineering: Vision for an Earthquake Resilient Society*, Geotechnical, Geological and Earthquake Engineering 32,
DOI 10.1007/978-94-017-8875-5_28, 

conditioned on a specified ground motion intensity level). Theoretical arguments are provided to support these findings, and implications for performance-based earthquake engineering are discussed.

**Keywords** Ground motion selection • Seismic risk assessment • Nonlinear analysis

## 28.1 Introduction

Recent work has illustrated that scaling up arbitrarily selected ground motions to a specified spectral acceleration (*Sa*) level at period *T* can produce overly conservative structural responses, because a single extreme *Sa*(*T*) level of interest for engineering analysis does not imply occurrence of equally extreme *Sa* levels at all periods. The Conditional Mean Spectrum (CMS) and Conditional Spectrum (CS) have been developed to describe the expected response spectrum associated with a ground motion having a specified *Sa(T)* level (Baker and Cornell 2006; Baker 2011; Lin et al. 2013). The Conditional Mean Spectrum for a rare (i.e., positive $\varepsilon$) *Sa*(*T*) level has a relative peak at *T* and tapers back towards the median spectrum at other periods. The Conditional Spectrum differs from the Conditional Mean Spectrum in that it also considers the variability in response spectra at periods other than the conditioning period (which by definition has no variability).

The Conditional Spectrum approach for selecting and scaling ground motions requires the user to specify a conditioning period (denoted here as $T^*$) that is used to compute corresponding distributions of spectral values at all other periods. Ground motions can then be selected to match these spectral values, and used as inputs to dynamic structural analysis, to compute Engineering Demand Parameters, or *EDP*s. When calculating Peak Story Drift Ratio (*PSDR*) in buildings, $T^*$ is often chosen to be the building's elastic first-mode period ($T_1$). This is done because $Sa(T_1)$ is often a "good" predictor of that *EDP*, so scaling ground motions based on this parameter can lead to reduced scatter in resulting response predictions and thus minimizes the required number of dynamic analyses (Shome et al. 1998).

There are situations where the application of the CS concept is not yet straight-forward. One such situation is for prediction of *EDP*s which are not dominated by the first-mode structural response, due to contributions from higher modes, such as peak floor accelerations, or to longer periods associated with reduced-stiffness nonlinear response such as the onset of collapse (Haselton and Baker 2006). A second situation is for selection of ground motions prior to identification of a single conditioning period, because the structure is not yet designed or because multiple designs having multiple periods are being considered. In both cases, one is faced with the possibility of using a Conditional Spectrum that is not conditioned on *Sa* at the period that most efficiently predicts structural response. And even in cases where one does know an effective conditioning period for computing the CS, the question still arises as to whether a comparable-intensity *Sa* at some other conditioning period might produce a larger level of *EDP*s than the primary CS being considered.

Here we will demonstrate that, *if* the analysis objective is to compute the annual rate of the structure experiencing $EDP > y$, and *if* the ground motions are selected to match the Conditional Spectrum, then the resulting answer is relatively insensitive to the choice of conditioning period. While some researchers have previously suggested that the choice of conditioning period may not be critical to estimates of reliability (e.g., Abrahamson and Yunatci 2010; Shome and Luco 2010), those efforts did not perform a full risk-based assessment using nonlinear dynamic analyses, and did not consider spectral variability at periods other than the conditioning period. Here we do repeated risk-based assessments (i.e., compute the rate of $EDP > y$ using nonlinear dynamic analysis at multiple *Sa* levels) to demonstrate this statement empirically. We also present theoretical arguments and intermediate results to support these findings.

## 28.2 Demonstration Analysis

To illustrate the effect of conditioning period, we first perform two parallel performance assessments, using ground motions selected to match Conditional Spectra conditioned on two periods. We will later look at the effect of repeating the procedure using other conditioning periods. The test case and analysis procedure is described in this section.

### *28.2.1 Building Site and Structural Model*

The structure being studied is assumed to be located in Palo Alto, California, approximately 10 km from the San Andreas Fault. The structure is a 20-story reinforced concrete special moment frame with the perimeter frame designed to resist lateral forces. This building was designed for the recent FEMA P695 project (Federal Emergency Management Agency 2009; Haselton and Deierlein 2007), and is denoted Building 1020 in that study. It is modeled in OpenSEES (2011), with strength deterioration (both cyclic and in-cycle) and stiffness deterioration that is believed to reasonably capture the responses up to the point of dynamic-instability collapse. The first three elastic modal periods are 2.6, 0.85 and 0.45 s. The building was designed per the ICC (2003), for a site with a slightly lower design ground motion level than the site being utilized in this study (by approximately 20 %). Estimating the annual rate of exceeding various thresholds of Peak Story Drift Ratio in this building is not trivial, as the *PSDR* is affected by multiple modes excited at multiple periods, and experiences effective period lengthening as it behaves nonlinearly up to the collapse level for high intensity ground motions.

### 28.2.2 Seismic Hazard Analysis and Ground Motion Selection

We perform seismic hazard analysis to obtain ground motion hazard curves for spectral accelerations at three periods (0.85, 2.6 and 5 s), corresponding to the first two modal periods of the building and a lengthened period that may be a good predictor of nonlinear response. For each spectral period and amplitude of interest, we obtain the rate of exceeding that amplitude and a deaggregation distribution providing the causal magnitudes, distances and $\varepsilon$ values associated with spectral accelerations of that amplitude. All of this data comes from the U.S. Geological Survey online tools (USGS 2008). Ten rates of exceedance are considered for each conditioning period, ranging from 0.023 to 0.00005 per year (i.e., 50 % in 30 years to 1 % in 200 years probability of exceedance), as those are the exceedance rates for which the USGS provides the needed hazard and deaggregation information. Hazard curves and deaggregation results are not provided for exactly the conditioning periods used here, so interpolation between hazard results at adjacent periods is utilized.

Using the hazard curve and deaggregation information for a particular conditioning period, the Conditional Spectrum calculation is used to compute the mean and standard deviation of logarithmic response spectral values at all other periods, conditioned on an amplitude of $Sa(T^*)$. The mean and standard deviation of ln$Sa$ are given by the following equations (Baker and Cornell 2005a; Baker 2011)

$$\mu_{\ln Sa(T_i)|\ln Sa(T^*)} = \mu_{\ln Sa}\left(M, R, T_i\right) + \rho\left(T_i, T^*\right)\varepsilon\left(T^*\right)\sigma_{\ln Sa}\left(T_i\right) \qquad (28.1)$$

$$\sigma_{Sa(T_i)|Sa(T^*)} = \sigma_{\ln Sa}\left(T_i\right)\sqrt{1-\rho^2\left(T_i, T^*\right)} \qquad (28.2)$$

where $\mu_{\ln Sa}(M, R, T_i)$ and $\sigma_{\ln Sa}(T_i)$ are the predicted mean and standard deviation from a ground motion prediction model (Boore and Atkinson 2008 in this case), $\rho(T_i, T^*)$ is the correlation between the spectral values at period $T$ and the conditioning period $T^*$ (obtained from Baker and Jayaram 2008), and $M$, $R$ and $\varepsilon(T^*)$ come from the deaggregation distributions described in the previous section. In this case the $M$, $R$ and $\varepsilon(T^*)$ values used are the mean values from deaggregation at the given $Sa(T^*)$ level; this is an approximation relative to the use of the full distributions of potential $M$, $R$ and $\varepsilon(T^*)$ values, and performing a more exact calculation is possible and important to do in some cases as discussed in detail by Lin (2012).

For each conditioning period and spectral amplitude, 40 recorded ground motions were selected and scaled such that their spectra matched the target mean and standard deviations computed using Eqs. 28.1 and 28.2. Figure 28.1 shows the target spectra and selected ground motions' spectra for 0.85 and 2.6 s conditioning periods, at $Sa$ amplitudes with 2 % probability of exceedance in 50 years. Ground motions were selected from the PEER NGA database (Chiou et al. 2008). No further constraints were placed on the ground motion selection (e.g., magnitudes and distances) other than limiting scale factors to less than four, with the primary selection focus being on the match of the ground motion spectra to the target spectra. This was done because the structure response parameter of interest in this

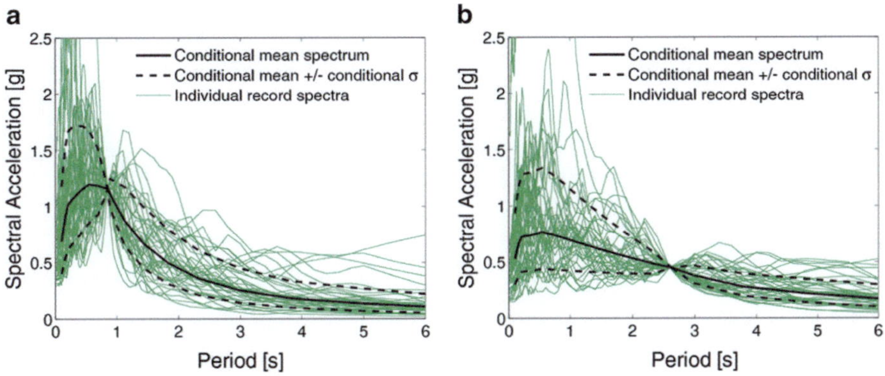


**Fig. 28.1** Conditional spectra and spectra of selected ground motions for a site at Palo Alto, California, with spectral acceleration at the conditioning period having a 2 % probability of exceedance in 50 years, (**a**) conditioned on $Sa(0.85\ \text{s}) = 1.2$ g, (**b**) conditioned on $Sa(2.6\ \text{s}) = 0.45$ g

case is thought to be most closely related to spectral values, and that earthquake magnitude and distance affect this structural response primarily as they relate to spectral values (which are accounted for directly through the Conditional Spectrum) rather than other ground motion parameters such as duration. Details regarding the ground motion selection algorithm and its implications are provided by Jayaram et al. (2011).

### 28.2.3 *Structural Analysis and Risk Assessment*

With the selected ground motions (40 motions at each of 10 intensity levels, for a given conditioning period), dynamic analysis of the structure described above was performed. Results are shown in Fig. 28.2 for the ground motions selected conditioned on two periods, and the fraction of ground motions causing collapse at each conditioning period and *Sa* level are shown in Fig. 28.3 (in this figure, results from ground motions with a third conditioning period of 5 s are also shown). A collapse fragility curve was obtained using a maximum likelihood approach to fit a lognormal fragility function to those observed fractions of collapse (Baker and Cornell 2005b, Appendix D).

For our risk-based assessment the structural analysis results are combined with the hazard curve for the corresponding conditioning *Sa*, to compute the annual rate of exceeding a given *PSDR* level as follows:

$$\lambda\left(PSDR > y\right) = \int_x P\left(PSDR > y|Sa\left(T^*\right) = x\right)\left|d\lambda\left(Sa\left(T^*\right) > x\right)\right| \quad (28.3)$$

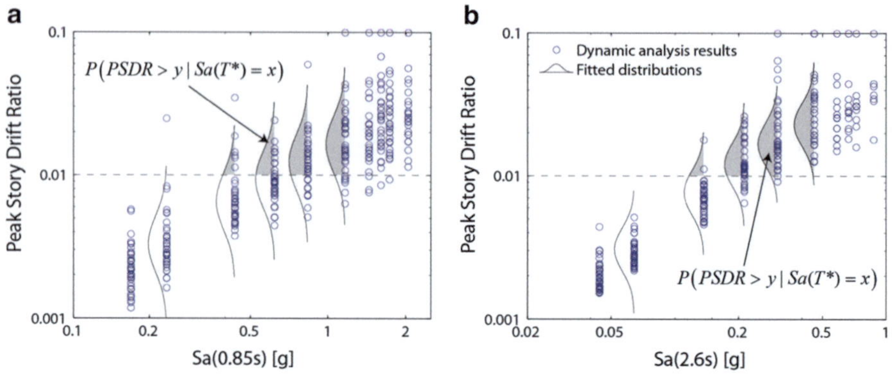


**Fig. 28.2** Peak Story Drift Ratios from non-collapse dynamic structural analysis, and fitted probability distributions, for ground motions selected to match Conditional Spectra with (**a**) $T^* = 0.85$ s and (**b**) $T^* = 2.6$ s

**Fig. 28.3** Observed fractions of analyses causing collapse from ground motions selected to match Conditional Spectra with three conditioning periods, and fitted fragility functions

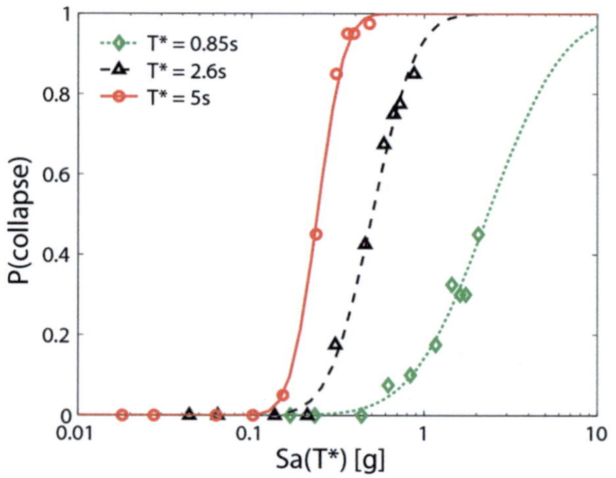


where $d\lambda(Sa(T^*) > x)$ is the derivative of the hazard curve for $Sa(T^*)$, $P(PSDR > y|Sa(T^*) = x)$ is the probability of Peak Story Drift Ratio exceeding $y$ given a ground motion with $Sa(T^*) = x$, and $\lambda(PSDR > y)$ is the rate of Peak Story Drift Ratio exceeding $y$. The $P(PSDR > y|Sa(T^*) = x)$ term is computed as follows

$$P\left(PSDR > y|Sa\left(T^*\right) = x\right) = P(C) + (1 - P(C))\left(1 - \Phi\left(\frac{\ln y - \mu_{\ln PSDR}}{\beta_{\ln PSDR}}\right)\right) \tag{28.4}$$

where $P(C)$ is the probability of collapse given $Sa(T^*) = x$ estimated from the collapse fragility function in Fig. 28.3, $\mu_{\ln PSDR}$ and $\beta_{\ln PSDR}$ are the mean and

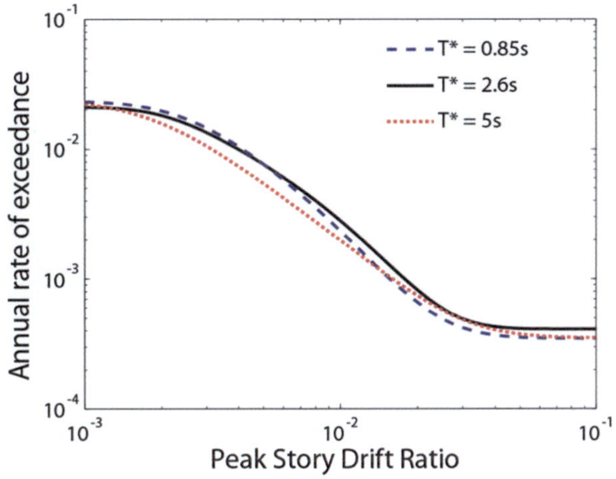


**Fig. 28.4** Risk assessment results showing annual rates of exceedance for various Peak Story Drift Ratios, obtained using hazard curves and ground motions with three different conditioning periods

standard deviation, respectively, of ln*PSDR* values given $Sa(T^*) = x$ in Fig. 28.2, and $\Phi(\ )$ is the standard normal distribution cumulative distribution function. This approach assumes that all collapse cases exceed *y*, and fits a lognormal distribution to the non-collapse *PSDR*s, following procedures proposed elsewhere (e.g., Shome and Cornell 1999).

The calculation in Eq. 28.3 is referred to here as a risk-based assessment, though it is also referred to elsewhere as the first step of the "PEER Integral," (Cornell and Krawinkler 2000), a "drift hazard" calculation (Krawinkler and Miranda 2004), or a "time-based assessment" (Applied Technology Council 2011). Equation 28.3 was evaluated using the three sets of hazard curves, ground motions and resulting structural responses associated with each of the considered $T^*$ values, and the resulting risk assessment results are shown in Fig. 28.4. The predictions of the rates of exceeding a given *PSDR* are very consistent regardless of the conditioning period.

The relative consistency of results in Fig. 28.4 may be surprising at first, so let us examine the data underlying these results more closely. In Fig. 28.5a, b, we see the response spectra of the ground motions selected and scaled to match *Sa*(2.6 s) and *Sa*(0.85 s) at the ten amplitudes considered; we see the "pinched" shapes of the spectra at 0.85 and 2.6 s in Fig. 28.5a, b, respectively, at those ten conditioning amplitudes. At other periods, the spectra are more varied, as the amplitudes at other periods have variability even when $Sa(T^*)$ is known with certainty. But the careful way in which these ground motions were selected, to maintain proper conditional means and variances, ensures that the distributions of spectra at all periods are still consistent with all known hazard information for the site being considered. It is difficult to see this consistency visually in Fig. 28.5a, b, because there are 40 ground motions at each *Sa* amplitude, while the real site will have many more low-amplitude ground motions than high-amplitude motions (and Eq. 28.3 captures this by incorporating weights from the hazard curve).

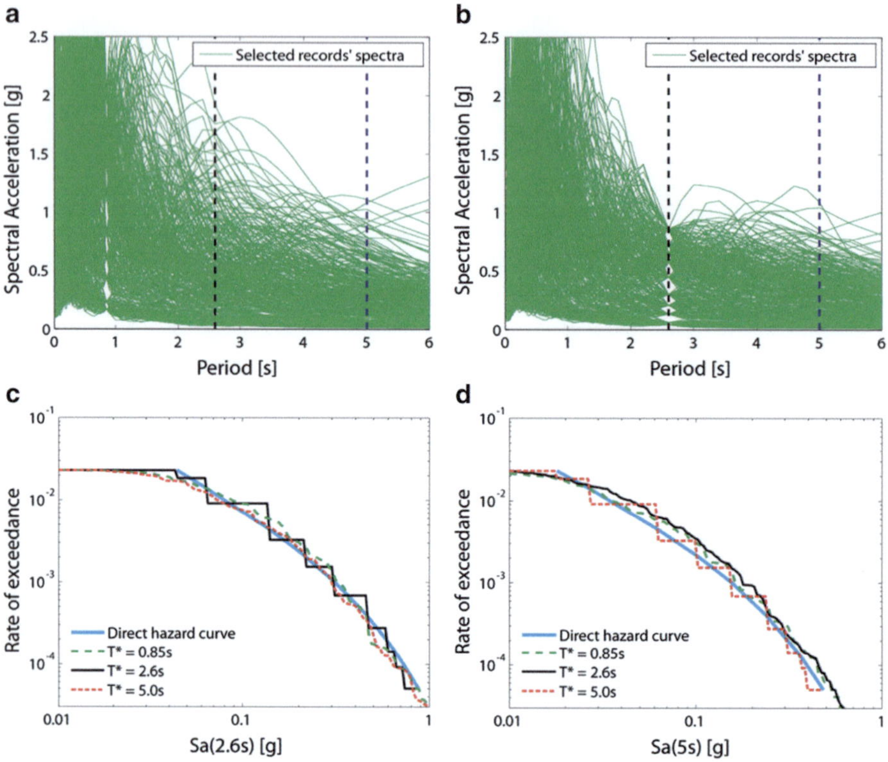


**Fig. 28.5** (**a**) Response spectra of ground motions selected using $T^* = 0.85$ s. (**b**) Response spectra of ground motions selected using $T^* = 2.6$ s. (**c**) Rate of $Sa(2.6\text{ s}) > y$ implied by each of the selected ground motion sets, plus the original ground motion hazard curve for reference. (**d**) Rate of $Sa(5\text{ s}) > y$ implied by each of the selected ground motion sets, plus the original ground motion hazard curve for reference

To make a quantitative comparison of the sets of response spectra, the rate of *Sa* exceeding a given amplitude *y* at an arbitrary period *T*, implied by the ground motions selected conditional on $Sa(T^*)$, is computing using an equation very similar to Eq. 28.3

$$\lambda\left(Sa(T) > y\right) = \int_x P\left(Sa(T) > y | Sa\left(T^*\right) = x\right) \left| d\lambda\left(Sa\left(T^*\right) > x\right)\right| \quad (28.5)$$

where $P(Sa(T) > y|Sa(T^*) = x)$ is the probability that a ground motion in the selected record set to have $Sa(T^*) = x$ has an *Sa* at period *T* that is greater than *y*. Here this probability is estimated as simply the fraction of the 40 ground motions with $Sa(T^*) = x$ that have $Sa(T) > y$. The multiplication of these probabilities by the derivative of the hazard curve reweights the results according to the rate of observing ground motions with $Sa(T^*) = x$, as was done in Eq. 28.3.

Figure 28.5c shows the implied rate of $Sa(2.6\ \text{s}) > y$ for the ground motions in Fig. 28.5a, b, plus comparable ground motions with $T^* = 5$ s. Additionally, the hazard curve for $Sa(2.6\ \text{s})$ is plotted, as this is the correct answer from hazard analysis that we are trying to represent using a suite of ground motions. The ground motions selected using $T^* = 2.6$ s have a stepped plot in Fig. 28.5c, due to the ten discrete $Sa(2.6\ \text{s})$ amplitudes that were considered when selecting motions. The ground motions with other $T^*$ values have smoother curves, but all of the curves are in good general agreement, indicating that even though the other sets of ground motions did not scale ground motions to match $Sa(2.6\ \text{s})$, they have the proper distribution of $Sa(2.6\ \text{s})$ as specified by the hazard curve at that period. Thus, if $Sa(2.6\ \text{s})$ is a good predictor of structural response, then the ground motions selected to match $Sa(0.85\ \text{s})$ will still do a good job of capturing the distribution of structural response values that might be observed for the given site and structure considered. A similar plot is shown in Fig. 28.5d for the rate of exceeding $Sa(5\ \text{s})$; in this case the ground motions with $T^* = 5$ s have the stepped curve, and the other $T^*$ cases are smooth. Again the curves are in relatively good agreement with each other, and with the true ground motion hazard curve.

## 28.3 Discussion

In principle, Eq. 28.3 is correct regardless of the value used to quantify intensity, but a few assumptions inherent in the application of this equation place practical constraints on this evaluation. First, Eq. 28.3 assumes that $P(PSDR > y|Sa(T^*) = x)$ is not dependent upon other ground motion properties besides the one quantified by the intensity measure, although this is never true for structures other than elastic single-degree-of-freedom systems (Luco and Cornell 2007). Here we have addressed that problem by further accounting for the effect of spectral values at other periods through ground motion selection with Conditional Spectrum targets, such that spectral values at all periods in the selected ground motions are consistent with hazard curves for the site, regardless of the spectral period used for conditioning. Nonetheless, we have only considered spectral values and not other ground motion properties (e.g. velocity pulses not fully captured in the spectral acceleration values, duration, etc.). If non-spectral ground motion parameters were also deemed important for predicting the *EDP* of interest, the approach above can be generalized to account for those parameters (Bradley 2010).

Another limitation of the approach used here is that Eqs. 28.1 and 28.2 used for computing the target Conditional Spectra are approximate if only a single magnitude and distance value is input, or only a single ground motion prediction model is used, because the calculations that produced the hazard curves use multiple magnitudes, distances and ground motion prediction models. That approximation was reasonable for the cases considered here, but is known to be unreasonable for many other cases. More exact uses of Eqs. 28.1 and 28.2 are available in (Lin et al. 2013), and the impact of this refinement is the subject of more detailed discussion in (Lin 2012).

## 28.4 Conclusions

We have presented risk-based assessment results for Peak Story Drift Ratios in a 20-story concrete frame structure located in Palo Alto, California, using a structural model with strength and stiffness deterioration that is believed to reasonably capture the responses up to the point of dynamic-instability collapse. The assessment was performed three times, using ground motions selected and scaled to match Conditional Spectra at three conditioning periods from 0.85 to 5.0 s (i.e., the second-mode structural period up to twice the first-mode period). For each case, the risk-based assessment results were similar. This similarity stems from the fact that the careful record selection ensures that the distributions of response spectra at all periods are consistent with targets specified by hazard analysis, so the distribution of resulting story drifts should also be comparable (to the extent that response spectra describe the relationship between the ground motions and structural responses).

From these results, we observe if the analysis goal is to perform a full "risk assessment" calculation, then one should be able to obtain an accurate result using any conditioning period, as long as careful ground motion selection ensures proper representation of spectral values and other ground motion parameters of interest. Here "proper representation" refers to consistency with the site ground motion hazard curves at all relevant periods, and this is achieved by using the Conditional Spectrum approach to determine target response spectra for the selected ground motions. The reproducibility of the results with varying conditional periods then results from the fact that the ground motion intensity measure used to link the ground motion hazard and the structural response is not an inherent physical part of the seismic reliability problem we are considering; it is only a useful link to decouple the hazard and structural analysis. If this link is maintained carefully then one should get a consistent answer (the correct answer) for the risk assessment in every case. The consistency in risk assessments achieved here is in contrast to some previous speculation on this topic, because this study utilizes the recently developed Conditional Spectrum target for ground motion selection, and uses a newly available algorithm for selecting ground motions to match this target.

Is the choice of conditioning period still important? Choice of a "good" conditioning period does still serve several useful purposes. A good conditioning period helps because the spectral accelerations at the conditioning period will be a good predictor of structural response; this makes any inaccuracies in representing spectral values at other periods have a less severe impact on the resulting calculations. Additionally, use of a good conditioning period reduces the variability in structural responses and thus reduces the number of dynamic analyses that are required to accurately estimate distributions of structural response. Luco and Cornell (2007) referred to these two properties as "sufficiency" and "efficiency," respectively. We take those concepts further here, acknowledge that there is no intensity measure with perfect efficiency and sufficiency, and so perform careful ground motion selection to compensate for shortcomings that are inherent in any intensity measure.

This document has presented a relatively simple illustration of the concept that hazard consistency in ground motions will lead to consistent risk-based assessment results. This work is part of a larger project on ground motion selection (NIST 2011), and the PhD thesis of Ting Lin (2012) provides a much more extensive set of analyses of this type, including studies of permutations on the target spectrum used, the *EDP* parameter of interest, and the type of structure being analyzed. Those results provide a more complete picture of the relationship between careful ground motion selection and robust structural response results.

**Acknowledgements** This work was supported in part by the NEHRP Consultants Joint Venture (a partnership of the Consortium of Universities for Research in Earthquake Engineering and Applied Technology Council), under Contract SB134107CQ0019, Earthquake Structural and Engineering Research, issued by the National Institute of Standards and Technology, for project ATC-82. Any opinions, findings and conclusions or recommendations expressed in this material are those of the authors and do not necessarily reflect those of the NEHRP Consultants Joint Venture. The authors also acknowledge the contributions of Jared DeBock and Fortunato Enriquez in conducting the structural analyses used in this study.

## References

Abrahamson NA, Yunatci AA (2010) Ground motion occurrence rates for scenario spectra. In: Fifth international conference on recent advances in geotechnical earthquake engineering and soil dynamics, San Diego, paper no. 3.18, 6p

Applied Technology Council (2011) ATC-58, guidelines for seismic performance assessment of buildings, 75% draft. Applied Technology Council, Redwood City, 266p

Baker JW (2011) Conditional mean spectrum: tool for ground motion selection. J Struct Eng 137(3):322–331

Baker JW, Cornell CA (2005a) A vector-valued ground motion intensity measure consisting of spectral acceleration and epsilon. Earthq Eng Struct Dyn 34(10):1193–1217

Baker JW, Cornell CA (2005b) Vector-valued ground motion intensity measures for probabilistic seismic demand analysis (Report no. 150). John A. Blume Earthquake Engineering Center, Stanford, 321p

Baker JW, Cornell CA (2006) Spectral shape, epsilon and record selection. Earthq Eng Struct Dyn 35(9):1077–1095

Baker JW, Jayaram N (2008) Correlation of spectral acceleration values from NGA ground motion models. Earthq Spectra 24(1):299–317

Boore DM, Atkinson GM (2008) Ground-motion prediction equations for the average horizontal component of PGA, PGV, and 5 %-damped PSA at spectral periods between 0.01 s and 10.0 s. Earthq Spectra 24(1):99–138

Bradley BA (2010) A generalized conditional intensity measure approach and holistic ground-motion selection. Earthq Eng Struct Dyn 39(12):1321–1342

Chiou B, Darragh R, Gregor N, Silva W (2008) NGA project strong-motion database. Earthq Spectra 24(1):23–44

Cornell CA, Krawinkler H (2000) Progress and challenges in seismic performance assessment. PEER Center News 3(2):1–3

Federal Emergency Management Agency (2009) Quantification of building seismic performance factors (FEMA P695, ATC-63). FEMA P695, prepared by the Applied Technology Council, 421p

Haselton C, Baker JW (2006) Ground motion intensity measures for collapse capacity prediction: choice of optimal spectral period and effect of spectral shape. In: Proceedings, 8th national conference on earthquake engineering, San Francisco, p 10
Haselton CB, Deierlein GG (2007) Assessing seismic collapse safety of modern reinforced concrete moment frame buildings. Pacific Earthquake Engineering Research Center, Berkeley
ICC (2003) International building code 2003. International Code Council, ICC (distributed by Cengage Learning)
Jayaram N, Lin T, Baker JW (2011) A computationally efficient ground-motion selection algorithm for matching a target response spectrum mean and variance. Earthq Spectra 27(3):797–815
Krawinkler H, Miranda E (2004) Performance-based earthquake engineering. In: Bozorgnia Y, Bertero VV (eds) Earthquake engineering: from engineering seismology to performance-based engineering. CRC Press, Boca Raton
Lin T (2012) Advancement of hazard consistent ground motion selection methodology. PhD thesis, Dept. of Civil and Environmental Engineering, Stanford University, Stanford
Lin T, Harmsen SC, Baker JW, Luco N (2013) Conditional spectrum computation incorporating multiple causal earthquakes and ground motion prediction models. Bull Seismol Soc Am 103(2A):1103–1116.
Luco N, Cornell CA (2007) Structure-specific scalar intensity measures for near-source and ordinary earthquake ground motions. Earthq Spectra 23(2):357–392
NIST (2011) Selecting and scaling earthquake ground motions for performing response-history analyses. NIST GCR 11-917-15, Prepared by the NEHRP Consultants Joint Venture for the National Institute of Standards and Technology, Gaithersburg
OpenSEES (2011) Open system for earthquake engineering simulation. Pacific Earthquake Engineering Research Center, http://opensees.berkeley.edu/. Accessed 20 Jun 2011
Shome N, Cornell CA (1999) Probabilistic seismic demand analysis of nonlinear structures (Report no. RMS35). PhD thesis, RMS program, Stanford, p 320
Shome N, Luco N (2010) Loss estimation of multi-mode dominated structures for a scenario of earthquake event. In: 9th US National and 10th Canadian conference on earthquake engineering, Toronto, p 10
Shome N, Cornell CA, Bazzurro P, Carballo JE (1998) Earthquakes, records, and nonlinear responses. Earthq Spectra 14(3):469–500
USGS (2008) Interactive deaggregation tools. United States Geological Survey, https://geohazards.usgs.gov/deaggint/2008/. Accessed 20 Jun 2011

# Chapter 29
# Reliability Considerations in the Seismic Capacity Design Requirements for Force-Controlled Components

**Victor K. Victorsson, Jack W. Baker, and Gregory G. Deierlein**

**Abstract** This chapter describes factors to consider in developing a methodology to establish capacity-design criteria for force-controlled elements in seismic force resisting systems. The focus is on capacity-designed connections in steel concentrically braced frames, but the concepts can be generally applied to other structural components and systems. The proposed methodology is an adaptation of the load and resistance factor design (LRFD) methodology, where the load effects are defined by the force demands from yielding components of the system. Demand and capacity factors (analogous to load and resistance factors) are determined considering the variability in inelastic earthquake demands and component capacities, along with a target reliability. The target reliability is based on a comprehensive collapse risk assessment that is evaluated using nonlinear dynamic analyses and benchmarked to the collapse safety of modern code-conforming buildings.

**Keywords** Seismic design • Capacity design • Reliability • Steel structures • Collapse safety • Load and resistance factor design

---

V.K. Victorsson
Global Engineering, Swiss Reinsurance Company Ltd., Mythenquai 50/60, 8002 Zürich, Switzerland
e-mail: victorknutur@gmail.com

J.W. Baker • G.G. Deierlein (✉)
Department of Civil and Environmental Engineering, John A. Blume Earthquake Engineering Center, Stanford University, 473 Via Ortega MC 4020, Stanford, CA, USA
e-mail: bakerjw@stanford.edu; ggd@stanford.edu

M. Fischinger (ed.), *Performance-Based Seismic Engineering: Vision for an Earthquake Resilient Society*, Geotechnical, Geological and Earthquake Engineering 32,
DOI 10.1007/978-94-017-8875-5_29, 

## 29.1 Introduction

Most modern building codes employ capacity design principles to help ensure ductile response and energy dissipation capacity in seismic force resisting systems. The design provisions are geared toward restricting significant inelastic deformations to those structural components that are designed to sustain large inelastic deformations. Such elements are often referred to as deformation-controlled components. Other structural components, referred to as force-controlled components, are designed with sufficient strength to remain essentially elastic, even under large earthquake ground motions.

The 2010 AISC *Seismic Provisions* (AISC 2010a) for brace connections, columns and beams in steel Special Concentrically Braced Frames (SCBFs) are one example of where capacity design principles are used to design force-controlled elements. The design provisions aim to confine significant inelastic deformation in the braces while the brace connections, columns and beams remain essentially elastic. The design intent is achieved by requiring that the design strengths of brace connections, columns and beams exceed the expected strength of the braces by an appropriate margin, considering the inherent variability in the force demands and component strengths. In concept, the capacity design requirement is given by the following equation:

$$\phi C_n \geq \gamma D_n \tag{29.1}$$

where $C_n$ is the nominal strength of the force-controlled component, $D_n$ is the nominal force demand, imposed by the yielding component; and $\gamma$ and $\phi$ are demand and capacity factors (similar to load and resistance factors), which are determined based on a target reliability for the force-controlled component.

As the primary goal of seismic building code provisions is to ensure that buildings have adequate collapse safety, the safety margins for capacity design should be determined in the context of the overall system safety. Thus, the establishment of capacity design requirements should consider the following questions:

1. What is the likelihood that the imposed force demand will exceed the strength of capacity designed force-controlled components?
2. How does the failure of a capacity designed component impact the collapse safety of the overall structural system?
3. What are the appropriate demand and capacity factors, and $\gamma$ and $\phi$, to ensure that the system meets the target collapse safety for new buildings.

In this chapter, methods to address these questions will be illustrated through an application to evaluate design requirements for braced connections in a six-story SCBF building. The example is based on a more comprehensive study of the reliability of capacity-designed components by Victorsson et al. (2012) (Fig. 29.1).

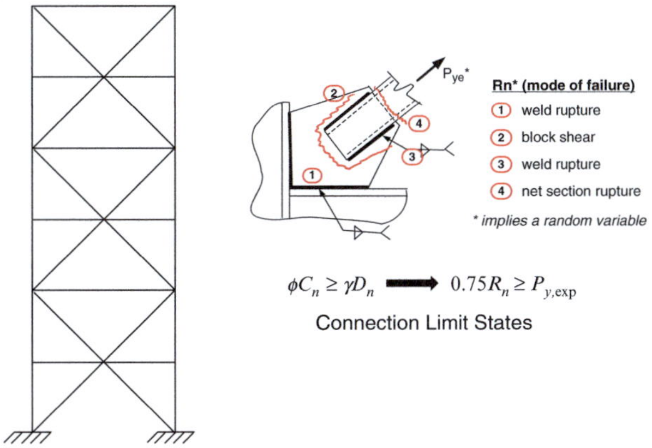


**Fig. 29.1** Force-controlled limit states design for brace connections in steel special concentrically braced frame (SCBF)

## 29.2 Seismic Collapse Safety of Modern Buildings

The FEMA P695 report on *Quantification of Building Seismic Performance Factors* (FEMA 2009) provides a framework to evaluate the collapse probability of building seismic systems. The framework provides a basis to establish minimum seismic design forces and related design requirements for seismic systems that helps ensure consistent collapse safety among the alternative building systems and materials permitted by modern building codes. The FEMA P695 framework employs nonlinear dynamic analyses to evaluate collapse probabilities, taking into account (1) variability in earthquake ground motions, (2) uncertainties in the design, quality assurance and nonlinear analysis, and (3) incomplete knowledge of the structural behavior.

The FEMA P695 framework assesses the reliability of structural systems by nonlinear dynamic analysis of structural archetype models, which are designed to generally represent the characteristics of the building system designation in the building code (e.g., steel SCBF). FEMA P695 specifies a set of 22 ground motion pairs, which are applied to the nonlinear analysis models with increasing intensity, i.e. using an Incremental Dynamic Analysis (IDA), until structural collapse is detected. The analysis data are used to determine the median ground motion collapse intensity, from which a collapse fragility curve is developed assuming a lognormal cumulative distribution function with a specified dispersion (logarithmic standard deviation) and an adjustment to account for ground motion spectral shape effects. The resulting collapse fragility curve (see Fig. 29.2) relates the ground motion intensity, described in terms of spectral acceleration (*Sa*), to the probability of collapse, i.e. $P(Collapse|Sa)$. Based on judgment informed by benchmark studies of several code-conforming systems, FEMA P695 specifies a maximum tolerable collapse risk of 10 % under *maximum considered earthquake* (MCE) ground motion intensities, i.e., $P(Collapse|Sa_{MCE}) \leq 10\,\%$.

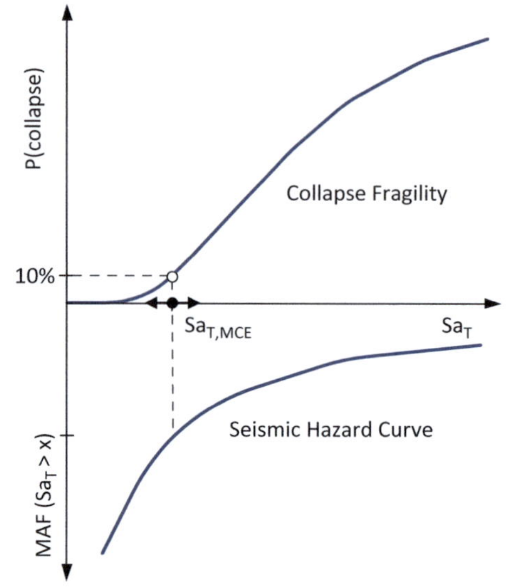


**Fig. 29.2** Integration of collapse fragility and seismic hazard curves

Building on the collapse fragilities defined in FEMA P695, the MCE seismic design maps for the United States have recently been revised to provide more consistent collapse risk safety throughout various regions of the United States (Luco et al. 2007). These new MCE design maps are predicated on achieving a maximum uniform risk of collapse less than a 1 % chance of exceedance in 50 years. This is in contrast to the prior definition of MCE maps, which were associated with ground motion intensities that had a 2 % chance of exceedance in 50 years. This recalibration of the MCE maps represents a change from the previous uniform-hazard ground motion intensity to uniform-risk ground motion intensity. As illustrated in Fig. 29.2, the new MCE design map intensities are obtained by integrating site ground motion hazards with a generic collapse fragility curve with a lognormal distribution and an assumed dispersion of 0.6, which is reasoned to be a conservative estimate based on FEMA P695 procedures. With the fixed dispersion of 0.6, the lognormal collapse capacity curve can be fully described by the assumed 10 % probability of collapse at the MCE intensity (as specified in the FEMA P695 procedures). Thus, given the default collapse fragility and the ground motion hazard curve for a specific site, the MCE intensity is then calculated for each map location, such that the integration of the two yields the target collapse risk of 1 % in 50 years, i.e., $P(Collapse)_{50\ years} \leq 1\ \%$. The resulting uniform risk MCE design maps have been adopted into the 2010 edition of the ASCE 7 (ASCE 2010) standard for seismic design in the United States. These developments are significant as they establish procedures and target collapse safety risk that provide the basis for establishing seismic design guidelines for new buildings.

## 29.3 Probability of Demand Exceeding Capacity of Force-Controlled Components

The nonlinear dynamic analyses used to establish the median collapse capacity in the FEMA P695 and similar procedures are typically performed using models that are calibrated to the expected values (central values) of the structural response parameters. As such, these collapse analyses do not directly account for the risk of failure in force-controlled components, since the expected properties of the force-controlled components are, by design, larger than the expected demands from yielding elements. Therefore, additional measures are needed to evaluate the failure risk in force-controlled components and how it may impact the collapse risk to the overall structural system. Assuming that the risk of collapse can be evaluated separately for the overall system, where force-controlled components are assumed to remain intact, $P(Coll_{D\leq C})_{50\,years}$, and the additional risk of collapse due to failure of force-controlled components, $P(Coll_{D>C})_{50\,years}$, then the total collapse risk is simply the sum of these two, where the probability is calculated based on a mean annual frequency over a 50-year time horizon:

$$P(Collapse)_{50\ years} = P\left(Coll_{D\leq C}\right)_{50\ years} + P(Coll_{D>C})_{50\ years} \tag{29.2}$$

The first term in Eq. 29.2, $P(Coll_{D\leq C})_{50years}$, can be determined by procedures similar to those of FEMA P695 where the capacity-designed components are assumed to remain intact. The focus of this study is on the second term, corresponding to collapse risk due to failure of the force-controlled components, $P(Coll_{D>C})_{50\,years}$.

Shown in Fig. 29.3 are nonlinear analysis results for a six-story SCBF that has been designed using the ASCE 7 and AISC Seismic Provisions for an MCE spectral intensity of Sa(T1) equal to 1.1 g and a system response factor of R equal to 6. The nonlinear analyses incorporate the effects of brace yielding, buckling and fracture, degrading flexural hinging in the beams and columns, and large deformation (P-$\Delta$) effects. As such, the analyses do a reasonably good job at capturing nonlinear behavior up to the onset of collapse. Figure 29.3a, b show results of an incremental dynamic analysis and the resulting collapse fragility calculated following the FEMA P695 procedures, where the risk of collapse under MCE ground motion intensity is about 10 %. Figure 29.3c, d show how the maximum brace forces develop under increasing ground motion, where the brace force is normalized by the expected tension strength of the braces. Points to note from these figures are (1) that the brace forces increase very rapidly and saturate at their maximum values at ground motion intensities significantly below the MCE intensities, and (2) in contrast to the large variability in drift response (Fig. 29.3a) the variability of the maximum brace forces (Fig. 29.3c) is well constrained about the expected brace yield strength.

Referring to Fig. 29.4a, b, the variable brace force demand (D) can be compared to the brace connection capacity (C) to determine the probability that the demand exceeds the capacity at variable ground motion intensities. As indicated, the failure probability can be controlled by the ratio of demand to capacity factors, $\gamma$ and

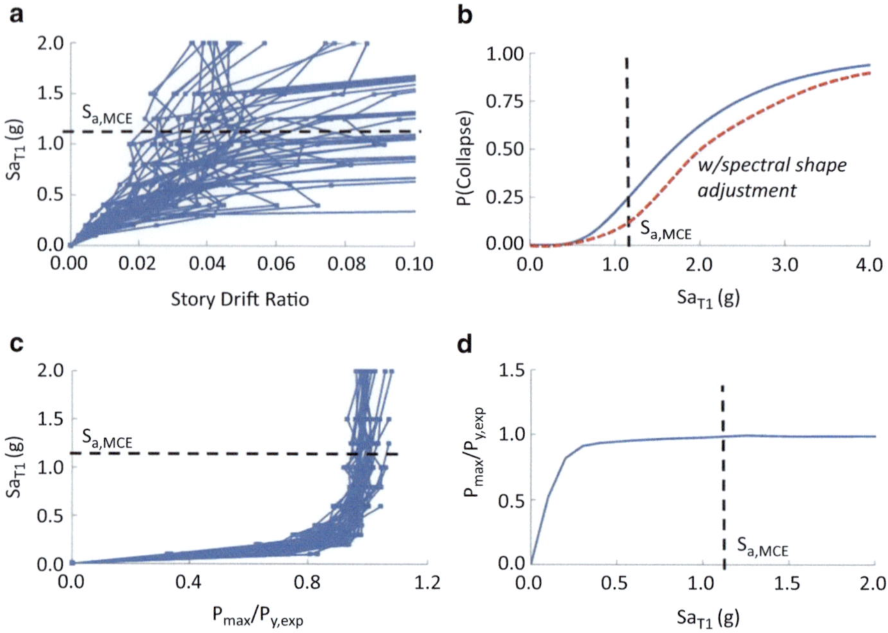


**Fig. 29.3** Nonlinear analysis results of 6-story SCBF (**a**) incremental dynamic analysis – spectral ground motion intensity versus story drift ratio, (**b**) collapse fragility curve assuming brace connections intact, i.e., D < C, (**c**) normalized brace force demands versus ground motion spectral intensity, (**d**) median normalized brace force demands versus ground motion spectral intensity

ϕ. Much like the brace force demand, the conditional probability of connection failure, $P(D > C)|Sa$, plotted in Fig. 29.4c, increases rapidly and saturates well below the MCE ground motion intensity. Thus, when integrated with the seismic hazard curve (Fig. 29.2), the early rise in $P(D > C)|Sa$ would lead to rather frequent expectations of connection failures. The steep increase in the plot of Fig. 29.4c further suggests that the calculations could be simplified by approximating the curve with a step function, which increases from zero to the expected $P(D > C)$ at a ground motion intensity corresponding to the point of significant yielding, $Sa_{,yield}$, in the structure. This approximation can simplify calculations for the risk occurrence of connection failure, i.e., the mean annual frequency $MAF(D > C)$, by replacing the integration to a simple product of $P(D > C|Sa > Sa_{,yield})$ and the mean annual frequency $MAF(Sa > Sa_{yield})$, which can be obtained from the ground motion seismic hazard curve. Mathematically, this is as follows:

$$MAF\,(D > C) \cong P\left(D > C\,\middle|Sa > Sa_{yield}\right) * MAF\left(Sa > Sa_{yield}\right) \quad (29.3)$$

In this example, $Sa_{,yield}$ is equal to about 0.25 g (about one quarter of the MCE intensity) and has a MAF of exceedance of 0.01/year for the chosen building site. When multiplied by the risk of connection failure (D > C, assuming a 0.09 failure

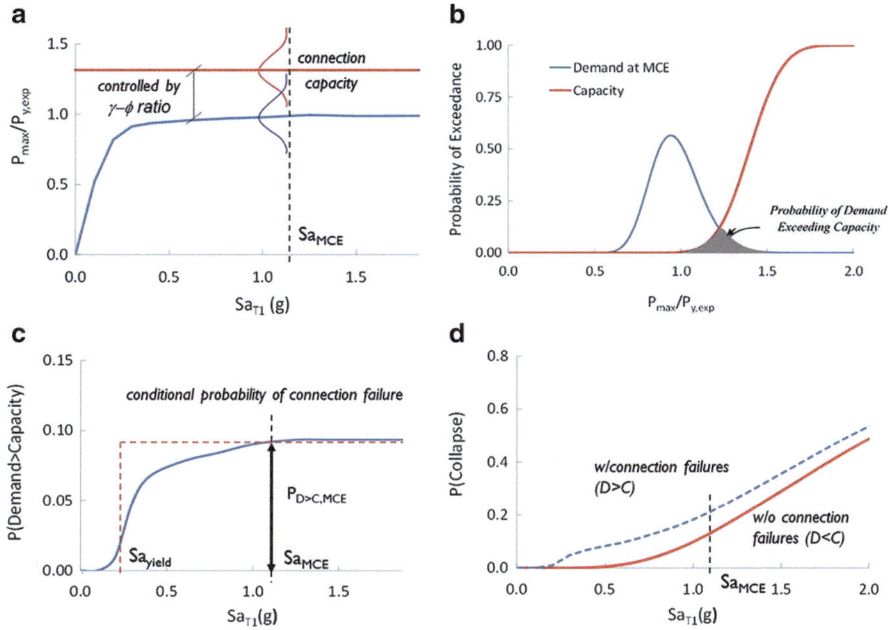


**Fig. 29.4** Connection failure data for six-story SCBF: (**a**) normalized brace force (**b**) Elevation of frame, (**c**) Maximum brace forces, $P_{max}$, recorded in each analysis normalized by expected yield strength, $P_{y,exp}$, (**d**) Probability of connection failure vs. spectral acceleration for a given connection capacity and dispersion

probability for $Sa > Sa_{,yield}$) the result is about a 4.5 % chance of connection failure in 50 years. This 4.5 % probability of connection failure is over four times the maximum target risk of building collapse of 1 % in 50 years.

## 29.4 Collapse Due to Failure of Force-Controlled Components

As shown in Fig. 29.4d, if one conservatively assumes that brace connection failure triggers frame collapse, then the probability of brace connection failure (Fig. 29.4c) would simply add directly to the probability of system collapse, obtained from the incremental dynamic analyses of the overall system (Figs. 29.3a, b). If judged by the change in collapse probability at the MCE intensity, the risk of connection failure would increase the probability of collapse, $P(Collapse)_{MCE}$, by about 1.8 times, from the original collapse probability of about 12 % (w/o connection failure) to 21 % (with connection failure). However, when integrated with the ground motion hazard curve to determine the annual rate of failure (e.g., as illustrated in Fig. 29.2), the addition of the connection failure probability to the collapse fragility curve

(Fig. 29.4d) has a much more dramatic effect on the collapse risk. This occurs because of the rapid increase in probability of connection failure at the low and frequent ground motion intensities. For example, when integrated with a hazard curve for the high seismic region of coastal California, the dashed fragility curve of Fig. 29.4d that includes connection failure would result in a $P(Collapse)_{50\,years}$ of 5.5 %, which is over six times larger than the than the 0.9 % probability calculated for the base fragility without connection failures. This example demonstrates how it can be misleading to evaluate collapse risk only at the MCE intensity as compared to integrating the full range of intensities with the seismic hazard curve. This has obvious implications on current engineering practice, where it is not uncommon to evaluate force-controlled limit states only at MCE level intensities, which can give misleading impressions as to the risk of failure.

While the simple addition of connection failure probability to the overall collapse probability is a logical first approximation, especially for systems with low redundancy such as the braced frame considered here, closer analysis shows that this can be a very conservative assumption. To more carefully assess how connection failures impact the overall frame stability, we conducted additional nonlinear response history analyses where connection failure was simulated directly. Since the connection failure criteria are uncertain, the analyses were conducted using a Monte Carlo type assessment where the brace connection strengths were assumed as uncorrelated random variables.

The Monte Carlo nonlinear analyses are initially performed with brace connection fracture excluded, and then the probability of brace demand exceeding the connection capacity is calculated for the non-collapsed cases. With an assumed median connection capacity of 1.35 times the median brace yield strength and dispersion of 0.15, the probability of demand exceeding capacity is calculated using the component reliability concepts described in the previous section. The connection strengths of the Monte Carlo realization are then incorporated in the model and the dynamic analyses are re-run for the cases where the connection capacity is less than the brace demand. The number of additional collapses due to connection failure is then incorporated into the collapse fragility curve.

Figure 29.5a demonstrates that the added probability of collapse due to connection fractures is not constant and initially increases as the ground motion intensity $Sa_{T1}$ increases. In other words, $P(Coll_{D>C})|D>C)$ varies with the ground motion intensity, $Sa_{T1}$. No new collapses are recorded at $Sa_{T1}=0.40$ g, suggesting that at this ground motion intensity, the frame is robust enough that it can survive even if connections fracture. As the ground motion intensity increases, the frame's inherent collapse resistance decreases and $P(Coll_{D>C})|D>C, Sa)$ increases. These results tend to agree with conclusions from Luco and Cornell (2000) on the effects of brittle connection fractures in steel special moment resisting frames, i.e. that the effect of connection fractures is less pronounced at lower ground motion intensities than at higher ones. These results greatly reduce the influence of brace connections on the system reliability as even if braces are likely to fracture at low spectral accelerations, i.e. close to $Sa_{y,exp}$, the probability of frame collapse is low.

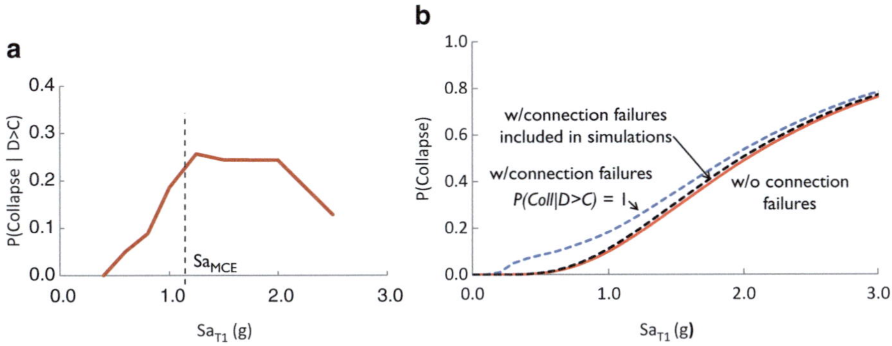


**Fig. 29.5** Collapse probabilities for six-story SCBF: (**a**) change in collapse probability conditioned on connection failure (**b**) collapse fragility curves with and without connection failures

Using the plot of the probability of collapse conditioned on connection failure for non-collapsed frames, $P(Coll_{D>C})|D > C)$, from Fig. 29.5a, combined with the previous data on the probability of connection failure, $P(D > C)|Sa$, from Fig. 29.4c, the total collapse fragility curve is calculated as shown in Fig. 29.5b. The lowest curve (solid red line) and the upper curve (blue dashed line) are the two cases shown previously (Fig. 29.4d) without and with connection failures; and the middle curve (black dashed line, close to the solid red line) represents the case with connection failures and including the conditional probability from Fig. 29.5a. As indicated, by considering the data on conditional collapse probabilities, the resulting collapse fragility indicates that connection failure has a very modest influence on the final collapse fragility. When the three fragility curves from Fig. 29.5b are integrated with the seismic hazard curve, the resulting collapse probabilities, $P(Collapse)_{50\ years}$, are 0.85 %, 0.90 % and 5.50 %, respectively. Thus, the additional probability of collapse due to connection fractures is only 0.05 % in 50 years, which is dramatically less than the value calculated when the conditional collapse probability (Fig. 29.5a) is ignored. It is important to note that the data in Fig. 29.5a are based on analyses where the variability in connection strength is assumed to be uncorrelated. Correlation between uncertainties in connection strengths will generally worsen the performance, though not to the extent as when connection failure is assumed to be synonymous with collapse.

## 29.5 Reliability-Based Method to Determine Capacity-Design Factors for Force-Controlled Components

The analyses presented above demonstrate how the risk of failure of force-controlled components is related to the overall risk of collapse to the structure. Ultimately, the target probability of failure (or reliability index) of the force controlled components depend on the following factors:

*$P(D > C)|Sa > Sa_{,yield})$:* the probability that the force demand D imposed by yielding components will exceed the capacity C, conditioned on the structure having experienced ground motions to initiate yielding.

*MAF $(Sa > Sa_{,yield})$:* the mean annual frequency that the structure will experiences ground motions that initiate significant yielding in the members that generate forces in the force-controlled components.

*$P(Coll_{D>C}|D > C,\ Sa)$:* the probability of collapse caused by failure of force-controlled components. As illustrated in Fig. 29.5a, this probability depends on the ground motion intensity and conditioned on the subset of cases where the structure has not collapsed due to other factors (e.g., sidesway collapse where the force-controlled components are intact).

*Target MAF $(Collapse_{D>C})$ or $P(Collapse_{D>C})_{50\,year}$:* the maximum permissible mean annual frequency of structural collapse, due to failure of the force controlled components. As described per Eq. 29.2, this target probability is constrained by the target limit on structural collapse from all causes, assumed to be on the order of 1 % in 50 years, per Luco et al. (2007), and the probability of collapse due to factors other than failure of the force-controlled components, which is assumed to be the main contributor to collapse.

Of these four probabilities, the first, $P(D > C)|Sa > Sa_{,yield})$, can be described in a design-sense in terms of an LRFD-like formulation (Ravindra and Galambos 1978; AISC 2010b) in which the reliability index, $\beta$, can be calculated as follows:

$$\beta = \frac{\ln\left(\frac{D_n}{D_m}\frac{C_m}{C_n}\frac{\gamma}{\varphi}\right)}{\sqrt{V_C^2 + V_D^2}} \tag{29.4}$$

where $D_n$ and $D_m$ are the nominal and median force demands, $C_n$ and $C_m$ are the nominal and median component capacities, $V_D$ and $V_C$ are variances in the force demands and capacities, $\gamma$ and $\phi$ are the demand and capacity factors, and $\beta$ is the resulting reliability index. Assuming that the force demands and capacities can be described by lognormal distributions, $\beta$ can be related to the probability of failure (i.e., that $D > C$, conditioned on $S_a > S_{a,yield}$) as shown in Fig. 29.6. In the case of brace connections in steel SCBFs, the connection capacity terms ($C_n$, $C_m$ and $V_C$) are the same as those assumed in the standard *AISC Specification* (AISC 2010b) requirements, the nominal demand $D_n$ is the expected yield strength of the brace, $P_{y,exp}$, and the median demand and variability in demand ($D_m$ and $V_D$) can be developed through nonlinear analysis of SCBFs (e.g., Fig. 29.3) and brace tests. Given this information to characterize the demands and capacities, once a target reliability index is known, then the $\gamma$ and $\phi$ factors can be used to adjust the probability of failure of the force-controlled components (e.g., $P(D > C)|Sa > Sa_{,yield})$, as shown in Fig. 29.4.

| Reliability Index β | Probability of Failure |
|---|---|
| 0.0 | 0.5 |
| 0.5 | 0.31 |
| 1.0 | 0.16 |
| 1.5 | 0.07 |
| 2.0 | 0.02 |
| 2.5 | 0.006 |
| 3.0 | 0.001 |
| 3.5 | 0.0002 |
| 4.0 | 0.00003 |
| 4.5 | 0.000003 |
| 5.0 | 0.0000003 |

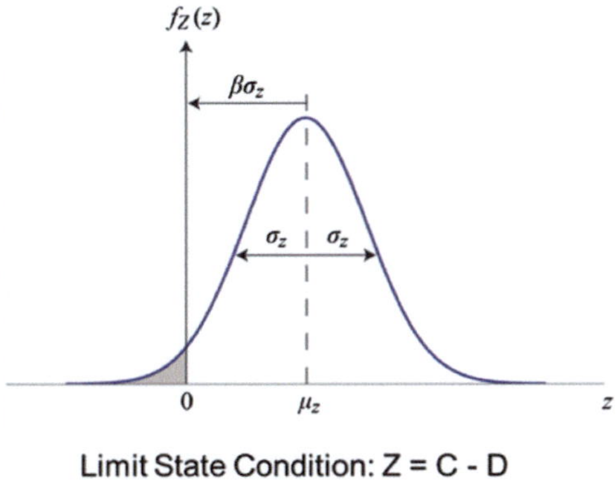


**Fig. 29.6** Collapse probabilities for six-story SCBF: (**a**) change in collapse probability conditioned on connection failure (**b**) collapse fragility curves with and without connection failures

The main challenge in the reliability assessment is to determine the target reliability index, β, which is equivalent to the establishing target failure probability $P(D > C)|Sa > Sa_{,yield})$. The appropriate target reliability (component failure probability) depends on the other three components of the analysis, i.e., $MAF\ (Sa > Sa_{,yield})$, $P(Coll_{D>C}|D > C,\ Sa)$, and $P(Collapse_{D>C})_{50year}$. The first of these, $MAF\ (Sa > Sa_{,yield})$, depends to a large extent on the seismic response factor that is used to define the required strength (e.g., the R-factor in United States practice), which is based on the inelastic deformation characteristics of the system. The second, $P(Coll_{D>C}|D > C,\ Sa)$, depends on the dynamic response characteristics, redundancy of the system, and the effect that failures of the force-controlled components have on the overall system behavior. The final term, $P(Collapse_{D>C})_{50\ year}$, should probably be limited to about 0.1–0.2 % in 50 years (MAF of 0.00002 to 0.00004/year), assuming that the total $P(Collapse)_{50\ year}$ is limited to 1 % in 50 years and that only a small portion (<10 %) of this should be attributed to failure of the force-controlled components.

In studies of SCBFs of the type described in this paper, the authors found that reliability indices, β, on the order of 2.5 provided acceptable performance. From Fig. 29.6, this β corresponds to a $P(D > C)|Sa > Sa_{,yield})$ of about 0.006, or 0.6 %. When combined using Eq. 29.4 with available statistical data on force demands and capacities ($D_n$, $D_m$, $C_n$, $C_m$, $V_D$ and $V_C$), this β of 2.5 implies that the demand and capacity factors of $\gamma = 1.0$ and $\phi = 0.75$, as specified for bracing connection components by the current AISC Provisions (AISC 2010a, b) are slightly conservative. Of course, while the underlying methodology outlined in this paper can be generally applied, the specific numerical results depend on data and assumptions that are specific to the SCBFs considered in this study.

## 29.6 Concluding Remarks

While the basic principle of capacity-design is straightforward, its implementation is complicated by uncertainties in the force demands and capacities, which introduce ambiguities as to how strong to make the force-controlled components. The calculation of appropriate demand and capacity factors for force-controlled components requires consideration of the overall system reliability, in order to maintain a reasonable balance between the achieving the idealized inelastic mechanism (as envisioned by capacity-design approach) and practical and economic limits on design. The proposed reliability-based methodology to establish capacity design requirements incorporates the main factors believed to influence the reliability of force-controlled components. While further work is needed to quantify the constituent components of the reliability assessment, the proposed methodology is intended to provide a framework that will enable the calculation of risk consistent capacity-designed components for structural components and systems.

**Acknowledgments** The authors would acknowledge the contributions of the late Professors Allin Cornell and Helmut Krawinkler to this research. Both of these individuals were involved in the early stages of this project and generously shared their knowledge and expertise. The authors would also acknowledge the advice and encouragement of Tom Schlafly and others on the specification committee of the American Institute of Steel Construction, who reviewed this project. The financial support of the American Institute of Steel Construction, the National Science Foundation (CMMI-1031722, Program Director M.P. Singh), and the Blume Earthquake Engineering Center at Stanford University is also acknowledged. The findings and conclusions made in this paper are those of the authors and do not necessarily reflect the views of the sponsors.

## References

AISC (2010a) Seismic provisions for structural steel buildings. American Institute of Steel Construction, Chicago

AISC (2010b) Specification for structural steel buildings. American Institute of Steel Construction, Chicago

ASCE (2010) ASCE-7-10: minimum design loads for buildings and other structures. American Society of Civil Engineers, Reston

FEMA (2009) Quantification of building seismic performance factors. FEMA P695/June 2009, Federal Emergency Management Agency, Washington, DC. http://www.fema.gov/media-library-data/20130726-1716-25045-9655/fema_p695.pdf

Luco N, Cornell CA (2000) Effects of connection fractures on SMRF seismic drift demands. ASCE J Struct Eng 126(1):127–136

Luco N, Ellingwood BR, Hamburger RO, Hooper JD, Kimball JK, Kircher CA (2007) Risk-targeted versus current seismic design maps for the conterminous United States, SEAOC 2007 convention proceedings, Squaw Creek, CA, 26–29 September 2007. Structural Engineers Association of California, 13 pp. http://books.google.com/books/about/SEAOC_2007_Convention_Proceedings.html?id=BIZUYgEACAAJ, http://www.seaoc.org/bookstore/proceedings-76th-annual-2007-squaw-creek

Ravindra MK, Galambos TV (1978) Load and resistance factor design for steel. J Struct Div ASCE 1978; 104(STD9):1337–1353

Victorsson VK, Deierlein GG, Baker JW, Krawinkler H (2012) The reliability of capacity-designed components in seismic resistant systems, J.A. Blume technical report 177, Stanford University, Stanford, CA

# Chapter 30
# Reassessing ACI 318 Shear Wall Provisions Based on Recent Earthquake and Test Observations

**John W. Wallace**

**Abstract** Observed wall damage in recent earthquakes in Chile (2010) and New Zealand (2011), where modern building codes exist, exceeded expectations. In these earthquakes, structural wall damage included boundary crushing, reinforcement fracture, and global wall buckling. Recent laboratory tests also have demonstrated inadequate performance in some cases, particularly for slender walls with thin boundary regions. These observations indicate a need to review code provisions, identify shortcomings, and make necessary revisions. Use of simple performance-based design approaches provides an ideal framework to incorporate code changes that balance performance expectations and impact/cost.

**Keywords** Earthquake damage • Laboratory testing • Field observations • Chile 2010 • Christchurch 2011 • Structural wall • Shear wall • ACI 318 • Detailing • 90-degree hooks • Confinement • Displacement-based design • Wall slenderness • Wall instability • Biaxial loading • Tension-controlled • Compression-controlled • Shaking table test • Full-scale • Shear • Sliding shear • Rebar buckling • Spectra • Displacement spectra • NEES • E-Defense

## 30.1 Introduction

Design and construction practice for special structural walls (ACI 318 designation) has evolved significantly since the system was introduced in the 1970s. Throughout the 1970s and 1980s, it was common to use so-called barbell-shaped wall cross

J.W. Wallace (✉)
Department of Civil Engineering, University of California, Los Angeles,
5731 Boelter Hall, Los Angeles, CA 90095-1593, USA
e-mail: wallacej@ucla.edu

M. Fischinger (ed.), *Performance-Based Seismic Engineering: Vision for an Earthquake Resilient Society*, Geotechnical, Geological and Earthquake Engineering 32,
DOI 10.1007/978-94-017-8875-5_30, 

sections, where a "column" was used at each wall boundary to resist axial load and overturning along with a narrow wall web. In the late 1980s and early 1990s, use of rectangular wall cross sections became common to produce more economical designs. Use of walls with rectangular cross sections is common in many countries, including Chile and New Zealand. Although use of walls with boundary columns is still common in Japan, based on information available in the literature, the AIJ Standard for "Structural Calculations of Reinforced Concrete Buildings" was revised in 2010 to show RC walls with rectangular cross-sections. Engineers around the world have pushed design limits in recent years, optimizing economy and design, and in many practices producing walls with higher demands and more slender profiles than have been verified in past laboratory testing or field experience. The trend towards more slender profiles has been accelerated by use of higher concrete strengths.

Observed wall damage in recent earthquakes in Chile (2010) and New Zealand (2011), where modern building codes exist, exceeded expectations. In these earthquakes, structural wall damage included boundary crushing, reinforcement fracture, and global wall buckling. Recent tests of isolated structural walls in the US and tests of two, full-scale 4-story buildings with high-ductility (or special) structural walls at E-Defense in December 2010 provide vital new data. A particularly noteworthy aspect of these recent tests is the failure of relatively thin wall boundaries to develop ductile behavior in compression, even though they complied with building code provisions and recommendations of ACI (ACI 318-95 1995, ACI 318-99 1999, ACI 318-11 2011) and AIJ (Architectural Institute of Japan 2010).

The observed performance following recent earthquakes and in recent laboratory tests suggests strongly that the problems observed are not isolated and that analysis and design provisions need to be reassessed. In particular, the quantity and configuration of transverse reinforcement required at wall boundaries needs to be reassessed to address issues associated with wall thickness, slenderness, axial load, and configuration, as well as expected displacement demands and load history. Preliminary studies indicate that greater amounts of transverse reinforcement may be required for thin walls or walls with large cover and that tighter spacing of transverse reinforcement may be required to suppress buckling of vertical reinforcement, especially for walls with light axial load or walls with flanges. These issues apply to both high ductility (ACI Special) and moderate ductility (ACI Ordinary) walls.

Given this background, the objectives of this paper are to review the performance of slender structural walls in recent earthquakes and laboratory tests, to review code provisions, to identify possible shortcomings, and to suggest approaches that could be implemented to address these issues. Use of simple performance-based design approaches are emphasized here because they provide an ideal framework to incorporate code changes that balance performance expectations and cost, with the added benefit of impacting the majority of buildings that are designed and constructed.

## 30.2 Observed Performance of Slender Structural Walls

### *30.2.1 Recent Earthquake Reconnaissance*

Recent earthquakes in Chile ($M_w$ 8.8, February 2010, EERI 2010), New Zealand (February 2011, $M_L = 6.3$), and Japan ($M_w$ 9.0, March 2011) have provided a wealth of new data on the performance of modern buildings that utilize structural walls for the primary lateral-force-resisting system. Although complete building collapse was rarely observed, damage was widespread and generally exceeded expectations.

In 1996, Chile adopted a new code (NCh 433.Of96 1996) based on ACI 318-95 and produced an immense inventory of progressively more slender buildings corresponding essentially to the US reinforced concrete code provisions, except boundary element confinement was not required. The 2010 $M_w$ 8.8 earthquake caused serious damage to many of these buildings, including crushing/spalling of concrete and buckling of vertical reinforcement, often over a large horizontal extent of the wall (Fig. 30.1). Damage tended to concentrate over a relatively short height of one to three times the wall thickness, apparently because buckling of vertical bars led to concentration of damage. Closer inspection of the wall boundary regions (Fig. 30.1) revealed the relatively large spacing of hoops (20 cm) and horizontal web reinforcement (20 cm), as well as the 90-degree hooks used on hoops and horizontal web reinforcement, which may have opened due to concrete crushing and/or buckling of vertical reinforcement (Fig. 30.1d). Some of the failures are attributable to lack of closely-spaced transverse reinforcement at wall boundaries, which was

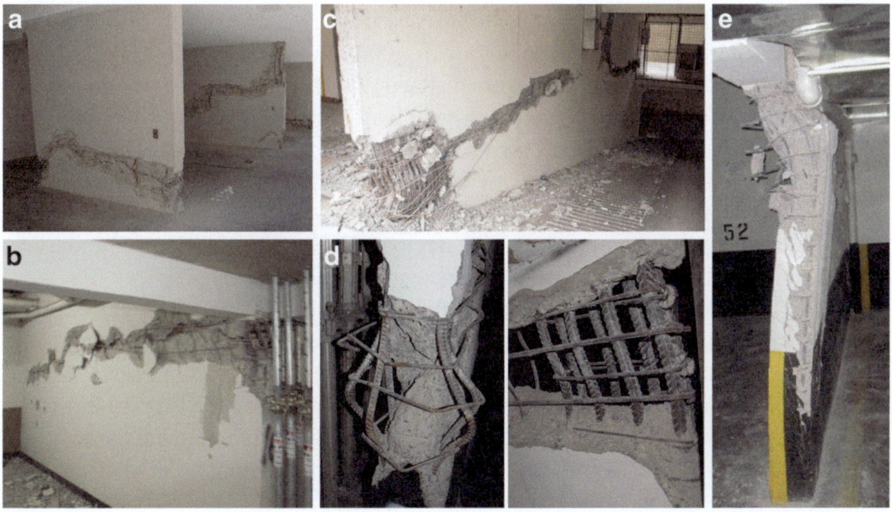


**Fig. 30.1** Typical wall damage in Chile earthquake. (**a**) Vina del Mar. (**b**) Santiago. (**c**) Concepcion. (**d**) Boundary zone details. (**e**) Wall lateral instability

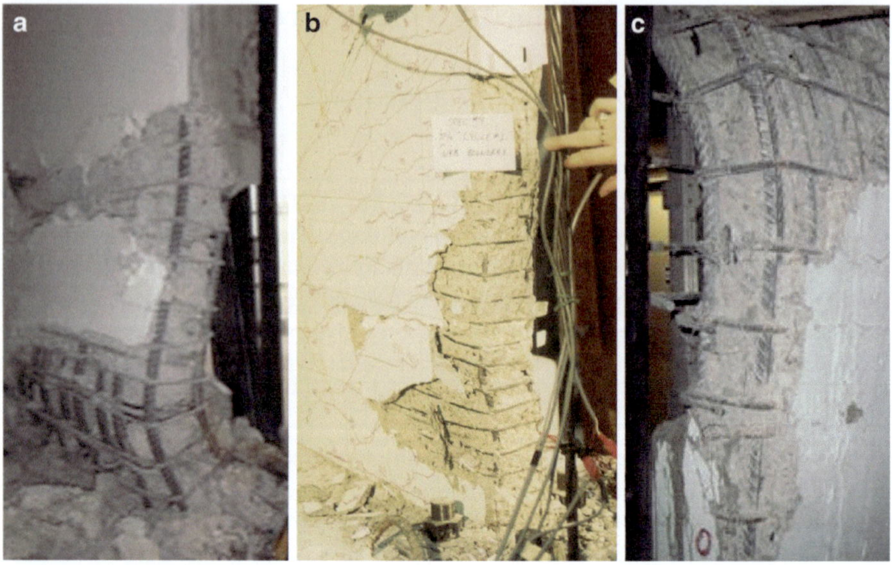

**Fig. 30.2** (**a**) Wall failure in 2011 Christchurch earthquake (Elwood 2011). (**b**) Specimen TW2 web boundary failure (Thomsen and Wallace 2004). (**c**) Wall failure in 2010 Chile earthquake

not required by the Chilean code based on the good performance of buildings in the 1985 M7.8 earthquake; however, many of the failures are not yet understood, and many suggest that there are deficiencies in current US design provisions (Wallace 2011; Massone and Wallace 2011). In some cases, lateral instability (buckling) of a large portion of a wall section was observed (Fig. 30.1e); prior to the Chile and New Zealand earthquakes, this global buckling failure had been primarily observed in laboratory tests (e.g., Thomsen and Wallace 2004). Detailed surveys conducted as part of ATC-94 (2011) indicate that global wall buckling was not driven by prior yielding in tension (as had originally been suspected based on past research, e.g., Corley et al. 1981; Paulay and Priestley 1993; Chai and Elayer 1999) but instead was the result of lateral instability of previously crushed boundary zones. Laboratory testing is required to understand these behaviors; preliminary studies are underway in Chile and the US to investigate these issues.

The 2011 Christchurch earthquake (EERI 2011a, b; NZRC 2011) shows many similar wall failures, suggesting the deficiencies observed in the 2010 Chile earthquake are not isolated (Fig. 30.2a). All of the walls depicted in Figs. 30.1 and 30.2 have either T-shaped (Figs. 30.1e and 30.2b) or L-shaped (Fig. 30.2a) cross sections, which leads to large cyclic tension and compressive demands at the wall web boundary (Wallace 1996). The wall web boundaries are apparently susceptible to out-of-plane buckling following cover concrete spalling. Although current ACI 318-11 provisions require consideration of an effective flange width, the provisions do not restrict use of narrow walls and do not address this out-of-plane failure mode, i.e., there are no restrictions on wall thickness or wall slenderness.

### 30.2.2 Recent Laboratory Studies of Conventional Walls

Recent laboratory testing of structural walls in the US has focused on addressing concerns related to behavior of walls with rectangular and T-shaped cross sections subjected to uniaxial and biaxial loading (Waugh et al. 2008; Waugh and Sritharan 2010; Brueggen 2009; Brueggen and French 2010), with couplers and splices in the plastic hinge region (Johnson 2010; Birely et al. 2010), with higher shear demands (Birely et al. 2008, 2010; Sriram and Sritharan 2010), and with coupling beams (Naish and Wallace 2010; Parra-Montesinos et al. 2012; Lehman and Lowes 2011). All of these studies involved quasi-static testing. Shake table testing of walls has been limited, except for 7-story "building slice" tests of walls with rectangular and T-shaped cross sections conducted by Panagiotou and Restrepo (2007) and the H-shaped cross section coupled wall tested by Fischinger et al. (2006). The overwhelming majority of quasi-static and shake table tests conducted in Japan have been conducted on barbell-shaped walls and low-rise buildings with "wing walls" (e.g., Kabeyasawa et al. 2008, 2010a, b), which are not common in the US. Only recently have the Japanese Building Standard Law and Architectural Institute of Japan (AIJ 2010) recommendations been modified to allow the use of rectangular walls with boundary elements, but their use is not yet widespread.

Johnson (2010) reports test results isolated, slender ($h_w/l_w$ and $Mu/V_u l_w = 2.67$) cantilever walls to investigate the behavior of anchorage details. Three walls were tested, one each with continuous (RWN), coupled (RWC), and spliced (RWS) vertical reinforcement. The wall cross sections were 6 in. $\times$ 90 in. (152.4 mm $\times$ 2.29 m), and the walls were subjected to horizontal lateral load 20 ft (6.1 m) above the bases. Although the wall cross-sections were rectangular, different amounts of boundary vertical reinforcement were used to simulate the behavior of T-shaped wall cross sections; 4-#6 ($d_b = 19$ mm) and 2-#5 ($d_b = 15.9$ mm) at one boundary and 8-#9 ($d_b = 28.7$ mm) at the other boundary. Horizontal wall web reinforcement, of #3 @7.5 in. or $\rho_t = 0.0049$ ($d_b = 9.5$ mm @ 19 cm), was selected to resist the shear associated with the expected moment strength (including overstrength). Wall web vertical reinforcement consisted of #4 @18 in. or $\rho_v = 0.0037$ ($d_b = 12.7$ mm @ 45.7 cm). It is noted that the 18 in. (45.7 cm) spacing of vertical web reinforcement is the maximum spacing allowed by ACI 318-11 §21.9.2.1. Lateral load versus top lateral displacement relations for RWC and RWS are plotted in Fig. 30.3a; since results for RWC and RWN are very similar. For RWC, the wall reached rotations exceeding +0.035 (#5 in tension) and −0.02 (#9 in tension), whereas for RWS, the wall reached rotations of approximately +0.02 (#5 in tension) and −0.012 (#9 in tension). Damage was concentrated at the foundation-wall interface, which accounted for about 0.015 rotation at a top rotation of 0.02. Significant horizontal cracking also was observed for specimens RWN and RWC, suggesting that the quantity (and large spacing) vertical web reinforcement was insufficient to restrain vertical crack opening between the boundary zones (Fig. 30.3b). However, the test results do indicate that the presence of the splice significantly reduced the wall lateral deformation capacity.

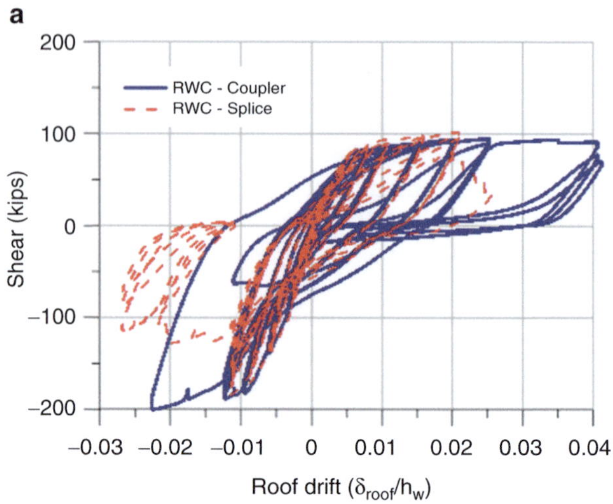

**Fig. 30.3** (**a**) Load-displacement relations. (**b**) Wall damage at end of test (RWS)

Tests of walls with splices also were conducted by Birely et al. (2010). The test specimens were roughly one-half scale replicas of the bottom three stories of a ten-story wall (Fig. 30.4a). Base shear versus 3rd story (top) displacement plots are shown in Fig. 30.4b for three of the tests, PW1 (splice, $M_b = 0.71 h_w V_b$), W2 (splice, $M_b = 0.50 h_w V_b$), and W4 (no splice, $M_b = 0.50 h_w V_b$). Design wall shear stresses were 0.23, 0.33, and 0.33 $\sqrt{f_c' \, MPa} \, MPa$ for W1, W2, and W4, respectively (equivalent to 0.7, 0.9, and $0.9V_n$). The #4 ($d_b = 12.7$ mm) boundary bars were lapped 0.61 m, with spacing of boundary transverse reinforcement of 51 mm ($s/d_b = 4$). The test with lower shear stress was reasonably ductile, achieving

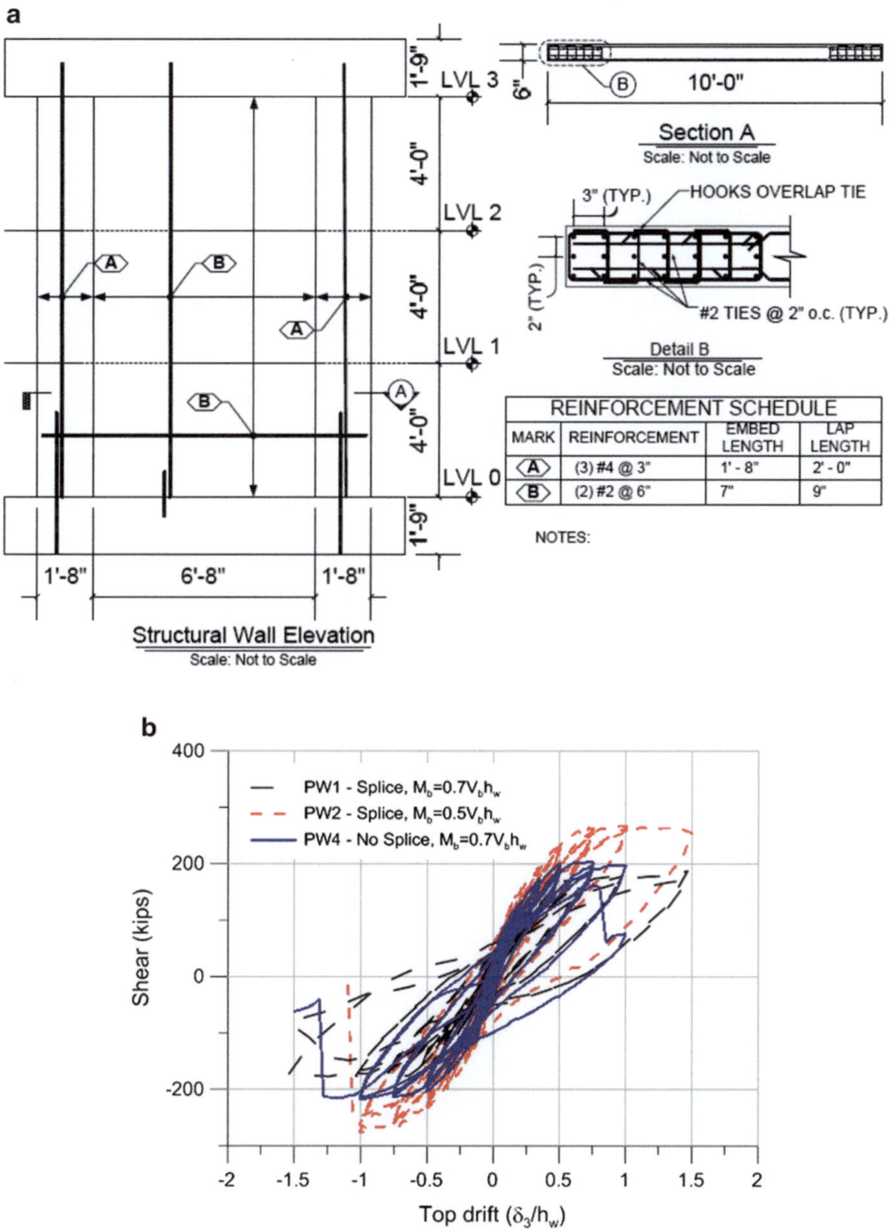


**Fig. 30.4** (**a**) NEESR UW wall tests (Birely et al. 2010). (**b**) Base shear vs. drift

$1.08M_n$ and a 3rd story lateral drift of 1.5 % prior to strength loss; however, test PW4, with no splice, reached only 1.0 % lateral drift at the third story (top) prior to strength loss. For all tests with splices, damage initiated with buckling of the interior bar at the wall edge (Fig. 30.5a) and then concentrated at the top of

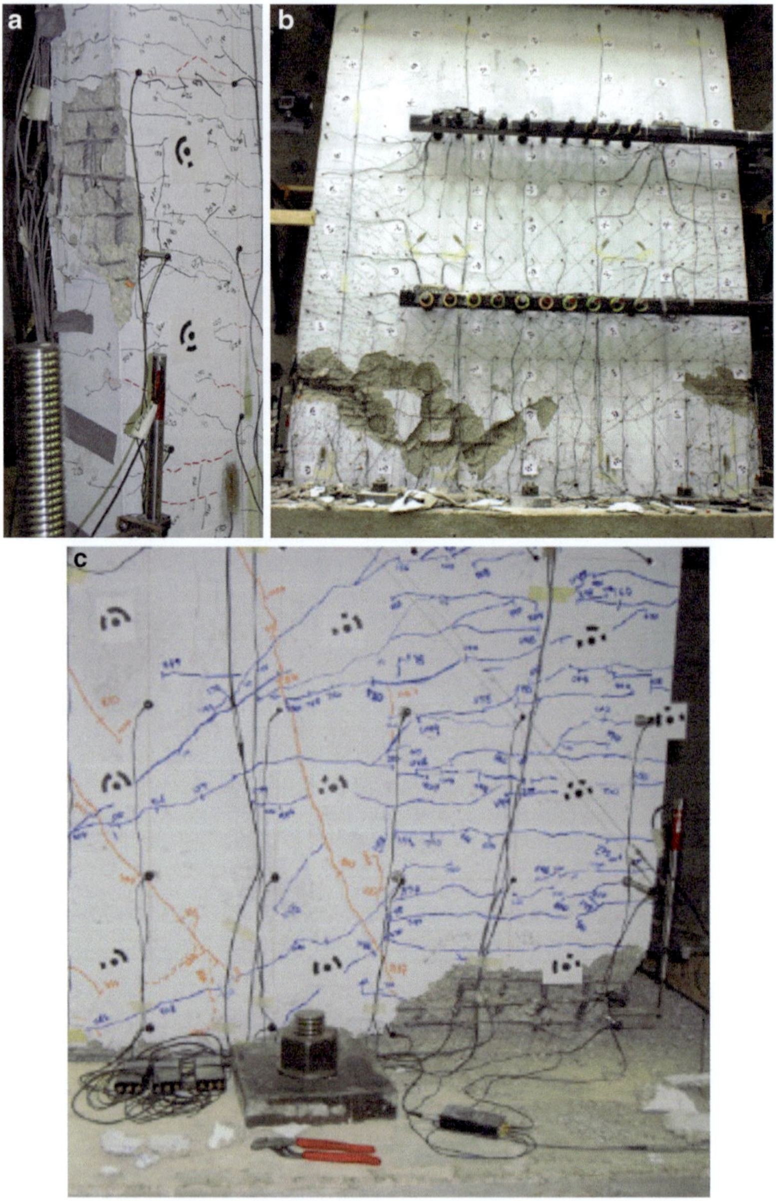

**Fig. 30.5** Wall damage: (**a**) PW2 @ 1.0 % drift; (**b**) PW2 end of test; (**c**) PW4 @ 1.0 % drift

the splices (Fig. 30.5b), whereas damage was concentrated at the foundation-wall interface for test PW4 with no splice (Fig. 30.5c). Even without consideration of the elastic deformations over the top seven stories not included in the test, deformation capacities of the walls are less than expected, especially for PW4, with no splice.

**a**

| | | | | | |
|---|---|---|---|---|---|
| 1Fl. | Section | 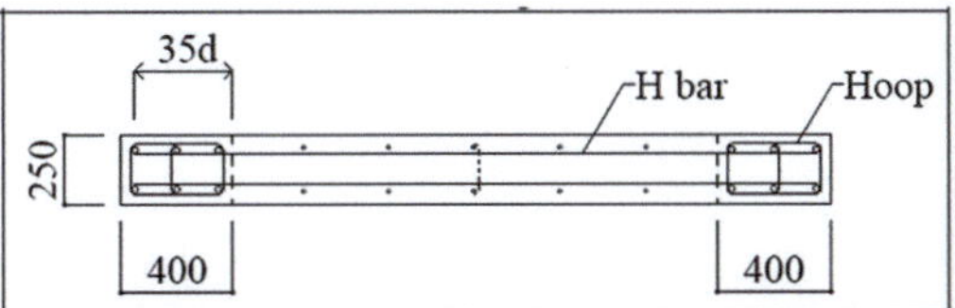  | | | |
| | B x D | 2,500 x 250 | | | |
| | Rebar | 2 x 6-D19 | Vertical | D13@300 (W) | |
| | Hoop | Ⓐ 2,3-D10@80 | Horizontal | Ⓐ D10@125 (W) | |
| | | Ⓒ 2,3-D10@100 | | Ⓒ D10@200 (W) | |
| | Joint | 2,2-D10@150 | | | |

**b**

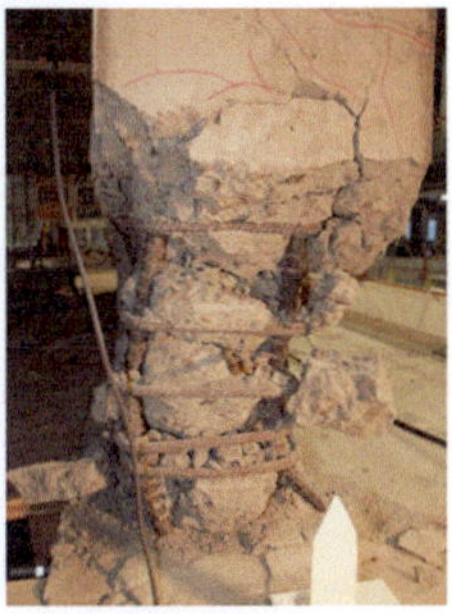

**Fig. 30.6** (**a**) RC conventional wall (Nagae et al. 2011). (**b**) Wall damage

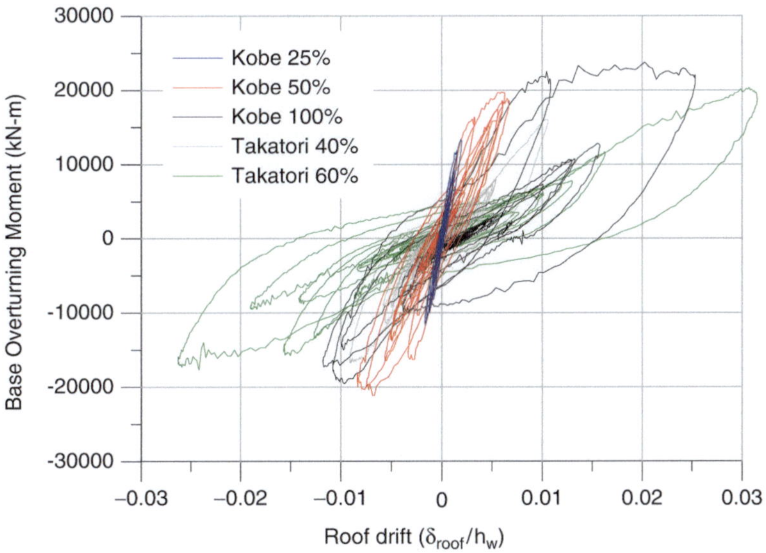


**Fig. 30.7** Base moment versus roof drift

Nagae et al. (2011) reports E-Defense tests on two 4-story buildings, one conventionally reinforced and the other using high-performance RC construction, both with rectangular wall cross sections (Fig. 30.6a). The conventionally reinforced wall had confinement exceeding U.S. requirements, with axial load of approximately $0.03A_g f_c'$, yet the compression boundary zone sustained localized crushing and lateral buckling ((Fig. 30.6b), following Kobe 100 % motion). The base overturning moment versus roof displacement responses are plotted in Fig. 30.7; base rotations are slightly less than the roof drift ratio (e.g., for Kobe 100 %, the base rotation measured over $0.27l_w$ is a little more than 0.02). Following crushing of boundary regions, sliding shear responses increased substantially during the Kobe 100 % test (Fig. 30.8). Sliding displacements in the Takatori 60 % test reached the limits of the

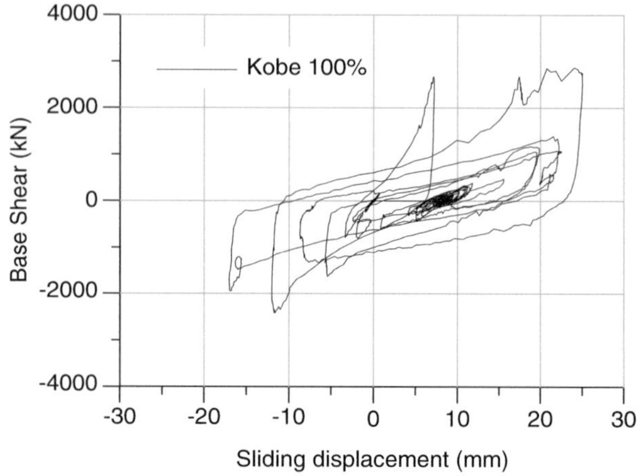


**Fig. 30.8** Shear versus sliding shear responses

sensor, +45 mm and −60 mm with peak shear of +/−2,000 kN. It is noted that the relatively large clear cover over the boundary longitudinal bars was used (∼40 mm) and the boundary transverse reinforcement was insufficient to maintain the boundary compressive load following cover spalling. It is noted that the crushing/spalling of the boundary region was accompanied by lateral buckling of the compression zone, as was observed in Chile and New Zealand (Figs. 30.1e and 30.2).

### *30.2.3 Recorded Ground Motions*

Response Spectra computed using ground motions recorded in recent earthquakes have significantly exceeded values used for design. For example, spectra for records in Chile and Christchurch significantly exceed values used for design (Fig. 30.9). For Chile, a vast majority of buildings were designed for the Soil II spectrum, whereas spectral ordinates are generally 2–6 times the values for Soil II over a broad period range. Given such large demands it is important to re-evaluate how displacement demands influence design requirements for structural walls and the associated consequences.

### *30.2.4 Summary*

Wall performance in recent earthquakes and laboratory tests raise a number of concerns. In Chile, brittle failures at wall boundaries were likely influenced by the level of axial stress (possibly leading to compression failures), the larger than expected displacement demands, the use of unsymmetric wall cross sections, and the lack of closely-spaced transverse reinforcement at wall boundaries. A particularly

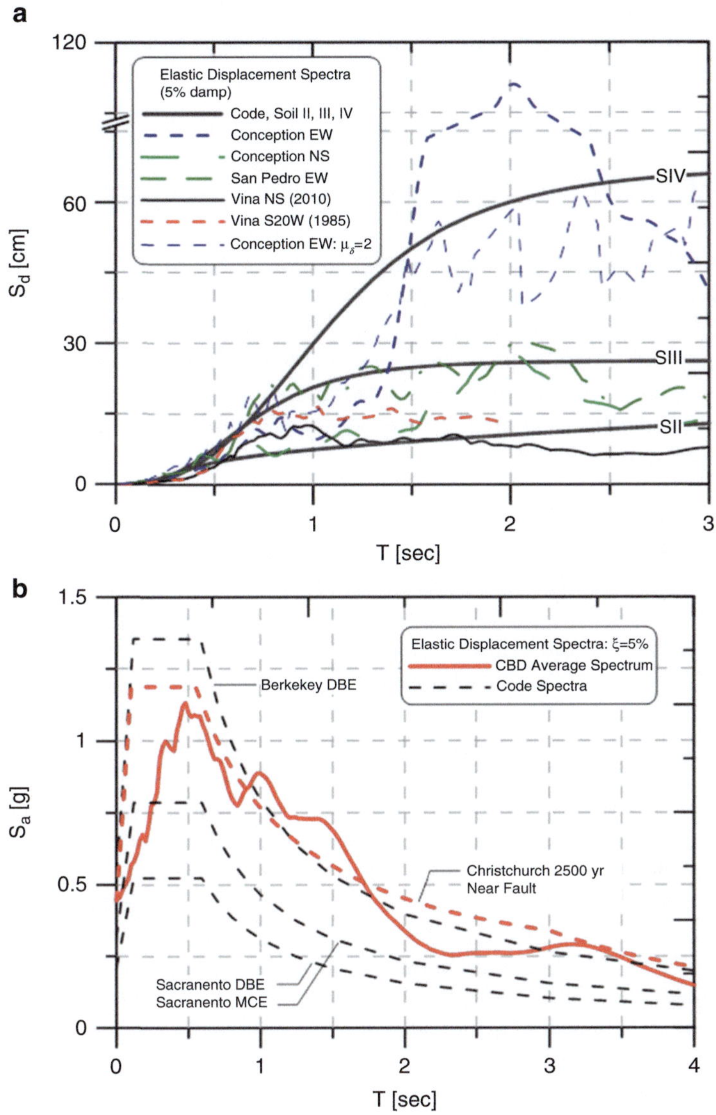


**Fig. 30.9** (**a**) Chile displacement spectra; (**b**) Christchurch acceleration spectra

noteworthy aspect of recent tests (Nagae et al. 2011; Lehman and Lowes 2011; Moehle et al. 2010) is the failure of relatively thin wall boundaries to develop ductile behavior in compression, even though they complied with ACI 318 special boundary element requirements, as well as Japan Standard Building Law and AIJ (Architectural Institute of Japan 2010) requirements. Recent tests to investigate the role of splices within the plastic hinge region of structural walls suggest that splices will substantially reduce wall inelastic deformation capacity. Given these observations, current ACI 318-11 code provisions for Special Structural Walls are reviewed to identify possible concerns and to suggest changes that could be implemented to address these concerns.

## 30.3 ACI 318 Code Provisions – Special Structural Walls

Provisions for "Special Structural Walls" are contained in ACI 318-11 §21.9 and include provisions for Reinforcement (§21.9.2), Shear Strength (§21.9.4), Design for Flexural and Axial Loads (§21.9.5), and Boundary Elements of Special Structural Walls (§21.9.6). In light of the prior discussion, key aspects of these provisions are reviewed and areas of concern are noted. In many cases, insufficient information is available to develop comprehensive (code) requirements and comments provided here are intended to provide guidance.

### *30.3.1 Reinforcement and Splices*

A single curtain of web reinforcement is allowed if the wall shear stress is less than $0.17\sqrt{f_c'MPa}\ MPa$. This provision is acceptable for squat walls with low shear stress (e.g., walls with aspect ratio less than 1.5); however, for slender walls where buckling of boundary vertical reinforcement and lateral instability are more likely due to significant tensile yielding of reinforcement under cyclic loading, two curtains should always be used. This recommendation applies to both Special Structural Walls (high ductility) and Ordinary Structural Walls (moderate ductility).

Recent laboratory tests have identified that wall deformation capacity may be compromised in cases where splices exist within the wall critical section (plastic hinge region) because nonlinear deformations are concentrated outside of the splice region, either at the wall-foundation interface (large moment gradient; Johnson 2010) or above the splice (nearly uniform wall moment; Birely et al. 2010). Given these results, it is questionable whether boundary vertical reinforcement should be lapped spliced within the plastic hinge region. Test results did indicate that use of ACI 318-11 Type II couplers performed adequately. The option of staggering splices is not addressed here.

### *30.3.2 Design Displacement and Plastic Hinge Length*

The model used to develop ACI 318-11 §21.9.6.2 provisions is shown in Fig. 30.10. Given this model, the design displacement $\delta_u(\text{ACI}) \equiv \delta_x = C_d\delta_e/I$ (ASCE 7-05; American Society of Civil Engineers 2005) is related to local plastic hinge rotation and extreme fiber compressive strain as:

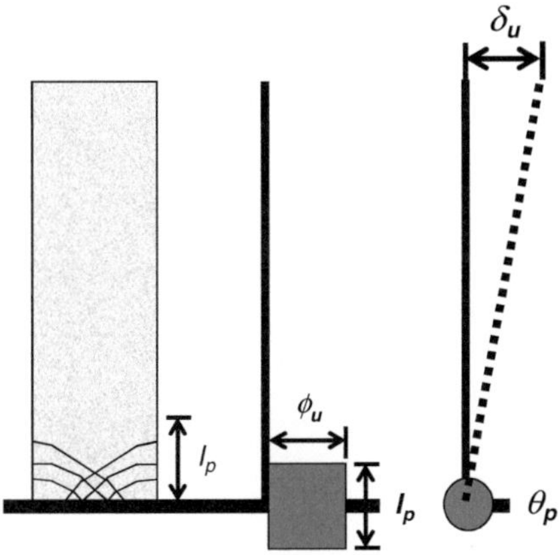


**Fig. 30.10** ACI 318-11 §21.9.6.2 model

$$\theta_p = \frac{\boldsymbol{\delta_u}}{h_w};\quad \theta_p = \left(\phi_u = \frac{\varepsilon_c}{c}\right)\left(\boldsymbol{l_p} = \frac{\boldsymbol{l_w}}{\mathbf{2}}\right) \therefore\ \varepsilon_c = 2\left[\frac{\delta_u}{h_w}\right]\left[\frac{c}{l_w}\right] \tag{30.1}$$

If the compressive strain exceeds a limiting value, typically taken as 0.003, then special transverse reinforcement is required. In ACI 318-11 Equation (21.8), this approach is rearranged to define a limiting neutral axis depth versus a limiting concrete compressive strain as:

$$c_{\text{limit}} = \frac{0.003 l_w}{2\left(\delta_u / h_w\right)} = \frac{l_w}{667\left(\delta_u / h_w\right)} \approx \frac{l_w}{600\left(\delta_u / h_w\right)} \tag{30.2}$$

In this approach, it is obvious that the result is sensitive to the values used for the design displacement and the plastic hinge length. Revised formulations, using a detailed displacement-based design approach (Wallace and Orakcal 2002) and a plastic hinge length that varies with wall thickness (Wallace 2011), produces the following relation:

$$\frac{\delta_u}{h_w} = \varepsilon_{cu}\left(\alpha\frac{\boldsymbol{t_w}}{\boldsymbol{l_w}}\frac{l_w}{c}\right)\left(1 - \frac{\alpha}{2}\frac{\boldsymbol{t_w}}{\boldsymbol{h_w}}\right) + \frac{\varepsilon_{sy}}{(1 - c/l_w)}\left(\frac{11}{40}\frac{\boldsymbol{h_w}}{\boldsymbol{l_w}} - \alpha\frac{\boldsymbol{t_w}}{\boldsymbol{l_w}} + \alpha^2\frac{\boldsymbol{t_w}}{\boldsymbol{h_w}}\frac{\boldsymbol{t_w}}{\boldsymbol{l_w}}\right) \tag{30.3}$$

Where $t_w$ is the wall thickness, $c$ is the neutral axis depth, $h_w$ is the wall height, $l_w$ is the wall length, and $\varepsilon_{sy}$ is the tensile reinforcement yield strain. The constant 11/40 results based on the assumed distribution of lateral force over the height of the wall (Wallace and Moehle 1992). In (30.3), the relationship between the wall neutral axis depth, concrete compressive strain, and drift is computed for various ratios of $l_w/t_w$ and $h_w/l_w$ for the three assumed values of plastic hinge length. For this preliminary study, wall aspect ratio $h_w/l_w$ is set to 3.0 and the ratio of $l_w/t_w$ is set to 13.3 for U.S. construction. Concrete compressive strain is set to 0.003; results presented in Fig. 30.11 define when special transverse reinforcement would be required at wall boundaries for three plastic hinge lengths.

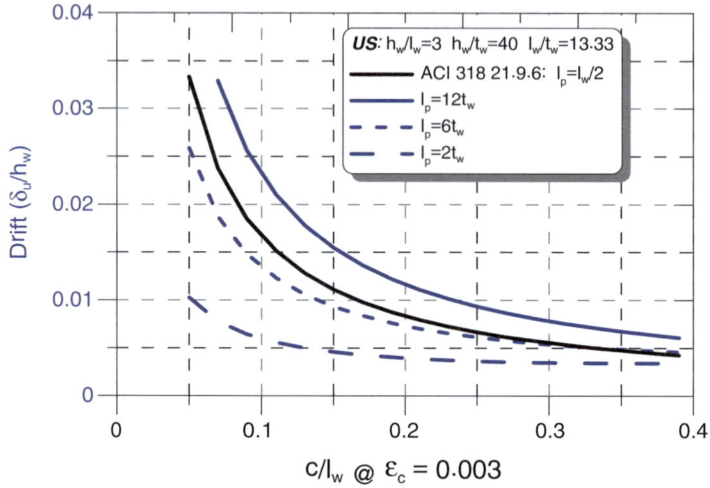


**Fig. 30.11** Influence of plastic hinge length on need for SBEs

According to Fig. 30.11, if the drift ratio is 0.01, the neutral axis must exceed $0.17l_w$ before SBE are required by ACI 318-11. However, for the same neutral axis depth of $0.17l_w$, if inelastic deformations are concentrated over a short height ($l_p = (a = 2)t_w$), only less than one-half of this drift ratio (0.005), can be tolerated before SBEs are required. The sensitivity of the results suggests that measures are needed to ensure appropriate spread of plasticity by requiring walls to be tension-controlled or by ductile yielding of concrete in compression.

In current US codes the intent is to provide 90 % confidence of non-collapse for MCE shaking. In contrast, the current ACI confinement trigger is based on 50 % confidence of not exceeding the concrete crushing limit in the Design Basis Earthquake (which is much lower shaking intensity than the MCE). To address this issue, it is necessary to adjust ACI Equation (21.8) to be more consistent with the building code performance intent. Three factors need to be considered: (1) MCE exceeds DBE. (2) There is dispersion about the median response. (3) Damping is likely to be lower than the 5 % value assumed in the ACI provisions. To address these issues, the coefficient of 600 in the denominator of Equation (21.8) in ACI 318-11 should be increased by a factor of approximately 1.5 to adjust to MCE level shaking and to consider dispersion, and by approximately 1.2–1.3 to account for potential lower damping ratios; therefore, a coefficient of about 1,200 is needed.

### 30.3.3 Axial Load and Compression-Controlled Walls

As noted above, the provisions of 318-11 §21.9.6.2 assume that nonlinear deformations within the critical (plastic hinge) region of the wall will spread out over a distance equal to one half the wall length. ACI 318-11 §9.4 defines tension- and compression-controlled sections; however, no guidance is provided on how these requirements should be applied to special (or ordinary) structural walls. In addition, ACI 318 and ASCE 7 do not place limits on wall axial stress. The performance of walls in Chile suggests that higher axial stresses and wall cross section shape (e.g., T-shaped) may lead to cases where concrete compressive strain reaches 0.003 prior to yield of tension steel.

Various approaches could be used to address this issue, such as placing limit on axial stress or requiring wall critical sections to be tension-controlled. In the 1997 version of the Uniform Building Code, wall axial load was limited to $0.35P_0$; for higher axial loads the lateral strength and stiffness of the wall could not be considered. An alternative to neglecting the lateral-force-resistance of compression-controlled walls would be to impose more stringent design requirements, such as always requiring SBEs to maintain a stable compressive zone as the concrete yields in compression. Even with more stringent design requirements, it might be prudent to place a limit on concrete compressive strain, e.g., 0.01, as it is not prudent to expect significant inelastic deformation capacity (rotation) can be achieved through compression yielding. This objective can be accomplished using displacement-based design using Eq. (30.1). For $c/l_w \geq 3/8$, the value at which a section is roughly no longer tension-controlled per ACI 318-08 9.4, Eq. (30.1) gives: $(\delta_u/h_w)_{\text{limit}} = 0.010l_w/(2 * 3l_w/8) = 0.0133$, whereas for $c/l_w \geq 0.6$, where a section is compression-controlled per ACI 318-08 9.4, Eq. (30.1) gives: $(\delta_u/h_w)_{\text{limit}} = 0.010l_w/(2 * 0.6l_w) = 0.0083$. If the drift limit is exceeded, then redesign of the wall section would be required.

### *30.3.4 Boundary Element Detailing*

ACI 318-11 detailing requirements for SBEs are based on requirements that were developed for columns; these provisions may be insufficient for SBEs of thin walls. The review of recent wall damage in earthquakes and laboratory tests provides sufficient evidence to raise concerns related to detailing of thin walls. For example, although the quantity of transverse reinforcement provided at the boundaries of the conventional RC wall tested at E-Defense were 1.4 and 2.1 times that required by ACI 318-11 §21.9.6.4 (for the larger spacing of 100 mm used at Axis C), concrete crushing and lateral instability (Fig. 30.6b) occurred earlier in the Kobe 100 % test, followed by substantial sliding (Fig. 30.7). Inspection of the damaged boundary zone revealed that relatively large clear cover was used, on the order of 40 mm (larger than the code minimum in ACI 318, which is 19 mm), suggesting that the confined core was incapable of maintaining stability of the compression zone following loss of concrete cover. For smaller columns, ACI 318-11 Equation (21.4), which is based on maintaining column axial load capacity after cover concrete spalling, typically governs the selection of transverse reinforcement for smaller columns where cover makes up a larger percentage of the gross concrete

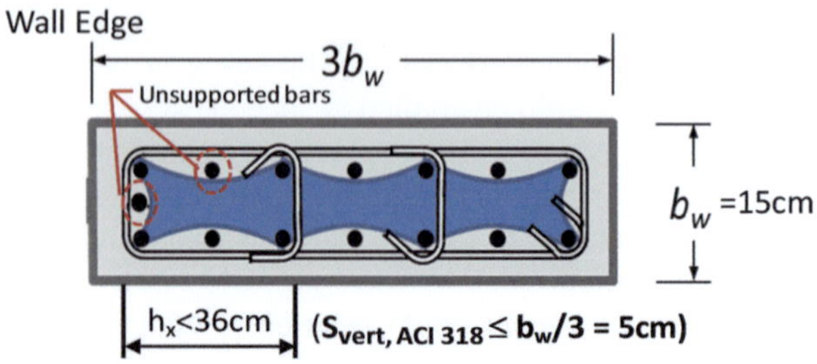


**Fig. 30.12** Wall special boundary element (ACI 31-11): length = $3b_w$

section. This equation also was required for wall SBEs prior to ACI 318-99; it was dropped because it rarely controlled for the thicker walls that were commonly used at that time. For the E-Defense conventional RC wall, the provided transverse reinforcement was only 0.34 and 0.45 times that required by ACI 318-11 Equation (21.4), suggesting that improved performance may have resulted had this relation been required. Additional testing is needed to determine if reinstating (21-4) is sufficient to ensure ductile behavior of thin boundary zones.

ACI 318-11 §21.6.6.2 allows a distance of 14″ (356 mm) between adjacent hoops or ties. Use of such a large spacing for thin SBEs is unlikely to provide sufficient confinement (Fig. 30.12) and is incompatible with use of a vertical spacing one-third the wall thickness. For example, for a 10 in. (254 mm) thick wall, such as used in the E-Defense test, the vertical spacing per ACI 318-11 is limited to 3.33″ (84.6 mm); however, the horizontal spacing along the wall can reach 356 mm (356/84.6 = 4.2). An additional limit should be considered for wall SBEs, similar to that used for vertical spacing, where the horizontal distance between legs of hoops or ties is limited to a fraction of the wall thickness, e.g., $2/3t_w$ or a value less than 356 mm, e.g., 200 mm. Not allowing intermediate, unsupported bars at the wall edge (Fig. 30.12), which initiated the section failure for test PW2 (Fig. 30.5a), also should be considered.

### *30.3.5 Wall Slenderness and Lateral Stability*

To limit instability failures, limits on wall slenderness should be considered, similar to what was done in the Uniform Building Code (1997), which imposed a slenderness limit of $t_w \geq h_s/16$. Lateral instability failures at wall boundary regions are influenced by a number of factors, including: slenderness, cross section shape, quantity of vertical reinforcement, detailing, axial load, design displacement, and load history. Introduction of a limit based on slenderness alone is unlikely to provide a robust solution to this problem; however, until a comprehensive study is available, use of $l_u/b \leq 16$ is recommended, although this limit may not be sufficient to preclude lateral instability failures for asymmetric wall cross sections (T- or L-shaped sections), where a lower limit of $l_u/b \leq 10$ might be appropriate at the web boundary opposite the flange given the large cyclic demands that may occur at this location (Wallace 2012). This issue is currently under study by ATC 94 (2011).

## 30.4 Conclusions

Wall performance in recent earthquakes and laboratory tests is reviewed and American Concrete Institute 318 provisions are reassessed to identify possible shortcomings. The findings suggest a number of issues require more in-depth study, particularly for thin walls, as well as approaches that could be implemented to address these issues. Changes are needed to increase the design displacement used in ACI 318-11 Equation (21.8), a factor of two is suggested, and to ensure spread of plasticity consistent with the derivation of Equation (21.8). To address this latter issue, walls should either be tension controlled or be designed and detailed to ensure ductile compression yielding by requiring that walls be thicker and by imposing a limit on wall slenderness. Limiting wall compression strain for compression-controlled walls also might be prudent. A simple limit on slenderness is suggested until more detailed studies are conducted.

**Acknowledgements** This research described in this paper was carried out with funding from various sources, including the EERI Learning from Earthquakes program (NSF CMMI-**0758529), NSF RAPID projects to enhance US-Japan collaboration related to the E-Defensetests in December 2010** (CMMI-1110860 and CMMI-1000268; Program Director Joy Pauschke), NSF NEES REU (CMMI-0927178), as well as support provided to the first author by the Japan Society for the Promotion of Science (JSPS) Invitation Fellowship Program during the fall 2010. This support is gratefully acknowledged. The author would like to express his appreciation to those researchers who have contributed their research results to NEEShub, which provides an invaluable resource, and to Japanese collaborators working on the December 2010 E-Defense tests for sharing their ideas and results, including: T Nagae (NIED), K Tahara (NIED), T Matsumori (NIED), H Shiohara (U Tokyo), T Kabeyasawa (U Tokyo ERI), S Kono (Kyoto U), M Nishiyama (Kyoto U); M. Nakashima (NIED, Kyoto U). The opinions, findings, conclusions, and recommendations in this paper are those of the author, and do not necessarily represent those of the sponsors or other individuals mentioned here.

## References

ACI 318-95 (1995) American Concrete Institute, Building code requirements for structural concrete and commentary (ACI 318R-95). American Concrete Institute, Farmington Hills

ACI 318-99 (1999) American Concrete Institute, Building code requirements for structural concrete and commentary (ACI 318R-99). American Concrete Institute, Farmington Hills

ACI 318-11 (2011) American Concrete Institute, Building code requirements for structural concrete and commentary (ACI 318R-11). American Concrete Institute, Farmington Hills

American Society of Civil Engineers (2005) ASCE7-05: minimum design loads for buildings and other structures. American Society of Civil Engineers, Reston

Architectural Institute of Japan (2010) Standard for structural calculation of reinforced concrete structures, first published in 1933

ATC-94 (2011) Analysis of seismic performance of reinforced concrete buildings in the 2010 Chile earthquake, including effects of non-seismic-force-resisting building structural elements – task order 21. https://www.atcouncil.org/Projects/nehrp-jv.html

Birely A, Lehman D, Lowes L, Kuchma D, Hart C, Marley K (2008) Investigation of the seismic behavior and analysis of reinforced concrete structural walls. In: Proceedings, 14th world conference on earthquake engineering, Beijing, China, 12–17 Oct 2008

Birely A, Lehman D, Lowes L, Kuchma D, Hart C, Marley K (2010) Investigation of the seismic response of planar concrete walls. In: Proceedings paper 773, 9th U.S. national conference and 10th Canadian conference on earthquake engineering, Ontario, Canada, 25–29 July 2010

Brueggen BL (2009) Performance of T-shaped reinforced concrete structural walls under multidirectional loading. PhD dissertation, University of Minnesota, Department of Civil Engineering, 498 pp

Brueggen BL, French CW (2010) Simplified modeling of non-rectangular RC structural walls. In: Proceedings paper 1713, 9th U.S. national conference and 10th Canadian conference on earthquake engineering, Ontario, Canada, 25–29 July 2010

Chai YH, Elayer DT (1999) Lateral stability of reinforced concrete columns under axial reversed cyclic tension and compression. ACI Struct J 96(5):780–789. ACI

Corley WG, Fiorato AE, Oesterle RG (1981) Structural walls, Publication SP-72, ACI, pp 77–131

EERI (2010) The Mw 8.8 Chile earthquake of February 27, 2010, EERI special earthquake report, Earthquake Engineering Research Institute, Oakland, California

EERI (2011a) NEES/E-Defense collaboration on tests. EERI Newsl 45(2). Earthquake Engineering Research Institute, Oakland, California

EERI (2011b) The M 6.3 Christchurch, New Zealand, earthquake of February 22, 2011, EERI special earthquake report, Earthquake Engineering Research Institute, Oakland, California, May 2011

Elwood KJ, (2011), Personal Communication. [see Also EERI Christchurch Earthquake Clearing House: http://www.eqclearinghouse.org/2011-02-22-christchurch/]

Fischinger M, Isakovic T, Kante P (2006) Shaking table response of a thin H-shaped coupled wall. In: 8th US national conference on earthquake engineering, San Francisco, 18–22 Apr 2006, pp 6619–6628

Johnson B (2010) Anchorage detailing effects on lateral deformation components of R/C shear walls. MS thesis, University of Minnesota, Department of Civil Engineering, 353 pp

Kabeyasawa T, Kabeyasawa T, Matcumori T, Kabeyasawa T, Kim Y (2008) Full-scale dynamic collapse tests of three-story reinforced concrete buildings on flexible foundation at E-defense. In: Proceedings of the 14th world conference on earthquake engineering, Beijing, China, October 2008, Paper ID: S15-002

Kabeyasawa T, Kabeyasawa T, Kim Y (2010a) Collapse simulation of reinforced concrete buildings with ASFI approach. In: Proceedings paper 816, 9th U.S. national conference and 10th Canadian conference on earthquake engineering, Ontario, Canada, 25–29 July 2010

Kabeyasawa T, Kabeyasawa T, Kim Y, Kabeyasawa T, Bae K, Quang PV (2010b) Strength and deformability of reinforced concrete columns with wing walls. In: Proceedings, paper 813, 9th U.S. national conference and 10th Canadian conference on earthquake engineering, Ontario, Canada, 25–29 July 2010

Lehman DE, Lowes LN (2011) Personal communication about laboratory test results for NEESR project seismic behavior, analysis, and design of complex wall systems. [see also nees.org/warehouse/project/104]

Massone LM, Wallace JW (2011) Lessons from Chile: impacts of earthquake engineering of RC buildings in the US, EERI/NEES Webinar. http://nees.org/resources/3192

Moehle JP, Acevedo C, Creagh A (2010) Exploratory tests of wall boundary elements subjected to alternating tensile and compressive loadings. Poster and oral presentations at 2010 PEER annual meeting, Pacific Earthquake Engineering Research Center (PEER), Berkeley, California

Nagae T, Tahara K, Matsumori T, Shiohara H, Kabeyasawa T, Kono S, Nishiyama M (Japan), Wallace JW, Ghannoum W, Moehle JP, Sause R, Keller W, Tuna Z (2011) Design and instrumentation of the 2010 E-defense four-story reinforced concrete and post-tensioned concrete buildings, Report No 2011-XX, Pacific Earthquake Engineering Research Center, Department of Civil & Environmental Engineering, University of California, Berkeley, February 2011, 235 pp

Naish D, Wallace J (2010) EERI 2010 paper – testing and modeling of diagonally reinforced concrete coupling beams. http://nees.org/resources/6998

NCh 433.Of96 (1996) Diseño Sismico de Edificios (Chilean Building Code), Instituto Nacional de Normalizacion, Chile

NZRC (2011) Canterbury earthquakes royal commission – interim report. http://canterbury.royalcommission.govt.nz/Interim-Report

Panagiotou M, Restrepo JI (2007) Practical lessons learned from the full-scale 7-story building shake table test at UC San Diego, 2007 SEAOC convention, Squaw Creek, CA, 26–29 Sept 2007, pp 57–74

Parra-Montesinos G, Wight JK, Lequesne RD, Seekit M (2012) A summary of ten years of research on HPFRC coupling beams. In: Parra-Montesinos GJ, Reinhardt HW, Naaman AE (ed) High performance fiber reinforced cement composites 6, 560p

Paulay T, Priestley MJN (1993) Stability of ductile structural walls. ACI Struct J 90(4):385–392. American Concrete Institute

Sriram A, Sritharan S (2010) Nonlinear fiber-based analysis of rectangular concrete walls designed with different anchorage details. In: Proceedings, paper 123, 9th U.S. national conference and 10th Canadian conference on earthquake engineering, Ontario, Canada, 25–29 July 2010

Thomsen JH IV, Wallace JW (2004) Displacement-based design of slender RC structural walls – experimental verification. J Struct Eng ASCE 130(4):618–630

UBC (1997) Uniform building code. International Council of Building Code Officials, Whittier

Wallace JW (1996) Evaluation of UBC-94 provisions for seismic design of RC structural walls. Earthq Spectra 12(2):327–348

Wallace JW (2011) February 27, 2010 Chile earthquake: preliminary observations on structural performance and implications for U.S. building codes and standards. ASCE structures congress, paper 1159, Las Vegas, 14–16 Apr 2011

Wallace JW (2012) Behavior, design, and modeling of structural walls and coupling beams – lessons from recent laboratory tests and earthquakes. Int J Concr Struct Mater 6(1):3–18

Wallace JW, Moehle JP (1992) Ductility and detailing requirements of bearing wall buildings. J Struct Eng ASCE 118(6):1625–1644

Wallace JW, Orakcal K (2002) ACI 318-99 provisions for seismic design of structural walls. ACI Struct J, American Concrete Institute 99(4):499–508

Waugh JD, Sritharan S (2010) Nonlinear analysis of T-shaped concrete walls subjected to multidirectional loading. In: Proceedings, paper 1506, 9th U.S. national conference and 10th Canadian conference on earthquake engineering, Ontario, Canada, 25–29 July 2010

Waugh J, Aaleti S, Sritharan S, Zhao J (2008) Nonlinear analysis of rectangular and T-shaped concrete walls, ISU-ERI-Ames report ERI-09327, Iowa State University, Department of Civil, Construction and Environmental Engineering, Ames, Iowa, December 2008, 351 pp

# Chapter 31
# Collapse Probability of Existing Concrete Buildings: The Evolution of Seismic Rehabilitation in North America

**Kenneth J. Elwood, Majid Baradaran Shoraka, and Tony Y. Yang**

**Abstract** Existing reinforced concrete buildings lacking details for ductile response during earthquake shaking represent a significant life safety risk in high seismic zones around the world. The poor seismic performance of these non-ductile concrete buildings is evident from recent earthquakes in Chile, New Zealand and Japan. Seismic rehabilitation of these existing buildings plays an important role in reducing urban seismic risk; however, with the massive inventory of existing concrete buildings and the high costs of seismic rehabilitation, it is necessary to start by identifying and retrofitting those buildings which are most vulnerable to collapse. Numerous sources of uncertainty complicate the ability to identify buildings which are vulnerable to collapse. For this reason, it is important to develop estimates of collapse probability to account for all significant sources of uncertainties. This chapter will introduce the concept of *collapse indicators*, design and response parameters that are correlated with "elevated" collapse probability. The methodology for identifying collapse indicators is based on results of comprehensive collapse simulations. Appropriate collapse indicators and corresponding limits are evaluated by seeking trends between probability of collapse and collapse indicators. This chapter will discuss significant challenges which pose a barrier to the assessment of collapse indicators that are applicable for the wide range of existing concrete buildings.

**Keywords** Reinforced concrete frames • Seismic assessment • Nonlinear analysis • Collapse assessment • Probabilistic analysis

---

K.J. Elwood (✉)
Civil and Environmental Engineering Department, University of Auckland, Auckland, New Zealand
e-mail: k.elwood@auckland.ac.nz

M. Baradaran Shoraka • T.Y. Yang
Civil Engineering Department, University of British Columbia, Vancouver, BC, Canada
e-mail: majidbs@gmail.com; yang@civil.ubc.ca

M. Fischinger (ed.), *Performance-Based Seismic Engineering: Vision for an Earthquake Resilient Society*, Geotechnical, Geological and Earthquake Engineering 32,
DOI 10.1007/978-94-017-8875-5_31, 

## 31.1 Introduction

In the mid-1990s new seismic rehabilitation guidelines (e.g. FEMA 1997) were introduced, providing the structural engineering profession with the first generation of 'performance-based' procedures for seismic assessment and rehabilitation design. These documents revolutionized the assessment of existing buildings by encouraging the use of nonlinear analysis, by enabling the engineer to select project-specific performance objectives, and perhaps most importantly, by recognizing that structural collapse was limited by both strength and deformation capacity. The past 15 years has seen modest improvements to this first-generation performance-based design procedure as FEMA 273 has evolved into of ASCE/SEI 2006, including updated acceptance criteria for concrete components based on new experimental data (Elwood et al. 2007). However, the overall framework remains essentially deterministic and inconsistent conservatism in specified deformation capacities throughout the document may impact the reliability of the predicted performance of a building structure. Furthermore, the component-based assessment procedures (i.e. once one component is determined to have exceeded a performance level, the entire structure is deemed to have exceeded the performance level) ignore the ability of a structural system to redistribute loads as damage accumulates and will tend to lead to conservative assessments of collapse vulnerability. Seismic evaluation documents based on checklist assessments (e.g. ASCE/SEI 2003) are also generally conservative to ensure dangerous buildings are not misdiagnosed. As currently formulated, ASCE/SEI 31 and ASCE/SEI 41 are not capable of reliably determining the relative collapse risk between different non-ductile concrete buildings. From a public policy standpoint, the ability to economically make this distinction across the large inventory of existing concrete buildings is a critical need and a necessary next step in the evolution of seismic rehabilitation documents.

ATC-76-5 (NIST 2010) identified the following critical needs for addressing the collapse risk associated with older concrete construction:

- Improved procedures for identifying building systems vulnerable to collapse, including simple tools that do not require detailed analysis.
- Updated component acceptance criteria based on latest research results.
- Identification of cost-effective mitigation strategies to reduce collapse risk.

To address the first critical need, ATC-76-5 proposed a methodology for identifying parameters, termed collapse indicators, correlated with an elevated probability of collapse based on results of comprehensive collapse simulations and estimation of collapse probabilities for prototypical concrete buildings. Ideally there should be a variety of collapse indicators, ranging from those appropriate for rapid assessment to others used to identify collapse potential based on results of detailed nonlinear analysis. Collapse indicators for rapid assessment must be very simple parameters which can be established based on basic information available from a quick survey of the building or engineering drawings. Collapse indicators for detailed collapse prevention assessment can make use of the results from building analyses. This chapter will demonstrate a methodology to identify collapse indicators and discuss challenges to be overcome in the identification of collapse indicators representative of a general inventory of buildings. The methodology discussed below was initially proposed as part of the ATC-76-5 project funded by NIST, and will be further developed in ongoing projects (ATC-78 and ATC-95) funded by NIST and FEMA. This chapter does not provide specific recommendations for collapse indicator limits, but instead focuses on the development of the methodology. The goal of the ongoing ATC projects is to refine the methodology and ultimately lead to simplified practical guidance for engineers on how to identify collapse vulnerable buildings based on limits for selected collapse indicators.

## 31.2 Potential Collapse Indicators

The identification of collapse indicators appropriate for engineering practice, and establishing limits on these indicators, is best accomplished through detailed analytical studies such as those described in Sect. 31.3. However, before embarking on the analytical studies, it is essential to come up with a list of potential collapse indicators from which the recommended collapse indicators will be selected. Engineering judgment and experience with collapse analyses were used to select the preliminary list of potential collapse indicators below. It is anticipated that this list will evolve as further experience is gained from the analyses described in Sect. 31.3.

Ideally there should be a spectrum of collapse indicators, ranging from those appropriate for rapid assessment to others used to identify collapse potential based on results of detailed nonlinear analysis. Collapse indicators for rapid assessment must be very simple parameters which can be established from basic information available from a quick survey of the building or engineering drawings. Conversely, collapse indicators for detailed collapse prevention assessment can make use of the results from nonlinear analyses. It is proposed to categorize collapse indicators into two fundamental types:

Design parameter collapse indicators: These collapse indicators are determined based on design features of a concrete building, including reinforcement details, structural system layout, and relative strength and stiffness of members. These indicators can be further sub-categorized as "rapid assessment" (RA) or "engineering calculation" (EC) collapse indicators, where the former can be determined from a quick survey of the building or engineering drawings and the latter requires some calculation of capacities and demands based on engineering drawings. RA and EC collapse indicators will be useful for refining the seismic evaluation procedures in ASCE/SEI 31.

Response parameter collapse indicators: These collapse indicators reflect the response of the structure based on results from building analysis (BA). Generally the most refined collapse indicators are expected to be derived from results of nonlinear analysis and provide system-level acceptance criteria for the Collapse Prevention performance level.

Table 31.1 provides a list of potential collapse indicators to be considered for investigation in the analytical studies discussed in Sect. 31.3. These collapse

**Table 31.1** Examples of collapse indicators

| Type of collapse indicator | | System-level | | Component-level | |
|---|---|---|---|---|---|
| Design parameters | **Rapid Assessment (RA)** *Quantities that can be determined from a quick survey of the building or engineering drawings* | *RA-S1.*[a] | Maximum ratio of column-to-floor area ratios for two adjacent stories | *RA-C1.* | Average minimum column transverse reinforcement ratio for each story |
| | | *RA-S2.* | Maximum ratio of horizontal dimension of the SFRS in adjacent stories | *RA-C2.* | Minimum column aspect ratio |
| | | *RA-S3.* | Maximum ratio of in-plane offset of SFRS from one story to the next to the in-plane dimension of the SFRS | *RA-C3.* | Misalignment of stories in adjacent buildings |
| | | *RA-S4.* | Plan configuration (L or T shape versus rectangular) | | |
| | | *RA-S5.* | Minimum ratio of column area to wall area at each story[b] | | |
| | **Engineering Calculations (EC)** *Quantities that require some calculation of capacities and demands based on engineering drawings, but do not require structural analysis results from computer modeling* | *EC-S1.* | Maximum ratio of story stiffness for two adjacent stories | *EC-C1.* | Maximum ratio of plastic shear capacity ($2M_p/L$) to column shear strength, $V_p/V_n$ |
| | | *EC-S2.* | Maximum ratio of story shear strength for two adjacent stories | *EC-C2.* | Maximum axial load ratio for columns with $V_p/V_n > 0.7$ |
| | | *EC-S3.* | Maximum ratio of eccentricity (distance from center of mass to center of rigidity or center of strength) to the dimension of the building perpendicular to the direction of motion | *EC-C3.* | Maximum ratio of axial load to strength of transverse reinforcement (45 deg truss model) |
| | | *EC-S4.* | Portion of story gravity loads supported by columns with ratio of plastic shear demand to shear capacity > 0.7. | *EC-C4.* | Maximum ratio of joint shear demand (from column bar force at yield) to joint shear capacity for exterior joints |
| | | | | *EC-C5.* | Maximum gravity shear ratio on slab-column connections |

| | | | | | |
|---|---|---|---|---|---|
| Response parameters | **Building Analysis (BA)** *Quantities for detailed collapse prevention assessment using the results from nonlinear building analyses* | *BA-S1.* | Maximum degradation in base or story shear resistance | *BA-C1.* | Maximum drift ratio |
| | | *BA-S2.* | Maximum fraction of columns at a story experiencing shear failures | *BA-C2.* | Maximum ratio of deformation demands to ASCE/SEI 41 limits for columns, joints, slab-column connections and walls |
| | | *BA-S3.* | Maximum fraction of columns at a story experiencing axial failures | | |
| | | *BA-S4.* | Minimum strength ratio (as defined in ASCE/SEI 41) | | |

[a]Collapse indicator notation: *RA* Rapid Assessment, *EC* Engineering Calculation, *BA* Building Analysis, *S* System, *C* Component

[b]May not result in collapse but could help prevent collapse if a mechanism forms

indicators have been grouped based on the classification described above, and further grouped as component or system-level parameters. Component Building Analysis indicators shown in Table 31.1 (BA – C#) can be interpreted as equivalent to component acceptance criteria in ASCE/SEI 41. It is anticipated that relationships may exist among the indicators, as vectors of indicators may be found to provide a better indication of collapse potential than any one indicator. For example, if the average minimum transverse reinforcement ratio (RA-C1) is less than a specific value and the maximum ratio of column-to-floor area ratios for two adjacent stories (RA-S1) is greater than a specific value, then collapse potential is expected to be high.

## 31.3 Collapse Simulation and Probability of Collapse

In order to better understand and define the collapse potential for concrete buildings, model building prototypes are developed for nonlinear analyses using OpenSees (2010). These models, capable of capturing onset of structural collapse, enable the selection of collapse indicators having strong correlation with collapse potential. Using the prototypes, the Design Parameter collapse indicators listed above, including geometric and mass properties (e.g. plan dimensions, building height), can be varied to explore the effects on collapse probability. Sophisticated building prototype models allow explicit consideration of collapse probability considering both loss of vertical load carrying capability and lateral dynamic instability; uncertainty in modeling and ground motion are also accommodated.

Collapse of real buildings is highly dependent on the complex behavior and interaction among individual components. The analyses to refine the selection of collapse indicators utilize building models to explore characteristic behavior and the effects of parametric variations.

In contrast to ductile structural systems, non-ductile concrete buildings will typically experience gravity-load collapse resulting from loss of vertical load carrying capacity in critical components prior to experiencing a "side-sway" collapse resulting from degradation in lateral resistance (Baradaran Shoraka et al. 2013). The nonlinear building models used for this study incorporate elements capable of approximately capturing the loss of vertical load carrying capacity for critical gravity-load supporting components (e.g. columns (Elwood 2004; Baradaran Shoraka and Elwood 2013) and slab-column connections (Kang et al. 2009)) and account for P-delta effects. It should be noted that these models provide a relatively simple representation of a very complex phenomenon at the point of gravity-load failure, and hence, may lack some sophistication required to accurately capture the behavior of the building to the point of total collapse. In particular, given the lack of data available for validation, modeling of three dimensional gravity load redistribution through a slab floor system after gravity load failure of a single component should be considered approximate at best. Despite possible inaccuracies,

**Table 31.2** Four and seven story non-ductile RC frame structures

| Num. of stories | Num. of bays | Story height (m) | Bay width (m) | Framing system | Period (s) |
|---|---|---|---|---|---|
| 4[a] | 3 | 3.96 | 7.62 | Space | 1.98 |
| 7[b] | 8 | 4.12 | 5.72 | Perimeter | 1 |

[a]Liel et al. (2011)
[b]Krawinkler (2005)

the failure models provide good estimates of observed collapse behavior of simple 2D frames (e.g. Yavari et al. 2009).

One significant challenge that must be overcome to undertake this analysis is the distinction between the "true" point of gravity-load collapse and non-convergence due to numerical instability of the model. In this study, collapse is detected based on a comparison of floor-level gravity load demands and capacities (adjusted at each time step to account for member damage and allow for load redistribution) (Elwood and Moehle 2008; Baradaran Shoraka et al. 2013). The first point when floor-level vertical load demands exceed the total vertical load capacity at that floor is defined as gravity-load collapse for the building system. Non-convergence of the analysis prior to significant degradation in the capacity to resist gravity loads and not associated with large lateral displacements (side-sway collapse) will not be considered as collapse, and the results of such analysis will be omitted from the calculation of collapse probabilities.

The following sections briefly describe the scope for the analytical studies. First, the key elements of the numerical models are introduced. In this chapter, four- (Liel et al. 2011) and seven-story (Krawinkler 2005) non-ductile reinforced concrete frame buildings located in Los Angeles, California, are used as example structures to illustrate the methodology. Next, the model and ground motion uncertainties are briefly explained. Finally, the two approaches used to establish limits on the collapse indicators (design and response parameters) are discussed.

### *31.3.1 Numerical Model*

Two-dimensional finite element models are used to simulate the seismic response of the buildings using OpenSees . A fixed-base model is used in the analysis; as a result soil-structure foundation interaction is neglected. The frame elements are modeled using the force-based beam–column model with distributed nonlinear fiber sections. The joints are modeled using the Alath and Kunnath model (Alath and Kunnath 1995) which includes the pinching hysteric behaviour to account for the degradation usually seen in these non-ductile elements. Axial failure of joints is not considered in this example. The shear and axial response in the columns are modeled using the modified Limit State material model (Baradaran Shoraka and Elwood 2013) recently implemented in OpenSees. Key characteristics of the structures are summarized in Table 31.2.

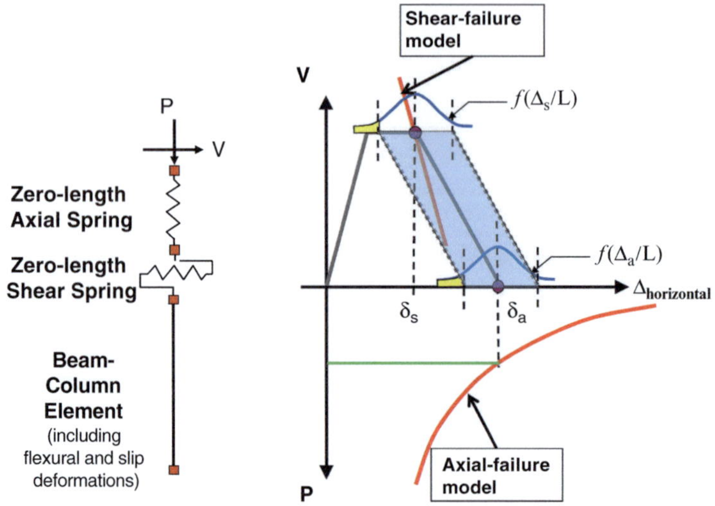


**Fig. 31.1** Structural model uncertainty in shear and axial failure models (Elwood 2004)

## *31.3.2 Record- to-Record and Model Uncertainty*

Performance-based earthquake engineering enables probabilistic prediction of structural response, incorporating key sources of uncertainty in the process. By using a suite of earthquake records, nonlinear dynamic analyses (via incremental dynamic analysis, IDA (Vamvatsikos and Cornell 2002), and/or multiple stripe analysis, MSA (Jalayer and Cornell 2009)) directly incorporates information about variability in ground motions in the collapse performance assessment. However, nonlinear dynamic analysis alone does not account for how well the structural model represents the collapse behavior of the building; hence, model uncertainties should also be accounted for in the process of collapse simulation. These modeling uncertainties are especially important in predicting collapse, because of the high degree of empiricism and uncertainty in predicting deformation capacity and other critical parameters.

In this methodology the uncertainty for each random variable is explicitly considered in the analysis and reflected in the final probabilities of collapse. The random variables selected with the respective probability distribution should have the capability of capturing the major uncertainties inherent in non-ductile reinforced concrete frames. Uncertainty in the shear and axial failure models for non-ductile columns are considered as the main sources of model uncertainty for the example buildings. The shaded area shown in Fig. 31.1, presents the entire outcome of failure model variability considered for these non-ductile columns. The variability in the drift at column shear and axial load failure is represented by a lognormal distribution with a mean equivalent to the limit-state material failure models and a coefficient of variation of approximately 0.3 (Elwood and Moehle 2008).

In addition to model variability, record to record variability is considered in the process. Accounting for record to record uncertainty is possible by selecting an appropriate suite of ground motions specific for the building site and the anticipated hazard level with a good fit to the conditional mean spectrum. Ten ground motions for each hazard level are considered using MSA in this example.

### *31.3.3 Assessment Procedure for Design Parameters*

The nonlinear models will be used to estimate the probability of collapse considering ground motion and model uncertainties. The procedure is to first build the complete nonlinear model of a building prototype and evaluate the building collapse fragility (Baradaran Shoraka et al. 2013). Two- and three-dimensional analytical models can be used. Then a selected collapse indicator parameter (e.g. RA-L1, average minimum column transverse reinforcement ratio) will be altered in the building model and the collapse fragility will be reassessed for each realization of the collapse indicator (Fig. 31.2a). Figure 31.2a shows the variation of the collapse fragilities for the four-story example building with the average column transverse reinforcement ratio, RA-L1. As expected, collapse probability increases as the average column transverse reinforcement ratio decreases. The same figure can be represented by grouping the collapse fragility into different bins of hazard intensities as shown in Fig. 31.2b. The variation of the collapse fragilities can be multiplied with the slope of the hazard curve shown in Fig. 31.2c to estimate the mean annual frequency of collapse ($\lambda_{collapse}$) as a function of the design parameter (Fig. 31.2d). The assessment illustrated in Fig. 31.2a through Fig. 31.2d can be repeated several times to access the sensitivity of the design parameters for several different building types (Fig. 31.2e).

Limits on the collapse indicators can be selected based on a “suitable” mean annual frequency of collapse ($\lambda_{collapse}$) (Fig. 31.2e). For instance, $\lambda_{collapse}$ could be selected to be consistent with the new risk-targeted ground motions used for new design (Luco et al. 2007), i.e. a uniform collapse risk of 1 % in 50 years. To be consistent with current seismic rehabilitation practice and to encourage economical retrofits, it is reasonable to relax the target collapse risk compared to that used for new buildings; for example, a mean annual of frequency of collapse of 0.001, equivalent to a probability of collapse of 5 % in 50 years. To achieve this selected risk level, Fig. 31.2e indicates that the transverse reinforcement ratio should not be below, approximately 0.0043 and 0.012 for the seven- and four-story example buildings, respectively. It should be noted that an ideal collapse indicator would have only limited variation in the collapse indicator limit suggested by the different building types. The results shown in Fig. 31.2 suggest that transverse reinforcement ratio by itself may not be an ideal collapse indicator for frame buildings as the number of columns and stories also influence the results. Future research will consider how to determine appropriate limits for combinations of collapse indicators.

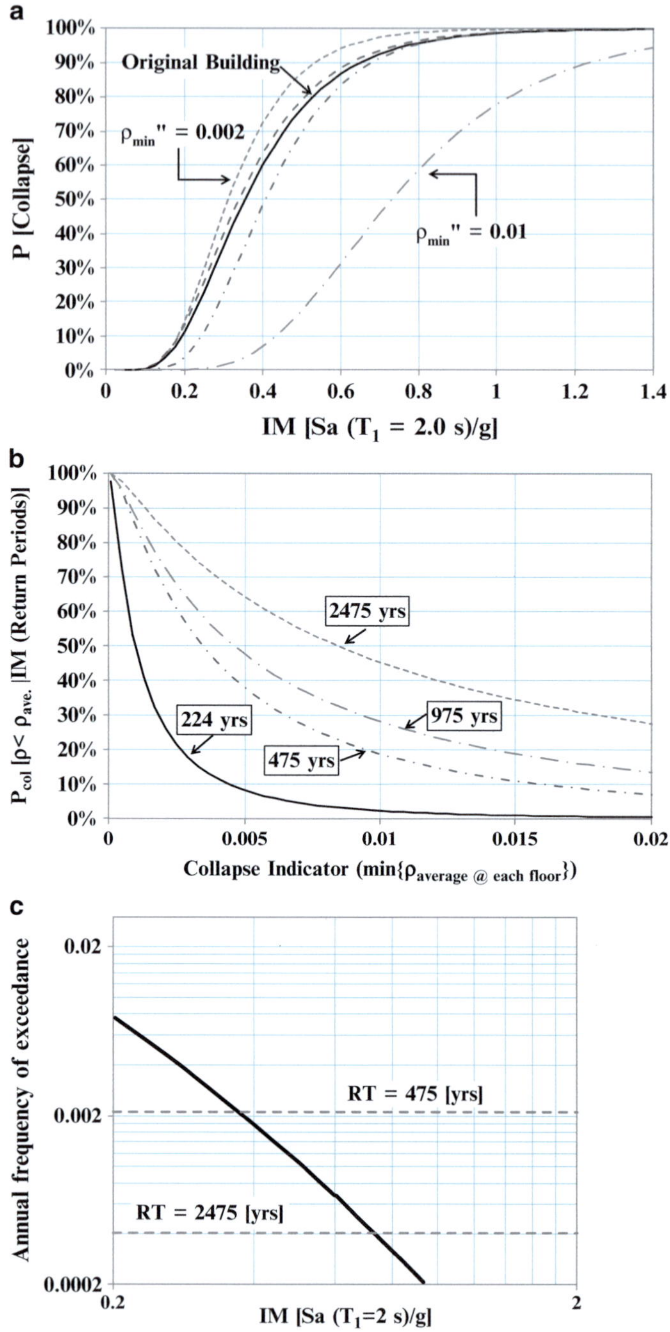


**Fig. 31.2** Procedure for establishing collapse indicator limits, design parameters. (**a**) Develop collapse fragilities for a range of the selected collapse indicator (*e.g. RA-L1*). (**b**) Develop collapse fragilities for a range of selected return periods (*e.g. 475 years, …*). (**c**) Estimate mean annual rate of collapse ($\lambda_{collapse}$) integrating the collapse fragility curves with the hazard curve. (**d**) Seek trends in $\lambda_{collapse}$ for changes in collapse indicator. (**e**) Repeat for "several" building prototypes and choose an appropriate risk and determine the range for the collapse indicator

d

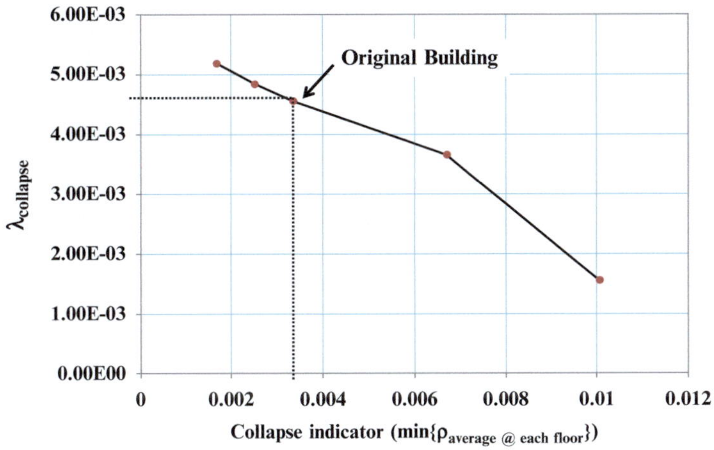


e

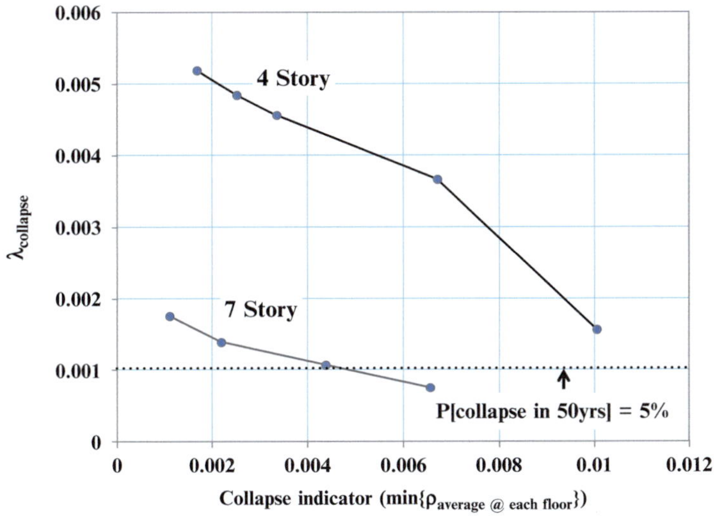


**Fig. 31.2** (continued)

### *31.3.4 Assessment Procedure for Response Parameters*

As shown in Table 31.1, response parameters considered as collapse indicators could be related to the deformations (e.g. global/interstory drift ratios) or forces (e.g. minimum strength ratio) extracted from the nonlinear analysis. These response parameters are also referred as engineering demand parameters (EDPs). Since the performance level considered in this study is collapse prevention, the damage states are discrete and binary and it is assumed that the collapse observation is an ordinary

Bernoulli random variable (i.e., a value of unity whenever the structure sustains collapse and zero in all other cases).

The variability in building responses can be accounted for using cumulative distribution functions (CDFs) to approximate the probability of each response parameter (e.g. maximum interstory drift ratio) occurring. For each building response (BR) and intensity measure (in this study the return period, RT, is used to represent the intensity measure), cumulative distribution functions are developed based on nonlinear dynamic analysis. The objective is to develop a CDF for the probability of collapse given BR and RT, P(Collapse| BR, RT). The probability of exceeding the collapse state conditioned on a particular building response and return period is modeled using a lognormal probability distribution, given by the following equation:

$$P\left(Collapse\middle| BR, RT\right) = \Phi\left[\frac{Ln(BR) - Ln\left(\overline{BR}\right)}{\sigma_{Ln\,BR}}\right] \quad (31.1)$$

where P(Collapse| BR, RT) is the probability of achieving the collapse state, $\overline{BR}$ is the median of the BRs at which the probability of collapse is observed, and $\sigma_{\mathrm{Ln}BR}$ is the standard deviation of the natural logarithm of the BRs. As suggested by Ramirez (2008) different methods could be used to determine the statistical parameters of the lognormal distribution for the CDF, for example, least square methods and the maximum likelihood method. The maximum likelihood method is used in this study to estimate the median $\left(\overline{BR}\right)$ and standard deviation ($\sigma_{\mathrm{Ln}BR}$). CDFs for building prototypes are developed and related to a selected collapse indicator (e.g. BA-S1 – maximum degradation in base or story shear resistance at collapse). Figure 31.3a shows a fitted curve of the response parameter. As shown in this figure, the results of the collapse indicator are either collapse or no collapse from response history analyses.

Using the methodology presented in the previous section, CDFs (Fig. 31.3b) are multiplied with the slope of the hazard curve shown in Fig. 31.2c to estimate the annual frequency of collapse ($\lambda_{collapse}$). Similar collapse fragilities would be determined for different building prototypes and trends in the probabilities of collapse are compared. Potential limits for the collapse indicator are estimated, similar to the design parameters, for a selected mean annual of frequency of collapse, such as 0.001. Using this limit, Fig. 31.3c indicates the maximum interstory drift should not exceed 4.5 % for the seven story building and 4 % for the four story building.

## 31.4 Summary and Future Challenges

The risk associated with older non-ductile concrete buildings internationally is significant, and the development of improved technologies for mitigating that risk is a large and costly undertaking. Considering the limited funding available for

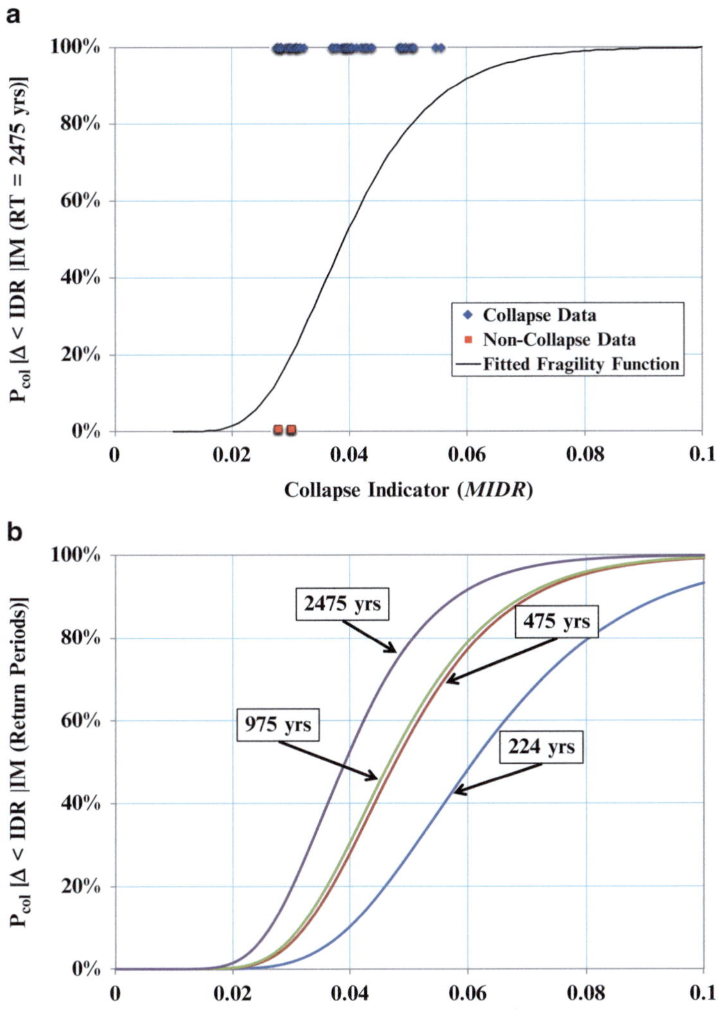


**Fig. 31.3** Procedure for establishing collapse indicator limits, response parameters. (**a**) Determine *CDF* for prototype building. (**b**) Develop *CDF*s for range of selected return periods (*e.g. 475 years*, ...). (**c**) Repeat for "several" building prototypes and choose an appropriate risk and determine the range for the collapse indicator

seismic retrofit, to achieve a meaningful reduction in the collapse risk it is essential to be able to identify the very worst buildings and fix these first. A potential methodology for identifying collapse indicators based on results of comprehensive collapse simulations and estimation of collapse probabilities for a collection of building prototypes is described.

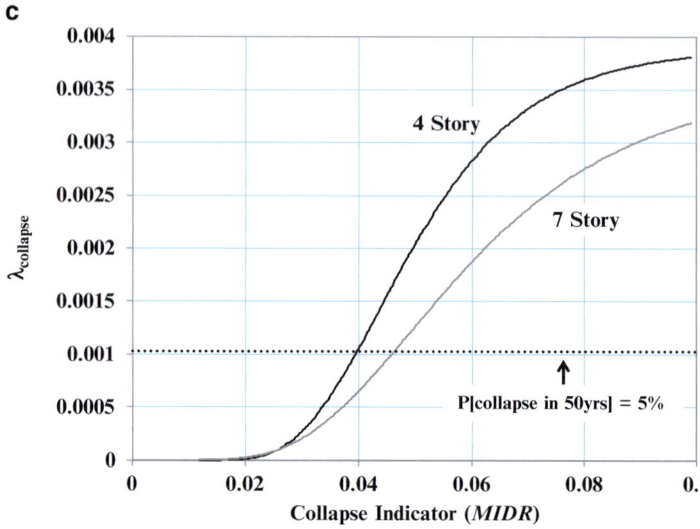


**Fig. 31.3** (continued)

Although only demonstrated here for frames, the probabilities of collapse must be considered for a broad cross section of building types to ensure the selected limits for the collapse indicators are appropriate for the large varied inventory of existing buildings. Methodologies for selecting sample design (e.g. average column transverse reinforcement ratio) and response (e.g. maximum interstory drift ratio) parameter are explained. Limits on the collapse indicators can be selected based on a suitable mean annual frequency of collapse ($\lambda_{collapse}$). In this chapter, collapse is selected based on a target collapse risk. An alternative to this method could be to compare collapse for the prototype buildings with the collapse of a "good" existing building, for which seismic rehabilitation is not required to achieve a collapse prevention performance level.

Additional research is required to establish limits for use in design practice and to improve the methodology to address the interaction of multiple collapse indicators. Ongoing studies funded by FEMA and NIST through the ATC-78 and ATC-95, respectively, are expected to result in specific guidance for practicing engineers based on some of the concepts presented in this chapter.

**Acknowledgments** This work is supported in part by the National Institute of Standards and Technology through the NEHRP Consultants Joint Venture Program and National Sciences and Engineering Research Council of Canada through the Canadian Seismic Research Network. Any opinions, findings, and conclusion or recommendations expressed in this work are those of the authors and do not reflect those of the organizations noted here.

# References


Alath S, Kunnath SK (1995) Modeling inelastic shear deformation in RC beam-column joints. Part 1 (of 2), 21–24 May, ASCE, Boulder, CO, USA, pp 822–825

ASCE/SEI (2003) Seismic evaluation of existing buildings, ASCE standard ASCE/SEI 31-03. American Society of Civil Engineers, Reston

ASCE/SEI (2006) Seismic rehabilitation of existing buildings, ASCE standard ASCE/SEI 41-06. American Society of Civil Engineers, Reston

Baradaran Shoraka M, Elwood KJ (2013) Mechanical model for non ductile reinforced concrete columns. J Earthq Eng 17(7):937–957

Baradaran Shoraka M, Yang TY, Elwood KJ (2013) Seismic loss estimation of non-ductile reinforced concrete buildings. Earthq Eng Struct Dyn 42:297–310. doi:10.1002/eqe.2213

Elwood KJ (2004) Modeling failures in existing reinforced concrete columns. Can J Civ Eng 31(5):846–859

Elwood KJ, Moehle JP (2008) Dynamic collapse analysis for a reinforced concrete frame sustaining shear and axial failure. Earthq Eng Struct Dyn 37(7):991–1012

Elwood KJ, Matamoros A, Wallace JW, Lehman DE, Heintz JA, Mitchell A, Moore MA, Valley MT, Lowes L, Comartin C, Moehle JP (2007) Update of ASCE/SEI 41 concrete provisions. Earthq Spectra 23(3):493–523. Earthquake Engineering Research Institute

FEMA (1997) NEHRP guidelines for the seismic rehabilitation of buildings, FEMA 273. Federal Emergency Management Agency, Washington, DC

Jalayer F, Cornell CA (2009) Alternative non-linear demand estimation methods for probability-based seismic assessments. Earthq Eng Struct Dyn 38(8):951–972

Kang T, Wallace J, Elwood K (2009) Nonlinear modeling of flat-plate systems. J Struct Eng 135(2):147–158

Krawinkler H (2005) Van Nuys hotel building testbed report: exercising seismic performance assessment. PEER Report 2005/11. College of Engineering, University of California, Berkeley

Liel AB, Haselton CB, Deierlein GG (2011) Seismic collapse safety of reinforced concrete buildings. II: comparative assessment of nonductile and ductile moment frames. J Struct Eng 137(4):492–502

Luco N, Ellingwood B, Hamburger RO, Hooper JD, Kimball JK, Kircher CA (2007) Risk-targeted vs. current seismic design maps for the coterminous United States. In: Proceedings of annual SEAOC 2007 conference, Structural Engineers Association of California

NIST GCR 10-917-7 (ATC-76-5) (2010) Program plan for the development of collapse assessment and mitigation strategies for existing reinforced concrete buildings. National Institute of Standards and Technology (NIST), Washington, DC

OpenSees (2010) Open system for earthquake engineering simulation. _OpenSees_ framework-Version 2.1.0. Pacific Earthquake Engineering Research Center, University of California, Berkeley. http://opensees.berkeley.edu. 10 Sept 2010

Ramirez CM (2008) Building-specific loss estimation methods & tools for simplified performance-based earthquake engineering. Doctoral dissertation, Stanford University, Stanford, CA

Vamvatsikos D, Allin Cornell C (2002) Incremental dynamic analysis. Earthq Eng Struct Dyn 31(3):491–514

Yavari S, Elwood KJ, Wu CL (2009) Collapse of a nonductile concrete frame: evaluation of analytical models. Earthq Eng Struct Dyn 38(2):225–241

# Chapter 32
# Earthquake-Resilient Communities: A Look from Mexico

**Sergio M. Alcocer and Roberto Meli**

**Abstract** It is the aim of this chapter to assess the general situation of earthquake resilience in communities in Mexico. This evaluation is performed from a public policy point of view. From the diagnosis presented, challenges and areas of opportunity for implementing programs aimed at reducing risk and attaining more resilience are discussed. It is conjectured that some conclusions and recommendations aimed at achieving resilient communities in the developing countries are also applicable to the developed world.

**Keywords** Resilience • Risk mitigation • Disaster prevention • Non-engineered construction • Engineered construction

## 32.1 Introduction

Mexico is a country subjected to different types of natural hazards. Seasonal hurricanes, intense convective rains, landslides, volcanic eruptions and earthquakes are the most prevalent hazards that inflict damage to Mexican communities. According to the National Center for Disaster Prevention of Mexico (CENAPRED), there has been considerable reduction in the number of human lives lost on the average, with a reduction to one-third of that recorded in the previous two decades (CENAPRED 2001, 2010). It is also clear the very significant increase of material losses. This phenomenon is attributed to the prevalence of damage due to hydro-meteorological events, which may very well be attributed to changes in climate patterns due to global warming. In contrast, in the past 10 years or so, besides

S.M. Alcocer (✉) • R. Meli
Instituto de Ingeniería, Universidad Nacional Autónoma de México, UNAM,
Ciudad Universitaria, México DF 04510, Mexico
e-mail: salcocerm@ii.unam.mx; rmep@ii.unam.mx

M. Fischinger (ed.), *Performance-Based Seismic Engineering: Vision for an Earthquake Resilient Society*, Geotechnical, Geological and Earthquake Engineering 32,
DOI 10.1007/978-94-017-8875-5_32, 

the 2003 earthquake in Tecomán, Colima (with only local damage consequences), Mexico has not been subjected to any significant ground motion shaking (EERI and SMIS 2006; EERI 2012a).

It is the aim of this chapter to assess the general situation of communities in Mexico from the perspective of resilience to earthquakes. This evaluation is performed from a public policy point of view. From the diagnosis presented, challenges and areas of opportunity for implementing programs aimed at reducing risk and attaining more resilience will be discussed. Reader is cautioned that several statements in this chapter are based on the authors' expertise and judgment, for which literature references and quantitative data may not exist.

## 32.2 Diagnosis of Earthquake Risk in Mexico

### *32.2.1 Seismic Hazard in Mexico*

Since the early 1960s, Mexico has devoted considerable efforts, organizational skills and resources to better characterize the seismic environment to which population is subjected. As a result, a reasonably accurate seismic zonation for the country has been developed. Highest seismic hazard is located along the coast in the Pacific Ocean where large subduction earthquakes take place, and at the northwest part of the country, where the system of faults from the West coast of the United States extends down into Mexico. The last population census in 2010 reported 112.3 million people, of which 95.5 million people live in urban and rural communities exposed to moderate to very high seismic hazard.

### *32.2.2 Seismic Risk in Mexican Communities*

The following are statements that have been developed on the basis of expert opinions and authors' expertise and judgment.

1. Seismic risk reduction is not a priority neither in the national nor local agendas

Soon after the 1985 earthquakes, public media, design and construction professionals and society in general became reasonably informed and were aware of the realities of seismic hazard, risk management and risk reduction. In the aftermath of the great 1985 Michoacán earthquakes, the National System for Civil Protection (SINAPROC is the acronym in Spanish) was established aiming at coordinating efforts at the federal, state and local levels. As part of this strategy, CENAPRED was organized as a means to develop applied research and deploy a training and dissemination strategy in all natural and man-made hazards. CENAPRED was also designed to serve as a link with academics, especially those at UNAM.

For over 15 years after the quakes, CENAPRED contributions to advancing the concept of disaster prevention, preparedness, and resilience were outstanding. Design and construction professionals, including both engineers and architects, maintained a close look at the development and enforcement of code provisions, thus leading to conclude that lessons learned after the events had been permanently assumed. Also, in this period, earthquake and evacuation drills were conducted in large urban centers with a positive attitude of population, mainly children. At the universities, a large number of good and motivated students became interested in graduate studies in structural and geotechnical engineering, engineering seismology, earth sciences and social sciences related to earthquake disaster prevention.

Unfortunately, consciousness and interest about the earthquake phenomena in general, and on risk reduction, in particular, has eroded in the last 10 or so years. It is the authors' opinion that reduction in earthquake awareness can attributed to four facts. Firstly, for the new generations, the large and devastating earthquakes, such as those of 1985 in Mexico City, are just part of history, they are not part of a present reality, and much less part of the future. This is coupled with the long-term dream of incorporating solid curricula on disasters at the elementary and high school levels that never took place.

Secondly, SINAPROC and CENAPRED leadership on disaster prevention has unfortunately declined. Progressively more attention has been given to activities related to emergency management, thus leaving little time and resources to think and develop new disaster prevention strategies.

In third place, Mexico has experienced a lack of significant earthquakes; the only exception is the event that occurred in 2003 in Tecomán, Colima. This earthquake caused damage but at the local level, thus leaving the idea that such event was an isolated one. The last earthquake of March 20, 2012 caused alarm in Mexico City and an overreaction of the population to an event whose intensity, in terms of peak ground acceleration, was about one-fifth the intensity of the 1985 events (EERI 2012a). Since 1985 there has been a good number of moderate earthquakes that have produced minimal damage. This has led the population and authorities to publicly indicate that the country and Mexico City are safe thanks to the code changes implemented after 1985.

In fourth place, with a very significant role, is that other topics became an everyday priority for Mexicans: employment, security, satisfaction of basic needs (health, water, sewage) and quality of education.

2. Policy makers and government officials are not, in general, aware of seismic risk

In general terms, government officials and policy makers, including legislators at the federal, state and municipal levels, are not cognizant of the seismic risk of the country, and much less, of the region where they inhabit. To make the situation worse, civil protection authorities at the municipal level most often change every 3 years, coinciding with the election cycle. Therefore, every 3 years new people have to be trained with the evident negative effects on the learning curve.

Overall, authorities are better trained to cope with emergencies caused by natural disasters, than for implementing preventive measures. Most officials ascertain to indicate that Mexico is an earthquake country, but little knowledge on seismic risk is present. Moreover, possibilities for seismic risk reduction are still too far from being considered, and therefore, implementation of risk reduction measures is unimportant.

3. Earthquake experts are often ill coordinated, and very seldom perform research on subtleties

Since the early 1950s, Mexico has developed a good reputation in the earthquake engineering community worldwide. The quality and diversity of its research groups, as well as the successful transfer and implementation of knowledge to engineering practice are distinctive characteristics. However, in recent times, research projects are routinely devoted to refine analytical methods or methodologies that have minimal impact on reducing vulnerability and risk. The loss of leadership that CENAPRED once exerted by establishing the "big-picture" topics certainly contributes to this. Also, scarcity of research funding has forced new researchers to embark on low-risk projects devoted to improve already existing knowledge. Therefore, the real issues that adversely affect the seismic vulnerability of the Mexican building inventory are not, in general, discussed and studied.

4. Lack of a continued and comprehensive program for seismic risk reduction

As it has been common in other countries, seismic risk reduction programs in Mexico have been typically implemented in the aftermath of damaging earthquakes. Examples are the school, housing and bridge rehabilitation programs carried out after the 1985 Mexico City, 1995 Colima and 2003 Tecomán earthquakes. In effect, after 1985, a comprehensive program aimed at reducing the seismic vulnerability of public schools was carried out. Such buildings have performed outstandingly in subsequent earthquakes. Just minor damage of nonstructural components has been reported. Although experts are aware of the need to upgrade critical facilities (such as hospitals), efforts to implement a massive program have failed. Reasons are mostly related to the lack of sound cost-benefit analysis to convince authorities to approve investments. Also, the lack of proper design guidelines, especially for hospital contents and special systems, is a deterrent for establishing a vulnerability reduction agenda. Similar assertions may be made for other types of infrastructure.

5. Not clear trend observed towards correction of inadequate construction practices

Earthquake after earthquake, similar lessons are learned and re-learned throughout the world. Deficient performance of buildings with soft stories, short columns, and non-ductile detailing, among others, is well documented and, in many cases, well understood. In effect, for most of these conditions, analysis, design and detailing requirements are available in building codes. However, for these requirements to lead to an adequate seismic behavior, code enforcement is essential. Although the latter sentence is obvious, reality is different. One caveat of such statement is that sometimes the optimal solution is to enforce minimal, simple requirements, rather than complex and detailed guidelines.

The lack of code enforcement finds its roots on many topics. Indeed, one may think that code compliance is just a direct consequence of a system organized to promote adherence to the law by discovering and punishing persons who violate the rules and norms. That is, code compliance may be thought to be the result of a societal attitude towards respecting the law. In the case of Mexico, code misinterpretation and the existence of strong commercial interests must be added to the picture.

Taking Mexico City as an example, the Advisory Committee for Structural Safety of the city's government recently financed a study to assess the degree of compliance of recent projects built in the city. Main deficiencies encountered were:

- Ill-conceived structural layouts, for example, large mass concentration and stiffness irregularities in plan and elevation
- Inadequate interpretation of code requirements, e.g. members with smaller dimensions and/or reinforcement than minimum prescribed values, as well as non-ductile detailing in zones where plastic deformations are anticipated
- Lax of code enforcement, characterized by gross errors that should have been avoided from design or corrected during inspection.

Findings reinforce the need to enhance the teaching-learning process on structural engineering and foundations, to improve skills and knowledge of practicing professionals through continuing education programs, as well as to develop simple, yet robust, technical requirements that may be easily understood.

The other issue that contributes to lack of full code compliance is the large pressure exerted by real-estate developers on reducing design and construction time, as well as to reduce the cost of foundations and structural system and nonstructural elements to the minimum possible. In this process, it is not rare for design and construction professional to perform unethically by bridging code requirements to increase revenues.

6. Are the abovementioned issues valid in other developing countries?

The diagnosis of the Mexican situation on community resilience and seismic risk reduction was broadly described in the past five statements. The question posed is now if such diagnosis is only valid for Mexico or may be assumed to be correct for other developing countries. The following are two excerpts from reports on reconnaissance visits to Haiti and Chile in 2009, respectively:

> The massive human losses can be attributed to a lack of attention to earthquake-resistant design and construction practices, and the poor quality of much of the construction. (USGS and EERI 2010)
>
> Many of the Chilean standards are adopted from standards in use in the United States (U.S.). In some cases, these standards, where implemented, resulted in buildings and infrastructure that performed well. In other cases, observed performance was less satisfactory, suggesting there may be shortcomings in the available standards and programs for earthquake risk reduction. (EERI 2010)

The statement above related to Chile may not correctly describe what occurred. For many years, a conservative design approach had been implemented, in which buildings with plenty of walls were built. Due to pressures exerted by housing and office developers, designers were pushed to apply the codes to their limits. Such norms lacked of requirements for achieving ductile behavior, thus leading to failures and collapses of recent buildings. In any event, Haiti and Chile are two distinctly different countries in the Latin American region. However, as indicated by the excerpts provided, damage in both countries is related to some, or all, of the statements in which the Mexican diagnosis was based. Therefore, it may be assumed that the overall diagnosis for Mexico is correct for other developing countries. Evidently, the degree of participation or relative weight that the five statements have is different for each country.

## 32.3 Challenges and Areas of Opportunity for Implementing of a Seismic Risk Reduction Program

Once the diagnosis of the prevalent situation on earthquake resilient communities in Mexico is given, some challenges and areas of opportunity are identified in this section. Discussion on how strategies and measures should be implemented is presented.

### *32.3.1 Change of Paradigm: Emphasis on Prevention of Loss of Functionality and on Repairability*

As it was indicated above, societal response to disasters starts with managing emergencies and evolves to prevent and reduce the consequences of natural phenomena. As the disaster management system progresses, societies become more acquainted on the importance of prevention, as a rational process to invest public funding to increase its resilience. Disregarding the background of the civil protection or disaster prevention systems, one may state that seismic risk reduction should be attained through damage control and explicit resilience strategies. Such reduction implicitly assumes that in order to achieve a sustainable development, within the framework of earthquake engineering, investments on prevention ought to be made.

In the case of Mexico, the unforeseeable impact and damage on the engineering profession should a large seismic disaster occurs, is another significant factor. In the aftermath of the 1985 Mexico City earthquakes, because of the large number of casualties and damage to the city, groups from different sectors of the society were largely critical against the engineering profession. Engineers responded in a unified manner by rationally explaining the motions and effects, as well as by improving the building code and standards in a swiftly, coherent manner. Furthermore, the vast and complex process of building rehabilitation exhibited the technical capacity, social commitment and professional performance on the Mexican engineering community. Engineers convinced media and society at large that profession has correctly responded to this crisis.

After the diagnosis presented, if a large disaster occurs with significant damage, it is very likely that Mexican society (and of course, public media) will be much more critical to the engineering profession than in 1985. The quality of the profession at large (design, construction, inspection, quality assurance) will be severely questioned.

### *32.3.2 Vision*

In order to achieve resilient communities, it is advisable to develop a vision to be shared at large. The following is a vision developed on the basis of the authors' expertise and judgment:

- Seismic risk reduction should be a societal need so that appropriate protection against earthquakes is provided through damage control. This is particularly relevant to housing, in which repairable minor damage is expected for largest earthquakes.
- Technologies and their applications should be consistent with the country's level of development. Competences of design and construction professionals and personnel should be consistent with the degree of complexity and refinement of technology and requirements. Care should be exercised when using technologies (design requirements) that are far from being correctly interpreted and implemented in practice.
- Research and outreach should be enhanced and performed through some sort of coordination. In all cases, research and outreach efforts should be conducted within a multidisciplinary framework.

One example of a coordination setup was the Consejo Consultivo sobre Sismos, CoCoS, (Advisory Council on Earthquakes) which was proposed for the 20th anniversary of the Mexico City earthquakes in 2005. CoCoS membership consists of the largest research universities and institutes, as well as of professional societies interested in the earthquake phenomenon. It was purposely decided that all members were non-government agencies. Main purpose of this group is to define the "big-picture" topics in which research, outreach, training, innovation and public policy should be conducted in Mexico, all from the point of view of experts, academic and users' communities.

### *32.3.3 Specific Challenges and Areas of Opportunity*

In this section, specific challenges and areas of opportunities to improve the seismic resilience of Mexican communities are proposed. Suggestions are based on the authors' judgment and experience and are ranged as some of the most significant to greatly improve resilience.

1. Quality assurance systems

As it has been indicated, code compliance in urban constructions is one of the challenges to be fully attained in countries like Mexico. Positive experiences in different countries (United States, Japan, Chile, among others) on implementing and operating compulsory and independent peer reviews for particular buildings should be revised and adapted to local practices. In the case of Mexico City, as a result of the code compliance assessment discussed before, a similar scheme has been established. The system comprises a closed-loop process in which besides peer reviews, independent construction inspections are to be carried out. Implementation has been programmed in different phases to assess effectiveness and fine tune as required. Candidates to be peer reviewed are all essential facilities and a random sample of typical buildings. Revision will consist of evaluating structural layout, design criteria, geometry and reinforcement schedule, some connections and critical zones.

One limitation of the system adopted in Mexico City is that housing or office projects, up to 10,000 $m^2$ of built area, are exempted from obtaining a construction license. Such practice should be avoided because these buildings, although small in nature, are too many and pose a large risk to the population. Other shortcoming is that all structures will be revised for ultimate limit state, while damage control would make more sense for certain type of buildings (e.g. hospitals).

2. Qualified professionals and experts must talk about seismic risk

The best way to convey accurate and timely information to society at large is to have experts talking about earthquakes and their impact to the society. Very often, regardless of their inaccuracies and faults, journalists and science reporters become the "experts" to the eyes of the society. This is particularly the case when a vacuum of information exists; if experts are unavailable or uninterested in disseminating earthquake knowledge at large, anyone who starts doing it, will become the society's expert.

3. Organize joint technical seminars, press conferences and interviews

It is largely desirable to have experts from different fields (structural, geotechnical, seismology, psychology, etc.) talking to public media and society at large in joint meetings. When people representing a group of stakeholders talk to public media and the society, it is obvious that only such interested party will be promoted. In contrast, when several groups of stakeholders are represented, a more balanced and comprehensive message may be conveyed. The Earthquake Engineering Research Institute is one of those very few success stories of engaging a multidisciplinary group advocating, in a coherent and unified fashion, on reducing earthquake risk.

4. Marketing professionals to be part of earthquake resilience

Earthquake experts are often illiterate with regards to marketing issues. Therefore, it is advisable to bring marketing professionals to aid experts so that their message and approach make earthquakes a national/local priority. Marketing professionals have shown the advantages of their profession in making indispensable many of our everyday commodities. The question is how we could benefit from their training and experience in making earthquake risk an everyday matter of utmost importance to our lives and communities sustainability.

5. Qualified professionals and experts to advocate on community resilience

Following the idea of statement 1 above, professionals and experts must advocate for attaining resilient communities. Firstly, a vision on resilient communities should be developed in a coordinated manner, through a consensus groups like CoCoS or comprehensive professional societies like the Earthquake Engineering Research Institute in the United States. It should be clear that community resilience could only be obtained if changing the paradigm from emergency attention to disaster prevention is accepted, thus implying seeing public disbursements as investments on reducing seismic risk.

6. Experts must not look for funding for their personal research

As it has been mentioned, experts (in Mexico, at least) are not coordinated. Furthermore, academics from universities and research centers are heavily motivated by internal and external personal grants that become part of their monthly income. Although the system is organized to enhance the scientific and technological output through stimulus and grants, researchers have found that such awards are given if peer-reviewed papers are published regardless of their impact on solving national or regional problems. Therefore, earthquake-engineering researchers, in general, have taken the easy route of working on refinements of existing knowledge, rather than on topics related to actual vulnerability of Mexican buildings. A shift in this state of mind is needed, but must be developed within the research community itself. Again, consensus groups may become an excellent vehicle to promote a serious discussion and assumption of better attitudes towards improving resilience through coordinated research and outreach.

7. Researchers and experts must learn to talk to politicians

Much of the success of implementing a resilient community advocacy program has to do with convincing politicians from the executive and legislation branches of the government, at all levels –federal, state, municipal- of the benefits of investing in this theme. Typically, the decision making process of politicians is marked by the clock of the next election; thus, experts must convince them that time needed for seismic risk reduction is far greater that their term in office, but that the rewards to do it are of paramount importance for the community's and country's future. Also, experts must understand the avidness of politicians to have solutions in the short term, as they are uninterested in academic products and activities, i.e. papers and attendance to conferences. In this process, patience from both parties is required.

8. Research and outreach should be performed in a coordinated manner

Commonly, experts are mostly interested in performing research, but little time, if any, is devoted to outreach activities. In general, society is strongly interested and becomes motivated and enthusiastic when accurate and timely knowledge is provided. Therefore, for any program on resilient communities to be effective, a strong outreach strategy should be included. Again, earthquake experts must develop and implement coordinating mechanisms to improve outreach.

9. International collaboration

Advances and lessons learned in other countries should be considered as important as if one's has acquired them directly. Therefore, the engineering community of a country should be open to the exchange of experiences and knowledge from and to other countries. This attitude should be expressed in the form of more active participation in international academic and technical events and projects, even if such contributions are limited at the beginning.

## 32.4 Recommendations for Earthquake Mitigation Programs

A description of some recommendations for establishing successful earthquake mitigation programs (EMPs) in developing countries is presented herein. In this section the term mitigation is used as a synonymous of risk reduction. The term "beneficiary" defines the target population of the mitigation program. Differences are made between non-engineering and engineering constructions, since they are relevant in the context of those types of countries. Some recommendations are directly applicable to developed nations.

### *32.4.1 Premises for a Successful EMP*

Three basic ideas to be assumed in planning and executing successful EMPs are proposed:

- Premise A. In developing countries, safety is not a concept easily accepted by beneficiaries who lack basic infrastructure; rather, safety should be fostered along improvements in the quality of life (literacy, water supply, sewage, health services, etc.).
- Premise B. Recognition of the different levels of scholarship, expertise, interests and cultural background among parties involved is needed to achieve success.
- Premise C. Admit that code compliance depends on the complexity of the code itself as it relates to the level of expertise and socioeconomic situation.

These premises have been developed after reviewing what worked well and what did not perform as intended in different EMPs implemented in distinct countries. Premise A is fundamental; it is the core of a successful EMP in areas where most vulnerable groups and communities exist. Rural communities are a prime example of vulnerable groups where earthquake safety is far from being important when compared to everyday needs that have been unmet for many years.

Premise B assumes that a successful EMP cannot be designed and implemented without proper consideration of local characteristics, resources, culture, and idiosyncrasy. Several EMPs have failed to fulfill its objectives because a government office had designed them far from where the program was going to be implemented.

Finally, Premise C implies the need for codes with distinctly different level of complexity and detail built in. For urban construction, well-detailed and state-of-the-art codes may be needed. In rural areas, it does not make economic sense to require a complex code. Rather, a simple code that establishes minimum level of safety should be developed. Its compliance would strongly depend on the breadth and depth of its dissemination and training among groups (engineers, architects, masons) involved in construction.

### 32.4.2 *Topics to Be Considered for Non-engineered Construction*

Non-engineering construction comprise structures built without construction permits, formal code compliance, and the participation of qualified professionals. A particular case of non-engineered construction is that built by the owners; this case is quite rare as most non-engineered constructions are built through local masons. Typical examples of non-engineered construction are houses in the rural areas. Such houses are typically made of some type of unreinforced masonry, including adobe, or made of confined masonry. The latter construction type has shown to perform very well under severe ground shaking, when confinement elements (i.e. tie-columns and bond-beams) are properly located and constructed. Interestingly, adequate performance is attributed to good reinforcing schemes in the practice of local masons rooted on repeating them for many years, rather than based on technical knowledge or training. Confined masonry has drawn considerable attention from the international engineering community. The Confined Masonry Network has been established and is dedicated to promote seismically safe and economical housing worldwide by bringing quality confined masonry into the design and construction mainstream (EERI 2012b). A seismic design guide for low-rise confined masonry buildings has been developed with contributions from different countries (Meli et al. 2011).

In the following, main issues to be taken into account when designing and implementing an EMP for non-engineered construction are discussed.

1. Solutions should be compatible to local practice

Structural solutions, whether for new construction or rehabilitation schemes, should use materials and technologies familiar to the local workforce available. Care must be exercised so that local practice is applied if it provides adequate seismic safety. Extraneous solutions, not known and understood by local beneficiaries and construction workers, very often lead to unsafe results.

2. Education and training at the local level

Implementation of an EMP provides a unique opportunity to improve the quality of life of benefited people. Enhancing their competence as construction laborers through local education and training directly increases the quality of constructions made under the EMP. It also allows beneficiaries to apply for jobs better paid than those typically accessible to people in isolated or poor communities.

3. Participation of local technical groups and professional societies

Sound EMPs seldom benefit from the technical assistance and support provided by local universities and professional societies, at least in the case of Mexico. In the proposed strategy, senior engineering and architecture students may be very helpful as the first line of contact with beneficiaries. Coordinators belonging to professional societies and local governments would oversee students. All levels of support should get specific training depending upon their responsibilities. Participation of local technical groups, universities and professional societies should be properly stimulated and recognized.

4. Financial incentives

Evidently, EMPs require funding to be implemented. In most cases, national or state governments provide resources either to fully pay for the program, or to heavily subsidize construction materials and labor force. In the latter, successful EMPs required all activities, decisions and resources being accountable and transparent.

5. Foster the participation of beneficiaries

As it was indicated, habitually benefited people have some skills to implement the construction phase of the EMP. It is wise to include temporary employment as part of the EMP because it allows beneficiaries to receive money and wages during critical times, especially in the aftermath of an earthquake. In all cases, success of an EMP will depend on the acceptance of the target population. Probabilities of success will increase if beneficiaries take part of the planning process of the EMP.

6. Disseminate solutions and achievements

Dissemination of strategies, objectives and targets at all stages of an EMP is as important part as the implementation itself. Of particular relevance is the communication of milestones or specific achievements. Fulfillment of goals improves the self-esteem and pride of all parties involved; this is quite significant if the EMP was implemented after a seismic disaster.

7. Carry out demonstration projects

Demonstration projects have shown to improve the likelihood of success of an EMP. In India, for example, small shaking table have been constructed on truck platforms to perform simple tests of vulnerable and retrofitted houses. Through direct evidence, people observe and understand the advantages of using the retrofit scheme suggested. A similar objective is looked for when pilot constructions, i.e. houses, are built following the proposed materials and technologies. Local population should visit such pilot buildings in order to get their feedback aimed at improving and correcting any deficiency. One of such EMPs is that implemented in Oaxaca, Mexico, after the 1999 earthquake. Some reasons for its success were the following:

- Program design considered the local practice, as well as the availability and quality of materials and workmanship
- Technical information was conveyed in a simple manner
- Solutions did not depend on complex design and construction details.

In recent years, the Mexican Society of Structural Engineering, with the support of the Mexico's National Housing Commission developed two guidelines, 16-page each, one for new construction and the other for rehabilitation of non-engineered construction (SMIE 2011a, b). Both documents are directed to rural and suburban areas where vulnerable low-rise houses are common. Design and construction recommendations are embedded within cartoons. Each guideline is based on a story of a typical Mexican family eager to properly build their new house, or to safely rehabilitate their existing one. A critical message conveyed during the stories is that rules provided come from good engineering and that engineering professionals should be sought for advice.

### *32.4.3 Topics to Be Considered for Engineered Construction*

Over the last years, enforcement of building codes in developing countries has been a major concern of academics, technical societies, practicing engineers and government officials. In Mexico City, for example, aiming at simplifying the process to obtain a building permit, in 2004, the city government decided to implement a "Notification of construction process" applicable for buildings up to 10,000 $m^2$ of built area. In this process, the building owner is only required to inform the local authority about general characteristics of a building to be constructed. This process, in lieu of a formal building permit, has then omitted the formal revision of building drawings and calculations that was typically made and filed by local building officials. The outcome of such process is buildings with evident irregularities in stiffness, strength and/or mass distribution, and excessive lateral flexibility. In some cases, deficient designs can be tracked to engineers with limited skills, very often underpaid, who do a poor job in their designs. Buildings with such characteristics were the most affected during the 1985 Mexico City earthquakes, such as soft stories.

The case of taller structures, although different in nature, has led to similar concerns. Poor workmanship and evidence of a systematic misinterpretation of code requirements are often found. Up to now, reasons for this phenomenon are not clear. Lack of understanding of code requirements, ignorance about the effects on structural behavior of assumptions made during analysis and design, and corruption have been discussed as possible reasons. In the opinion of the authors, the first two reasons prevail.

Although reasons may be different, similar deficiencies and vulnerable characteristics have been recorded in other countries subjected to significant earthquake hazard.

Vulnerability reduction of engineered construction is founded on the following proposed issues:

1. Structural systems with built-in large seismic capacity

Earthquakes have evidenced the superior performance of structural systems with inherent large seismic capacity. In general, buildings comprising wall systems have superiorly performed when compared to buildings relying only on frame action. This statement should not be interpreted as the authors' rejection of systems based on frame behavior. However, wall behavior is less sensitive to deviations on design and detailing when compared to that of frame structures. Moreover, damage of some nonstructural components and contents has shown to be dependent on lateral drift, which in turn is better controlled through walls.

2. Avoid foreign engineering concepts without local review and assimilation

The Izmit earthquake in Turkey in 1999 was key for understanding the negative consequences of importing engineering solutions and concepts, mainly developed for non-seismic areas, without a local review of technical experts. Through this reviewing process, weaknesses can be identified so that improvements and corrections can be developed. Once the foreign system has been checked and improved, it can be assimilated to practice reducing the likelihood of improper behavior under local conditions. In the case of Turkey, buildings designed abroad exhibited large open spaces at the ground story, thus leading to soft-story failure with the obvious consequences.

3. Design and rehabilitation requirements for critical facilities

Critical facilities, such as hospitals and telecommunications centers, should be designed according to their expected performance and vital role for the community during and after an earthquake. As a consequence of their significance, most codes have implemented an importance factor, larger than 1.0, aimed at increasing the seismic demands (i.e. lateral forces) from those applicable to normal constructions. However, current procedures fail to reconcile that building capacity (including structural and nonstructural components) should be tailored to exhibit the intended performance. The performance-based seismic design approach, when fully implemented, is a step towards the right direction. Indeed, in this procedure it is correctly acknowledged that capacity requirements (mostly reflected on member detailing) should correspond to distinct performance objectives. So far, detailing rules are relevant for ultimate design approach, roughly equivalent to collapse-prevention performance objective. Design and detailing requirements for attaining more stringent performance objectives (v.gr. immediate occupancy) are lacking.

Existing critical facilities pose a challenge because many of them had been designed and constructed using codes and specifications that are substandard compared to today's knowledge. To improve their expected performance, specific rehabilitation requirements need to be developed through phases. As it was indicated above, structural and nonstructural components should be included. Guidelines should be implemented as they become available, understanding that the development process will take some time.

4. Codes with procedures and requirements of different levels of complexity

For many countries, it is of vital importance to recognize that technical expertise in the engineering community is heterogeneous and, therefore, that they should implement means to improve the quality of engineering education, as well as the technical expertise of professionals. Very often, engineering community comprises design professionals who perform their practice following state-of-the-art knowledge and approaches. But, also, it includes professionals with limited skills and knowledge. In this environment, small yet simple buildings should be designed by those professionals with the minimum of skills and knowledge. More complex and important structures should, then, be designed by the most knowledgeable engineers. According to this, codes should be developed to reflect procedures and requirements of different levels of complexity, consistent with the building importance, type and size. Refined analysis and design procedures should be favored, but optional simple yet conservative approaches should be offered for those design professional with limited expertise to follow refined methods.

The idea behind having levels of complexity embedded in codes and requirements is applicable to the design of buildings depending upon their importance. Buildings for normal occupancy could be designed according to simplified procedures; important buildings and critical facilities should be designed following more elaborate procedures.

## 32.5 From the Topics Discussed: Is There Anything Applicable to the Developed World?

The aim of the paper has been to present challenges and areas of opportunity for improving community resilience under earthquakes in Mexico. A diagnosis and suggested recommendations to overcome the present state of affairs has been discussed. Basics for a successful implementation of EMPs were presented. Specific suggestions for non-engineered and engineered construction were made. In the discussion it was made apparent that conclusions and recommendations are applicable to other developing countries. Before closing, authors would like to pose a challenge to the reader of this chapter. Based on the known damage characteristics and consequences of earthquakes in the developed world (2011 Japan and New Zealand, for example), are there similarities to damage (reasons and consequences) in developing countries? Authors argue that some suggestions for the Mexican case are also applicable to the developed world.

**Acknowledgments** Authors wish to dedicate this paper to the memory of Prof. Helmut Krawinkler, whose inspiring leadership and insightful research ideas contributed to improving earthquake resilience worldwide.

## References

Centro Nacional de Prevención de Desastres, CENAPRED (2001) Características del impacto socioeconómico de los principales desastres ocurridos en México en el período 1980–99 (in Spanish). Mexico. ISBN 970-628-5911. http://www.cenapred.unam.mx/es/Publicaciones/archivos/2892006Impacto1.pdf. Accessed 3 July 2012

Centro Nacional de Prevención de Desastres, CENAPRED (2010) Impacto socioeconómico de los principales desastres ocurridos en la República Mexicana en el año 2009 (in Spanish). Mexico. ISBN 978-607-7558-19-4

Earthquake Engineering Research Institute, EERI (2010) Chile research needs workshop report. http://www.eqclearinghouse.org/20100227-chile/wp-content/uploads/2010/11/Chile-Workshop-Report_FINAL.pdf. Accessed 3 July 2012

Earthquake Engineering Research Institute, EERI (2012a) The March 20, 2012, Ometepec, Mexico, earthquake. http://www.eeri.org/wp-content/uploads/Ometepec-2012-eq-report.pdf. Accessed 3 July 2012

Earthquake Engineering Research Institute, EERI (2012b) The confined masonry network. http://www.confinedmasonry.org/. Accessed 3 July 2012

Earthquake Engineering Research Institute, EERI, and Sociedad Mexicana de Ingeniería Sísmica, SMIS (2006) The January 21, 2003 Tecomán, México, earthquake. ISBN 1-932884-08-4. www.eeri.org/products-page/reconnaissance-reports/the-tecom-2/. Accessed 3 July 2012

Meli R et al (2011) Seismic design guide for low-rise confined masonry buildings. The confined masonry network. http://www.confinedmasonry.org/wp-content/uploads/2009/09/ConfinedMasonryDesignGuide82011.pdf. Accessed 3 July 2012

Sociedad Mexicana de Ingeniería Estructural, SMIE (2011a) Guía de autoconstrucción de vivienda (in Spanish). México

Sociedad Mexicana de Ingeniería Estructural, SMIE (2011b) Guía de reparación de vivienda (in Spanish). México

United States Geological Survey, USGS, and Earthquake Engineering Research Institute, EERI (2010) 7.0 Haiti earthquake advanced reconnaissance team report. http://www.eqclearinghouse.org/20100112-haiti/wp-content/uploads/2010/02/USGS_EERI_HAITI_V1.1.pdf. Accessed 3 July 2012

# Index



M. Fischinger (ed.), *Performance-Based Seismic Engineering: Vision for an Earthquake Resilient Society*, Geotechnical, Geological and Earthquake Engineering 32,
DOI 10.1007/978-94-017-8875-5, © Springer Science+Business Media Dordrecht 2014

MIX
Papier aus verantwortungsvollen Quellen
Paper from responsible sources
FSC® C105338

If you have any concerns about our products,
you can contact us on
**ProductSafety@springernature.com**

In case Publisher is established outside the EU,
the EU authorized representative is:
**Springer Nature Customer Service Center GmbH**
**Europaplatz 3, 69115 Heidelberg, Germany**

Printed by Libri Plureos GmbH
in Hamburg, Germany